Green Metathesis Chemistry

NATO Science for Peace and Security Series

This Series presents the results of scientific meetings supported under the NATO Programme: Science for Peace and Security (SPS).

The NATO SPS Programme supports meetings in the following Key Priority areas: (1) Defence Against Terrorism; (2) Countering other Threats to Security and (3) NATO, Partner and Mediterranean Dialogue Country Priorities. The types of meeting supported are generally "Advanced Study Institutes" and "Advanced Research Workshops". The NATO SPS Series collects together the results of these meetings. The meetings are co-organized by scientists from NATO countries and scientists from NATO's "Partner" or "Mediterranean Dialogue" countries. The observations and recommendations made at the meetings, as well as the contents of the volumes in the Series, reflect those of participants and contributors only; they should not necessarily be regarded as reflecting NATO views or policy.

Advanced Study Institutes (ASI) are high-level tutorial courses intended to convey the latest developments in a subject to an advanced-level audience

Advanced Research Workshops (ARW) are expert meetings where an intense but informal exchange of views at the frontiers of a subject aims at identifying directions for future action

Following a transformation of the programme in 2006 the Series has been re-named and re-organised. Recent volumes on topics not related to security, which result from meetings supported under the programme earlier, may be found in the NATO Science Series.

The Series is published by IOS Press, Amsterdam, and Springer, Dordrecht, in conjunction with the NATO Public Diplomacy Division.

Sub-Series

A.	Chemistry and Biology	Springer
B.	Physics and Biophysics	Springer
C.	Environmental Security	Springer
D.	Information and Communication Security	IOS Press
E.	Human and Societal Dynamics	IOS Press

http://www.nato.int/science
http://www.springer.com
http://www.iospress.nl

Series A: Chemistry and Biology

Green Metathesis Chemistry

Great Challenges in Synthesis, Catalysis and Nanotechnology

edited by

Valerian Dragutan

Romanian Academy, Institute of Organic Chemistry
Bucharest, Romania

Albert Demonceau

University of Liège, Institute of Chemistry
Liège, Belgium

Ileana Dragutan

Romanian Academy, Institute of Organic Chemistry
Bucharest, Romania

and

Eugene Sh. Finkelshtein

Russian Academy of Sciences
Topchiev Institute of Petrochemical Synthesis
Moscow, Russia

Springer

Published in cooperation with NATO Public Diplomacy Division

Proceedings of the NATO Advanced Study Institute on
Green Metathesis Chemistry: Great Challenges in Synthesis, Catalysis and Nanotechnology
Bucharest, Romania
21 July – 2 August 2008

Library of Congress Control Number: 2009933585

ISBN 978-90-481-3432-8 (PB)
ISBN 978-90-481-3431-1 (HB)
ISBN 978-90-481-3433-5 (e-book)

Published by Springer,
P.O. Box 17, 3300 AA Dordrecht, The Netherlands.

www.springer.com

Printed on acid-free paper

Table of Contents

PART III. NEW MATERIALS BY METATHESIS POLYMERIZATION AND RELATED CHEMISTRY

Preface

Preface

Green Metathesis Chemistry: Great Challenges in Synthesis, Catalysis and Nanotechnology

For the last 2 decades NATO ASI meetings on Metathesis Chemistry have acted as promoters of excellence in research on and valorization of this fascinating scientific area. Five such events organized previously (Akçay, Turkey, 1989, 1995; Polanica-Zdrój, Poland, 2000; Antalya, Turkey, 2002, 2006) have known a well-deserved success. In the context of the spectacular advances in the field culminating in the Nobel Prize for Chemistry awarded to alkene metathesis and the follow-up research, a sixth NATO ASI of the kind was organized in Bucharest, Romania (July 21–August 2, 2008), with a focus on green metathesis chemistry.

Over 70 scientists, a blend of top international experts in the field, academics and young researchers or students, from 17 countries gathered to unveil and debate the utmost new progress in this domain. During the intense 11 days of activities, four main themes were repeatedly addressed in plenary lectures and invited short contributions: (i) Catalyst design and development of cost-efficient and user-friendly processes (Grela, Fogg, Delaude, Clavier, Balcar); (ii) Metathesis-related fine chemical synthesis (Demonceau, Marciniec); (iii) Architecturally complex assemblies and nanostructures (Astruc); (iv) Tailored polymers and new technologies for smart materials (Khosravi, Finkelshtein). In correlation, these topics, indicative of the vibrant research and imaginative use of metathesis in new realms, offered ideas for solving acute contemporary problems such as environmental issues and growing demand for active, selective and recyclable catalysts. Being an atom efficient catalytic reaction, olefin metathesis is, not surprisingly, frequently the key step in multifarious synthetic protocols.

Newly alkene metathesis has been called upon to promote green chemistry applications as finely illustrated during the NATO ASI (Grela: ionic liquids and “greener” solvents; Fogg: catalyst lifetime; Malacea: vegetal oils). The sophisticated chemistry ongoing in organic synthesis and in the polymer arena was masterfully illuminated from perspectives presented by Demonceau and Khosravi, respectively. The material science theme saw interdisciplinary bridges shrink further with presentations on metallodendrimers and precisely controlled nanomaterials (Astruc). A special feature of the Bucharest workshop, happening for the first time in a metathesis NATO ASI, namely a theoretical perspective on the reaction through *in silico* experiments (Cavallo et al.) generated a vivid interest in the audience.

Two poster sessions and a Round Table, liberated from limitations in time, provided perfect settings for deeper delving into the four themes. As remarked by many of the participants, there was an exceptionally enthusiastic, unbridled discussion stimulated by the session chairpersons (Fogg, Khosravi) and the eager, young participants. Starting at the sessions, debate continued at breakfast, over lunch and dinner and, occasionally, during the social programme organized in a friendly style so as to facilitate contacts and stimulate exchange of ideas and information between research groups.

For this scientific event, we warmly thank the lecturers, the discussion leaders and, most of all, the participants. The Bucharest NATO ASI will be remembered as a forum for intense and free scientific interaction among individuals of different nationalities. We hope that this volume, while reflecting the multifaceted new achievements in metathesis, transmits to the readers this spirit and will be of real help to scientists and engineers active in this field.

Our special thanks are due to NATO for generously providing the financial support as well as to the Polytechnic University of Bucharest and the Institute of Organic Chemistry of the Romanian Academy who jointly hosted the scientific sessions.

The Editors

List of Contributors

AFANASIEV Vladimir	United Research and Development Center, 55/1, b.2, Leninskii pr., 119333 Moscow, Russia
ASTRUC Didier	Nanoscience and Catalysis Group LCOO, Institute of Molecular Science, UMR CNRS No. 5802 University of Bordeaux I, 33405 Talance Cedex, France, E-mail: d.astruc@ism.u-bordeaux1.fr
BALCAR Hynek	J. Heyrovsky Institute of Physical Chemistry, Academy of Sciences of the Czech Republic, Dolejskova 3, 18223 Prague 8, Czech Republic, E-mail: balcar@jh-inst.cas.cz
BARTHA Emeric	Institute of Organic Chemistry, Romanian Academy, 202B Spl. Independentei, Bucharest, Romania
BEK David	J. Heyrovsky Institute of Physical Chemistry, Academy of Sciences of the Czech Republic, Dolejskova 3, 18223 Prague 8, Czeck Republic
BELYAEV Boris A.	Topchiev Institute of Petrochemical Synthesis, Russian Academy of Sciences, Leninskii Prospect 29, Moscow 119991, Russia
BENCZE Lajos	Muller Laboratory, Department of Organic Chemistry, University of Veszprem, Egyetem utca 10, H-8200, P.O. Box 158, 8201 Veszprem, Hungary, E-mail: ben016@almos.vein.hu
BERMESHEV Maxim	Topchiev Institute of Petrochemical Synthesis, Russian Academy of Sciences, 29 Leninskii pr., 119991 Moscow, Russia
BESPALOVA Natalia	United Research and Development Center, 55/1, b.2, Leninskii pr., 119333 Moscow, Russia

BICCHIELLI Dario	Center for Education and Research on Macromolecules, Institute of Chemistry (B6a), University of Liege, Sart Tilman, B-4000 Liege, Belgium
BORGUET Yannick	Center for Education and Research on Macromolecules, Institute of Chemistry (B6a), University of Liege, Sart Tilman, B-4000 Liege, Belgium
BUTENKO Tamara A.	Topchiev Institute of Petrochemical Synthesis, Russian Academy of Sciences, Leninskii Prospect 29, Moscow 119991, Russia
BUZDUGAN Emil	Institute of Chemical Research, 202 Spl. Independentei, Bucharest, Romania
BYKOV Victor	Topchiev Institute of Petrochemical Synthesis, Russian Academy of Sciences, Leninskii Prospect 29, Moscow 119991, Russia
CALDARARU Monica	Institute of Physical Chemistry of the Romanian Academy, 202A Spl. Independentei, Bucharest, Romania
CAPROIU Teodor	Institute of Organic Chemistry, Romanian Academy, 202B Spl. Independentei, Bucharest, Romania
CASTARLENAS Ricardo	University of Zaragoza, Zaragoza, Spain
CASTLE Lyle W.	Chemistry Department, Idaho State University, Campus Box 8023, Pocatello, Idaho 83209, USA
CAVALLO Luigi	University of Salerno, University of Salerno, via Ponte don Melillo, Fisciano, I-84084, Italy, E-mail: lcavallo@unisa.it
ČEJKA Jiří	J. Heyrovsky Institute of Physical Chemistry, Academy of Sciences of the Czech Republic, Dolejskova 3, 18223 Prague 8, Czech Republic
CETINKAYA Sevil	Kyrykkale University, Kyrykkale, Turkey

CLAVIER Hervé	University of St-Andrews, School of Chemistry, Purdie Building, North Haugh, St Andrews, Fife KY16 9ST, UK, E-mail: hc31@st-andrews.ac.uk Institute of Chemical Research of Catalonia, (ICIQ), Av. Països Catalans 16, 43007 Tarragona, Spain
CORREA Andrea	University of Salerno, via Ponte don Melillo, Fisciano, I-84084, Italy, E-mail: acorrea@unisa.it
DAVIDENKO Lyudmyla	Institute of Surface Chemistry, National Academy of Sciences, Kiev, Ukraine
DE CANCK Els	University of Ghent, Krijgslaan 281 (S4), B-9000 Ghent, Belgium
DELAUDE Lionel	Center for Education and Research on Macromolecules, Institute of Chemistry (B6a), University of Liege, Sart Tilman, B-4000 Liege, Belgium, E-mail: L.Delaude@ulg.ac.be
DELEANU Calin	Institute of Organic Chemistry, Romanian Academy, 202B Spl. Independentei, Bucharest, Romania
DELFOSSE Sebastien	Center for Education and Research on Macromolecules, Institute of Chemistry (B6a), University of Liege, Sart Tilman, B-4000 Liege, Belgium
DEMONCEAU Albert	Center for Education and Research on Macromolecules, Institute of Chemistry (B6a), University of Liege, Sart Tilman, B-4000 Liege, Belgium, E-mail: A.Demonceau@ulg.ac.be
DIALLO Abdou	Nanoscience and Catalysis Group LCOO, Institute of Molecular Science, UMR CNRS No. 5802 University of Bordeaux I, 33405 Talance Cedex, France
DIMONIE Mihai	Department of Macromolecular Chemistry, Polytechnic University of Bucharest, Romania

DIXNEUF Pierre H.	Catalyse et Organometalliques UMR 6226 CNRS, Universite de Rennes, Institut Sciences Chimiques de Rennes, Bat 10C, Campus de Beaulieu, 35042 Rennes Cedex, France, E-mail: Pierre.Dixneuf@univ-rennes1.fr
DOLGINA Tatyana	United Research and Development Center, Moscow, Russia
DONESCU Dan	Institute of Chemical Research, 202 Spl. Independentei, Bucharest, Romania
DRAGUTAN Ileana	Institute of Organic Chemistry, Romanian Academy, 202B Spl. Independentei, Bucharest, Romania, E-mail: idragutan@yahoo.com
DRAGUTAN Valerian	Institute of Organic Chemistry, Romanian Academy, 202B Spl. Independentei, Bucharest, Romania, E-mail: vdragutan@yahoo.com
DROZDZAK Renata	University of Ghent, Department of Inorganic and Physical Chemistry, Krijgslaan 281 (S4), B-9000 Ghent, Belgium, E-mail: renatamatusiak@yahoo.fr
DUGNE Esra	Hacettepe University, Faculty of Science, Department of Chemistry, 06800 Beytepe, Ankara, Turkey
ECE Abdulilah	Hacettepe University Ankara, Beytepe, Ankara, Turkey
ELMAATY Tarek A.	Mansoura University, Mansoura, Egypt
FILIP Petru	Institute of Organic Chemistry, Romanian Academy, 202B Spl. Independentei, Bucharest, Romania
FINKELSHTEIN Eugene	Topchiev Institute of Petrochemical Synthesis, Russian Academy of Sciences, Leninskii Prospect 29, Moscow 119991, Russia, E-mail: fin@ips.ac.ru
FISCHER Helmut	University of Konstanz, Universitat Strasse 10, Konstanz, Germany

FOGG Deryn E.	Center for Catalysis Research and Innovation, Department of Chemistry, University of Ottawa, 10 Marie Curie KZN 6N5 Ottawa, ON, Canada, E-mail: dfogg@science.uottawa.ca
GAVAT Inge	Polytechnic University of Bucharest, Department of Electronics and Telecommunications, Romania
GAVIN Rafal	Institute of Organic Chemistry, Polish Academy of Sciences, Kasprzaka 44/52, 01-224Warsaw, Poland
GHIOCA Paul	Institute of Chemical Research, 202 Spl. Independentei, Bucharest, Romania
GHITA Stefan	Polytechnic University of Bucharest, Department of Electronics and Telecommunications, Romania
GRELA Karol	Institute of Organic Chemistry, Polish Academy of Sciences, Kasprzaka 44/52, 01-224Warsaw, Poland, E-mail: grela@icho.edu.pl
GRINGOLTS Maria	Topchiev Institute of Petrochemical Synthesis, Russian Academy of Sciences, Leninskii Prospect 29, Moscow 119991, Russia, E-mail: gringol@ips.ac.ru
GUŁAJSKI Łukasz	Institute of Organic Chemistry, Polish Academy of Sciences, Kasprzaka 44/52, 01-224Warsaw, Poland
HENDRICKX Pieters MS.	University of Ghent, Krijgslaan 281 (S4), B-9000 Ghent, Belgium
IANCU Lorena	Institute of Chemical Research, 202 Spl. Independentei, Bucharest, Romania
IMAMOGLU Iavuz	Hacettepe University, Faculty of Science, Department of Chemistry, 06800 Beytepe, Ankara, Turkey, E-mail: imamoglu@@hacettepe.edu.tr

IOVU Horia	Department of Macromolecular Chemistry, Polytechnic University of Bucharest, Romania
JOSSIFOV Christo	Institute of Polymers, Bulgarian Academy of Sciences, Acad. G. Bonchev str. 103A, 1113 Sofia, Bulgaria, E-mail: jossifov@polymer.bas.bg
KALINOVA Radostina	Institute of Polymers, Bulgarian Academy of Sciences, Acad. G. Bonchev str. 103A, 1113 Sofia, Bulgaria, E-mail: kalinova@polymer.bas.bg
KARABULUT Solmaz	Hacettepe University, Faculty of Science, Department of Chemistry, 06800 Beytepe, Ankara, Turkey, E-mail: solmazk@hacettepe.edu.tr
KHARLAMOV Oleksiy	Institute for Problems of Materials Science, National Academy of Science of Ukraine, 3 krjijanovskogo str., 03680 Kiev, Ukraine
KHOSRAVI Ezat	IRC in Polymer Science and Technology, University of Durham, South Road Durham DHI 3LE, UK
LACHTIN Valentin	Topchiev Institute of Petrochemical Synthesis, Russian Academy of Sciences, Leninskii Prospect 29, Moscow 119991, Russia
LEDOUX Nele	University of Ghent, Department of Inorganic and Physical Chemistry, Krijgslaan 281 (S4), B-9000 Ghent, Belgium
LUNIKA Michael	National Academy of Sciences, Kiev, Ukraine
MACAEV Fliur	Academy of Sciences of Moldova, Institute of Chemistry, Chisinau, Moldova
MAUDUIT Marc	UMR CNRS 6226, Ecole Nationale Superieure Chimie de Rennes, 35700 Rennes, France
MAKOVETSKII Kiryl L.	Topchiev Institute of Petrochemical Synthesis, Russian Academy of Sciences, Leninskii Prospect 29, Moscow 119991, Russia

MALACEA Raluca	Catalyse et Organometalliques UMR 6226 CNRS, Universite de Rennes, Institut Sciences Chimiques de rennes, Bat 10C, Campus de Beaulieu, 35042 Rennes Cedex, France
MARCINIEC Bogdan	Faculty of Chemistry, Department of Organometallic Chemistry, Adam Mickiewicz University, Grunwaldzka 6, 60-780 Poznan, Poland, E-mail: bogdan.marciniec@amu.edu.pl
MARTINS Jose C.	University of Ghent, Krijgslaan 281 (S4), B-9000 Ghent, Belgium
MIAH Mannan	Hacettepe University, Faculty of Science, Department of Chemistry, 06800 Beytepe, Ankara, Turkey
MIAO Xiaowei	Catalyse et Organometalliques UMR 6226 CNRS, Universite de Rennes, Institut Sciences Chimiques de rennes, Bat 10C, Campus de Beaulieu, 35042 Rennes Cedex, France
MICHROWSKA Anna	Max Plank Institut fur Kohlenforschung, Germany
MONFETTE Sebastien	Center for Catalysis Research and Innovation, Department of Chemistry, University of Ottawa, 10 Marie Curie KZN 6N5 Ottawa, ON, Canada
MONSAERT Stijn	University of Ghent, Department of Inorganic and Physical Chemistry, Krijgslaan 281 (S4), B-9000 Ghent, Belgium
NOLAN Steven	University of St-Andrews, School of Chemistry, Purdie Building, North Haugh, St Andrews, Fife KY16 9ST, UK, E-mail: sn17@st-andrews.ac.uk Institute of Chemical Research of Catalonia, (ICIQ), Av. Països Catalans 16, 43007 Tarragona, Spain
NICOLAE Cristian	Institute of Chemical Research, 202 Spl. Independentei, Bucharest, Romania

NIZOVTSEV Alexey	United Research and Development Center, Moscow Russia
ORNELAS Catia	Nanoscience and Catalysis Group LCOO, Institute of Molecular Science, UMR CNRS No. 5802 University of Bordeaux I, 33405 Talance Cedex, France
PIETRASZUK Cezary	Faculty of Chemistry, Department of Organometallic Chemistry, Adam Mickiewicz University, Grunwaldzka 6, 60-780 Poznan, Poland
POATER Albert	University of Salerno, Department of Chemistry, via Ponte don Melillo, Fisciano, I-84084, Italy, E-mail: apoater@unisa.it
POP Mircea	Institute of Organic Chemistry, Romanian Academy, 202B Spl. Independentei, Bucharest, Romania
RAGONE Francesco	University of Salerno, Department of Chemistry, via Ponte don Melillo, Fisciano, I-84084, Italy
RAZUS Alexandru	Institute of Organic Chemistry, Romanian Academy, Bucharest, Romania
RIBKOVSKAIA Zinaida	Academy of Sciences of Moldova, Institute of Chemistry, Chisinau, Moldova
RIX Diane	UMR CNRS 6226, Ecole Nationale Superieure Chimie de Rennes, 35700 Rennes, France
ROGALSKI Szymon	Faculty of Chemistry, Department of Organometallic Chemistry, Adam Mickiewicz University, Grunwaldzka 6, 60-780 Poznan, Poland
ROGAN Yulia	IRC in Polymer Science and Technology, University of Durham, South Road Durham DHI 3LE, UK
ROSCA Sorin	Department of Organic Chemistry, Polytechnic University of Bucharest, Romania

RUIZ Jaime	Nanoscience and Catalysis Group LCOO, Institute of Molecular Science, UMR CNRS No. 5802 University of Bordeaux I, 33405 Talance Cedex, France
SEDLÁČEK Jan	Heyrovsky Institute of Physical Chemistry, Academy of Sciences of the Czech Republic, Dolejskova 3, 18223 Prague 8, Czech Republic
SPURCACIU Bogdan	Institute of Chemical Research, 202 Spl. Independentei, Bucharest, Romania
STANESCU Michaela	Department of Organic Chemistry, Polytechnic University of Bucharest, Romania
SUCMAN Natalia	Academy of Sciences of Moldova, Institute of Chemistry, Chisinau, Moldova
TEODORESCU Mircea	Department of Macromolecular Chemistry, Polytechnic University of Bucharest, Romania
TUDOSE Adriana	Center for Education and Research on Macromolecules, Institute of Chemistry (B6a), University of Liege, Sart Tilman, B-4000 Liege, Belgium
UDOVYK Oleg	Ukrainian Academy of Sciences, Kiev, Ukraine
VAN DER VOORT Pascal	University of Ghent, Department of Inorganic and Physical Chemistry, Krijgslaan 281 (S4), B-9000 Ghent, Belgium
VASILESCU Sorin	Department of Macromolecular Chemistry, Polytechnic University of Bucharest, Romania
VERPOORT Francis	University of Ghent, Department of Inorganic and Physical Chemistry, Krijgslaan 281 (S4), B-9000 Ghent, Belgium
VULUGA Mircea	Institute of Organic Chemistry, Romanian Academy, 202B Spl. Independentei, Bucharest, Romania

PART I. INNOVATIONS IN CATALYSTS DESIGN

Recent Advances in Ruthenium Catalysts for Alkene Metathesis

Lionel Delaude,[1] Albert Demonceau,[1] Ileana Dragutan,[2] Valerian Dragutan[2*]

[1]Laboratory of Macromolecular Chemistry and Organic Catalysis, Institut de Chimie (B6a), University of Liège, Sart-Tilman, 4000 Liège, Belgium

[2]Institute of Organic Chemistry of the Romanian Academy, 202B Spl. Independentei, 060023 Bucharest, P.O. Box 35-108, Romania

[*]E-mail: vdragutan@yahoo.com

Abstract Although ruthenium initiators currently available for alkene metathesis are endowed with many beneficial properties, there is still room for improvement and many research groups are actively pursuing the quest for the next generation of alkene metathesis catalysts. The present contribution aims at providing a critical survey of some of the most significant achievements accomplished toward this goal during the last few years. New ligands and complexes designed to achieve the appropriate balance between electronic and steric properties of the ruthenium active centres are depicted, and their stability, activity, and chemoselectivity are briefly discussed.

Keywords Alkylidene ligand · Arene ligand · Grubbs catalyst · Hoveyda–Grubbs catalyst · Indenylidene ligand · N-heterocyclic carbene · Phosphine ligand

1 Introduction

Thanks to the development of well-defined molybdenum- and ruthenium-alkylidene catalysts initiated by Schrock and Grubbs in the late 1990s, the various embodiments of alkene metathesis have acquired a central role in organic synthesis and in macromolecular chemistry [1–4]. Indeed, ring-closing metathesis (RCM) or cross-metathesis (CM) are now routinely used for the construction of small organic molecules and macrocycles [5–8], whereas ring-opening metathesis polymerisation (ROMP) is a method of choice for the controlled assembly of functionalised polymer chains [9–12]. Although ruthenium initiators currently available are endowed with many beneficial properties (good reactivity, air and moisture stability, ease of handling and storage, potential for immobilisation), there is still room for improvement and many research groups are actively pursuing the quest for the next generation

V. Dragutan et al. (eds.), *Green Metathesis Chemistry: Great Challenges in Synthesis, Catalysis and Nanotechnology*,

of alkene metathesis catalysts. The present contribution aims at providing a critical survey of some of the most significant achievements accomplished toward this goal during the last few years.

2 Alkylidene Complexes

The routine procedure used for phosphine substitution by N-heterocyclic carbene (NHC) ligands led to the thermally stable, halogenated NHC complex **1** and to the unsymmetrically substituted NHC complexes **2** and **3** containing, respectively, a perfluoroalkyl chain or a silylether derivative [13]. Both species were catalytically active in metathesis reactions. Unsymmetrically substituted complexes **4** could readily metathesise their own ligands to form chelated NHC-ruthenium complexes in which the NHC and the alkylidene unit were tethered by a variable length chain. As an example, the metallacyclic complex **5** was obtained in 75% yield by refluxing a solution of complex **4** (n = 2) in toluene. It was assumed that the catalytic species might be able to regenerate themselves after the productive metathesis is over and the substrate in solution has been quantitatively consumed [13]. A broad palette of chiral NHC ruthenium–benzylidene complexes (**6**) was reported and screened by Grubbs for achieving enantioselectivity in asymmetric

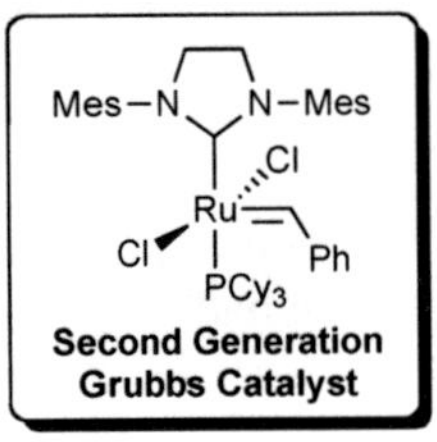

Second Generation Grubbs Catalyst

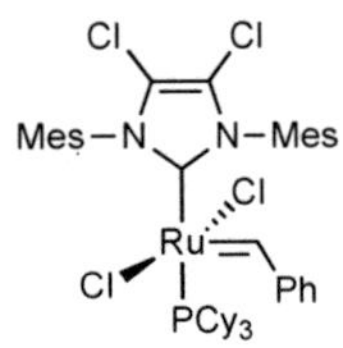

1

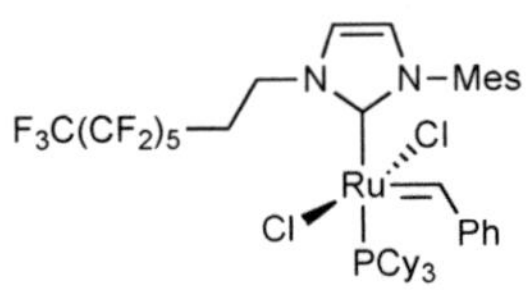

2

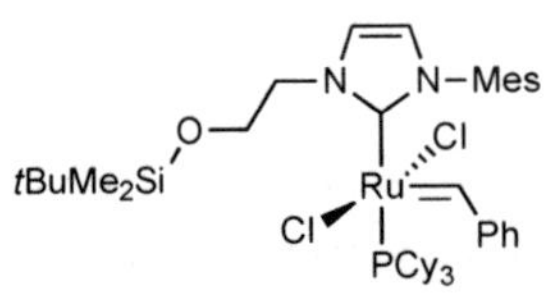

3

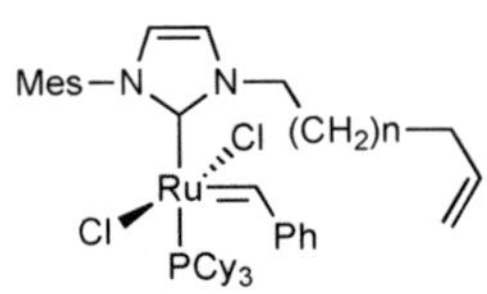

4

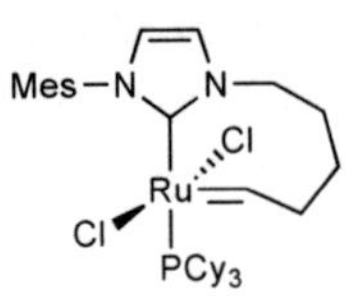

5

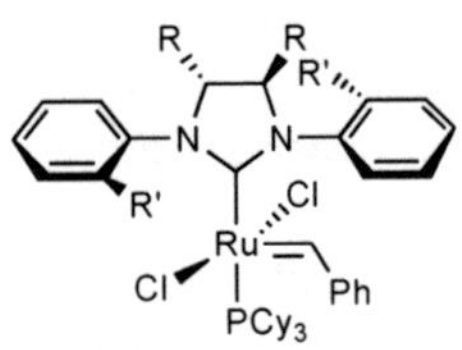

6

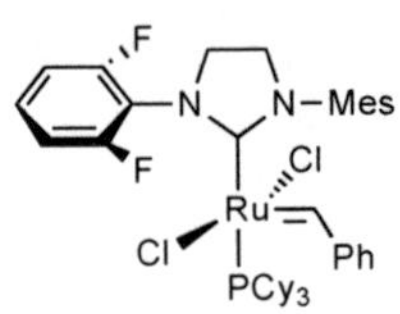

7

ring-closing metathesis (ARCM) [14]. Credit should also be given to the Grubbs group for making available catalysts **7** and its 2-isopropoxybenzylidene analogue, each bearing an unsymmetrical NHC ligand with a fluoro-substituted aromatic unit performing well in RCM, CM, and ROMP reactions [15].

A fresh development concerning Grubbs' benzylidene catalysts was brought up by Fogg et al. who introduced chelating aryloxides (pseudohalides) ligands onto complexes **8–10**. These catalysts performed RCM with exceptionally high efficiency, due to enhanced lability of the substituted pyridine ligand. Moreover, they were easily removed from the reaction media by one-run flash chromatography. This feature makes them promising candidates for application in biological areas or in pharmaceutical and agrochemical industries where colourless and metal-free products are required [16–18].

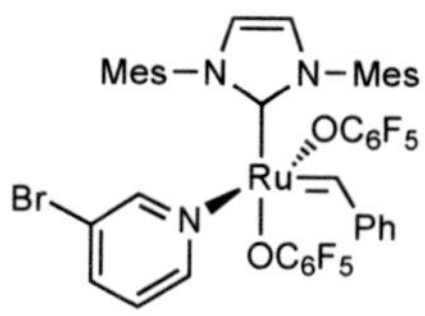

8

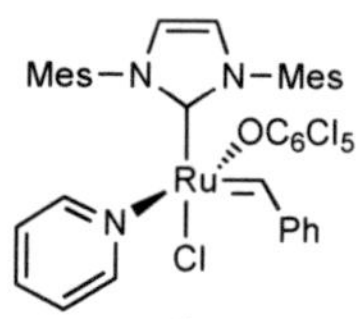

9

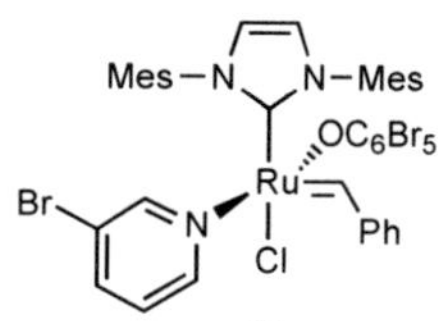

10

Interestingly, N-aryl thiazolin-2-ylidene ligands were proposed by Grubbs for synthesising a novel family of ruthenium complexes in one step, from either $(PCy_3)_2Cl_2Ru{=}CHPh$ or $(PCy_3)Cl_2Ru{=}CH(i\text{-}PrO\text{-}C_6H_4)$. In the solid state, the aryl substituents of the N,S-heterocyclic carbenes were located above the empty coordination site of the ruthenium centre. Despite the decreased steric bulk of these ligands compared to imidazole-based NHCs, complexes **11** and **12** efficiently promoted standard RCM reactions, ROMP of 1,5-cyclooctadiene and norbornene, CM of allylbenzene with *cis*-1,4-diacetoxy-2-butene, as well as the macrocyclic ring-closing of a 14-membered lactone. The phosphine-free catalysts **12** were more stable than complexes **11** and they exhibited pseudo first order kinetics in the RCM of diethyl 2,2-diallylmalonate. Removing steric bulk from the *ortho* positions in the N-aryl group of the thiazolin-2-ylidene decreased their stability. On the other hand, introduction of too bulky substituents resulted in prolonged induction periods. Among the ancillary ligands examined, 3-(2,4,6-trimethylphenyl)- and 3-(2,6-diethylphenyl)-4,5-dimethylthiazolin-2-ylidene afforded the most efficient and stable catalysts. In cross-metathesis of allylbenzene with *cis*-1,4-diacetoxy-2-butene increasing bulkiness at the *ortho* positions in the N-aryl substituents afforded more *Z*-selective catalysts [19].

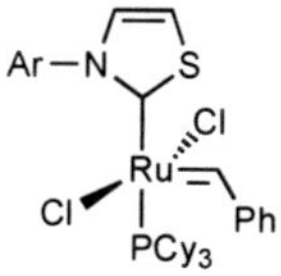

11

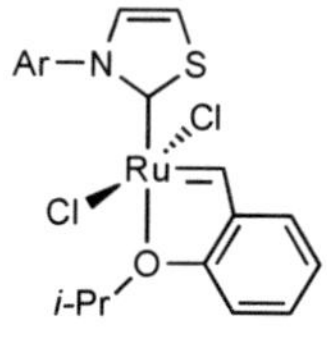

12

3 Oxygen-Chelated Alkylidene Complexes

Grela reported adequate modifications of the Hoveyda–Grubbs catalyst via electronic and structural manipulations of the chelating alkoxy ligand [20]. The resulting new complexes **13**–**15**, in which the alkoxybenzylidene ligand was doubly chelated to the metal centre, proved very stable and active in RCM and enyne metathesis reactions. Surprisingly, complex **15** displayed a latent behaviour in model RCM reactions, turning highly active only upon thermal or chemical treatment. The robust complex **16**, whose metathetical activity paralleled that of the parent Hoveyda–Grubbs catalyst, presented a remarkably high affinity for silica gel when dichloromethane was used as eluent, thereby enabling its efficient removal from the reaction media. Building on this observation, Grela also developed a new strategy for phase-separation and recovery of **16** that afforded products with excellent purities (containing less than 400 ppm of ruthenium) and allowed to recycle the catalyst up to nine times [21].

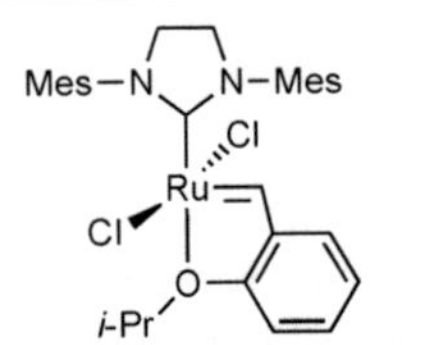

Second Generation Hoveyda–Grubbs Catalyst

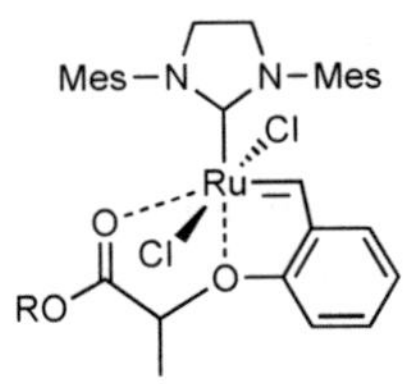

13 (R = H) **14** (R = Me)

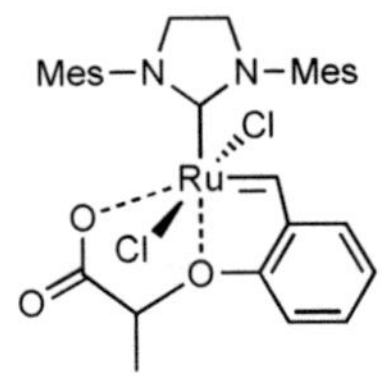

15

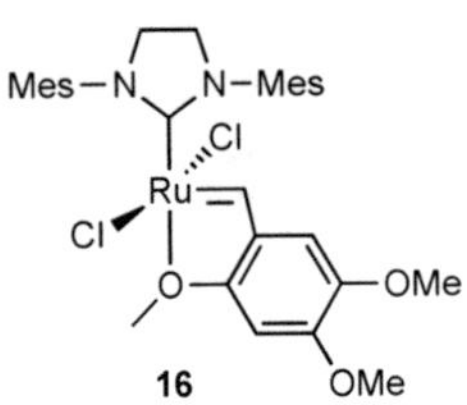

16

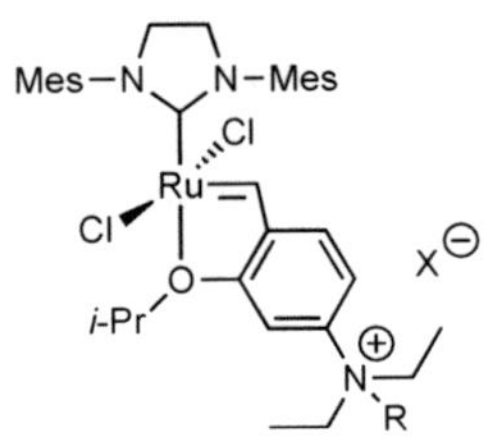

17 (R = H) **18** (R = Me)

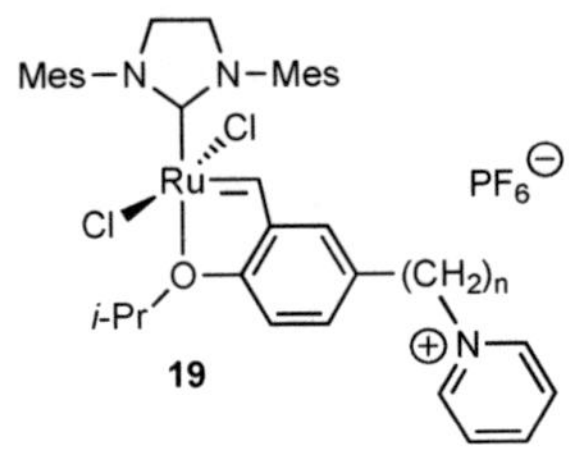

19

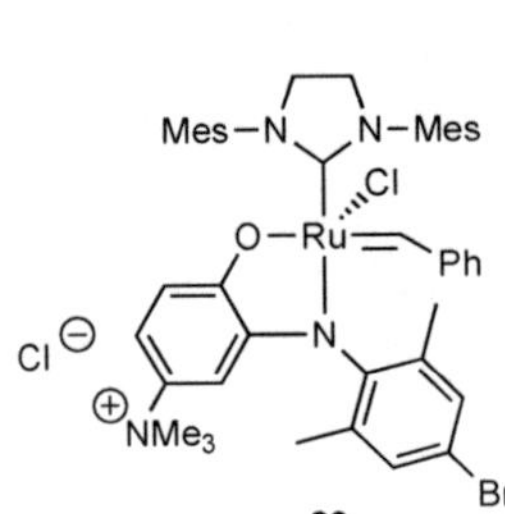

20

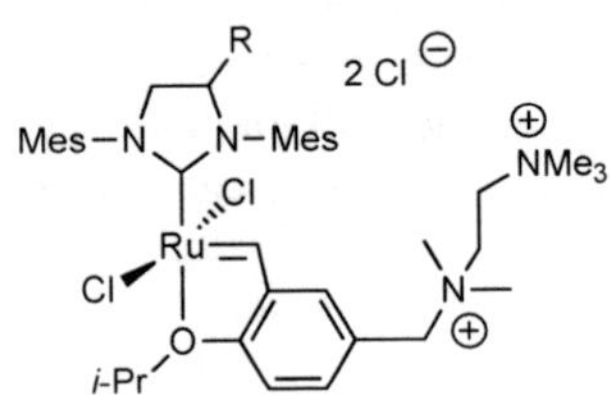

21

We also owe to Grela salt **17**, which is a polar alkene metathesis catalyst prepared *in situ* by neutralisation of the corresponding free amine with Brønsted acids, resulting in an electron-donating to an electron-withdrawing group switch. The quaternary ammonium function not only activated the ruthenium catalyst but also made it more hydrophilic. Consequently, salt **17** could be used for metathesis reactions in conventional solvents, such as dichloromethane or toluene, but also in alcohols, alcohol/water mixtures, or neat water, even in the presence of air. In addition, catalyst **17** acted as an "inisurf" (initiator and surfactant), thereby promoting metathesis in aqueous heterogeneous media [22]. Further valuable additions to the family of polar quaternary ammonium catalysts for metathesis in the "greenest" solvent of all, water, sometimes using ultrasound emulsification, include Grela's complex **18** [23–25], pyridinium salt **19** [26], as well as complexes **20** and **21** reported by Raines [27] and Grubbs [28], respectively.

A fairly recent technique for non-covalent immobilisation of homogeneous ruthenium metathesis catalysts onto liquid supports took advantage of room temperature ionic liquids (RTILs). Representatives from this class of green solvents were first investigated merely as reaction media, in particular in earlier work by Bruneau and Dixneuf using ruthenium–allenylidene precatalysts [29]. Their use was further extended to ionic liquid-tagged catalyst precursors. Thus, several Ru–NHC complexes, including the IL-tagged counterparts of the Hoveyda–Grubbs catalysts, were tested in various metathesis reactions conducted in ILs or IL/organic solvent mixtures (biphasic catalysis). These new complexes demonstrated convenient recyclability combined with high reactivity and led to extremely low residual levels of ruthenium in the reaction products [30–34]. Two Hoveyda-type catalysts containing an IL-tag linked either to the *ortho*-oxygen substituent or to the *meta*-position of the styrenylidene ligand were recently reported [35]. They were evaluated in the RCM of dimethyl 2,2-diallylmalonate and *N*,*N*-diallyltosylamide conducted in an ionic liquid medium, where they showed moderate recyclability, yet good activity for the first cycle. As an alternative method for tagging ruthenium complexes, the "light fluorous" versions of the first and second generation Hoveyda–Grubbs catalysts were also synthesised. They exhibited the expected reactivity profile of their illustrious parents and were readily recovered from reaction mixtures by fluorous solid-phase extraction. Furthermore, they could be recycled routinely five or more times and employed either in a stand-alone fashion, or supported on fluorous silica gel [36].

Ruthenium alkene metathesis catalysts **22–24** bearing cyclic(alkyl)(amino) carbenes (CAACs) were evaluated in the CM of *cis*-1,4-diacetoxy-2-butene with allylbenzene and in the ethenolysis of methyl oleate. Compared to most NHC-based complexes, the CAAC-substituted species afforded lower *E*/*Z* ratios (3:1 at 70% conversion) in the reactions under investigation. Additionally, they exhibited a good selectivity for the formation of terminal alkenes versus internal ones in the ethenolysis process. Indeed, with complex **22**, TONs of 35,000 – the highest recorded to date – were achieved. Importantly, CAAC-substituted ruthenium complexes exhibited markedly different kinetic selectivity from most NHC-substituted complexes [37].

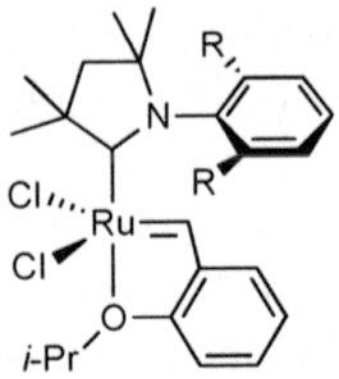

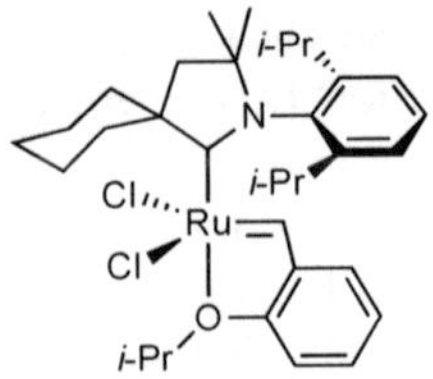

22 (R = *i*-Pr) **23** (R = Et) **24**

4 Other Chelated Alkylidene Complexes

The third generation Grubbs catalyst (**24**) enjoyed further transformation into the chelated complexes **25**–**27**, obtained in high yield (ca. 80%) by stoichiometric cross-metathesis with 2-(3-butenyl)pyridine. Compared to **24**, these new catalysts displayed lower initiation rates and were more latent in RCM and ROMP, especially for a *trans* arrangement of the NHC and pyridine ligands [38]. This behaviour is particularly beneficial for reaction injection molding (RIM) applications, because it allows handling of the monomer/catalyst mixture for longer periods of time before polymerisation begins. Starting again from **24**, other rather stable, yet thermally switchable Ru–NHC initiators of great interest for polymer chemistry (**28** and **29**), were introduced by Slugovc et al. taking advantage of carbene exchange with Schiff bases. Application of these ruthenium complexes as initiators in ROMP revealed considerable latency at room temperature, but very fast polymerisation rates were achieved around 110°C. Catalyst **28** with a six-membered chelate ring displayed a more pronounced latency than complex **29** containing a five-membered chelate ring. This difference translated into a higher switching temperature and a lower polymerisation rate for **28** as compared to **29.** These observations were rationalised taking into account the greater stability of the six-membered chelate ring in **28** compared to the five-membered ring in **29** [39]. Latency effects due to tuning of the bidentate Schiff base ligands were also recently reported by Verpoort and coworkers. For these systems, addition of a controlled amount of hydrochloric acid or trichlorosilane to the reaction media successfully triggered the transformation of dormant precatalysts into active species for the ROMP of cycloalkenes [40, 41].

Lemcoff et al. prepared a new series of sulphur-chelated latent ruthenium alkene metathesis catalysts **30**–**34** that possessed an uncommon *cis*-dichloro arrangement and were mostly inactive at room temperature. Modifications of the size of remote substituents on the sulphur atom significantly affected the catalytic activity at different temperatures. More bulky substituents raised activity at lower temperatures. Catalysts **30**–**34** were also stable in solution and retained their catalytic activity in RCM reactions even after being exposed to air for 2 weeks [42, 43].

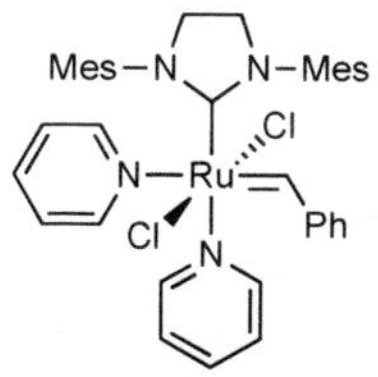

Third Generation Grubbs Catalyst (24)

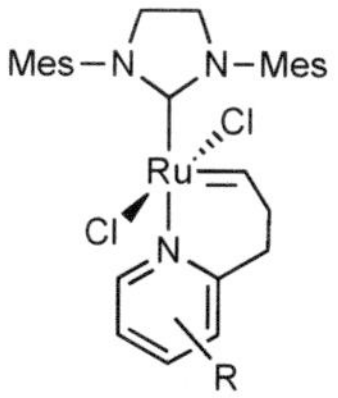

25 (R = H)
26 (R = 2-Me) **27** (R = 4-Me)

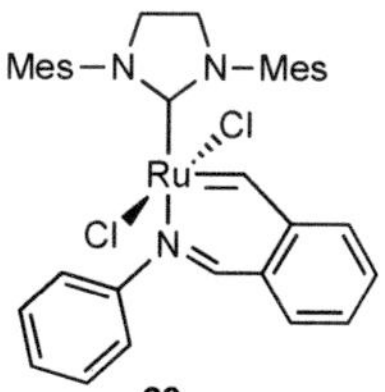

28

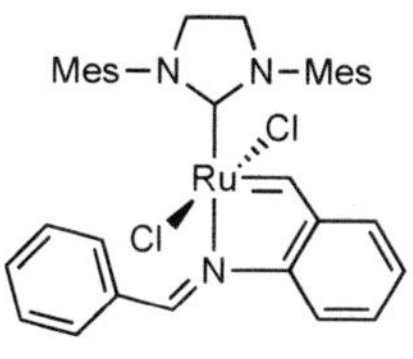

29

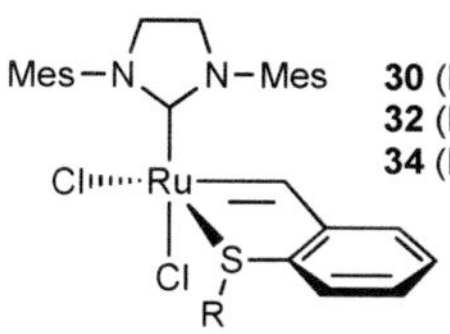

30 (R = Me) **31** (R = Et)
32 (R = *i*-Pr) **33** (R = *t*-Bu)
34 (R = Ph)

5 Indenylidene Complexes

Ruthenium–indenylidene complexes bearing different ancillary ligands have emerged as quite productive and versatile metathesis precatalysts [44, 45]. They proved to be rather robust and remained stable upon heating in solution, a highly desirable feature for various metathetical applications. As a special bonus, they made possible reactions that were not promoted by many earlier ruthenium–benzylidene catalysts. In particular, they allowed the convenient synthesis of tri- and tetra-substituted cycloalkenes, as well as RCM reactions involving highly substituted acrylates. Due to their straightforward access, enhanced activity, increased stability, and extended area of application, they successfully complement conventional ruthenium complexes currently employed in RCM of linear dienes, ADMET of α,ω-dienes, enyne metathesis, and ROMP of cycloalkenes.

The first generation of ruthenium–indenylidene complexes was easily obtained from [$RuCl_2(PPh_3)_4$] and 3,3-diphenylpropyn-3-ol [46, 47]. Upon coordination and rearrangement, this propargyl alcohol derivative afforded the 3-phenylindenylidene fragment. The two triphenylphosphine ligands of complex **35**, which is now commercially available, were subsequently replaced by better donating ligands such as tricyclohexylphosphine (PCy_3), thereby affording complex **36** with improved activity and stability [48]. Further substitution of one phosphine by an imidazolin-2-ylidene or imidazolidin-2-ylidene ligand bearing bulky substituents on its nitrogen atoms led to a second generation of 16-electron ruthenium–indenylidene complexes with excellent catalytic properties [49, 50]. In the most common protocol, addition of 1,3-dimesitylimidazolin-2-ylidene (IMes) or 1,3-dimesitylimidazolidin-2-ylidene (SIMes) to the 3-phenylindenylidene complexes **35** and **36** in toluene at room temperature afforded complexes **37**–**40** in high yields. These catalysts displayed an outstanding catalytic activity and selectivity in the RCM of diethyl 2,2-diallylmalonate and *N,N*-diallyltosylamide [49] and in the CM of *tert*-butyl(hex-5-enyloxy)dimethylsilane with methyl acrylate [50].

Increasing interest in this family of robust and convenient metathesis catalysts triggered synthesis of the third generation ruthenium–indenylidene complexes, e.g. **41–43** and **44** [51, 52]. An independent approach started from the diphosphine indenylidene complex **36** and an aromatic salicylaldimine affording the Schiff base-containing ruthenium–indenylidene complex **44** [53]. Other related Schiff base-ligated ruthenium–indenylidene complexes were also prepared according to the same procedure. Their structures were established by ^{1}H, ^{13}C, and ^{31}P NMR spectroscopies and their catalytic activity was evaluated in the ROMP of cycloalkenes and the atom transfer radical polymerisation (ATRP) of vinyl monomers [54].

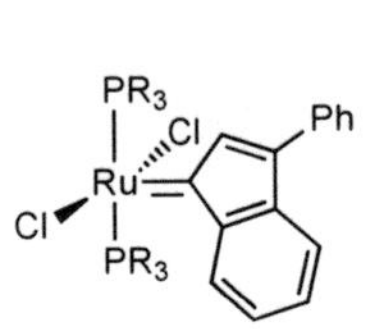

35 (R = Ph) **36** (R = Cy)

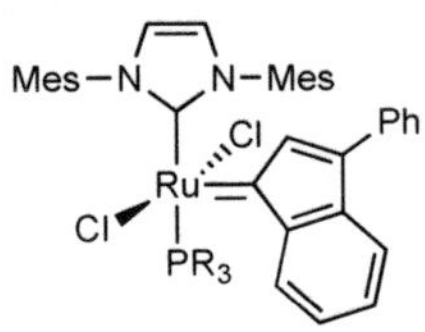

37 (R = Ph) **38** (R = Cy)

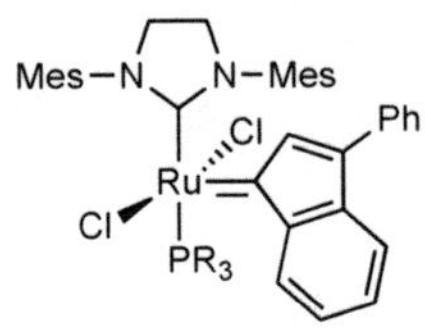

39 (R = Ph) **40** (R = Cy)

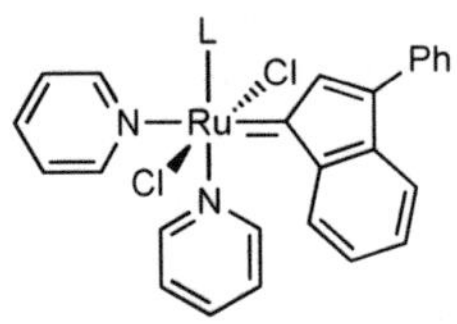

41 (L = PCy3) **42** (L = IMes) **43** (L = SIMes)

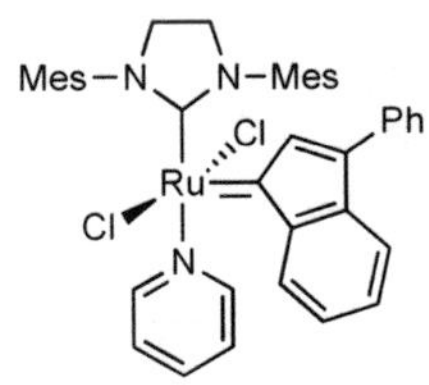

44

6 Arene Complexes

Delaude and Demonceau proposed a distinctive approach for the advancement of metathesis catalysts. It relied on ruthenium–arene complexes generated *in situ* from the $[RuCl_2(p\text{-cymene})]_2$ dimer and NHC ligands. A broad range of imidazolium and imidazolinium salts or 2-carboxylate inner salts were synthesised and their ability to act as stable NHC ligand precursors was thoroughly investigated in the photoinduced ROMP of norbornene and cyclooctene. Thus, 1,3-diarylimidazol(in)ium chlorides bearing diversely substituted, phenyl, 1-naphthyl, or 4-biphenyl groups on their nitrogen atoms were reacted with $[RuCl_2(p\text{-cymene})]_2$ and potassium *tert*-butoxide or sodium hydride to afford the corresponding ruthenium–arene complexes $[RuCl_2(p\text{-cymene})(NHC)]$ *in situ*, e.g. **45** and **46** [55, 56].

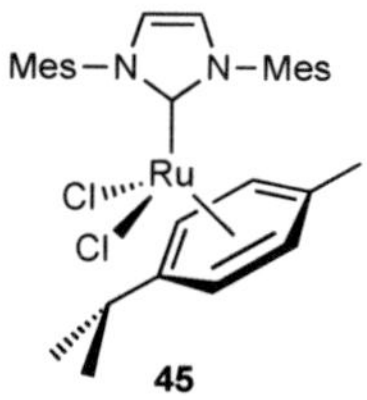

45

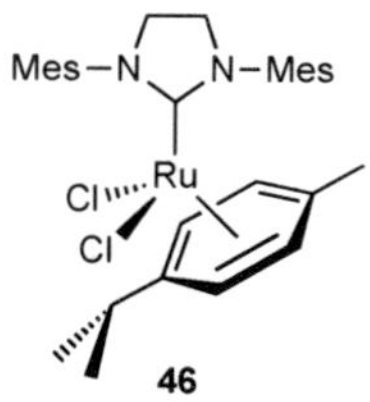

46

Compared to imidazol(in)ium salts, the 2-carboxylate betaines did not require the addition of a strong base to release free carbenes. The catalytic activity of the ruthenium–arene complexes generated *in situ* from $[RuCl_2(p\text{-cymene})]_2$ and five $NHC \cdot CO_2$ adducts was evaluated in the cyclopropanation of styrene and cyclooctene and in the photoinduced ROMP of cyclooctene and norbornene [57]. Irrespective of the ligand precursors used, these ruthenium catalytic systems afforded polynorbornenes that contained mostly *trans* double bonds, as determined by ^{13}C NMR spectroscopy, a feature shared with many other ruthenium precatalysts. A reaction mechanism was postulated, in which the initiating metal–alkylidene species could arise by a direct interaction between the monomer and a coordinatively unsaturated ruthenium species generated *in situ* under the influence of visible light [58]. Indeed, independent investigations showed that the active propagating species involved in the self-metathesis of ethyl oleate obtained from ruthenium–arene catalyst precursors and trimethylsilyldiazomethane were probably identical to those derived from the first generation Grubbs benzylidene complex [59].

7 Homobimetallic Complexes

Delaude and Demonceau obtained homobimetallic ruthenium–arene complexes **47** and **48** in high yields (70–80%) upon heating a toluene solution of $[RuCl_2(p\text{-cymene})]_2$ with 1 eq. of carbene ligand under ethylene atmosphere. Coupling reactions with various styrene derivatives confirmed the outstanding aptitude of complex **47** to catalyse alkene metathesis. Unlike its monometallic counterparts of formula $[RuCl_2(p\text{-cymene})(NHC)]$, the homobimetallic catalyst precursor did not require the addition of a diazo compound nor visible light illumination to initiate the ROMP of norbornene or cyclooctene. When α,ω-dienes were reacted with **47** or **48**, a mixture of cycloisomerization and RCM products was obtained in a non-selective way. Strikingly, addition of a terminal alkyne co-catalyst enhanced the metathetical activity, while completely repressing the cycloisomerisation process. Hence, quantitative conversions of diethyl 2,2-diallylmalonate and *N,N*-diallyltosylamide were achieved within 2 h at room temperature using 2 mol% of **47** and 6 mol% of phenylacetylene [60]. These observations prompted the Liege group to further investigate the role of the alkyne co-catalyst. Thus, a series of new homobimetallic ruthenium–arene complexes bearing vinylidene, allenylidene, and indenylidene ligands was prepared very recently starting from propargyl alcohol derivatives. Their catalytic activities toward several types of alkene metathesis reactions were investigated and they were found valuable intermediates for the safe and efficient one-pot synthesis of Hoveya–Grubbs alkoxybenzylidene catalyst $[Cl_2Ru(PCy_3)(=CH\text{-}o\text{-}O\text{-}i\text{-}PrC_6H_4)]$. In the RCM of diethyl diallylmalonate, the homobimetallic ruthenium–indenylidene complex **49** outperformed all the ruthenium–benzylidene complexes under investigation and was only slightly less efficient than its monometallic parent **36** [61].

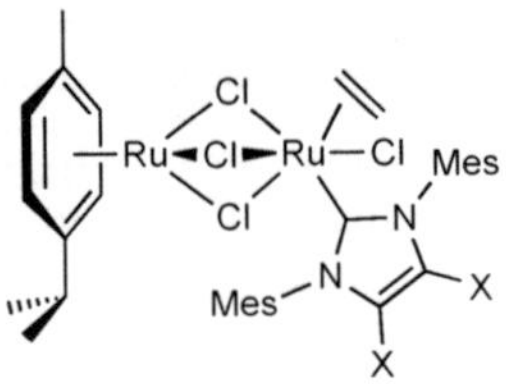

47 (X = H) **48** (X = Cl)

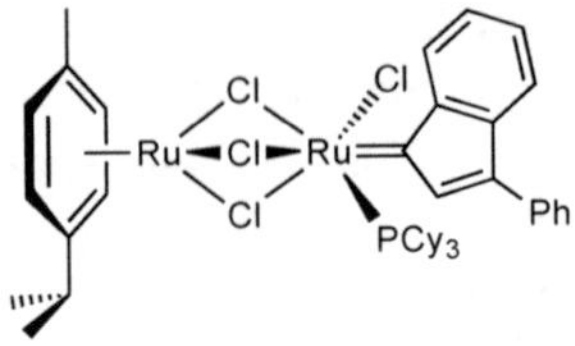

49

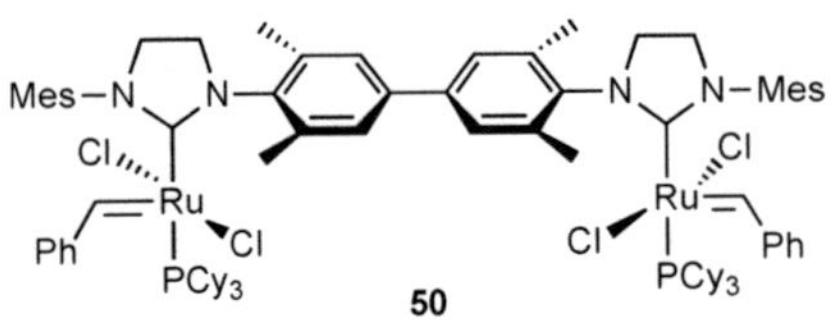

50

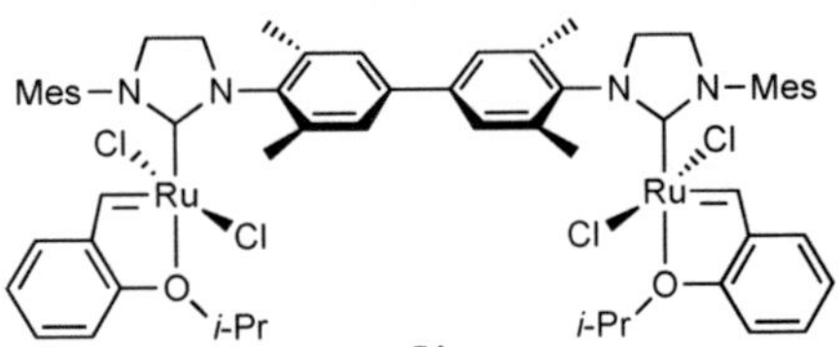

51

Lemcoff et al. designed a novel series of homodinuclear ruthenium catalysts to assist the dimer ring-closing metathesis (DRCM) of α,ω-dienes. Using species such as **50** and **51** to catalyse the metathesis of trideca-1,13-diene, they succeeded in directing selectively the coupling process toward the formation of the cyclic dimer, cyclododeca-1,12-diene, at the expense of usual RCM product, cycloundecene, or the linear ADMET polymers [62].

8 Conclusion and Perspectives

During the past few years, a rich assortment of new ruthenium catalysts were developed for alkene metathesis. Ligands on these species were carefully designed to achieve the appropriate balance between electronic and steric properties of the metal centre, thereby fine-tuning its stability, activity, and chemoselectivity. Future developments will undoubtedly further improve overall catalytic efficiencies. Increasingly so, a main research target will be also to increase tolerance of ruthenium promoters toward the widest range of unprotected functional groups, air, and moisture, widening accordingly their application profiles. In addition, catalyst recycling capability and the removal of metal traces from the products is likely to become a mandatory requirement for the next generation of alkene metathesis initiators [63].

Acknowledgments Financial support from the Romanian Academy and the Belgian C.G.R.I. is gratefully acknowledged.

References

[1] Grubbs RH (ed.) (2003) Handbook of metathesis. Wiley-VCH, Weinheim

[2] Astruc D (2005) The metathesis reaction: from a historical perspective to recent developments. New J Chem 29: 42–56

[3] Astruc D (2006) Answer to Katz's criticisms on the history of metathesis. New J Chem 30: 1848–1852

[4] Delaude L, Noels AF (2007) Metathesis. In: Seidel A (ed.), Kirk-Othmer encyclopedia of chemical technology. Wiley, New York, vol. 26, pp. 920–958

[5] Connon SJ, Blechert S (2003) Recent developments in olefin cross-metathesis. Angew Chem Int Ed 42: 1900–1923

[6] Nicolaou KC, Bulger PG, Sarlah D (2005) Metathesis reactions in total synthesis. Angew Chem Int Ed 44: 4490–4527

[7] Dragutan V, Dragutan I (2006) A resourceful new strategy in organic synthesis: tandem and stepwise metathesis/non-metathesis catalytic processes. J Organomet Chem 691: 5129–5147

[8] Kotha S, Lahiri K (2007) Synthesis of diverse polycyclic compounds via catalytic metathesis. Synlett 18: 2767–2784

[9] Frenzel U, Nuyken O (2002) Ruthenium-based metathesis initiators: development and use in ring-opening metathesis polymerization. J Polym Sci A: Polym Chem 40: 2895–2916

[10] Slugovc C (2004) The ring opening metathesis polymerisation toolbox. Macromol Rapid Commun 25: 1283–1297

[11] Dragutan V, Dragutan I, Fischer H (2008) Synthesis of metal-containing polymers via ring opening metathesis polymerization (ROMP). Part I: polymers containing main group metals. J Inorg Organomet Polym Mater 18: 18–31

[12] Dragutan I, Dragutan V, Fischer H (2008) Synthesis of metal-containing polymers via ring opening metathesis polymerization (ROMP). Part II: polymers containing transition metals. J Inorg Organomet Polym Mater 18: 311–324

[13] Fürstner A, Ackermann L, Gabor B, Goddard R, Lehmann CW, Mynott R, Stelzer F, Thiel OR (2001) Comparative investigation of ruthenium-based metathesis catalysts bearing N-heterocyclic carbene (NHC) ligands. Chem Eur J 7: 3236–3253

[14] Seiders TJ, Ward DW, Grubbs RH (2001) Enantioselective ruthenium-catalyzed ring-closing metathesis. Org Lett 3: 3225–3228

[15] Vougioukalakis GC, Grubbs RH (2007) Ruthenium olefin metathesis catalysts bearing an N-fluorophenyl-N-mesityl-substituted unsymmetrical N-heterocyclic carbene. Organometallics 26: 2469–2472

[16] Conrad JC, Parnas HH, Snelgrove JL, Fogg DE (2005) Highly Efficient Ru−pseudohalide catalysts for olefin metathesis. J Am Chem Soc 127: 11882–11883

[17] Conrad JC, Camm KD, Fogg DE (2006) Ru-aryloxide metathesis catalysts with enhanced lability: assessing the efficiency and homogeneity of initiation via ring-opening metathesis polymerization studies. Inorg Chim Acta 359: 1967–1973

[18] Monfette S, Fogg DE (2006) Ruthenium metathesis catalysts containing chelating aryloxide ligands. Organometallics 25: 1940–1944

[19] Vougioukalakis GC, Grubbs RH (2008) Synthesis and activity of ruthenium olefin metathesis catalysts coordinated with thiazol-2-ylidene ligands. J Am Chem Soc 130: 2234–2245

[20] Michrowska A, Bujok R, Harutyunyan S, Sashuk V, Dolgonos G, Grela K (2004) Nitro-substituted Hoveyda–Grubbs ruthenium carbenes: enhancement of catalyst activity through electronic activation. J Am Chem Soc 126: 9318–9325
[21] Michrowska A, Gułajski Ł, Grela K (2006) A simple and practical phase-separation approach to the recycling of a homogeneous metathesis catalyst. Chem Commun: 841–843
[22] Gułajski Ł, Michrowska A, Bujok R, Grela K (2006) New tunable catalysts for olefin metathesis: controlling the initiation through electronic factors. J Mol Catal A: Chem 254: 118–123
[23] Michrowska A, Gułajski Ł, Kaczmarska Z, Mennecke K, Kirschning A, Grela K (2006) A green catalyst for green chemistry: synthesis and application of an olefin metathesis catalyst bearing a quaternary ammonium group. Green Chem 8: 685–688
[24] Gułajski Ł, Sledz P, Lupa A, Grela K (2008) Olefin metathesis in water using acoustic emulsification. Green Chem 10: 271–274
[25] Gułajski Ł, Michrowska A, Naroznik J, Kaczmarska Z, Rupnicki L, Grela K (2008) A highly active aqueous olefin metathesis catalyst bearing a quaternary ammonium group. ChemSusChem 1: 103–109
[26] Rix D, Caïjo F, Laurent I, Gułajski Ł, Grela K (2007) Highly recoverable pyridinium-tagged Hoveyda–Grubbs pre-catalyst for olefin metathesis. Design of the boomerang ligand toward the optimal compromise between activity and reusability. Chem Commun: 3771–3773
[27] Binder JB, Guzei IA, Raines RT (2007) Salicylaldimine ruthenium alkylidene complexes: metathesis catalysts tuned for protic solvents. Adv Synth Catal 349: 395–404
[28] Jordan JP, Grubbs RH (2007) Small-molecule N-heterocyclic-carbene-containing olefin-metathesis catalysts for use in water. Angew Chem Int Ed 46: 5152–5155
[29] Bruneau C, Dixneuf PH (2006) Metal vinylidenes and allenylidenes in catalysis: applications in anti-Markovnikov additions to terminal alkynes and alkene metathesis. Angew Chem Int Ed 45: 2176–2203
[30] Yao Q, Zhang Y (2003) Olefin metathesis in the ionic liquid 1-butyl-3-methylimidazolium hexafluorophosphate using a recyclable Ru catalyst: remarkable effect of a designer ionic tag. Angew Chem Int Ed 42: 3395–3398
[31] Audic N, Clavier H, Mauduit M, Guillemin J-C (2003) An ionic liquid-supported ruthenium carbene complex: a robust and recyclable catalyst for ring-closing olefin metathesis in ionic liquid. J Am Chem Soc 125: 9248–9249
[32] Clavier H, Audic N, Mauduit M, Guillemin J-C (2004) Ring-closing metathesis in biphasic BMI·PF6 ionic liquid/toluene medium: a powerful recyclable and environmentally friendly process. Chem Commun: 2282–2283
[33] Yao Q, Sheets M (2005) An ionic liquid-tagged second generation Hoveyda–Grubbs ruthenium carbene complex as highly reactive and recyclable catalyst for ring-closing metathesis of di-, tri- and tetrasubstituted dienes. J Organomet Chem 690: 3577–3584
[34] Clavier H, Audic N, Guillemin J-C, Mauduit M (2005) Olefin metathesis in room temperature ionic liquids using imidazolium-tagged ruthenium complexes. J Organomet Chem 690: 3585–3599
[35] Thurier C, Fischmeister C, Bruneau C, Olivier-Bourbigou H, Dixneuf PH (2007) Ionic imidazolium containing ruthenium complexes and olefin metathesis in ionic liquids. J Mol Catal A: Chem 268: 127–133
[36] Matsugi M, Curran DP (2005) Synthesis, reaction, and recycle of light fluorous Grubbs–Hoveyda catalysts for alkene metathesis. J Org Chem 70: 1636–1642
[37] Anderson DR, Ung T, Mkrtumyan G, Bertrand G, Grubbs RH, Schrodi Y (2008) Kinetic selectivity of olefin metathesis catalysts bearing cyclic (alkyl)(amino)carbenes. Organometallics 27: 563–566
[38] Ung T, Hejl A, Grubbs RH, Schrodi Y (2004) Latent ruthenium olefin metathesis catalysts that contain an N-heterocyclic carbene ligand. Organometallics 23: 5399–5401

[39] Slugovc C, Burtscher D, Stelzer F, Mereiter K (2005) Thermally switchable olefin metathesis initiators bearing chelating carbenes: influence of the chelate's ring size. Organometallics 24: 2255–2258

[40] Allaert B, Dieltiens N, Ledoux N, Vercaemst C, Van Der Voort P, Stevens CV, Linden A, Verpoort F (2006) Synthesis and activity for ROMP of bidentate Schiff base substituted second generation Grubbs catalysts. J Mol Catal A: Chem 260: 221–226

[41] Ledoux N, Alaert B, Schaubroeck D, Monsaert S, Drozdzak R, Van Der Voort P, Verpoort F (2006) In situ generation of highly active olefin metathesis initiators. J Organomet Chem 691: 5482–5486

[42] Ben-Asuly A, Tzur E, Diesendruck CE, Sigalov M, Goldberg I, Lemcoff NG (2008) A thermally switchable latent ruthenium olefin metathesis catalyst. Organometallics 27: 811–813

[43] Kost T, Sigalov M, Goldberg I, Ben-Asuly A, Lemcoff NG (2008) Latent sulfur chelated ruthenium catalysts: steric acceleration effects on olefin metathesis. J Organomet Chem 693: 2200–2203

[44] Dragutan V, Dragutan I, Verpoort F (2005) Ruthenium indenylidene complexes. Metathesis catalysts with enhanced activity. Platinum Metals Rev 49: 33–40

[45] Boeda F, Clavier H, Nolan SP (2008) Ruthenium–indenylidene complexes: powerful tools for metathesis transformations. Chem Commun: 2726–2740

[46] Fürstner A, Hill AF, Liebl M, Wilton-Ely JDET (1999) Coordinatively unsaturated ruthenium allenylidene complexes: highly effective, well defined catalysts for the ring-closure metathesis of ,-dienes and dienynes. Chem Commun: 601–603

[47] Fürstner A, Guth O, Düffels A, Seidel G, Liebl M, Gabor B, Mynott R (2001) Indenylidene complexes of ruthenium: optimized synthesis, structure elucidation, and performance as catalysts for olefin metathesis – application to the synthesis of the ADE-ring system of Nakadomarin A. Chem Eur J 7: 4811–4820

[48] Jafarpour L, Schanz H-J, Stevens ED, Nolan SP (1999) Indenylidene–imidazolylidene complexes of ruthenium as ring-closing metathesis catalysts. Organometallics 18: 5416–5419

[49] Clavier H, Nolan SP (2007) N-Heterocyclic carbene and phosphine ruthenium indenylidene precatalysts: a comparative study in olefin metathesis. Chem Eur J 13: 8029–8036

[50] Boeda F, Bantreil X, Clavier H, Nolan SP (2008) Ruthenium–indenylidene complexes: scope in cross-metathesis transformations. Adv Synth Catal 350: 2959–2966

[51] de Fremont P, Clavier H, Montembault V, Fontaine L, Nolan SP (2008) Ruthenium–indenylidene complexes in ring opening metathesis polymerization (ROMP) reactions. J Mol Catal A: Chem 283: 108–113

[52] Monsaert S, Drozdzak R, Dragutan V, Dragutan I, Verpoort F (2008) Indenylidene–ruthenium complexes bearing saturated N-heterocyclic carbenes: synthesis and catalytic investigation in olefin metathesis reactions. Eur J Inorg Chem: 432–440

[53] Opstal T, Verpoort F (2002) Ruthenium indenylidene and vinylidene complexes bearing Schiff bases: potential catalysts in enol-ester synthesis. Synlett: 935–941

[54] Opstal T, Verpoort F (2003) Synthesis of highly active ruthenium indenylidene complexes for atom-transfer radical polymerization and ring-opening-metathesis polymerization. Angew Chem Int Ed 42: 2876–2879

[55] Delaude L, Szypa M, Demonceau A, Noels AF (2002) New in situ generated ruthenium catalysts bearing N-heterocyclic carbene ligands for the ring-opening metathesis polymerization of cyclooctene. Adv Synth Catal 344: 749–756

[56] Maj AM, Delaude L, Demonceau A, Noels AF (2007) Synthesis of N-heterocyclic carbene precursors bearing biphenyl units and their use in ruthenium-catalyzed ring-opening metathesis polymerization. J Organomet Chem 692: 3048–3056

[57] Tudose A, Demonceau A, Delaude L (2006) Imidazol(in)ium-2-carboxylates as N-heterocyclic carbene precursors in ruthenium-arene catalysts for olefin metathesis and cyclopropanation. J Organomet Chem 691: 5356–5365

[58] Delaude L, Demonceau A, Noels AF (2006) Synthesis and application of new N-heterocyclic carbene ruthenium complexes in catalysis: a case study. Curr Org Chem 10: 203–215

[59] Ahr M, Thieuleux C, Copéret C, Fenet B, Basset J-M (2007) Noels' vs. Grubbs' catalysts: evidence for one unique active species from two different systems! Adv Synth Catal 349: 1587–1591

[60] Sauvage X, Borguet Y, Noels AF, Delaude L, Demonceau A (2007) Homobimetallic ruthenium-N-heterocyclic carbene complexes: synthesis, characterization, and catalytic applications. Adv Synth Catal 349: 255–265

[61] Sauvage X, Borguet Y, Zaragoza G, Demonceau A, Delaude L (2009) Homobimetallic ruthenium vinylidene, allenylidene, and indenylidene complexes: synthesis, characterization, and catalytic studies. Adv Synth Catal 351: 441–455

[62] Tzur E, Ben-Asuly A, Diesendruck CE, Goldberg I, Lemcoff NG (2008) Homodinuclear ruthenium catalysts for dimer ring-closing metathesis. Angew Chem Int Ed 47: 6422–6425

[63] Clavier H, Grela K, Kirschning A, Mauduit M, Nolan SP (2007) Sustainable concepts in olefin metathesis. Angew Chem Int Ed 46: 6786–6801

New N-Heterocyclic Carbene Ligands in Grubbs and Hoveyda–Grubbs Catalysts

Stijn Monsaert, Nele Ledoux, Renata Drozdzak, Pascal Van Der Voort, Francis Verpoort*

Department of Inorganic and Physical Chemistry, Ghent University, Krijgslaan 281 (S3), B-9000 Ghent, Belgium

*Fax: (+) 32-9-264-4183; e-mail: Francis.Verpoort@UGent.be

Abstract A series of N-heterocyclic carbene (NHC) ligands bearing aliphatic amino side groups were synthesized and reacted with the Grubbs first generation catalyst. Reactions involving symmetrical, aliphatic NHCs did not allow the isolation of any pure NHC substituted complexes due to their instability. Unsymmetrical NHCs having a planar mesityl group on one amino side reacted with Grubbs catalyst in a favorable manner, and the resulting complexes were stable enough to be isolated. X-ray crystallographic analysis demonstrated that the mesityl group is co-planar with the phenyl ring of the benzylidene, which indicates that a π–π interaction between the mesityl arm and the benzylidene moiety might constitute an important structural element. Catalysts substituted with an NHC derived from a primary or secondary amino-group were found to surpass the parent-complex for the ROMP of cycloocta-1,5-diene. The catalyst substituted with an NHC derived from tBu-NH_2 was considerably less metathesis active. Also new N-alkyl-N′-(2,6-diisopropylphenyl) heterocyclic carbenes were synthesized. These NHC ligands revealed a different reactivity towards Grubbs complexes than the hitherto reported imidazolinylidenes: (i) facile bis(NHC) coordination was found, and (ii) both NHCs on the bis(NHC) complexes can be exchanged with a phosphine, thereupon regenerating the Grubbs first generation complex. Furthermore, a comparison between the classical Hoveyda–Grubbs complexes and complexes substituted with N-alkyl-N′-(aryl) heterocyclic carbenes demonstrates that the introduction of one aliphatic group into the NHC framework does not improve the catalytic activity in any of the tested metathesis reactions. The introduction of two aliphatic amino side groups enhances the reactivity in the ROMP reaction while the increase of steric interactions lowers the CM activity. The lower activity of the N-alkyl-N′-(2,6-diisopropylphenyl) heterocyclic carbene complexes compared with the N-alkyl-N′-mesityl heterocyclic carbene complexes, may analogously be attributed to a more demanding steric environment. While small differences in donor capacities might cause a significantly different catalytic behavior, it is thus plausible that subtle steric differences exert a more determining influence on the activity of the catalysts. In addition, the obtained results confirm that the NHC's amino side groups play a pivotal role in determining the reactivity, selectivity as well as the stability of the corresponding catalysts.

V. Dragutan et al. (eds.), *Green Metathesis Chemistry: Great Challenges in Synthesis, Catalysis and Nanotechnology*,

Keywords Ruthenium catalysts · NHC ligands · Bis(NHC) complexes · Cross-Metathesis · Ring-Closing Metathesis · ROMP · Cyclooctadiene

1 Introduction

Introduction of N-heterocyclic carbene ligands [1] in Ru olefin metathesis catalysts opened up a new era in the development of highly active and stable catalysts for the formation of carbon–carbon double bonds by the olefin metathesis transformation [2]. 4,5-dihydro-1,3-(aryl)-imidazol(in)-2-ylidenes (aryl = mesityl; 2,6-diisopropyl-phenyl) derived from the commercially available carbene precursors remain to be the only examples of symmetrical saturated N-heterocyclic carbene ligands [3]. While some literature reports describe the functionalization of these N-heterocyclic carbenes ligands, others strive for the fine tuning of these ligands, aiming at enhanced catalyst stability, activity and/or selectivity [4]. Regarding the fine tuning of these ligands, Mol et al. reported on the synthesis of symmetric and asymmetric N-heterocyclic carbenes precursors bearing respectively adamantyl (H_2IAd) and adamantyl mesityl (H_2IAdMes) substituents [5]. While the symmetrical H_2IAd ligand did not yield the desired catalyst assigned to the steric bulk of the adamantyl side groups, the asymmetrical H_2IAdMes afforded its second generation Grubbs' catalyst analogue. X-ray analysis further showed only one isomer with the mesityl group coplanar above the benzylidene phenyl ring. The described complex exhibits negligible activity ascribed to the steric blocking of the adamantyl group.

Therefore, we aimed at further fine tuning of these ligands, i.e. reduction of the steric bulk of the amino side groups was elaborated in order to succeed in isolation of these Grubbs' type catalysts bearing symmetric aliphatic and asymmetric aliphatic-aromatic N-heterocyclic carbene ligands and to improve their reactivity. Furthermore, application of these ligands in the Grubbs–Hoveyda type catalysts was fully examined [6].

2 Results and Discussion

2.1 N,N′-Dialkyl Heterocyclic Carbenes in Grubbs' Catalyst [6a]

The symmetric N,N′-dialkyl heterocyclic carbene precursors used for our study are summed in Figure 1. Whereas ligand precursors **b–d** bear respectively tertiary, secondary and primary alkyl substituents in order to alter the steric bulk of these side groups, carbene precursors **a** and **e** were chosen in order to induce some selectivity at the ruthenium carbene center.

Introduction of these ligands in the Grubbs' catalyst was achieved by in situ deprotonation of the carbene by addition of 1 eq. KHMDS relative to the dihydroimidazolinium chloride **a** and Grubbs' first generation catalyst, **1**. Since no reaction was observed, a more than twofold excess of the carbene precursor was added. At this point, the ^{31}P NMR spectrum showed a new signal at 20.14 ppm and a signal of free PCy_3 while all first generation catalyst had been consumed. Efforts to isolate this compound, i.e. precipitation of the complex or column chromatography on silica gel did not allow isolation of the NHC-substituted complex. Likewise, ligand precursors c and e led to the observation of the corresponding NHC-Ru complexes, with new ^{1}H benzylidene resonances at respectively 20.28 and 20.08 ppm, however without allowance for their isolation. In case of ligand precursors **b** and **d**, no reaction was observed.

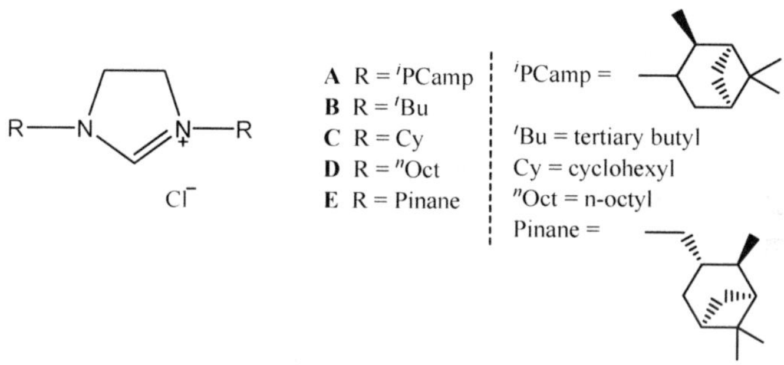

Figure 1 N,N′-dialkyl heterocyclic carbene precursors **a–e**

2.2 N-Alkyl-N′-Mesityl Heterocyclic Carbenes in Grubbs' Catalyst [6a]

Since these symmetric aliphatic NHC-ligands did not yield the desired second generation analogues, efforts were directed to their corresponding asymmetric N-alkyl-N′-mesityl heterocyclic carbene substituted analogues **3a–e** (Figure 2).

Their synthesis was straightforward, following a procedure previously reported in literature [7]. Using KHMDS to deprotonate the imidazolinium chlorides in presence of Grubbs first generation catalyst sufficed to give full conversion of the starting complex under mild conditions and to afford the isolated complexes in high yields. Single-crystal X-ray analysis of complex **3c** unambiguously exemplified the coordination of the NHC ligand in such a way that the mesityl group is nearly co-planar to and above the phenyl ring from the benzylidene ligand (Figure 3, Table 1). This specific coordination leaves the possibility of a π–π interaction between the discussed aromatic rings and allows for the assumption that such interaction might constitute a strong structural element, since the comparable symmetric bisalkyl NHC-substituted complexes did not allow for their isolation.

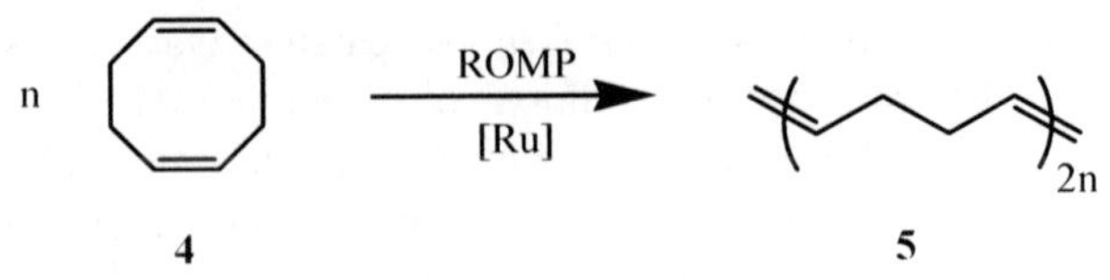

Scheme 1 Ring-opening metathesis polymerization of cis,cis-1,5-cyclooctadiene (4)

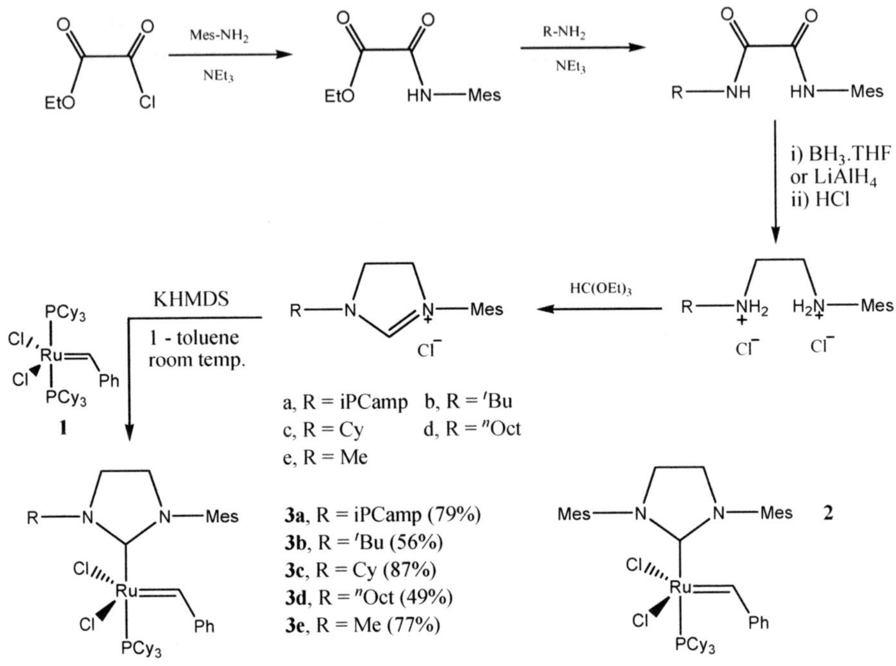

Figure 2 Synthesis of complexes 3a–e

In order to further examine the obtained complexes, their activity towards the Ring-Opening Metathesis Polymerization of COD, 4, in various solvents, i.e. $CDCl_3$ and C_6D_6, was compared to the activity of Grubbs' first and second generation (Scheme 1, Figure 4).

Whereas the activity of 3b is negligible, the incorporation of asymmetric NHC's stimulated the catalyst activity for all other complexes. Complex 3b, however, still managed to afford 72% conversion after 2 h at 70°C. While the catalytic activity of complex 2 is strongly enhanced in aromatic solvents compared to its activity in $CDCl_3$, activity of our complexes 3a and 3c–e proves slightly lower. At higher monomer to catalyst ratios (3,000/1), a slight increase of activity is noticed. The increase of catalyst activity in aromatic solvents has been described by Fürstner et al. for $Cl_2Ru(=CHPh)(IMes)(PCy_3)$ [9], and is assigned to π–π interactions between the mesityl group and solvent molecules, consequently reducing the intramolecular π–π interaction. This observation can be rationalized through the assumption that the solvent molecules interact with the mesityl group which is not co-planar to the benzylidene unit. Alternatively, it can be reasoned that rotation of the NHC ligand around the Ru-C bond allows for two populations stabilized by π–π interactions, compared to only one for the asymmetric NHC-substituted catalysts.

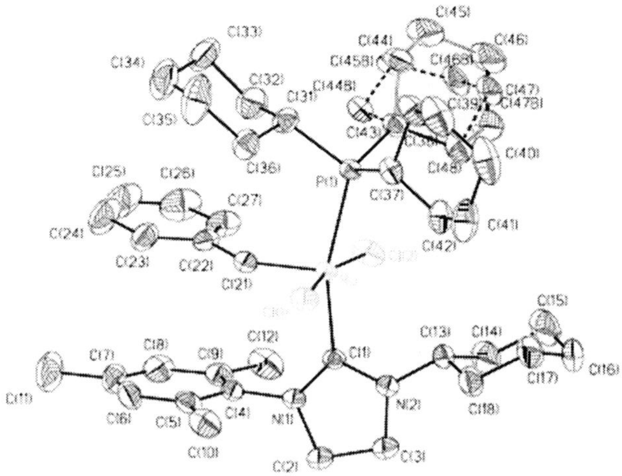

Figure 3 ORTEP diagram of **3c**. For clarity hydrogen atoms have been omitted

Table 1 Selected bond lengths (Å) and angles (°) for complexes **3c** and **2**

	3c[a]	**2**[b]
Ru=C	1.830(6)	1.835(2)
Ru-CNN	2.060(5)	2.085(2)
RuCl(1)	2.419(3)	2.3988(5)
RuCl(2)	2.376(3)	2.3912(5)
Ru-P	2.481(3)	2.4225(5)
Cl-Ru-Cl	167.73(6)	167.71(2)
N_2C-Ru-P	162.42(14)	163.73(6)
N_2C-Ru=C	98.9(2)	100.24(8)
P-Ru=C	98.63(18)	95.98(6)
N_2C-Ru-Cl(1)	85.58(15)	83.26(5)
N_2C-Ru-Cl(2)	89.18(16)	94.55(5)
Ru=C-Ph	137.2(4)	136.98(16)

[a]This work.
[b]Refs. [1i, 8].

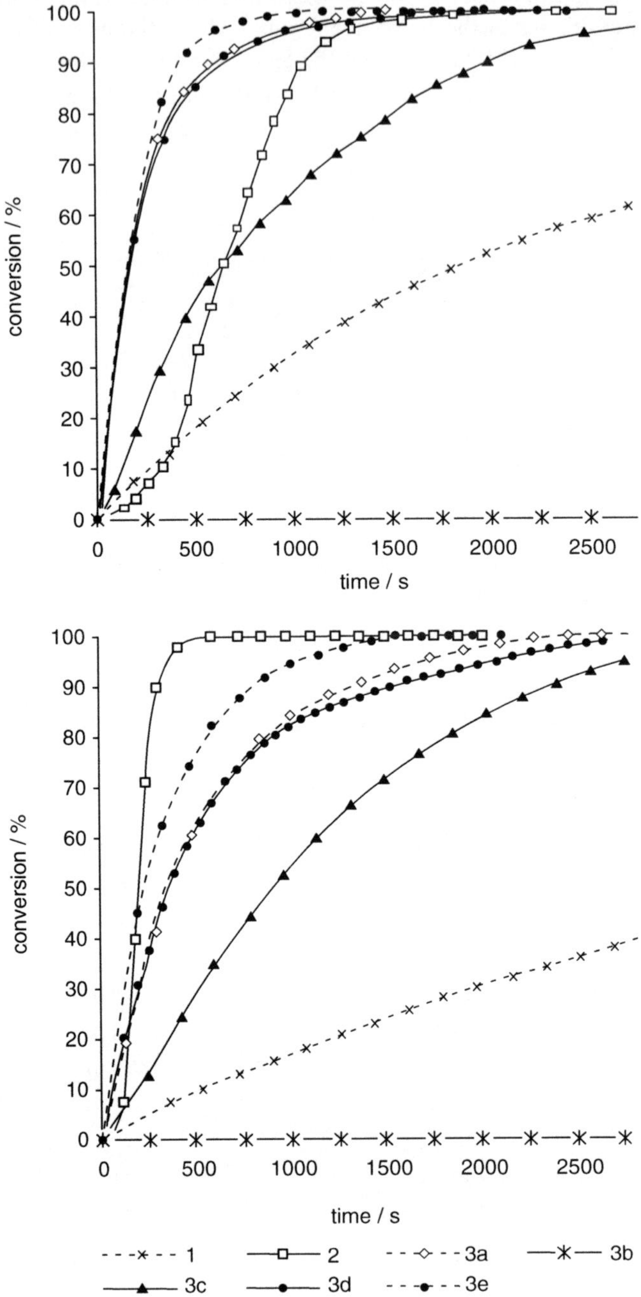

Figure 4 Monitoring ROMP of COD via ^{1}H NMR spectroscopy. COD/cat. = 300, cat. conc. = 4.52 mM, 20°C, solvent = $CDCl_3$ (*top*), C_6D_6 (*bottom*)

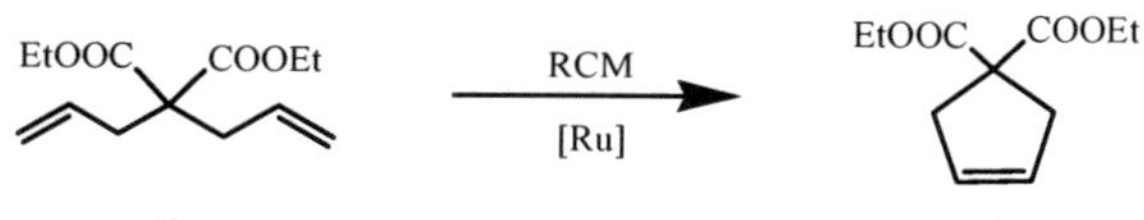

Scheme 2 Ring-closing metathesis of diethyl diallylmalonate (**6**)

Further investigation of these catalysts established their activity for the RCM of diethyl diallylmalonate, **6** (Scheme 2, Figure 5). It is clearly shown that the bulkiness of the amino side group has a profound influence on the catalyst activity, i.e. the reduction of the bulkiness of the aliphatic group affords higher activities. During the course of our research, Blechert et al. [10] reported on the synthesis of complex **3e** as well and the results show improved diastereoselectivities in RCM and significantly different E/Z ratios in cross metathesis.

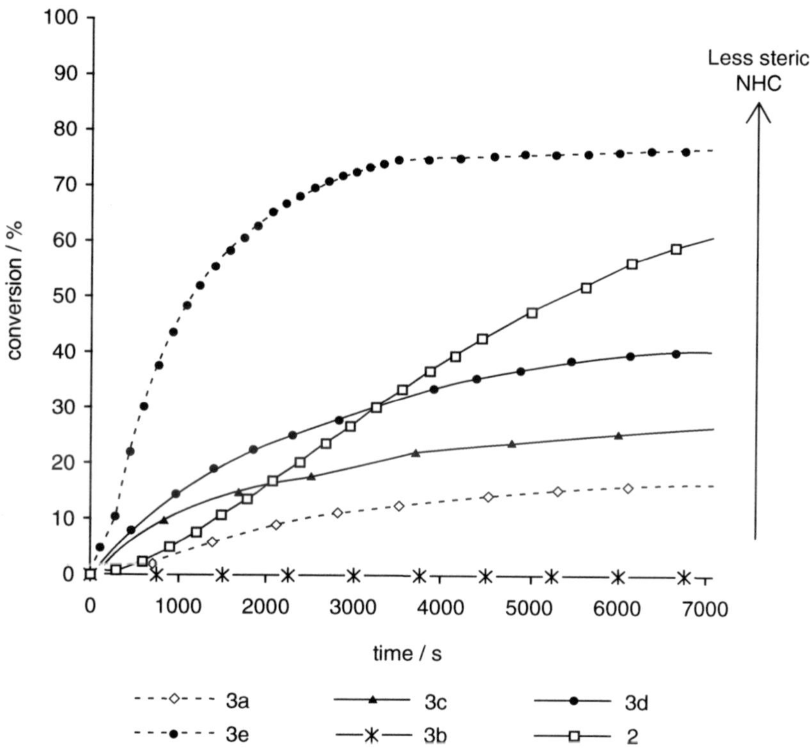

Figure 5 Monitoring RCM of diethyl diallylmalonate via ^{1}H NMR spectroscopy. substrate/cat. = 200, cat. conc. = 4.52 mM, 20°C, solvent = CD_2Cl_2

2.3 N-Alkyl-N'-(2,6-Diisopropylphenyl) Heterocyclic Carbenes [6b]

Using similar procedures as described in paragraph 1 and 2 for the synthesis of N-alkyl-N′-(2,6-diisopropylphenyl) heterocyclic carbene (alkyl = Me; Cy) substituted catalysts resulted in the observation of primarily bis(NHC)coordinated complexes, while only traces of the mono(NHC)substituted analogues were observed. Addition of 2.2 eq. of imidazolinium salt **a** or **b** (Figure 6) to **1** allowed for the full conversion of the starting complex respectively into the bis(NHC)-substituted complexes **8a** and **8b** (Figure 6).

To gain insight in the catalytic activity, the obtained complexes were tested for the ROMP of COD (Table 2). Both complexes showed low activity at room temperature (Entry 1 and 6), while activity increased substantially when heated to

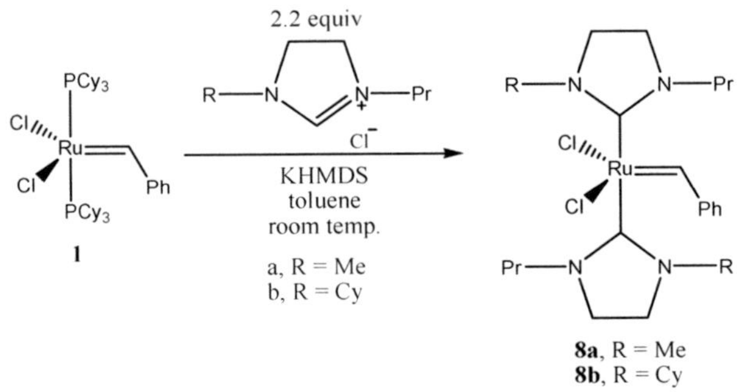

Figure 6 Synthesis of bis(NHC) complexes **8a** and **b**

Table 2 Ring-opening metathesis polymerization of 1,5-cyclooctadiene, **4**

Entry	Cat.	Temp (°C)	COD/cat.	Time (h)	Conv. (%)[a]	Cis- (%)[b]	Mn[c]	PDI[c]
1	**8a**	20	100	20	2	–	–	–
2	**8a**	80	100	1	100	43	31,200	1.4
3	**8a**	80	300	1	96	73	33,200	1.6
4	**8a**	80	3,000	20	100	60	42,800	1.7
5	**8a**	80	30,000	20	4	–	–	–
6	**8b**	20	100	20	24	91		
				40	63	80	48,100	1.5
7	**8b**	80	100	0.5	100	21	33,500	1.3
8	**8b**	80	300	0.5	100	31	39,000	1.6
9	**8b**	80	3,000	1	98	46	44,300	1.8
10	**8b**	80	30,000	20	60	81	80,000	2.5

[a]Conversion, determined by ^{1}H NMR spectroscopy.
[b]Percent olefin with *cis*-configuration in the polymer backbone – ratio based on ^{13}C NMR spectra (δ = 32.9 ppm: allylic carbon *trans*; δ = 27.6 ppm: allylic carbon *cis*).
[c]Determined by GPC ($CHCl_3$) analysis. Results are relative to polystyrene standards.

80°C (Entry 2–5 and 7–10). NHC decoordination, which is expected to induce catalyst initiation, proceeds more smoothly when temperature is raised. Indeed, a dissociative pathway for the Grubbs' type catalysts **1** and **2** is generally accepted and is strongly supported by experimental and computational studies, and it is within reason to assume that it holds for bis(NHC)-substituted olefin metathesis catalysts as well. While the lability of NHC ligands in organometallic complexes is typically low, our observations are in agreement to the generally accepted insights in the olefin metathesis mechanism using Grubbs' type catalysts [11].

2.4 New N-Heterocyclic Carbenes in Hoveyda–Grubbs' Catalyst [6c]

Whereas asymmetric NHC ligands bearing an alkyl and a mesityl side group are easily introduced using the well established procedure [12] for the synthesis of Hoveyda–Grubbs type catalysts, **9a–c**, an alternative procedure proved mandatory for the synthesis of the analogous complexes bearing N-alkyl-N′-(2,6-diisopropyl-phenyl)-, **10a–c**, or N,N′-dialkyl-heterocyclic carbenes, **11a**,**b**, since the first tend to form bis(NHC)-substituted complexes and the latter were unlikely to be isolated as such. Heating of the respective bis(NHC)-substituted complexes at 80°C in toluene for 2 h in the presence of an excess of 2-isopropoxystyrene afforded the desired complexes in good yields. Alternatively, an approach inspired by a procedure described by Blechert et al. [13] proved successful. Unmasking of the appropriate imidazolinium hydrochloride salts using LiHMDS in presence of the Hoveyda catalyst upon stirring in chloroform enabled the introduction of the NHC ligand. Complexes bearing N,N′-dialkyl-heterocyclic carbenes were obtained by an analogous procedure (Figure 7).

In order to depict the potential of the thus obtained complexes, their use as catalysts for ROMP of *cis*,*cis*-cycloocta-1,5-diene (**4**) and cross-metathesis of acrylonitrile (**14**) and allylbenzene (**13**) (Scheme 3) was elaborated.

Figure 8 depicts the results for the studied complexes towards the ROMP of COD. Complexes bearing symmetrical NHC ligands denote for full conversion before the first measurement could be performed. Furthermore, results show that complexes bearing an N-alkyl-N′-mesityl heterocyclic carbene ligand exhibit higher activity than their corresponding analogues bearing an N-alkyl-N′-(2,6-diisopropylphenyl) heterocyclic carbene.

Finally, we defined the potential of these complexes for their activity towards the cross-metathesis (CM) of allylbenzene, **13**, with the rather challenging acrylonitrile, **14**, at different catalyst loadings (2.5–5%) (Scheme 3, Table 3).

Whereas the Hoveyda catalyst, **12**, shows only poor activity, all catalysts bearing an NHC-ligand show substantially improved activity. All newly developed complexes display lower activities compared to the classical Hoveyda–Grubbs catalysts, **16**, **17** (Figure 8). Furthermore, a distinct influence on the selectivity can be observed. Complexes bearing a symmetric NHC-ligand yield higher Z selectivity. In general, asymmetric NHC-ligands induce a E/Z selectivity reversal. As an exception, complex **9c** bearing mesityl and isopinocamphyl amino side groups tends to favor E-selectivity.

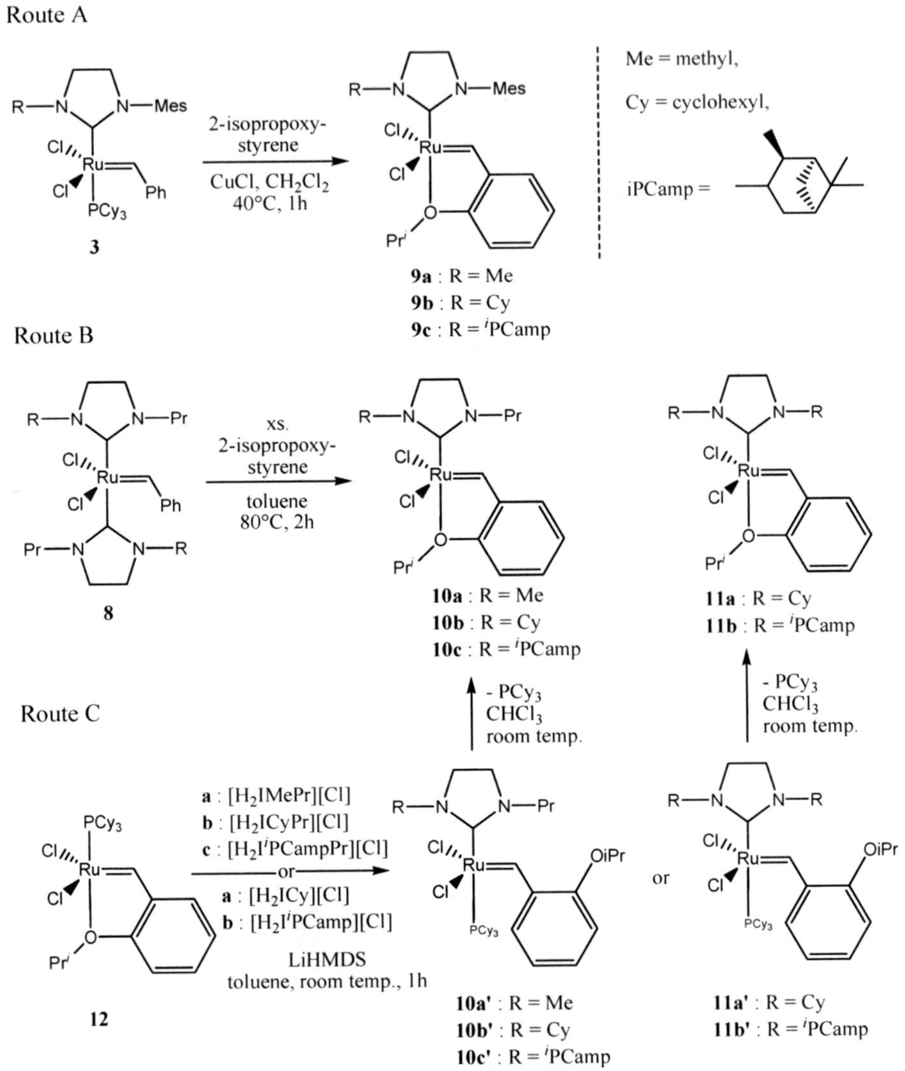

Figure 7 Synthesis of Hoveyda–Grubbs type complexes with modified N-heterocyclic carbene ligands

These observations stress out the importance of modified NHC-ligands. Besides catalytic activity, catalyst selectivity can be altered by a rigorous choice of the

NHC-ligand. Thus, new tools for specific applications in organic chemistry can be developed.

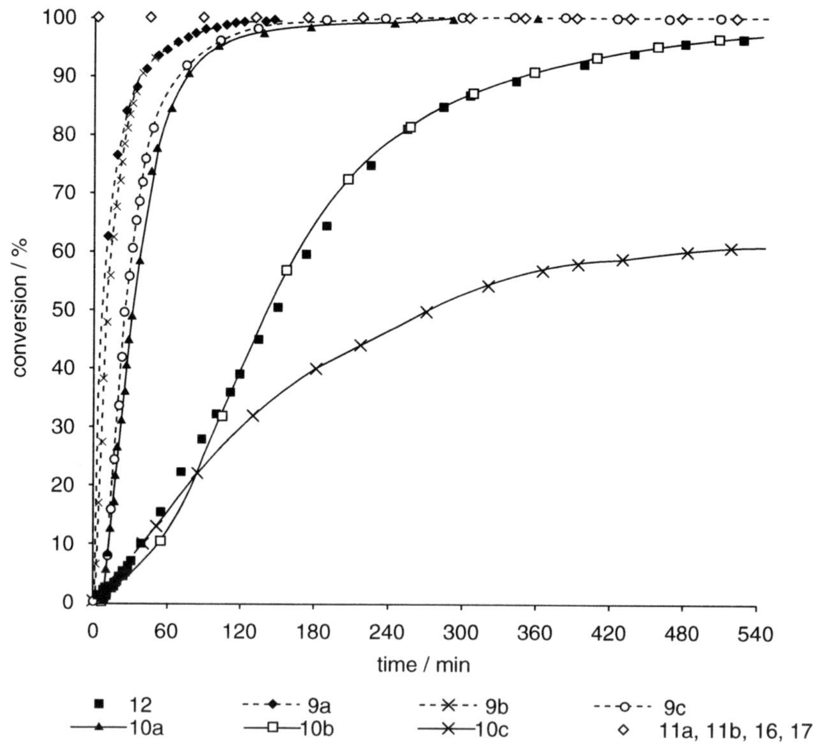

Figure 8 Monitoring ROMP of COD via ^{1}H NMR spectroscopy (20°C), COD/catalyst = 300, catalyst concentration = 4.52 mM, solvent = $CDCl_3$; **16** = SIMes-Hoveyda, **17** = SIPr-Hoveyda

3 Conclusion

In this paper, we give an overview of the research conducted in our group over the last years on the development on new N-heterocyclic carbene ligands and their application in Grubbs' and Hoveyda–Grubbs' catalysts. Although symmetrical NHC ligands could be introduced in Grubbs' catalyst, their isolation proved to be impossible, probably due to a lack of catalyst stability. Asymmetrical NHC ligands bearing a mesityl- and an alkyl-group were readily introduced in Grubbs' type complexes, and their activity for ROMP of cyclooctadiene and RCM of diethyl diallylmalonate increased as the steric bulk of the alkyl-group decreased. In case of NHC ligands bearing a 2,6-diisopropyl- and an alkyl-group, bis(NHC) coordination was observed. These complexes suffered from a lack of initiation efficiency and proved active only at elevated temperatures. Grubbs–Hoveyda type complexes bearing a wide variety of NHC ligands were readily prepared.

Complexes bearing symmetrical NHC ligands proved to exhibit higher activities during ROMP of cyclooctadiene. In addition, asymmetrical NHC ligands induced E/Z enantioselectivity during the CM of allylbenzene with acetonitrile.

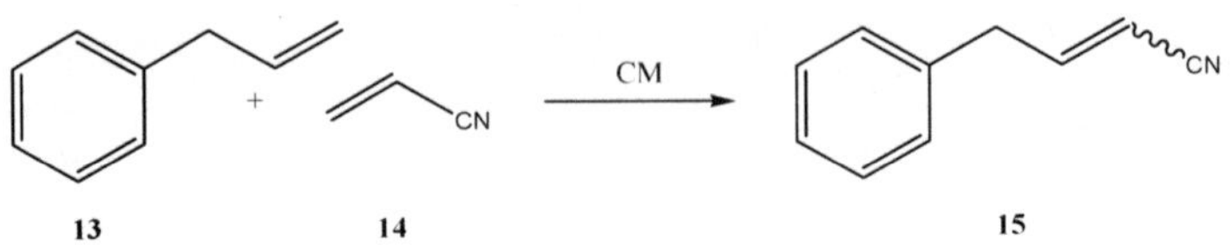

Scheme 3 Cross-metathesis of acrylonitrile (**14**) and allylbenzene (**13**)

Table 3 Cross-metathesis of acrylonitrile (**14**) and allylbenzene (**13**)

Catalyst	Catalyst loading (mol%)	Conversion (%)	E/Z ratio
12	5	<2	–
16	2.5	91	0.7/1
17	2.5	93	0.5/1
9a	2.5	20	1.9/1
9a	5	34	1.8/1
9b	2.5	33	1.5/1
9b	5	39	1/1
9c	2.5	43	0.6/1
9c	5	44	0.6/1
10a	2.5	15	2.5/1
10a	5	31	2.8/1
10b	2.5	12	3.2/1
10b	5	26	2.9/1
10c	2.5	21	2.2/1
10c	5	31	2.4/1
11a	2.5	5	0.8/1
11a	5	26	0.6/1
11b	2.5	7	0.5/1
11b	5	30	0.4/1

Reaction conditions: 40°C, 3 h, solvent = CH_2Cl_2. Conversion and E/Z ratios determined by ^1H NMR. ($ArCH_2R$ protons allylbenzene: δ 3.36 ppm, Z-isomer: δ 3.73 ppm, E-isomer: δ 3.51 ppm)

Acknowledgements S.M., N.L., P.V.D.V. and F.V. gratefully acknowledge the Research Fund of Ghent University (BOF) and the FWO for the financial support.

References

[1] (a) Nolan SP, Huang J, Stevens ED, Petersen JL (1999) J Am Chem Soc 121:2674–2678; (b) Westkamp T, Kohl FJ, Hieringer W, Gleich D, Herrmann WA (1999) Angew Chem Int Ed 38:2416–2419; (c) Scholl M, Ding S, Lee CW, Grubbs RH (1999) Org Lett 1:953–956; (d) Scholl M, Trnka TM, Morgan JP, Grubbs RH (1999) Tetrahedron Lett 40:2247–2250; (e) Huang J, Schanz H-J, Stevens ED, Nolan SP (1999) Organometallics 18:5375–5380; (f) Arduengo AJIII, Krafczyk R, Schmutzler R (1999) Tetrahedron 55:14523–14534; (g) Love JA, Morgan JP, Trnka TM, Grubbs RH (2002) Angew Chem Int Ed 21:4035–4037; (h) Choi T-L, Grubbs RH (2003) Angew Chem Int Ed 42:1743–1746; (i) Love JA, Sanford MS, Day MW, Grubbs RH (2003) J Am Chem Soc 125:10103–10109

[2] (a) Chauvin Y (2006) Angew Chem 118:3824–3831; (b) Chauvin Y (2006) Angew Chem Int Ed 45:3740–3747; (c) Schrock RR (2006) Angew Chem 118:3832–3844; (d) Schrock RR (2006) Angew Chem Int Ed 45:3748–3759; (e) Grubbs RH (2006) Angew Chem 118:3845–3850; (f) Grubbs RH (2006) Angew Chem Int Ed 45:3760–3765; For selected reviews on olefin metathesis, see: (g) Grubbs RH (ed.) (2003) Handbook of metathesis. Wiley-VCH, Weinheim; (h) Fürstner A (2000) Angew Chem Int Ed 39:3013–3043; (i) Schrock RR, Hoveyda AH (2003) Angew Chem Int Ed 42:4592–4633; (j) Grubbs RH (2004) Tetrahedron 60:7117–7140; (k) Astruc D (2005) New J Chem 29:42–56

[3] (a) Allaert B, Dieltiens N, Ledoux N, Vercaemst C, Van Der Voort P, Stevens CV, Linden A, Verpoort F (2006) J Mol Catal A 260:221–226; (b) Drozdzak R, Ledoux N, Allaert B, Dragutan I, Dragutan V, Verpoort F (2005) Cent Eur J Chem 3:404–416; (c) Drozdzak R, Allaert B, Ledoux N, Dragutan I, Dragutan V, Verpoort F (2005) Coord Chem Rev 249:3055–3074; (d) Drozdzak R, Allaert B, Ledoux N, Dragutan I, Dragutan V, Verpoort F (2005) Adv Synth Catal 347:1721–1743; (e) De Clercq B, Verpoort F (2003) J Organomet Chem 672:11–16; (f) Melis K, De Vos D, Jacobs P, Verpoort F (2003) J Organomet Chem 671:131–136; (g) Melis K, Verpoort F (2003) J Mol Catal A 201:33–41; (h) Melis K, Verpoort F (2003) J Mol Catal A 194:39–47; (i) Scholl M, Ding S, Lee CW, Grubbs RH (1999) Org Lett 1:953–956; (j) Dinger MB, Mol JC (2002) Adv Synth Catal 344:671–677

[4] Fürstner A, Ackermann L, Gabor B, Goddard R, Lehmann CW, Mynott R, Stelzer F, Thiel OR (2001) Chem Eur J 7:3236–3253

[5] Dinger MB, Nieczypor P, Mol JC (2003) Organometallics 22:5291–5296

[6] (a) Ledoux N, Allaert B, Pattyn S, Van der Mierde H, Vercaemst C, Verpoort F (2006) Chem Eur J 12:4654–4661; (b) Ledoux N, Allaert B, Linden A, Vander Voort P, Verpoort F (2007) Organometallics 26:1052–1056; (c) Ledoux N, Linden A, Allaert B, Vander Mierde H, Verpoort F (2007) Adv Synth Catal 349:1692–1700

[7] (a) Waltman AW, Grubbs RH (2004) Organometallics 23:3105–3107; (b) Clavier H, Coutable L, Guillemin J-C, Mauduit M (2005) Tetrahedron Asymm16:921–914

[8] Trnka TM, Morgan JP, Sanford MS, Wilhelm TE, Scholl M, Choi T-L, Ding S, Day MW, Grubbs RH (2003) J Am Chem Soc 125:2546–2558

[9] (a) Fürstner A, Ackermann L, Gabor B, Goddard R, Lehmann CW, Mynott R, Stelzer F, Thiel OR (2001) Chem Eur J 7:3236–3253; (b) Prühs S, Lehmann CW, Fürstner A (2004) Organometallics 23:280–287

[10] Vehlow K, Maechling S, Blechert S (2006) Organometallics 25:25–28

[11] (a) Herrmann WA (2002) Angew Chem Int Ed 41:1290–1309; (b) Huang J, Schanz H-J, Stevens ED, Nolan SP (1999) Organometallics 18:2370–2375

[12] (a) Garber SB, Kingsbury JS, Gray LB, Hoveyda AH (2000) J Am Chem Soc 122:8168–8179; (b) Wakamatsu H, Blechert S (2002) Angew Chem Int Ed 41:794–796; (c) Wakamatsu H, Blechert S (2002) Angew Chem Int Ed 41:2403–2405; (d) Dunne AM, Mix S, Blechert S (2003) Tetrahedron Lett 44:2733–2736; (e) Grela K, Harutyunyan S, Michrowska A (2002) Angew Chem Int Ed 41:4038–4040; (f) Michrowska A, Bujok R, Harutyunyan S, Sashuk V, Dolgonos G, Grela K (2004) J Am Chem Soc 126:9318–9325

[13] Gessler S, Randl S, Blechert S (2000) Tetrahedron Lett 41:9973–9976

Ruthenium–Indenylidene Complexes Bearing Saturated N-Heterocyclic Carbenes: Synthesis and Application in Ring-Closing Metathesis Reactions

Stijn Monsaert,[1] Els De Canck,[1] Renata Drozdzak,[1] Pascal Van Der Voort,[1] Pieter M.S. Hendrickx,[2] José C. Martins,[2*] Francis Verpoort,[1*]

[1]Department of Inorganic and Physical Chemistry, Ghent University, Krijgslaan 281 (S3), B-9000 Ghent, Belgium

[2]Department of Organic Chemistry, Ghent University, Krijgslaan 281 (S4), B-9000 Ghent, Belgium

[*]Fax: (+) 32-9-264-4183, (+) 32-9-264-4972; e-mails: Francis.Verpoort@UGent.be, Jose.Martins@UGent.be

Abstract The synthesis of complexes of the general formula $Cl_2Ru(SIMes)(L)$(3-phenylinden-1-ylidene), **3a** (L = PCy_3), **3b** (L = PPh_3), **3c** (L = py) and $Cl_2Ru(SIMe)(L)$(3-phenylinden-1-ylidene), **4c** (L = PCy_3), **4b** (L = PPh_3), **4c** (L = py) from commercially available $Cl_2Ru(PR_3)_2$(3-phenylinden-1-ylidene), **2a** (R = Cy), **2b** (R = Ph), is reported, and the summed complexes proved to be useful catalysts for various ring-closing metathesis reactions. The catalytic data reported furthermore demonstrate that the substitution pattern of the *N*-aryl group has a definite influence on the activity of the second generation indenylidene catalysts for a given metathesis reaction, i.e. catalysts containing the SIMe-ligand showed improved initiation compared to the more robust SIMes substituted catalysts. A strong temperature effect was observed on all of the reactions tested. It is shown that complexes containing the SIMe-ligand display higher initiation efficiency then their corresponding SIMes analogues.

Keywords Metathesis catalysts · Indenylidene complexes · N-Heterocyclic carbenes · Ring-closing metathesis · Diethyl diallylmalonate · Pyridine ligand

1 Introduction

The metathesis road of success has been endowed with the unravelling of the olefin metathesis mechanism [1a, b] and the development of well-defined Mo, W [1c, d] and Ru [1e, f] catalysts by Chauvin, Schrock and Grubbs respectively. Since the olefin metathesis method has proved to be an atom-efficient, synthetic

V. Dragutan et al. (eds.), *Green Metathesis Chemistry: Great Challenges in Synthesis, Catalysis and Nanotechnology*,

strategy for the formation of C–C double bonds, due diligence has been directed to the development of novel catalysts with improved activity, stability and selectivity [2]. Within the class of ruthenium complexes, efforts were directed towards the development of complexes bearing different ancillary ligands, among which benzylidene, phosphine and N-heterocyclic carbene ligands [3] (Scheme 1).

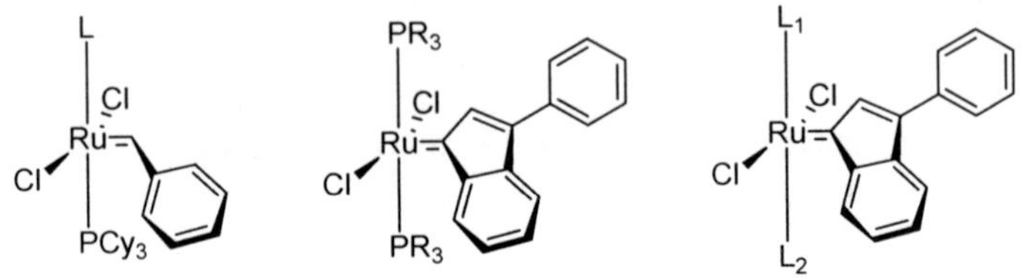

1a, L = PCy_3
1b, L = IMes
1c, L = SIMes

2a, R = Cy
2b, R = Ph

3a, L_1 = SIMes, L_2 = PCy_3
3b, L_1 = SIMes, L_2 = PPh_3
3c, L_1 = SIMes, L_2 = py

Scheme 1 Well-defined olefin metathesis catalysts **1**, **2** and **3**

Ruthenium indenylidene complexes were serendipitously discovered and exhibited remarkable stability and reactivity. In addition, the ease of their synthesis afforded a strong benefit over benzylidene type catalysts. Further research along these lines resulted in the development of ruthenium indenylidene type catalysts bearing various phosphine ligands and saturated and unsaturated N-heterocyclic carbene ligands [4].

The current research was focussed on the influence of the substitution pattern of the aryl moiety in the N-heterocyclic carbene ligand. Dinger and Mol [5] and more recently Stewart et al. [6] have shown that altering this substitution pattern can strongly affect the olefin metathesis activity of benzylidene catalysts. Preliminary research has shown that absence of the para-methyl substituent on the aryl moiety of the N-heterocyclic carbene results in a strongly improved initiation rate of ruthenium benzylidene type catalysts, however, at the expense of catalyst lifetime [7]. Our experience in the development of olefin metathesis indenylidene type catalysts moved us to research this effect in the N-heterocyclic carbene ligand for indenylidene type catalysts.

2 Results and Discussion

2.1 Synthesis of Second Generation Indenylidene Complexes

Other groups [8] and we [41] previously reported on the use of thermally decomposing chloroform adducts for the introduction of N-heterocyclic carbenes in organometallic complexes. This synthetic approach was adapted for the synthesis of **3a** starting from **2a** and **5a**. A similar approach using a pentafluorobenzene adduct **5b** was applied for the synthesis of **4a** from **2a** (Scheme 2). Monitoring the reaction of **2a** and **5a** using ^{31}P NMR showed that the reaction was complete after 1.5 h refluxing in THF. However, a rather large excess (2 eq.) of the carbene ligand precursor was found necessary. Likewise, **4a** was obtained from **2a** and **5b** after 1.5 h of reaction at 100°C in toluene. Surprisingly, 1.15 eq. of the pentafluorobenzene adduct **5b** proved satisfactory, suggesting a much less sterical hindrance thwarting the complex formation, and thus a much less sterically demanding ligand environment. The complexes were easily purified by evaporation of all volatiles and subsequent suspending in MeOH. Filtration and drying afforded **3a** as a red powder in good yield (82%). Complex **4a** was obtained as a red solid as well (56%). NMR analysis of **3a** and **4a** showed a single peak up-field to the starting complex in the ^{31}P NMR spectrum, respectively at δ 27.0 and δ 26.1 ppm. The ^{1}H NMR spectrum revealed a doublet at δ 9.13 ppm for **3a** and at δ 9.20 ppm for **4a**, which is typical for indenylidene complexes. The ethylene backbone of the imidazolin-2-ylidene ligand in **3a** and **4a** forms a complex multiplet (at δ 3.41–3.12 ppm for **3a** and at δ 3.35–3.06 ppm for **4a**) which indicates the complexes' asymmetry. The ^{13}C NMR spectrum further proved presence of the imidazolin-2-ylidene ligand, with doublets at δ 216.34 and δ 215.27 ppm for **3a** and **4a** correspondingly. Elemental analysis showed that the obtained compounds are indeed **3a** and **4a**, and that they were obtained as pure compounds. The complexes were found to be stable in solid state and were stable for days in dichloromethane.

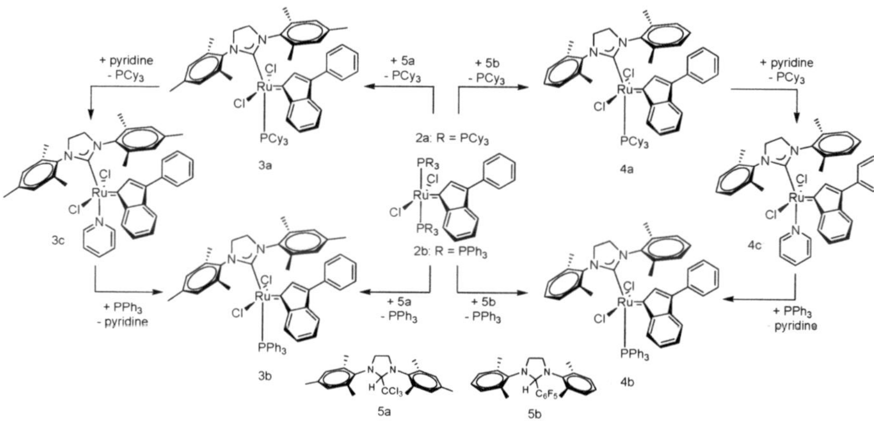

Scheme 2 Synthesis of new generations Ru indenylidene metathesis catalysts with saturated N-heterocyclic carbene ligands

Synthesis of **3c** proceeded easily by treatment of **3a** with an excess of pyridine. The indenylidene complex, **3c**, was isolated in good yield (70%) as an orange-brown powder. The analogous approach for the SIMe-ligated catalyst resulted **4c** as an orange-brown powder in 60% yield.

Complex **3b** and **4b** respectively were obtained from **3c** and **4c** by simple ligand exchange and isolated in 89% and 53% yield as clear red powders. In addition, they were straightforwardly obtained from reaction of **2b** with **5a**/**b** (1 h in refluxing THF) in good yields, respectively 86% and 74%. The high thermal stability of **2a** prevents decomposition under these conditions and thus provides a cheap and economical way to this second generation type catalysts.

2.2 Ring-Closing Metathesis Activity

A standard set of ring-closing metathesis reactions and more challenging diphenyl diallylsilane (Scheme 3) were used to depict the performance of complexes **2a**, **3a–c** and **4a–c**.

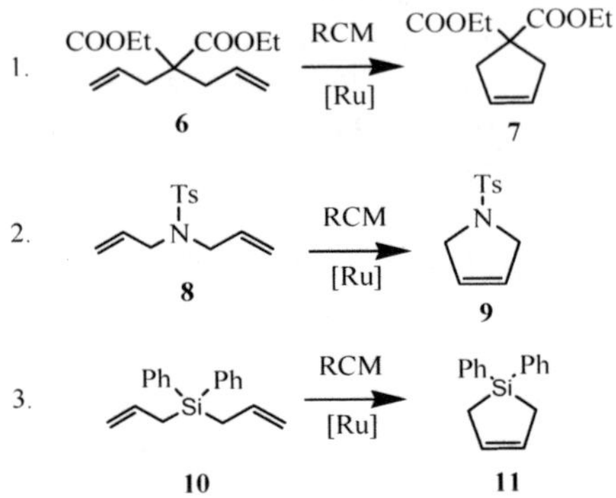

Scheme 3 Representative metathesis reactions

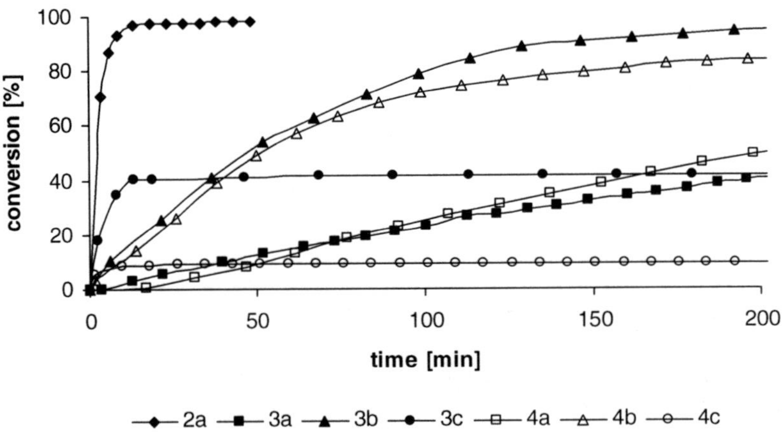

Figure 1 RCM of **6**, catalysts **2a**, **3a–c** and **4a–c**. Catalyst to substrate ratio = 1/200; solvent: $CDCl_3$; temperature: 20°C; conversion determined by ^1H NMR; lines are intended as visual aids

First generation indenylidene catalyst **2a** unequivocally displays the best activity for the RCM of diethyl diallylmalonate, converting 200 eq. of the substrate almost quantitatively within 30 min. Second generation type complexes with a PCy_3

ligand trans to the N-heterocyclic carbene moiety, **3a** and **4a**, suffer from a dramatic decrease in activity, which may be rationalized by a reduction of the catalysts initiation efficiency. Complex **3a** shows 50% conversion after 5 h for the RCM with **6** (Figure 1). The conversion proceeds to 89% after 24 h, which indicates a long lifetime of the complex. Full conversion could not be attained due to the low initiation rate. **4a** exhibits the same behaviour but exceeds the activity of **3a**; after 24 h **4a** reaches almost full conversion (99%). The stability of the PCy_3 ligand at room temperature is implied by the requirement for elevated temperatures in order to achieve high activity, as we reported earlier for indenylidene catalysts **3a**.

In case the complexes bear a weaker donating PPh_3 ligand, **3b** and **4b**, a definite increase in the catalyst's activities is observed. Complexes bearing a pyridine ligand trans to the NHC ligand, **3c** and **4c**, exhibit a high initial activity, succeeded by a strong activity drop-off which suggests dramatic catalyst decomposition. Although no discernable effect of the NHC ligand on the catalytic activity is observed for phosphine ligated complexes, a profound effect is perceived for the pyridine complexes **3c** and **4c**. While the SIMes ligated catalyst **3c** attains 40% conversion, SIMe ligated catalyst **4c** manages to convert only 9% of the substrate. A profound influence of the N-aryl substitution pattern in the N-heterocyclic carbene ligand on the stability of the active species can thus be assumed.

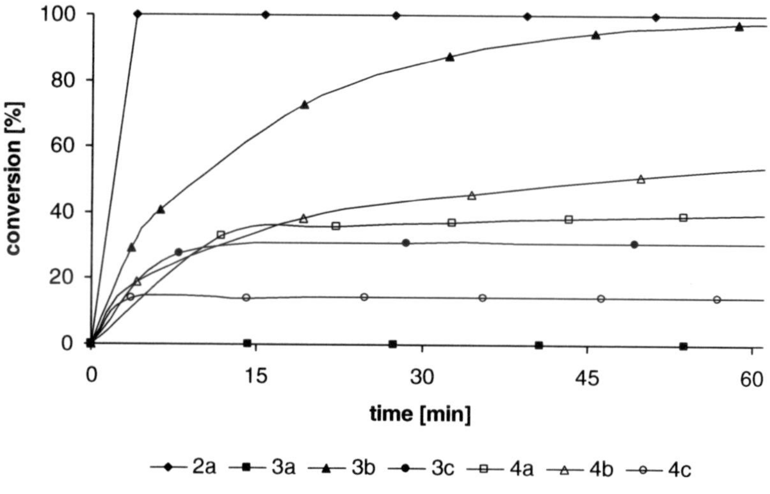

Figure 2 RCM of N,N-diallyltosylamide (**8**), catalysts **2a**, **3a–c** and **4a–c**. Catalyst/substrate ratio = 1/200; solvent: $CDCl_3$; temperature: 20°C; conversion determined by ^1H NMR; lines are intended as visual aids

Similar to the RCM of diethyl diallylmalonate, the RCM of N,N-diallyltosylamide proceeds smoothly using first generation type catalyst **2a**, affording quantitative conversion within 4 min. More strikingly is the negligible activity of **3a** towards the RCM of **8**. A comparable catalyst with the SIMe ligand, **4a**, still manages 40%

conversion. Catalysts **3b** and **4b** with a PPh_3 ligand again allow for higher activities, converting 98% and 58% of the substrate respectively within 1 h (Figure 2). The pyridine ligated complexes again exhibit a high initial activity followed by a dramatic decrease in catalyst activity, comparable to the results obtained for the RCM of diethyl diallylmalonate.

Schmidt and coworkers have proposed an RCM strategy for the synthesis of silacycloalkenes starting from diallylsilane derivatives [9]. The problems encountered during the ring-closing of **10** were mainly attributed to the large size of the silicon atom, disfavouring the transition state geometry for cyclization reactions. We were attracted by the sterically encumbered nature of this substrate to reveal peculiarities in the sterically demanding properties of our NHC ligands. Best performance is found for the PCy_3 ligand containing catalysts (**3a**, **4a**, and **1a**) but high catalyst loadings (5 mol%) are required (Table 1). A literature report for **1c** (5 mol%) describes only 70% formation of **11** after 16 h in refluxing CCl_4, suggesting much lower activity for the Grubbs' catalysts compared to indenylidene catalysts. PPh_3 and py containing catalysts **3b**, **4b**, **3c** and **4c** are less interesting catalysts regarding to this substrate. Furthermore, it should be noted that the SIMes containing catalysts are generally more active in RCM of **10**. Catalyst **3a** exhibits by far the highest activity, with nearly quantitative formation of **11** within 16 h in refluxing $CDCl_3$.

Table 1 Representative ring-closing and cross metathesis reactions

Entry	Substrate	Product	Catalyst	Time (h)	T(°C)	Solvent	Conversion (%)
1	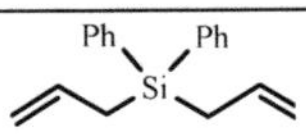 10	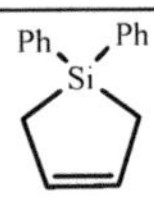 11	1c[a]	16	65	CCl_4	70
2			3a[b]	16	60	$CDCl_3$	95
3			3b[b]	2	60	$CDCl_3$	68
4			3c[b]	2	60	$CDCl_3$	62
5			4a[b]	2	60	$CDCl_3$	76
6			4b[b]	2	60	$CDCl_3$	62
7			4c[b]	2	60	$CDCl_3$	42

Conditions:
[a]See Ref. [9].
[b]See Ref. [4m].

3 Conclusions

In this article, we presented the synthesis and screening results for a series of second and third generation indenylidene olefin metathesis catalysts applied to a set of ring-closing metathesis transformations. The aim of this study was to reveal the relative efficacies of different catalysts containing a SIMes or a SIMe ligand. We have compared six of the ruthenium-indenylidene olefin metathesis catalysts in a set of metathesis reactions and described them in terms of their performance, selectivity, and stability. During this comparison it became evident that a small modification of the substituents on the NHC ligand influence the catalyst initiation rate. Nevertheless, as ligand (phosphine, pyridine) dissociation promotes catalyst decomposition, therefore that complexes bearing SIMe ligand decompose faster. It was evidenced that second generation type indenylidene catalysts suffer from low initiation efficiency. Therefore, first generation type catalyst **2a** often excels other studied catalysts for RCM transformations. Third generation type catalysts exhibit a high initial activity, ensued by a definite drop in activity, a fingerprint of their fast decomposition. Second generation type indenylidene catalysts bearing a SIMe ligand generally surpass the activity of those bearing a SIMes ligand, since the latter suffer from a more pronounced initiation period. Third generation type catalysts bearing a SIMe ligand suffer to a larger extent from decomposition, compared to their SIMes ligated counterparts. Therefore, their RCM activity is rather marginally.

Acknowledgements S.M. and F.V. gratefully acknowledge the Research Fund of Ghent University (BOF) for the financial support. The IWT Flanders is thanked for a Ph.D. grant to P.M.S.H. J.C.M. and F.V. thank the FWO for financial support and several NMR equipment grants (G.0036.00N, G.0365.03). The 700 MHz equipment of the Interuniversitary NMR Facility was financed by Ghent University, the Free University of Brussels (VUB) and the University of Antwerp via the 'Zware Apparatuur' Incentive of the Flemish Government. Umicore is acknowledged for a generous gift of compound **2a**.

References

[1] (a) Chauvin Y (2006) Angew Chem 118:3824–3831; (b) Chauvin Y (2006) Angew Chem Int Ed 45:3740–3747; (c) Schrock RR (2006) Angew Chem 118:3832–3844; (d) Schrock RR (2006) Angew Chem Int Ed 45:3748–3759; (e) Grubbs RH (2006) Angew Chem 118:3845–3850; (f) Grubbs RH (2006) Angew Chem Int Ed 45:3760–3765

[2] For selected reviews on olefin metathesis, see: (a) Grubbs RH (ed.) (2003) Handbook of metathesis. Wiley-VCH, Weinheim; (b) Fürstner A (2000) Angew Chem Int Ed 39:3013–3043; (c) Schrock RR, Hoveyda AH (2003) Angew Chem Int Ed 42:4592–4633; (d) Grubbs RH (2004) Tetrahedron 60:7117–7140; (e) Astruc D (2005) New J Chem 29:42–56

[3] (a) Schwab P, France MB, Ziller JW, Grubbs RH (1995) Angew Chem Int Ed 34:2039–2041; (b) Schwab P, Grubbs RH, Ziller JW (1996) J Am Chem Soc 118:100–110; (c) Nolan SP, Huang J, Stevens ED, Petersen JL (1999) J Am Chem Soc 121:2674–2678; (d) Westkamp T, Kohl FJ, Hieringer W, Gleich D, Herrmann WA (1999) Angew Chem Int Ed 38:2416–2419; (e) Scholl M, Ding S, Lee CW, Grubbs RH (1999) Org Lett 1:953–956; (f) Scholl M, Trnka TM, Morgan JP, Grubbs RH (1999) Tetrahedron Lett 40:2247–2250; (g) Huang J, Schanz H-J, Stevens ED, Nolan SP (1999) Organometallics 18:5375–5380; (h) Arduengo AJIII, R. Krafczyk R, Schmutzler R (1999) Tetrahedron 55:14523–14534; (i) Love JA, Morgan JP, Trnka TM, Grubbs RH (2002) Angew Chem Int Ed 21:4035–4037; (j) Choi T-L, Grubbs RH (2003) Angew Chem Int Ed 42:1743–1746; (k) Love JA, Sanford MS, Day MW, Grubbs RH (2003) J Am Chem Soc 125:10103–10109

[4] (a) Schanz H-J, Jafarpour L, Stevens ED, Nolan SP (1999) Organometallics 18:5187–5190; (b) Jafarpour L, Schanz H-J, Stevens ED, Nolan SP (1999) Organometallics 18:5416–5419; (c) Fürstner A, Hill AF, Liebl M, Wilton-Ely JDET (1999) J Chem Soc Chem Commun :601–602; (d) Fürstner A, Thiel OR, Ackermann L, Schanz H-J, Nolan SP (2000) J Org Chem 65:2204–2207; (e) Fürstner A, Guth O, Düffels A, Seidel G, Liebl M, Gabor B, Myonott R (2001) Chem Eur J 22:4811–4820; (f) Castarlenas R, Dixneuf PH (2003) Angew Chem Int Ed 42:4524–4527; (g) Castarlenas R, Vovard C, Fischmeister C, Dixneuf

PH (2006) J Am Chem Soc 128:4079–4089; (h) Forman GS, Bellabarba RM, Tooze RP, Slawin AMZ, Karch R, Winde R (2006) J Organomet Chem 691:5513–5516; (i) Clavier H, Petersen JL, Nolan SP (2006) J Organomet Chem 691:5444–5447; (j) Shaffer EA, Chen C-L, Beatty AM, Valente EJ, Schanz H-J (2007) J Organomet Chem 692:5221–5233; (k) Clavier H, Nolan SP (2007) Chem Eur J 28:8029–8036; (l) Monsaert S, Drozdzak R, Dragutan V, Dragutan I, Verpoort F (2008) Eur J Inorg Chem 3:432–440; (m) Monsaert S, De Canck E, Drozdzak R, Van Der Voort P, Verpoort F, Martins JC, Hendrickx PMS (2008) Eur J Org Chem accepted

[5] Dinger MB, Mol JC (2002) Adv Synth Catal 344:671–677

[6] Stewart IC, Ung T, Pletnev AA, Berlin JM, Grubbs RH, Schrodi Y (2007) Org Lett 9:1589–1592

[7] Monsaert S, Drozdzak R, Verpoort F, unpublished data

[8] (a) Arduengo AJ, Calabrese JC, Davidson F, Dias HVR, Goerlich JR, Krafczyk R, Marshall WJ, Tamm M, Schmutzler R (1999) Helv Chim Acta 82:2348–2364; (b) Trnka TM, Morgan JP, Sanford MS, Wilhelm TE, Scholl M, Choi T-L, Ding S, Day MW, Grubbs RH (2003) J Am Chem Soc 125:2546–2558; (c) Nyce GW, Csihony S, Waymouth RM, Hedrick JL (2004) Chem Eur J 10:4073–4079; (d) Blum AP, Ritter T, Grubbs RH (2007) Organometallics 26:2122–2124

[9] Schmidt B, Pohler M, Costisella B (2004) J Org Chem 69:1421–1424

Building Indenylidene–Ruthenium Catalysts for Metathesis Transformations

Hervé Clavier,[1,2] **Steven P. Nolan**[1,2]

[1]Institute of Chemical Research of Catalonia (ICIQ), Av. Països Catalans 16, 43007 Tarragona (Spain)

[2]University of St-Andrews, School of Chemistry, Purdie Building, North Haugh, St Andrews, Fife KY16 9ST (UK)

E-mails: hc31@st-andrews.ac.uk, sn17@st-andrews.ac.uk

Abstract Ruthenium-mediated olefin metathesis has emerged as an indispensable tool in organic synthesis for the formation carbon–carbon double bonds, attested by the large number of applications for natural product synthesis. Among the numerous catalysts developed to mediate olefin metathesis transformations, ruthenium–indenylidene complexes are robust and powerful pre-catalysts. The discovery of this catalyst category was slightly muddled due to a first mis-assignment of the compound structure. This report provides an overview of the synthetic routes for the construction of the indenylidene pattern in ruthenium complexes. The parameters relating to the indenylidene moiety construction will be discussed as well as the mechanism of this formation.

Keywords Olefin metathesis · Ruthenium catalysts · Indenylidene complexes · Allenylidene complexes · Fluorenylidene complexes

1 Introduction

Over the past decade, ruthenium-mediated olefin metathesis has emerged as an indispensable tool in organic synthesis for the formation carbon–carbon double bonds [1]. One fascinating feature of olefin metathesis is the access to numerous variations on the theme achieved as a function of both substrates and reaction conditions. As a testimony, this synthetic strategy is now employed for obtaining fine chemicals, biologically active compounds, new functionalised materials, and various polymers [2]. Diverse metal-based complexes can catalyse these transfomations. However, the rapid development of this area has been punctuated by groundbreaking developments particularly focusing on well-defined ruthenium–carbene complexes (Figure 1). Among them, benzylidene catalysts [3], are most widely used. Many efforts have been aimed at tuning these structures: L ligand screening (mainly phosphines [4] and N-heterocyclic carbenes [5]), pyridine

V. Dragutan et al. (eds.), *Green Metathesis Chemistry: Great Challenges in Synthesis, Catalysis and Nanotechnology*,

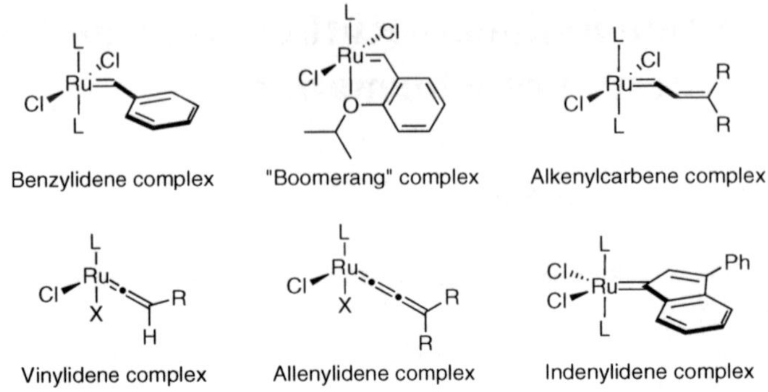

Figure 1 Ruthenium-based olefin metathesis catalysts (L = neutral ligand; X = anionic ligand)

derivatives [6] or anionic ligands [7], development of "boomerang" [8] and related supported catalysts [9], etc.

Nevertheless few structurally different complexes, such as alkenylcarbene [10], vinylidene [11], allenylidene [12] and indenylidene [13] complexes, have proven competent in mediating olefin metathesis transformations (Figure 1). As an alternative to benzylidene-catalysts, indenylidene–ruthenium catalysts are especially attractive due to their straightforward synthesis from $[RuCl_2(PPh_3)_3]$ and 1,1-diphenylpropargyl alcohol [14]. For this reason, several studies have been carried out on the synthesis of these complexes but also on their modification and applications.

2 Formation of Ruthenium–Indenylidene Complex

Originally, the product formed from the reaction of $[RuCl_2(PPh_3)_3]$ and 1,1-diphenylpropargyl alcohol was identified as the diphenylallenylidene complex **1** [14]. However, more detailed studies including an X-ray determination of a parent complex bearing an N-heterocyclic carbene, demonstrated that the stable product isolated was the rearranged indenylidene ruthenium complex **2** (Scheme 1) [15]. The possible origin of this misinterpretation might originate from the report of Selegue in 1982 stating the dehydration of 2-propyn-1-ols in presence of the ruthenium complexes which affords cumulenylidene complexes [16].

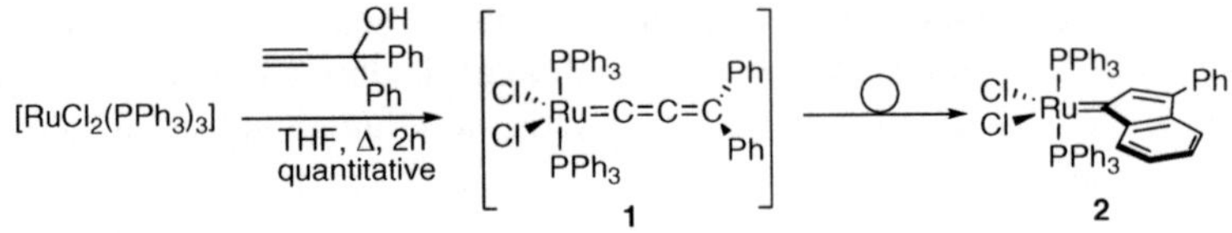

Scheme 1 First synthesis of ruthenium-indenylidene catalyst

As a consequence, some confusion between allenylidene and rearranged indenylidene complexes is encountered in the literature. For example, the product of the previous reaction was engaged with the dimer $[(p\text{-cymene})\text{-}RuCl_2]_2$ leading to the formation of a bimetallic compound. The structure of the complex was first presented to be the allenylidene–ruthenium **3**, before for a revision to the indenylidene complex **4** [17] (Figure 2).

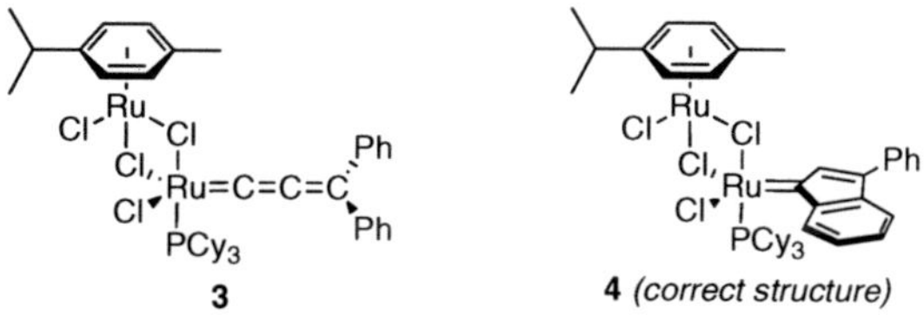

Figure 2 Structural revision

However such rearrangement into indenylidene moiety does not occur in numerous cases. When the reaction between $[RuCl_2(PPh_3)_3]$ and 1,1-diphenyl-propargyl alcohol is performed in the presence of tricyclohexylphosphine, the allenylidene **5** was isolated and no traces of the indenylidene **6** could be observed (Scheme 2) [18]. Of note, various attempts to convert **5** into **6** have been attempted using either thermal activation or acid conditions without any conclusive outcomes. The synthetic route leading to **6**, is to achieve the phosphine exchange after the formation of the indenylidene *i.e.* starting from complex **2**. When the dimer $[(p\text{-cymene})\text{-}RuCl_2]_2$ reacts with the propargylic alcohol in the presence of tricyclohexylphosphine **5** is isolated as the major product. Interestingly, slight amounts of the rearranged product 6 were detected (8%) (Scheme 3) [18].

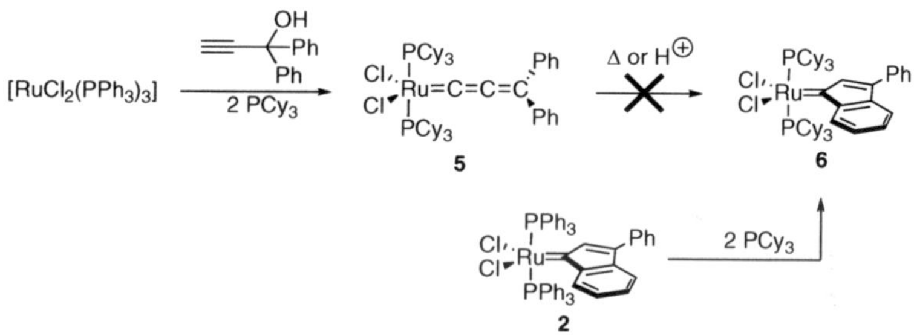

Scheme 2 Synthesis of the tricyclohexylphosphine complex **6**

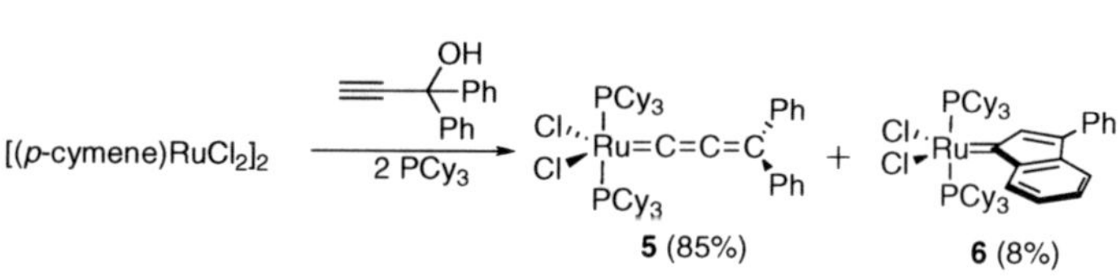

Scheme 3 Use of $[(p\text{-cymene})\text{-}RuCl_2]_2$ as ruthenium source

Such rearrangement does not seem to occur when the ruthenium metallic center contains an additional ligand. Indeed Schanz and coworkers have isolated and characterized by X-ray diffraction various complexes bearing chlorine, dimethyl-aminopyridine, water or alcohols **7-9** (Figure 3) [19].

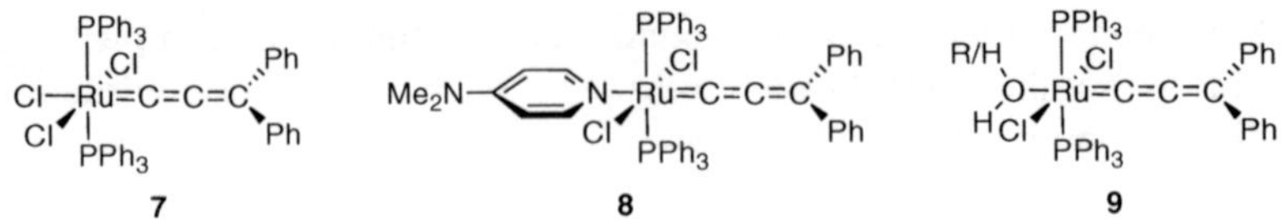

Figure 3 Stable ruthenium allenylidene complexes

Monomeric species [(*p*-cymene)$RuCl_2(PR_3)$] react also easily with propargylic alcohols in the presence of AgOTf (or $NaBPh_4$, $NaPF_6$ etc.) at ambient temperature to form the corresponding Ru-allenylidene complexes **11** in excellent yields [12c] (Scheme 4). According to the authors, these complexes are and can be stored under argon for several months without noticeable decomposition. While the catalytic performances of these complexes were investigated other organometallic species were observed and are assumed to act as the active catalyst [20]. The formation of indenylidene pattern being presumed, investigations on the related mechanism were carried out.

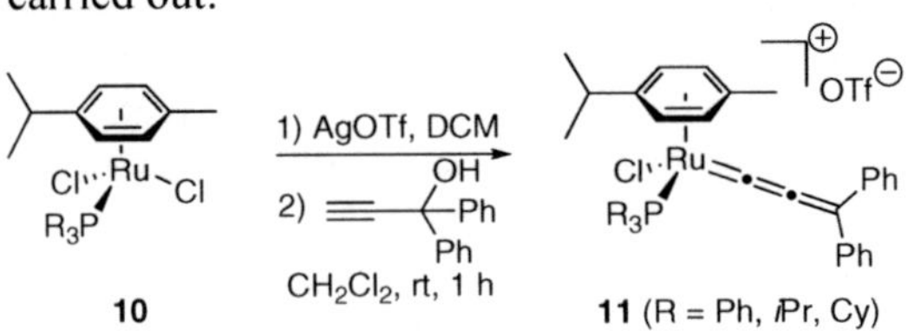

Scheme 4 Synthesis of the tricyclohexylphosphine complex **6**

3 Mechanistic Studies

It was observed that the addition of a strong acid such as trifluoromethanesulfonic acid to **11** led to a striking increase of the catalytic activity in several metathesis transformations [20]. To understand the true nature of the catalytic species and also control their synthesis, low temperature NMR studies were performed by Dixneuf and coworkers [20b, 21]. These demonstrated that after addition of 2 eq. of triflic acid to a CD_2Cl_2 solution of **11** at ~40°C, the ionic compound **12** can be clearly identified (Scheme 5). Then, upon slight warming to ~20°C, complex **12** was completely converted into phenylindenylidene arene–ruthenium complex **13**.

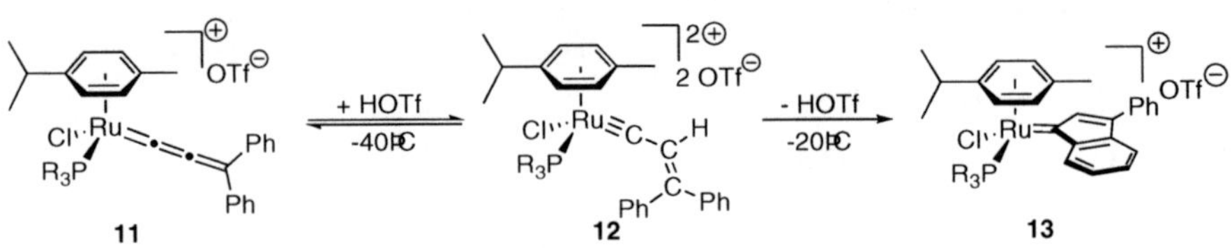

Scheme 5 Rearrangement of allenylidene into indenylidene for arene-ruthenium complex

Recently, Schanz and coworkers suggested that the formation of the allenylidene **1** is the rate-determining step of the synthesis of **2** [19]. The subsequent rearrangement is presumed to occur very rapidly, especially under acidic conditions and this pathway involves the formation of the carbynes **14** and **15**. It is of note that the intermediates **1** and **14** have been isolated and characterised by X-ray diffraction techniques.

The two mechanistic studies carried out are consistent and suggest that the α-carbon atom of the protonated intermediate **12** or **14** displays an elevated electrophilicity and thus facilitates the electrophilic ortho-substitution of the phenyl group (Scheme 6).

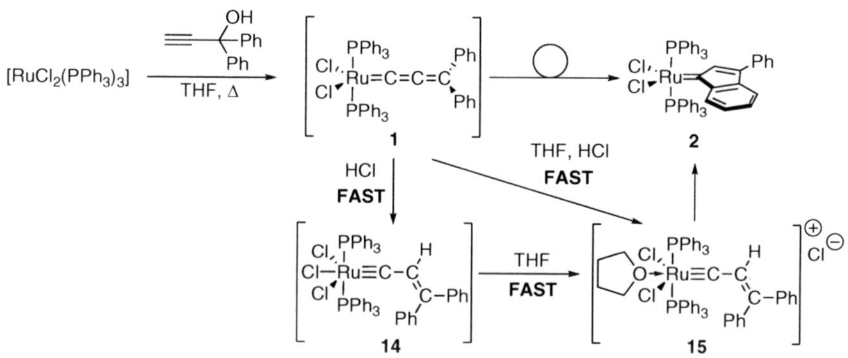

Scheme 6 Proposed mechanism for the indenylidene moiety generation

4 Derivatization of Indenylidene Complexes

As depicted in Scheme 7, from **3** one or both PCy_3 ligand were easily replaced by other phosphines such as alkyl-phoban yielding catalysts **16** [22], by N-heterocyclic carbenes (NHC) giving second generation catalysts **17** [15, 23] or by more labile pyridine ligands leading to the 18-electron adduct complexes **18** [24].

In order to tune the electronic properties around the metal center, Verpoort and coworkers investigated Schiff base as ligand [25]. Such ligands bear two donor atoms having opposite character. Indeed, the phenolate oxygen atom is a hard donor and is known to stabilise the higher oxidation states of ruthenium, whereas the imine nitrogen atom is in comparison soft and is a stabiliser of the lower oxidation states. Ruthenium indenylidene-based complexes bearing salicylaldimidato groups **19** were synthesized by treatment of salicylaldimine thallium salts with complex **6**. Upon substitution of an anionic chloride and a neutral phosphine, complexes **19** can be isolated in high yields. Recently Mauduit and Nolan demonstrated that complex **6** or **17** could be good precursors for the synthesis of "boomerang" type ruthenium catalysts **20** [26]. Excellent yields were obtained for differently substituted styrenyl ligands.

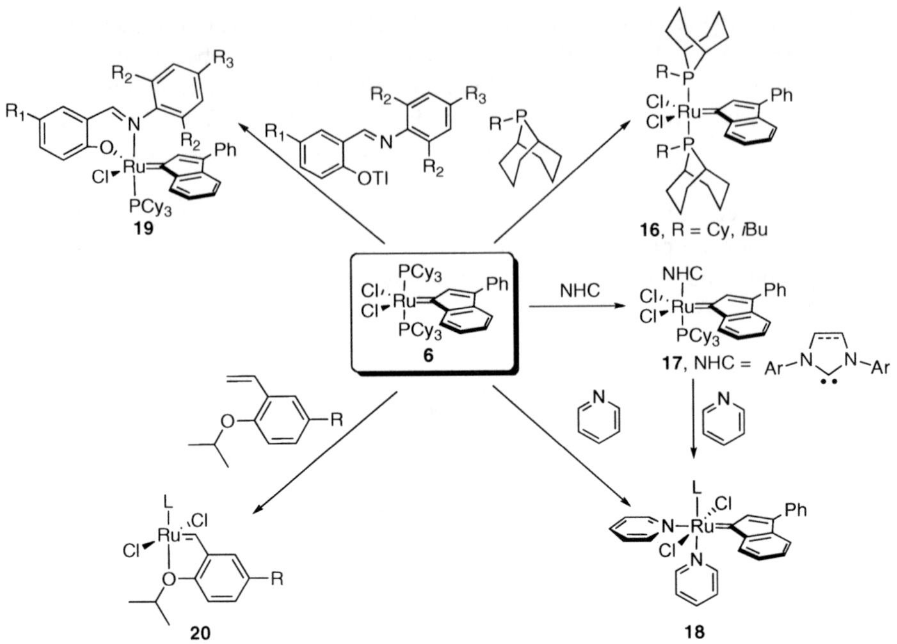

Scheme 7 Derivatization of indenylidene complex **3**

All these complexes were evaluated and found efficient in various olefin metathesis transformations including polymerization reactions (ring-opening metathesis polymerization and atom-transfer radical polymerization), ring-closing metathesis of diene and enyne, or cross metathesis of alkenes [22–26].

Surprisingly, whereas various modifications of the ligands bearing the metal center were achieved, to the best of our knowledge, no modification of the indenylidene framework has been investigated. We believed such variations using functionalized propargylic alcohols could lead to modification of the rearrangement process but also of the catalytic efficiency of resulting complexes.

5 Synthesis of Fluorenylidene–Ruthenium Complexes

Very recently, Milstein reported the synthesis of several fluorenylidene–ruthenium complexes [27]. Due to the close structural relationship between indenylidenes and fluorenylidenes, this approach has to be considered for further application to indenylidene–ruthenium formation. This synthetic route using a diazo compound is similar to the one developed for the formation of benzylidene complex [3], but so far no application to indenylidene–ruthenium catalysts has been reported. When $[RuCl_2(PPh_3)_3]$ was treated with 1 eq. of 9-diazofluorene at room and followed by the addition of a phosphine pincer ligand (2,6-bis(di-*tert*-butylphosphinomethyl-pyridine) complex **22** was isolated in good yields (Scheme 8). The formation of **22** may involved the passage through the intermediate **21** which is analogous to the indenylidene **2**. Stability tests have shown that **22** is stable in various solvent for several weeks.

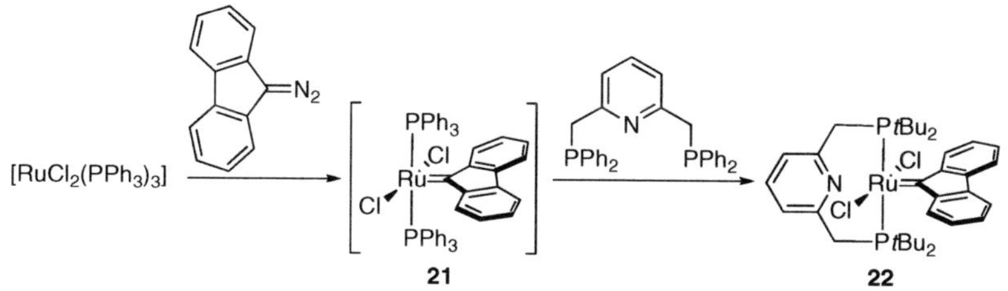

Scheme 8 Synthesis of fluorenylidene complex **22**

The stability towards thermal activation and the reactivity to silver salt was investigated. By simple heating to 110°C for 8 h, the cationic complex **23** (X = Cl) was generated (Scheme 9). Treatment with a silver salt such as $AgBF_4$ resulted in formation the cationic fluorylidene complex **23** (X = BF_4) confirmed by simple crystals X-ray determination. Analogous work has been conducted with phosphine- and N-heterocyclic carbene-containing indenylidene complexes, unfortunately no attempt to isolate the cationic complexes has been attempted so far and no characterization data are available [25c].

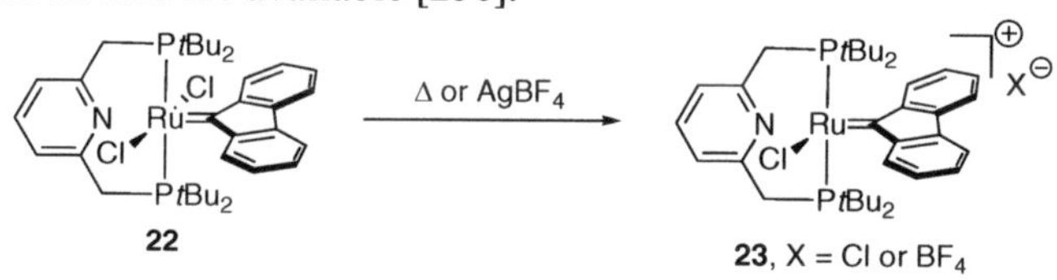

Scheme 9 Formation cationic fluorenylidene complexes **23**

6 Conclusions

As described in this paper, in spite of a confused discovery period, the ruthenium–indenylidenes are interesting complexes for olefin metathesis transformations, not only due to their straightforward syntheses but also because it has been shown that they catalyse efficiently various ruthenium-mediated metathesis reactions. As a result, a number of these catalysts are commercially available. Since these complexes were scarcely investigated so far, we believe that indenylidene-containing complexes should find new uses in the future.

Acknowledgments Funding for this work generously provided by the EC through the seventh framework program through grant (CP-FP 211468-2 EUMET). SPN (research grant) and HC (postdoctoral fellowship) both acknowledge the Ministerio de Educación y Ciencia (Spain) as well as the ICIQ Foundation for partial support of this work.

References

[1] For reviews on metathesis, see: (a) Fürstner A (2000) Angew Chem Int Ed 39: 3013–3043; (b) Trnka TM, Grubbs RH (2001) Acc Chem Res 34: 18–29; (c) Grubbs RH (ed.) (2003) Handbook of metathesis, p. 1204. Wiley-VCH, Weinheim; (d) Astruc D (2005) New J Chem 29: 42–56; (e) Deshmukh PH, Blechert S (2007) J Chem Soc, Dalton Trans: 2479–2491

[2] For reviews on synthetic applications, see: (a) Deiters A, Martin SF (2004) Chem Rev 104: 2199–2238; (b) McReynolds MD, Dougherty JM, Hanson PR (2004) Chem Rev 104: 2239–2258; (c) Nicolaou KC, Bulger PG, Sarlah D (2005) Angew Chem Int Ed 44: 4490–4527; (d) Van de Weghe P, Bisseret P, Blanchard N, Eustache J (2006) J Organomet Chem 691: 5078–5108; (e) Donohoe TJ, Orr AJ, Bingham M (2006) Angew Chem Int Ed 45: 2664–2670; (f) Gradillas A, Pérez-Castells J (2006) Angew Chem Int Ed 45: 6086–6100; (g) Hoveyda AH, Zhugralin AR (2007) Nature (London) 450: 243–251; (h) Kotha S, Lahiri K (2007) Synlett 18: 2767–2784; (i) Compain P (2007) Adv Synth Catal 349: 1829–1846

[3] (a) Schwab P, France MB, Ziller JW, Grubbs RH (1995) Angew Chem Int Ed 34: 2039–2041; (b) Schwab P, Grubbs RH, Ziller JW (1996) J Am Chem Soc 118: 100–110; (c) Roberts AN, Cochran AC, Rankin DA, Lowe AB, Schanz H-J (2007) Organometallics 25: 6515–6518

[4] For selected references, see: (a) Dias EL, Nguyen T, Grubbs RH (1999) J Am Chem Soc 119: 3887–3897; (b) Love JA, Sanford MS, Day MW, Grubbs RH (2003) J Am Chem Soc 125: 10103–10109

[5] (a) Beligny S, Blechert S (2006) In Nolan SP (ed.) N-heterocyclic carbenes in synthesis. Wiley-VCH: Weinheim, pp. 1–25; (b) Despagnet-Ayoub E, Ritter T (2007) Top Organomet Chem 21: 193–218; (c) Colacino E, Martinez J, Lamaty F (2007) Coord Chem Rev 251: 726–764

[6] (a) Sanford MS, Love JA, Grubbs RH (2001) Organometallics 20: 5314–5318; (b) Love JA, Sanford MS, Grubbs RH (2002) Angew Chem Int Ed 41: 4035–4037; (c) P'Pool SJ, Schanz H-J (2007) J Am Chem Soc 129: 14200–14212

[7] (a) Sanford MS, Henling LM, Day MW, Grubbs RH (2000) Angew Chem Int Ed 39: 3451–3453; (b) Conrad JC, Parnas HH, Snelgrove JL, Fogg DE (2005) J Am Chem Soc 127: 11882–11883; (c) Monfette S, Fogg DE (2006) Organometallics 25: 1940–1944

[8] For selected references, see: (a) Kingsbury JS, Harrity JPA, Bonitatebus PJ, Hoveyda AH (1999) J Am Chem Soc 121: 791–799; (b) Garber SB, Kingsbury JS, Gray BL, Hoveyda AH (2000) J Am Chem Soc 122: 8168–8179; (c) Wakamatsu H, Blechert S (2002) Angew Chem Int Ed 41: 794–796; (c) Wakamatsu H, Blechert S (2002) Angew Chem Int Ed 41: 2403–2405; (d) Grela K, Harutyunyan S, Michrowska A (2002) Angew Chem Int Ed 41: 4038–4040; (e) Michrowska A, Bujok R, Harutyunyan S, Sashuk V, Dolgonos G, Grela K (2004) J Am Chem Soc 126: 9318–9325; (e) Bieniek M, Bujok R, Cabaj M, Lugan N, Lavigne G, Arlt D, Grela K (2006) J Am Chem Soc 128: 13652–13653; (f) Rix D, Caijo F, Laurent I, Boeda F, Clavier H, Nolan SP, Mauduit M (2008) J Org Chem 72: 4225–4228

[9] Clavier H, Grela K, Kirschning A, Mauduit M, Nolan SP (2007) Angew Chem Int Ed 46: 6786–6800

[10] (a) Nguyen ST, Johnson LK, Grubbs RH, Ziller JW (1992) J Am Chem Soc 114: 3974–3975; (b) Nguyen ST, Grubbs RH, Ziller JW (1993) J Am Chem Soc 115: 9858–9859; (c) Wilhelm TE, Belderrain TR, Brown SN, Grubbs RH (1997) Organometallics 16: 3867–3869; (d) Gandelman M, Rybtchinski B, Ashkenazi N, Gauvin RM, Milstein D (2001) J Am Chem Soc 123: 5372–5373

[11] (a) Werner H, Stark A, Schulz M, Wolf J (1992) Organometallics 11: 1126–1130; (b) Grünwald C, Gevert O, Wolf J, González-Herrero P, Werner H (1996) Organometallics 15: 1960–1962; (c) Wolf J, Strüer W, Werner H, Schwab P, Schulz M (1998) Angew Chem Int Ed 37: 1124–1126; (d) Castarlenas R, Eckert M, Dixneuf PH (2005) Angew Chem Int Ed 44: 2576–2579
[12] (a) Fürstner A, Picquet M, Bruneau C, Dixneuf PH (1998) Chem Commun: 1315–1316; (b) Picquet M, Bruneau C, Dixneuf PH (1998) Chem Commun: 2249–2250; (c) Fürstner A, Liebl M, Lehmann CW, Picquet M, Kunz R, Bruneau C, Touchard D, Dixneuf PH (2000) Chem Eur J 6: 1847–1857
[13] For reviews, see: (a) Dragutan V, Dragutan I, Verpoort F (2005) Platinum Metals Rev 49: 33–40; (b) Boeda F, Clavier H, Nolan SP (2008) Chem. Commun: 2726–2740
[14] (a) Harlow KJ, Hill AF, Wilton-Ely JDET (1999) J Chem Soc, Dalton Trans: 285–291; (b) Fürstner A, Hill AF, Liebl M, Wilton-Ely JDET (1999) Chem Commun: 601–602
[15] Jafarpour L, Schanz H-J, Stevens ED, Nolan SP (1999) Organometallics 18: 5416–5419
[16] Selegue JP (1982) Organometallics 1: 217–218
[17] Fürstner A, Guth O, Düffels A, Seidel G, Liebl M, Gabor B, Mynott R (2001) Chem Eur J 7: 4811–4820
[18] Schanz H-J, Jafarpour L, Stevens ED, Nolan SP (1999) Organometallics 18: 5187–5190
[19] Shaffer EA, Chen CL, Beatty AM, Valente EJ, Schanz H-J (2007) J Organomet Chem 692: 5221–5233
[20] (a) Basseti M, Centola F, Sémeril D, Bruneau C, Dixneuf PH (2003) Organometallics 22: 4459–4466; (b) Castarlenas R, Dixneuf PH (2003) Angew Chem Int Ed 42: 4524–4527
[21] Castarlenas R, Vovard C, Fischmeister C, Dixneuf PH (2006) J Am Chem Soc 128: 4079–4089
[22] (a) Forman GS, McConnell AE, Hanton MJ, Slawin AMZ, Tooze RP, Janse van Rensburg W, Meyer WH, Dwyer C, Kirk MM, Serfontein DW (2004) Organometallics 23: 4824–4827; (b) Forman GS, Bellabarba RM, Tooze RP, Slawin AMZ, Karch R, Winde R (2006) J Organomet Chem 691: 5513–5516; (c) Boeda F, Clavier H, Jordaan M, Meyer WH, Nolan SP (2008) J Org Chem 73: 259–263
[23] Clavier H, Nolan SP (2007) Chem Eur J 13: 8029–8036
[24] (a) Clavier H, Petersen JL, Nolan SP (2006) J Organomet Chem 691: 5444–5447; (b) Clavier H, Nolan SP (2007) In Imamoglu Y, Dragutan V (eds.) Metathesis chemistry: From nanostructure design to synthesis of advanced materials, NATO Science Ser II.243: 29–38. Springer, Dordrecht; (c) de Frémont P, Clavier H, Montembault V, Fontaine L, Nolan SP (2008) J Mol Catal A: Chem 283: 108–113; (d) Monsaert S, Drozdzak R, Dragutan V, Dragutan I, Verpoort F (2008) Eur J Inorg Chem: 432–440
[25] (a) Opstal T, Verpoort F (2002) Synlett 6: 935–941; (b) De Clercq B, Verpoort F (2002) Adv Synth Catal 344: 639–648; (c) Opstal T, Verpoort F (2003) New J Chem 27: 257–262; (d) Opstal T, Verpoort F (2003) Angew Chem Int Ed 42: 2876–2879; (e) Drozdzak R, Ledoux N, Allaert B, Dragutan I, Dragutan V, Verpoort F (2005) Cent Eur J Chem 3: 404–416
[26] (a) Clavier H, Nolan SP, Mauduit M (2008) Organometallics 27: 2287–2292; (b) Rix D, Caijo F, Laurent I, Boeda F, Clavier H, Nolan SP, Mauduit M (2008) J Org Chem 73: 4225–4228
[27] Zhang J, Gandelman M, Shimon LJW, Milstein D (2008) Organometallics 27: 3526–3533

The Influence of the Anionic Counter-Ion on the Activity of Ammonium Substituted Hoveyda-Type Olefin Metathesis Catalysts in Aqueous Media

Łukasz Gułajski,[1] **Karol Grela**[1,2*]

[1]Department of Chemistry, Warsaw University, Pasteura 1, 02-093 Warsaw (Poland)

[2]Institute of Organic Chemistry, Polish Academy of Sciences, Kasprzaka 44/52, 01-224, Warsaw (Poland)

[*]E-mail: grela@icho.edu.pl

Abstract Polar olefin metathesis catalysts, bearing an ammonium group are presented. The electron withdrawing ammonium group not only activates the catalysts electronically, but at the same time makes the catalysts more hydrophilic. Catalysts can be therefore efficiently used not only in traditional media, such as methylene chloride and toluene, but also in technical-grade alcohols, alcohol–water mixtures and in neat water. Finally, in this overview the influence of the anionic counter-ion on the activity of ammonium substituted Hoveyda-type olefin metathesis catalysts in aqueous media is presented.

Keywords Catalysis · Metathesis · Tagged catalyst · Aqueous metathesis

1 Introduction

Olefin metathesis is now a widely used methodology in organic synthesis. The recent development of ruthenium catalysts, such as **1–3** (Figure 1) has an important impact in organic chemistry [1]. Using these catalysts chemists can now efficiently synthesize an impressive range of molecules that only a decade ago required significantly longer and tedious routes [1].

Despite the general superiority offered by this family of catalysts, they share some disadvantages. Since olefin metathesis reactions are expected to be used in pharmaceutical processes, the most undesirable feature of these complexes is that during the reaction they form ruthenium by-products which are difficult to remove from the reaction products [2]. In many cases, ruthenium levels of >2,000 ppm remain after chromatography of products prepared by RCM with 5 mol% of Grubbs

V. Dragutan et al. (eds.), *Green Metathesis Chemistry: Great Challenges in Synthesis, Catalysis and Nanotechnology*,

catalysts [3]. The ruthenium has to be removed prior to further processing [4]. Several protocols to solve problems associated with Ru contamination have been proposed but none is universally attractive so far [3, 5].

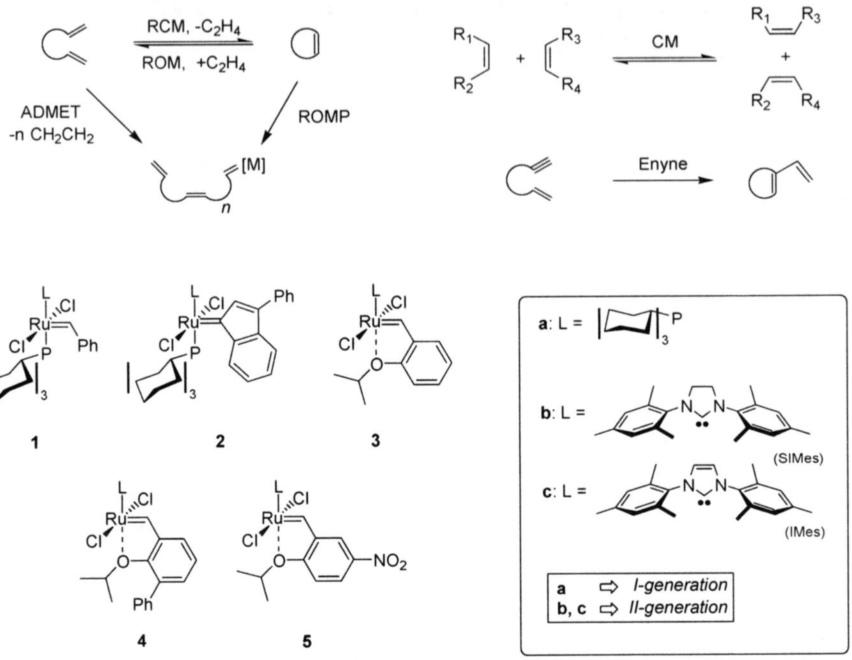

Figure 1 Selected modern catalysts for olefin metathesis and the transformation itself: RCM = ring-closing metathesis, ROM = ring-opening metathesis, ADMET = acyclic diene metathesis polymerization, ROMP = ring-opening metathesis polymerization, CM = cross-metathesis

In addition, due to environmental problems, there is a need to use water as a more safe, benign and cheap solvent [6, 7]. Furthermore, the aqueous olefin metathesis can be critical for some biological applications of olefin metathesis [8]. Classical ruthenium initiators were used in ring opening metathesis polymerization (ROMP) in mixed systems water/organic solvent, in the presence of dodecyltrimethylammonium bromide or sodium dodecylsulfate (SDS) [9]. Ring-closing metathesis (RCM) in water has been a challenge for a long time because of the instability of the formed carbene in water [10a]. However, the notion of "green chemistry" [7] has encouraged some groups to test precatalysts **1a, b** in methanol or water in the presence of surfactants, polydimethylsiloxane or without additives [10]. Unfortunately, in addition to the low solubility of **1a, b** in these media, a decrease of the reaction rate is often observed and, therefore, an increase of catalyst loading or reaction temperature is usually needed.

Onium-tagged catalysts are an alternative solution for problems with purification and application of metathesis reaction in aqueous media.

We demonstrated that the 5- and 4-nitro-substituted Hoveyda–Grubbs catalysts initiate olefin metathesis dramatically faster than the parent complex **3b** [11]. It was proposed that the electron-withdrawing (EWG) nitro group decreases the electron density on the chelating oxygen of the isopropoxy group and weakens the O→Ru coordination, facilitating faster initiation of the metathesis catalytic cycle [11, 12]. In accordance with this assumption, it was observed that complex **6** (Figure 2), bearing the electron-donating (EDG) diethylamino group, shows little or no activity in olefin metathesis [13]. However, in a striking contrast, the *in situ* formed salts 7, obtained by treatment of aniline 7 with Brønsted acids HA, are of high activity, surpassing the parent Hoveyda–Grubbs complex **3b** in terms of initiation speed [13]. The formation of a polar salt not only activates the catalyst but also changes its physical properties, such as solubility in polar media, and was used to create a site for non-covalent immobilisation on a solid phase [14, 15].

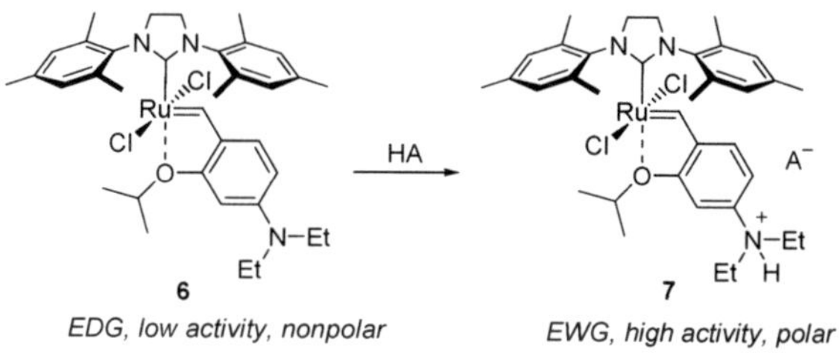

Figure 2 The concept of "electron donating to electron withdrawing activity switch"

The concept of "electron donating to electron withdrawing activity switch" [12] (Figure 2) was later extended by Grela to prepare complex **8** (Figure 3), bearing a quaternary ammonium group [16–19]. After this preliminary report, other catalysts bearing polar quaternary ammonium groups in the benzylidene fragment were reported by Raines (**9**) [20], Mauduit and Grela (**10–12**) [21] and by Grubbs (**13**, **14**) [22].

2 Results and Discussion

As illustrated in Scheme 1, we used commercially available aldehyde **15** as a starting material for the preparation of the corresponding quaternary salt **17**. Complex **8** was obtained in the reaction of **17** (1.1 eq.) with **2b** (1.0 eq.) and CuCl (1.5 eq.) as an air-stable deep green microcrystalline solid. A sample of this catalyst has been stored for a period of 6 months (+4°C, air) without any sign of decomposition. Interestingly, we found that iodide **8** is soluble in water at concentrations ≥0.002 M. Therefore, we decided to check the catalytic activity of **8** in metathesis of representative water-soluble substrates *in neat water*.

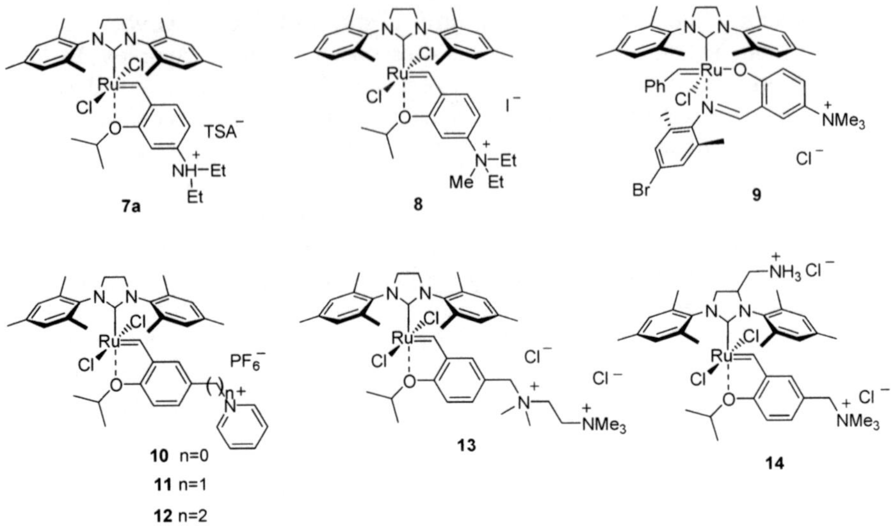

Figure 3 Ruthenium quaternary ammonium tagged catalysts (TSA = p-$CH_3C_6H_4SO_3$)

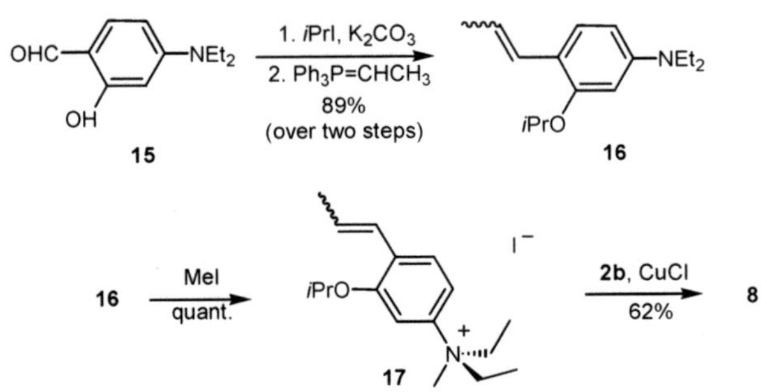

Scheme 1 Synthesis of complex **8** by the new methodology from complex **2b**

For activity comparison, we examined the homo-cross-metathesis of allyl alcohol **20** in the presence of catalyst **8** [17, 19]. As reported in Table 1 (entry 1), catalyst **8** showed an improved activity, leading -even at lower loadings- to complete conversion to **21**. When Grubbs **1b**, Hoveyda–Grubbs **3b** or nitro substituted Hoveyda **5** catalysts were used, no conversion was observed. The reason is that none of these catalysts is soluble in neat water [17, 19]. We also tested catalyst **10** and **11** but the results were not so satisfactory (Table 1, entries 6 and 7) [21].

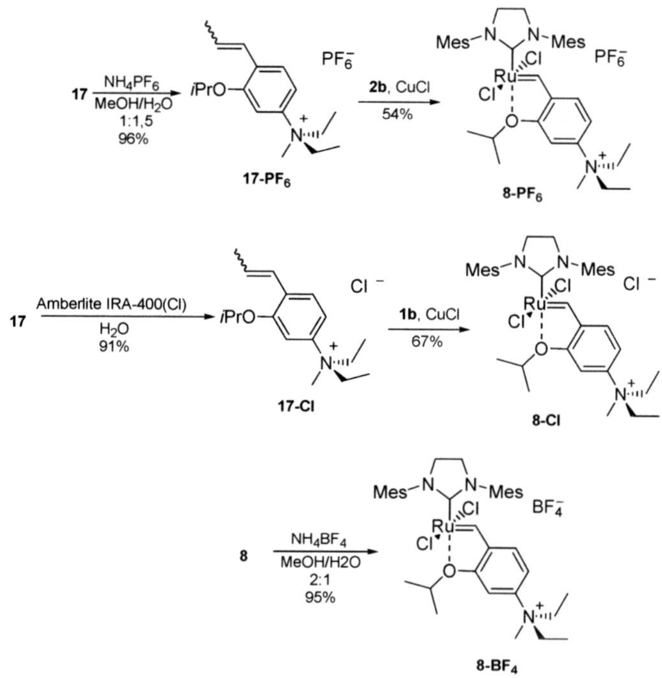

Scheme 2 Synthesis of analogues of complex **8** by using **1b** or **2b**

After completing this research we decided to check the influence of anionic counter-ions on the activity of ammonium substituted Hoveyda-type olefin metathesis catalysts in aqueous media. We prepared and tested the BF_4^-, PF_6^- and Cl^- containing analogues of **8**. Complexes **8-BF$_4$**, **8-PF$_6$** and **8-Cl** were conveniently obtained, according to routes presented in Scheme 2, as air stable green solids [19]. Within these complexes, we observed the expected relationship between the nature of the escorting counter anion and the catalyst's activity in water, since the more hydrophilic I and BF_4, and finally the best counter anion, Cl, lead to much better conversion as compared to **8-PF$_6$**. (Table 1, entries 1–5) [19]. Finally, we made an experiment in which to the reaction mixture containing catalyst **11** we added NH_4Cl and in this case conversion increased twice (entry 8) [21]. It suggests that in the reaction mixture the catalyst changes some of the counter-ion from PF_6 to Cl which is more soluble. Our experiments show that the best counter ion is Cl. Interestingly, two other quaternary nitrogen-containing catalysts, **10** and **11**, exhibited lower selectivity in the aqueous CM of **20** giving some amount of **22** (entries 6, 7) [22].

Table 1 Activity of ammonium substituted Hoveyda-type olefin metathesis catalysts in aqueous media in cross-metathesis of allylic alcohol [a] – reaction in presence of 450 eq. of NH_4Cl; [b] – data taken from [22]; [c] – catalysts not soluble in D_2O

entry	substrate	product	**cat** (mol %)	*t* (h)	temp (°C)	conv. (%)
1			**8** (2.5)	3.5	25	>99
2			**8** (1)	3.5	25	58
3			**8-PF$_6$**(1)	8	25	12
4			**8-BF$_4$** (1)	8	25	60
5			**8-Cl** (1)	1	25	46
6	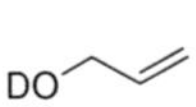	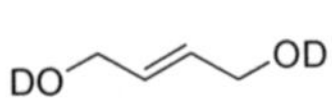	**10** (1)	8	25	14
7	**20**	**21**	**11** (1)	8	25	19
8		(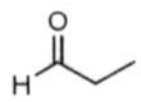)	**11** (1)	8	25	39[a]
9		**22**	**13** (5)	24	45	82 (+4)[b]
10			**14** (5)	6	45	69 (+12)[b]
11			**1b** or **3b** or **5** (1)	8	25	0[c]

3 Experimental

Representative procedure of olefin metathesis in water (homogeneous mixture). A reaction tube equipped with a magnetic stirring bar was charged with substrate **20** (0.13 mmol) and non-degassed D_2O (0.65 ml). To the solution of substrate catalyst **8** was added (5.5 mg, 0.007 mmol, 5 mol%). The reaction mixture was stirred at 25°C. After 3.5 h the reaction mixture was transferred to an NMR tube. Conversion was determined by ^{1}H NMR.

General procedure for preparation of complexes 8. Carbene complex **2b** (569.5 mg, 0.60 mmol), CuCl (89.1 mg, 0.9 mmol) and toluene (20 ml) were placed in a Schlenk flask equipped with a condenser. A solution of styrene **17** (280.3 mg, 0.72 mmol) in toluene (10 ml) was added and the resulting solution was stirred under argon at 80°C for 30 min. Next after reaction mixture (suspension) was filtered and the residue was washed on a Buchner funnel with acetone (20 ml) and dried under vacuum to give complex **8** as green microcrystalline solid (0.312 g, 62%). Complex **8**: IR (KBr) ν/cm 3,431, 2,928, 2,852, 1,629, 1,592, 1,484, 1,448, 1,430, 1,398, 1,384, 1,295, 1,262, 1,214, 1,168, 1,141, 1,098, 1,030, 980, 898, 852, 821, ^{1}H NMR (500 MHz, $CDCl_3$) δH/ppm: 1.05 (br.s, 6H), 1.34–1.55 (m, 6H), 2.40 (s, 6H), 2.46 (s,12H), 3.60 (s, 3H), 4.18 (s, 4H), 4.20–4.65 (m, 4H), 5.25–5.45 (m, 1H), 6.90–7.15 (m, 6H), 7.60 (s, 1H), 16.59 (s, 1H), ^{13}C NMR (125 MHz, $CDCl_3$) δC/ppm: 8.9, 19.4, 21.1, 22.0, 35.1, 51.5, 64.6, 78.3, 108.0, 115.7, 122.0, 126.8, 128.7, 129.4, 139.1, 139.5, 145.4, 153.5, 208.3, 290.1, HRMS

(ESI (+)): calcd for [M-126]+($C_{36}H_{50}N_3O^{35}Cl_2{}^{102}Ru$): 712.23744. found 712.2379. Complex **8-PF**$_6$: (ESI (−)): 144.9 (PF_6^-). Complex **8-BF**$_4$: (ESI (−)): 87.0 (BF_4^-). Complex **8-Cl**: (ESI (−)): 35.1 and 37.1 (Cl^-).

4 Conclusion

In conclusion, we report on homogeneous Hoveyda–Grubbs precatalysts tagged with a quaternary ammonium group and coordinating various counter anions. We have shown a relationship exists between the nature of the escorting counter anion and the catalyst activity in neat water.

Acknowledgments The KG thanks the Foundation for Polish Science for a 'Mistrz' professorship.

References

[1] General reviews: (a) Schrock RR, Hoveyda AH (2003) Angew Chem Int Ed 42:4592 (b) Trnka TM, Grubbs RH (2001) Acc Chem Res 34:18 (c) Fűrstner A (2000) Angew Chem Int Ed 39:3012 (d) Grubbs RH, Chang S (1998) Tetrahedron 54:4413 (e) Schuster M, Blechert S (1997) Angew Chem Int Ed 36:2037 (f) Dragutan V, Dragutan I, Balaban AT (2001) Platinum Met Rev 45:155 (g) For an industrial perspective: Thayer AM (2007) C&EN 85(7):37

[2] Nicola T, Brenner M, Donsbach K, Kreye P (2005) Org Process Res Dev 9:513

[3] (a) Conrad JC, Parnas HH, Snelgrove JL, Fogg DE (2005) J Am Chem Soc 127:11882; (b) For example, in a crude untreated product of the diethyl diallylmalonate RCM catalysed by 5 mol% of Grubbs I-generation catalyst the theoretical amount of Ru is 90 mg per 5 mg of product (18,000 ppm). After filtration of the crude reaction mixture, the Ru level was reduced to 59.7 ± 0.50 mg per 5 mg (12,000 ppm). Further purification of such crude metathesis products usually reduces ruthenium levels below 2,000 ppm, see ibid, and McEleney K, Allen DP, Holliday AE, Crudden CM (2006) Org Lett 8:2663

[4] Another solution to this problem might be based on the immobilisation of a metathesis catalysts in a separate liquid or solid phase. For recent reviews, see: (a) Hoveyda AH, Gillingham DG, Van Veldhuizen JJ, Kataoka O, Garber S B, Kingsbury JS, Harrity JPA (2004) Org Biomol Chem 2:1 (b) Buchmeiser RM (2004) New J Chem 28:549. For related systems developed in our laboratories, see: (c) Grela K, Mennecke K, Kunz U, Kirschning A (2005) Synlett 19:2948 (d) Grela K, Tryznowki M, Bieniek M (2002) Tetrahedron Lett 43:6425

[5] Cho J H, Kim BM (2003) Org Lett 5:531

[6] Cornils B, Hermann WA (eds.) (2004) Aqueous phase organometallic catalysis. Wiley-VCH, Weinheim, Germany

[7] For a review on sustainable aspects of olefin metathesis, see: Clavier H, Grela K, Kirschning A, Mauduit M, Nolan SP (2007) Angew Chem Int Ed 46:6786

[8] For example, see: (a) Gordon EJ, Sanders WJ, Kiessling LL (1998) Nature 392:30; (b) Kanai M, Mortell KH, Kiessling LL (1997) J Am Chem Soc 119:993; (c) Manning DD, Hu X, Beck P, Kiessling LL (1997) J Am Chem Soc 119:3161; (d) Manning DD, Strong LE, Hu X, Beck P, Kiessling LL (1997) Tetrahedron 53:11937

[9] (a) Lynn DM, Kanaoka S, Grubbs RH (1996) J Am Chem Soc 118:784; (b) Monteil V, Wehrmann P, Mecking S (2005) J Am Chem Soc 127:14568

[10] (a) Kirkland TA, Lynn DM, Grubbs RH (1998) J Org Chem 63:9904; (b) Davis KJ , Sinou D (2002) J Mol Catal A: Chem 177:173; (c) Mwangi MT, Runge MB, Bowden NB (2006) J Am Chem Soc 128:14434; (d) Connon SJ, Rivard M, Zaja M, Blechert S (2003) Adv Synth Catal 345:572; (e) Zarka MT, Nuyken O, Weberskirch R (2004) Macromol Rapid Commun 25:858; (f) For early examples of ROMP in aqueous media initiated by poorly defined ruthenium complexes such as $RuCl_2(H_2O)n$ or $Ru(H_2O)_6(TsO)_2$, see: Novak BM, Grubbs RH (1988) J Am Chem Soc 110:960; (g) Novak BM, Grubbs RH (1988) J Am Chem Soc 110:7542; (h) Hillmeyer MA, Lepetit C, McGrath DV, Novak BM, Grubbs RH (1992) Macromolecules 25:3345; (i) Mortell KH, Weatherman RV, Kiessling LL (1996) J Am Chem Soc 118:2297; (j) Lipshutz BH, Aguinaldo GT, Ghorai S, Voigtritter K (2008) Org Lett 10:1325; (k) Lipshutz BH, Ghorai S, Aguinaldo GT (2008) Adv Synth Catal 7–8:953; (l) For a review on aqueous olefin metathesis, see: Burtscher D, Grela K (2009) Angew Chem Int Ed 48:442

[11] (a) Grela K, Harutyunyan S, Michrowska A (2002) Angew Chem Int Ed 41:4038; (b) Michrowska A, Bujok R, Harutyunyan S, Sashuk V, Dolgonos G, Grela K (2004) J Am Chem Soc 126:9318

[12] (a) Grela K, Michrowska A, Bieniek M (2006) Chem Rec 6:144; (b) Michrowska A, Grela K (2008) Pure Appl Chem 80:31

[13] (a) Gułajski L, Michrowska A, Bujak R, Grela K (2006) J Mol Catal A: Chem 254:118; (b) Kirschning A, Gułajski Ł, Mennecke K, Meyer A, Busch T, Grela K (2008) Synlett 2692

[14] Michrowska A, Mennecke K, Kunz U, Kirschning A, Grela K (2006) J Am Chem Soc 128:13261

[15] Kirschning A, Harmrolfs K, Mennecke K, Messinger J, Schön U, Grela K (2008) Tetrahedron Lett 49:3019

[16] Michrowska A (2006) Ph.D. thesis, Institute of Organic Chemistry, Warsaw

[17] Michrowska A, Gułajski Ł, Grela K (2006) Chem Today 24(6):19

[18] Michrowska A, Gułajski Ł, Kaczmarska Z, Mennecke K, Kirschning A, Grela K (2006) Green Chem 8:685

[19] Gułajski Ł, Michrowska A, Narożnik J, Kaczmarska Z, Rupnicki L, Grela K (2008) Chem SusChem 1:103; Ł. Gułajski; K. Grela unpubliced results

[20] Binder JB, Guzei IA, Raines RT (2007) Adv Synth Catal 349:395

[21] (a) Rix D, Clavier H, Gułajski Ł, Grela K, Mauduit M (2006) J Organomet Chem 691:5397; (b) Rix D, Caïjo F, Laurent I, Gułajski Ł, Grela K, Mauduit M (2007) Chem Commun 3771

[22] Jordan J P, Grubbs R H (2007) Angew Chem Int Ed 46:5152

Ruthenium Catalysts Bearing Carboxylate Ligand

Rafał Gawin, Karol Grela

Institute of Organic Chemistry, Polish Academy of Sciences, Kasprzaka 44/52, 01-224, Warsaw, Poland

E-mails: gawin@icho.edu.pl, grela@icho.edu.pl

Abstract Novel class of ruthenium complex bearing carboxylate function is described in this highlight. Chemical activation of **14** providing highly active metathesis catalysts is discussed. New catalytic system due to high affinity to silica gel allow to efficient catalyst removal to obtain colorless products with low ruthenium contamination. Unique structural motif allows to introduce new carboxylate and sulfonate derived ligands without use of silver or thallium salts. Latest results of metathesis with derivative of **14** in ionic liquids and aqueous emulsions are presented.

Keywords Ruthenium · Metathesis · Carboxylate

1 Introduction

The tremendous success of olefin metathesis is largely due to the evolutionary development of active, well-defined ruthenium-based catalysts [1]. In the case of Grubbs' first generation catalyst, **1a**, major progress has been attained by optimizing the ligand sphere around the ruthenium center. Key advances include the development of *N*-heterocyclic carbene (NHC) complexes (e.g. Grubbs' second-generation catalyst **1b**) and derivatives containing a chelating 2-alkoxybenzylidene (e.g. Hoveyda's catalyst **2**) or indenylidene **3** [1] (Figure 1).

Despite many efforts still the most undesirable feature of modern homogeneous metathesis catalysts is that they often form deeply colored ruthenium by-products, which are difficult to remove from the desired products [2]. This can be a big problem when the product has a pharmaceutical destination, where ruthenium contamination has to be lower than 10 ppm.

A lot of work was made to answer the question: how conduct processes more economically, environment friendly and with the recoverable metathesis catalyst which price is important economic factor especially when loading of catalyst is high [2].

V. Dragutan et al. (eds.), *Green Metathesis Chemistry: Great Challenges in Synthesis, Catalysis and Nanotechnology*,

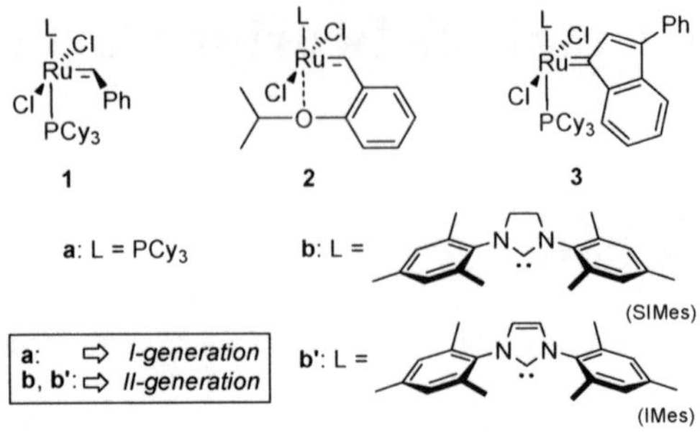

Figure 1 Modern olefin metathesis catalysts. PCy_3 – tricyclohexylphosphine

2 Ruthenium Catalysts Bearing Carboxylates and Sulfonates Ligand: Synthesis of New Ruthenium Complex Bearing Tridentate Carboxylate Function

It has been demonstrated that ruthenium alkylidene complexes bearing carboxylates and sulfonates ligands can reach metathesis efficiencies comparable or superior to those achieved by catalysts **1** and **2**.

The groups of Buchmeiser, Blechert, and Nuyken have prepared a series of complexes **4** and **5** bearing perfluorocarboxylic or triflic acid or derivatives of these ligands [3] (Figure 2). Fogg reported highly active catalysts **6** bearing bidentate sulfonate ligand [4]. He reported series of catalysts **7** bearing bidentate ligand with carboxylate and phosphine function [5]. Grubbs reported catalyst **8** bearing proline ligand [6].

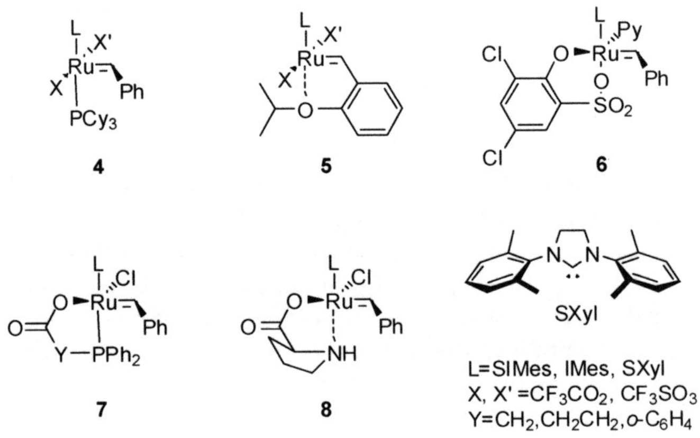

Figure 2 Ruthenium complexes bearing carboxylate ligand

Unfortunately in many protocols to exchange chlorine to new ligand the silver or thallium salts have to be employed [3, 4].

Bieniek et al. reported the highly active and stable catalyst **9** in which an alkoxybenzylidene ligand is doubly chelated to the ruthenium center through coordination of both the ether and an attached ester group [7] (Scheme 1).

In pursuit of this concept, we became interested in ruthenium complex **12** bearing a free carboxylic acid function, which might lend itself to non-covalent type immobilization [8]. During this investigation, however, an unprecedented bias of such a system to reorganize its ligand sphere was observed [9].

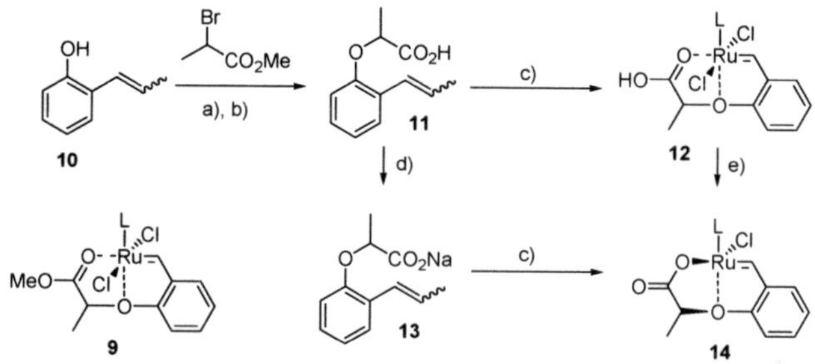

Scheme 1 Synthesis of new complexes **14** and **12**. Conditions: (a) K_2CO_3, Cs_2CO_3, DMF, rt, 24 h; (b) LiOH, THF, rt, 24 h; (c) **1b**, CuCl, DCM, reflux, 30 min or **3b**, CuCl, toluene, 80°C, 30 min; (d) NaOH; (e) SiO_2 (silica gel chromatography)

Starting from commercially available 2-propenylphenol **10** and methyl 2-bromopropionate ligand **11** was readily obtained in two steps. Reaction with Grubb's second generation catalyst **1b** or indenylidene second generation catalyst **3b** and standard purification procedure (chromatography and crystallization) provided unexpectedly new green complex **14** instead of **12** with yield 84%. NMR of crude reaction mixture and after column chromatography show that cyclization takes place during column chromatography (benzylidene proton shift 16.73 ppm

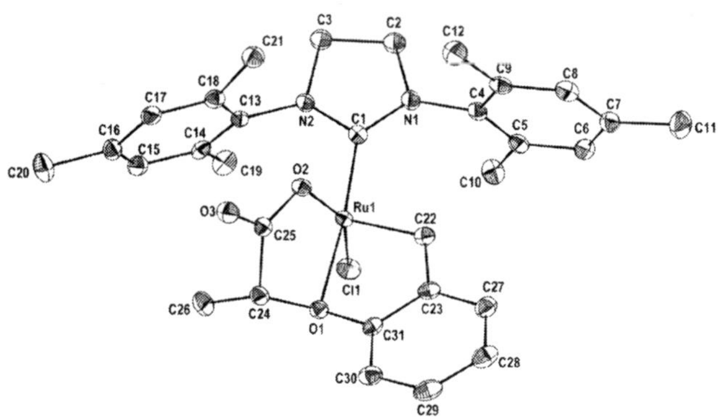

Figure 3 Solid state structure of **14** with thermal ellipsoids at 50% probability

before and 16.52 ppm after). Same reaction with sodium salt of acid **13** gave directly cyclized complex **14** (same benzylidene proton shift before and after chromatography: 16.52 ppm). Based on mass spectrometry, NMR spectroscopy, elemental analysis and X-ray analysis (Figure 3) we assigned structure of new complex to **14**.

The solid-state structure of **14** reveals some interesting properties. Analogous to most Hoveyda type catalysts, the ruthenium center in **14** is pentacoordinate and the geometry of the ligands is very close to square-pyramidal. In contrast to the structure of **9** in which an oxygen atom (O2) forms the additional coordination site of the ruthenium on one of the tops of the square bipyramid [7], in **14** the O2 oxygen atom replaces one of the chlorines in the coordination of the central metal ion in the *cis* position with respect to the O1 oxygen center. The O–Ru interactions are stronger in the case of **14**. Compared with **9** the O2–Ru1 distance is quite short while the O2–C25 distance is significantly elongated (Table 1).

Table 1 Comparison of bond length in **9**, **14** and Hoveyda catalyst **2b**

	Distance (Å)		
Bond	**9**	**14**	**2b**
Ru(1)-C(4)|C(22)	1.820(3)	1.832(2)	1.8286(15)
Ru(1)-C(1)	1.984(3)	1.9812(19)	1.9791(15)
Ru(1)-O(1)	2.207(2)	2.2286(14)	2.2562(10)
Ru(1)-Cl(1)	2.3583(10)	2.3222(5)	2.3279(4)
Ru(1)-Cl(2)	2.3560(11)	–	2.3380(4)
Ru(1)-O(2)	2.536(2)	2.0553(14)	–
O(2)-C(6)|C(25)	1.194(4)	1.291(2)	–
O(1)-C(36)|C(31)	1.371(4)	1.409(2)	1.3701(18)

The fact that the carboxylate moiety coordinates ruthenium by replacing one of the chlorine atoms influences the geometry around the ether oxygen. The O1 atom in **14** presents a geometry which suggests sp3 hybridization, with valence angles values closer to 110°. The O1–C31 bond length is significantly longer then the analogous bond in the Hoveyda catalyst **2b** or in **9** [7]. The coordination of the ruthenium by the oxygen atom O1 in compound **14** seems to take place by a single lone electron pair, which suggests a different strength and character of the interaction to that in **9** or other Hoveyda type complexes [10].

3 Acidic Activation of New Carboxylate Ruthenium Catalyst

After isolation and characterization of **14**, we attempted to convert it to parent complex **12**. Addition of 1 eq. of hydrochloric acid in diethyl ether to a solution of **14** in dichloromethane (DCM) and stirring for 5 min followed by precipitation with *n*-hexane, evaporation and drying, provided complex **12**. Other acids such as triflic acid and perfluorononanoic acid can be used. In same manner new complexes **15** and **16** were obtained (Scheme 2). After silica gel chromatography complex **14** was obtained irrespective of which acid was used.

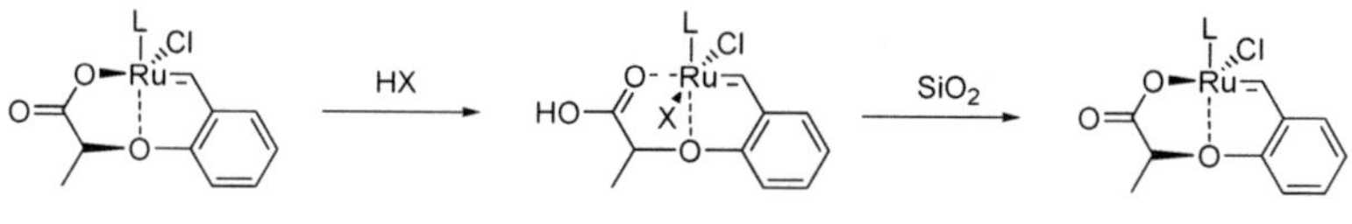

Scheme 2 Synthesis of catalysts **12**, **15** and **16**. Simple protocol to introduce new carboxylate and sulfonate derived ligands

Comparison of activity of the new complexes with **9** was investigated in standard model reaction – ring closing metathesis (RCM) of diethyl diallylmalonate **17a** (Scheme 3). It is worth mention that new complexes **12**, **15** and **16** were formed *in situ* by addition of 1 eq. of proper acid to solution of **17a** and **14** (Figure 4).

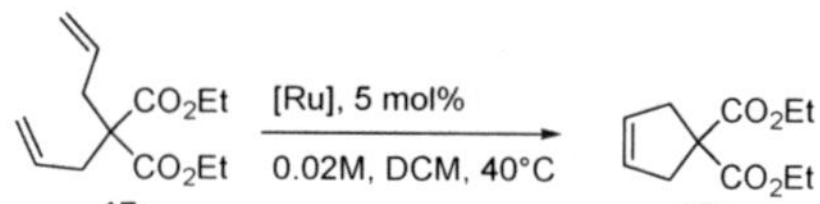

Scheme 3 Model metathesis reaction

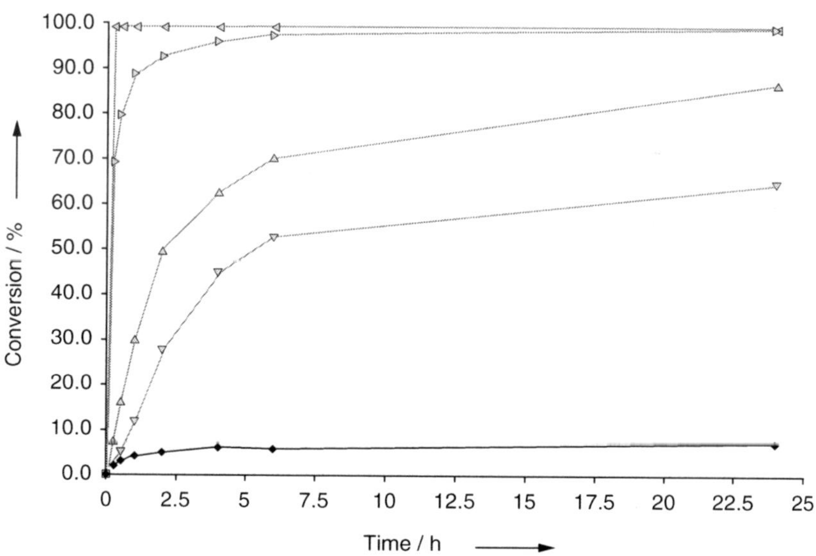

Figure 4 Catalytic activity of complexes **9** (◀), **12** (▶), **14** (■), **15** (▼) and **16** (▲) in RCM of **17a**. Conditions: 5 mol% of catalyst, 40°C, DCM, 24 h

Complex **12** with chlorine was the most active from new complexes. Complexes with perfluorononanoic **16** and triflic acid **15** were less active. It was not surprise that complex **14** bearing tridentate ligand was almost inactive.

Surprisingly complex **12** is much less active than corresponding **9**. The explanation for this fact may be a equilibrium which take place during reaction (Scheme 4). Although crystal structure of **12** was not obtained we believe that other reason may be a stronger bond (compared to **9**) between ruthenium and carboxylate moiety.

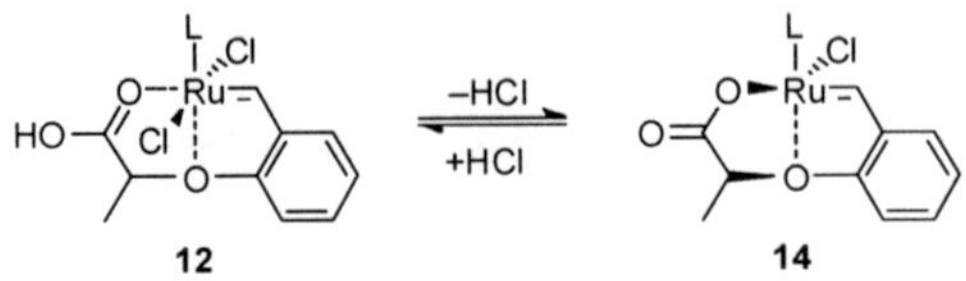

Scheme 4 Possible equilibrium between **12** and **14** during metathesis reaction. Proposed strong bond between carbonyl O and Ru in **12**

Because cyclization of **12** occurs even by treatment with silica, cyclization ability of **12** in reactions with bases is interesting. Reaction followed by NMR of **12** in CD_2Cl_2 and 1 eq. of sodium methoxide gives no **14** due to insolubility of base. Reaction with 1 eq. of pyridine gives mixture of **12** and cyclized **14** in ratio 1:1 whereas reaction with 1 eq. of potassium *t*-amylate gives only cyclized **14**.

4 Applications of New Carboxylate Catalysts in Metathesis Reactions

To investigate the scopes of new complexes standard set of metathesis reactions was used (Table 2). RCM of **19a** revealed that **14** has very low activity (entry 2). After refluxing reaction for 24 h only 14% of conversion was achieved. Worth to mention is fact that when after that time 1 eq. of hydrochloric acid was added and reaction was almost complete in next 1.5 h.

5 "Green" Aspects of New Carboxylate Catalysts

Removing catalyst and his by-products from reaction mixture especially where high loading of catalyst is necessary may be a problem [2]. Barrett reported synthesis of viridiofungin derivatives **23** where one of the key step was metathesis [11] (Figure 5).

Table 2 Set of standard metathesis reactions with new complexes **9**, **12**, **14**, **15** and **16**

Entry	Substrate	Product	Cat. (mol%), time, temp.	Yield (%)[a]	Regenerated 14 (%)
1			**9** (1), 1 h, 0°C	55	–
2			**14** (5), 24 h, 40°C	14	–
3	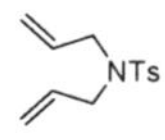 **19a**	 **19b**	**12** (5), 1 h, 40°C	96 (90)	95
4			**12** (1), 3 h, 40°C	96 (93)	–
5			**12** (0.6), 5.5 h, 40°C	92 (87)	–
6			**12** (0.3), 5.5 h, 40°C	93 (90)	–
7			**9** (5), 15 min, 40°C	99	–
8			**9** (1), 6 h, 0°C	70	–
9			**14** (5), 2 h, 20°C	0	–
10	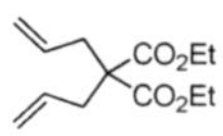 **17a**	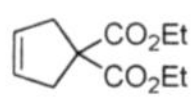 **17b**	**14** (5), 2 h, 40°C	10	–
11			**14** (5), 2 h, 100°C	88 (84)[b]	17
12			**12** (5), 24 h, 40°C	95 (86)	–
13			**15** (5), 24 h, 40°C	65	–
14			**16** (5), 24 h, 40°C	87	–
15			**9** (5), 15 min, 40°C	99	–
16			**14** (5), 24 h, 40°C	49	–
17	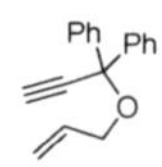 **20a**	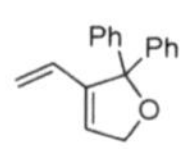 **20b**	**12** (5), 1 h, 40°C	99 (88)	96
18			**12** (1), 3 h, 40°C	96 (90)	–
19			**12** (0.6), 5.5 h, 40°C	96 (89)	–
20			**12** (0.3), 5.5 h, 40°C	99 (92)	–
21	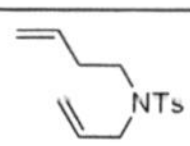 **21a**	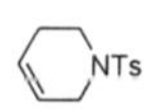 **21b**	**12** (1), 24 h, 40°C	98 (68)	–
22			**12** (0.3), 24 h, 40°C	98 (75)	–
23	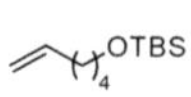 **22a**	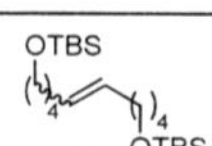 **22b**	**14** (5), 24 h, 40°C	0	97
24			**12** (5), 24 h, 40°C	96 (89)	–
25	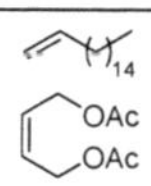 **23a**	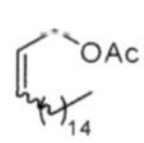 **23b**	**12** (5), 3 h, 40°C	98 (93)	–

[a]Isolated yields in parentheses.

[b]Reaction in toluene.

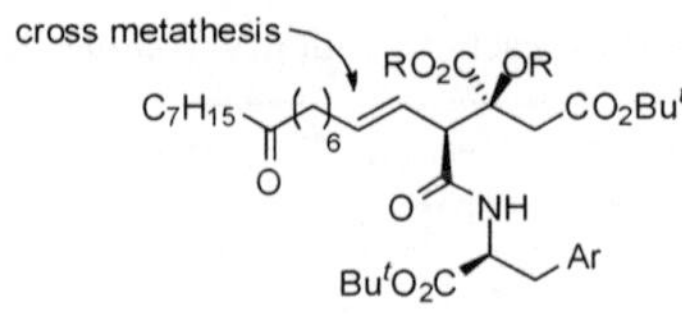

23

Figure 5 Structure of viridiofungin derivatives **23**

Despite almost complete conversion product was isolated with low yield due to extensive chromatography to remove metathesis catalyst and his by-products. The simplest solution to solve this problem is use metathesis catalysts with high affinity to silica gel. Use of highly polar catalysts not only provide product with low ruthenium contamination – it may be also useful in recovery protocol. The design of recoverable catalysts has become a central field of catalysis research, with an ideal recoverable catalyst having the following additional requirement to those listed above: it can be recovered either as the catalyst precursor or as a functionally equivalent resting state. In fact, catalyst decomposition associated with leaching of the active species and decomposition itself have to be taken into account. In practice, design efforts for effective recoverable catalysts must address the removal of these catalyst impurities from solution.

Catalysts **6** developed by Fogg (Figure 2) have shown a high affinity for silica, enabling their efficient removal in a single chromatographic pass. Michrowska et al. reported asarone derived catalysts **24** [12] and catalyst **25** bearing benzylidene unit with quaternary ammonium group (Figure 6) [13]. These catalyst posses high affinity to silica gel and can be recovered.

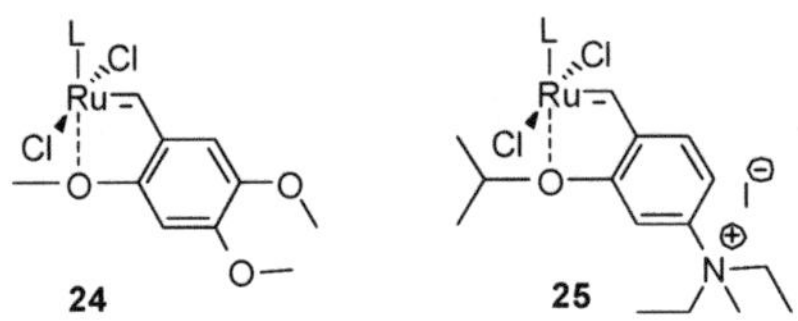

24 **25**

Figure 6 Highly polar ruthenium catalyst with good affinity to silica gel

Catalyst **14** also a high affinity to silica, enabling its efficient removal. For example, in the RCM of **17a** (5 mol% **12**, 1 h in refluxing dichloromethane, 0.2 mmol scale) filtration of the crude reaction mixture through a Pasteur pipette filled with 650 mg of silica gel yielded colorless product **17b** (Ru contamination of crude **17b** is 48 ppm, as determined by inductively coupled plasma mass spectrometry (ICP-MS)), while in the case of Grubbs' **1b** and Hoveyda's **2b** second generation catalyst removal of the catalyst was obviously incomplete under identical conditions (Figure 7). By washing of the Pasteur pipette containing **14** further with ethyl acetate it was possible to regenerate up to 95% of the catalyst (Table 1).

The regenerated complex shows identical analytical data to that of freshly prepared **14** and can be used, with similar results, in further metathesis reactions.

Although use of catalyst of high affinity to silica gel may be benefit in laboratory scale in large scale industrial processes using of chromatography is not practical and not economic. In these cases much more superior is use of immobilized and tagged catalyst.

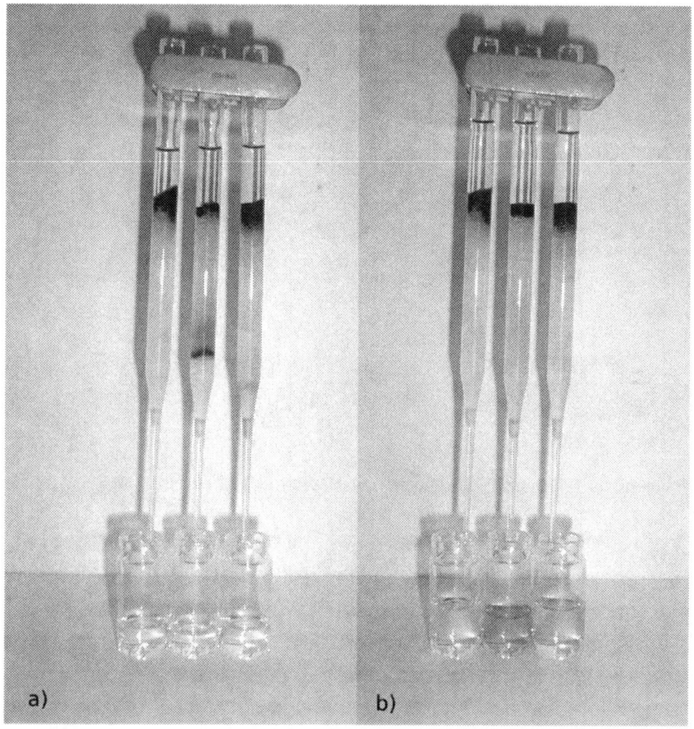

Figure 7 Different stage of DCM elution of crude reaction mixture containing **17b** and proper catalyst (from *left*: **12b**, **2b**, **1b**) trough silica contained Pasteur pipette

6 Immobilization of Ruthenium Carboxylate Catalyst in Ionic Liquid

Although chromatographic purification protocols have been optimized for decreasing the Ru content in the product to an almost acceptable level, the utilization of tagged or supported catalysts is a far superior strategy [2, 9, 14].

These strategies make use of the high affinity of these tags to alternative reaction media such as ionic liquids (ILs) or perfluorinated solvents that can be poorly miscible with the organic phase. The use of supported catalysts appears to be the most effective method to avoid contamination of product metal-containing catalysts. There is a lot of options to put the tag to ruthenium catalyst: NHC, benzylidene derivative or exchange of halogen ligand [2].

Almost in all cases exchange of halogen to carboxylate or sulfonate derived tag or support demand use of silver or thallium salts [3, 4].

In the case of complex **14** it is enough to use derivative of free acid to obtain tagged or supported catalyst.

Starting from 1-chlorohexadecane **26** and *N*,*N*-dimethylaminoethanol quaternary ammonium salt **27a** was obtained. Anion metathesis with KPF_6 provided **27b** (Scheme 5).

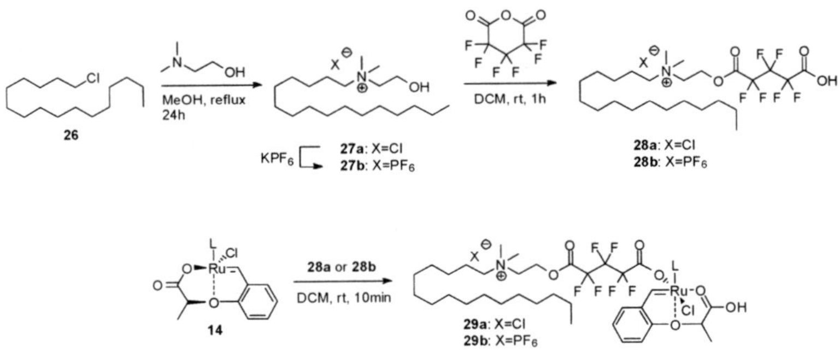

Scheme 5 Synthesis of ionically tagged catalyst **29a** and **29b**

In reaction with hexafluoroglutaric anhydride acids **28a** and **28b** were obtained which was reacted with **14** to obtain tagged catalyst **29a** and **29b**. Catalyst **29b** due to ionic group should have high affinity to polar media like ionic liquids. Use of ionic liquid immiscible with organic solvent is good strategy for recovery of catalyst and low ruthenium contamination in products [2].

Recovery of catalyst **29b** was investigated model RCM reaction of **17a** in mixture of butylmethylimidazole hexafluorophosphate (bmim·PF_6) and toluene (1:3) and 5 mol% of catalyst **29b** [15].

In first cycle total conversion was achieved after 30 min. Product **17b** was isolated by extraction of reaction mixture with toluene and evaporation. To reaction mixture was added second portion of **17a** in toluene and reaction was continued. Unfortunately only 72% conversion was achieved in second cycle after same isolation protocol and in following cycles conversion was even lower (Table 3).

Table 3 Conversions achieved in each cycle of RCM of **17a** in ionic liquid

Cycle	1	2	3	4	5	6
Conversion (%)	>98	72	61	39	31	31

Conditions: 5 mol% of **29b**, 0.2M, bmim·PF_6/toluene (1:3), 40°C, 30 min.

One of the reason of failure in recycling protocol may be a recyclization of catalyst **29b** to parent complex **14** (Scheme 4) which doesn't have a tag and can be extracted with toluene.

7 Reactions of Ruthenium Carboxylate Catalyst in Aqueous Emulsions

Recently using water as a solvent for metathesis reactions became of increasing interest to research [16]. Water not only can act like cheap and safe replacement for organic solvent, in many cases results obtained providing metathesis reaction in water are superior to these obtained in organic solvents.

Complex **29a** with long lipophilic chain, ionic function and polar carboxylate group is not soluble in water – it may act like surfactant and catalyst closed in one structure (inisurf catalyst). Standard set of metathesis reactions in non-degassed water in air at room temperature was conducted (Table 4). Products were isolated by solid phase extraction (Extrelut ®) with dichloromethane followed by silica gel chromatography. To be sure that reaction do not occur in organic phase, reaction mixtures were quenched by addition of vinyl ethyl ether.

8 Experimental Section

*Synthesis of ruthenium complex **14** bearing carboxylate ligand.*

Ligand **11** (49.5 mg; 0.24 mmol) or 13 (0.24 mmol), CuCl (23.8 mg; 0.24 mmol) and DCM (10 ml) were placed in a Schlenk flask. Afterward carbene complex **1b** (169.8 mg; 0.20 mmol) was added and the resulting solution was stirred under argon at 40°C for 40 min. From this point forth, all manipulations were carried out in air with reagent-grade solvents. The reaction mixture was concentrated under vacuum and resulting material was dissolved in ethyl acetate (ca. 10 ml), white solid was filtered off and the filtrate was concentrated under vacuum. The product was purified by column chromatography on silica gel (Merck, grade 9385, 230–400 mesh, no pre-treatment). Elution with *c*-hexane-ethyl acetate (2:1) (ca. 200 ml) then *c*-hexane-ethyl acetate (1:1) removes **14** as a green band. The solvent was evaporated and product dissolved in small amount of DCM, then *t*-butyl methyl ether was added until green crystals precipitated. The precipitate was filtered off, washed with *n*-pentane and dried in vacuum to afford complex **14** (104.8 mg, 84%) as a green solid. ^{1}H NMR (500 MHz, $CDCl_3$): δ = 16.52 (s, 1H), 7.49–7.52 (m, 1H), 7.11 (s, 2H), 7.06 (s, 2H), 7.02–7.05 (m, 1H), 6.97 (d, J = 8.23 Hz, 1H), 6.95 (d, J = 7.57 Hz, 1H), 4.65 (q, J = 7.02 Hz, 1H), 4.21 (s, 4H), 2.47 (bs, 6H, 2.39 (s, 6H), 2.37 (s, 6H), 1.19 (d, J = 7.02 Hz, 3H); ^{13}C NMR (125 MHz, $CDCl_3$): δ = 293.3, 210.3, 181.5, 154.6, 146.7, 139.3, 138.8, 130.0, 129.6, 129.4,126.6, 122.2, 119.1, 84.1, 51.4, 21.0, 19.00, 18.8, 18.5. IR (film from CH_2Cl_2): ν = 3483, 3044, 2977, 2919, 2861, 2736, 1662, 1607, 1594, 1571, 1483, 1451, 1401, 1376, 1318, 1297, 1269, 1185, 1151, 1100, 1069, 1017, 944, 918, 879, 846, 795, 752, 734, 700, 645, 617, 577, 501, 471, 438, 421 cm^{-1}; HRMS (FD/FI): m/z calcd for $C_{31}H_{35}N_2O_3{}^{35}Cl^{102}Ru$: [$M^+$] 620.1380 found, 620.1357; elemental analysis (%) calcd for $C_{31}H_{35}N_2O_3ClRu$ (620.16): C 60.04, H 5.69, N 4.52, Cl 5.72; found: C 59.92, H 5.69, N 4.55, Cl 5.48.

Table 4 Metathesis reactions in non degassed water on air catalyzed by **29a**

Entry	Substrate	Product	Time (h)	Yield (%)
1	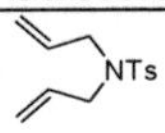 **19a**	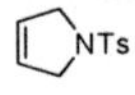 **19b**	1	98
2	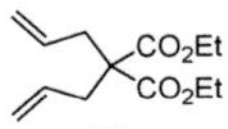 **17a**	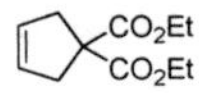 **17b**	3	94
3	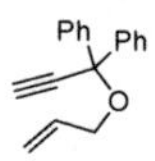 **20a**	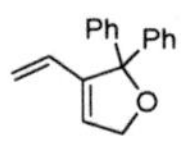 **20b**	1	98
4	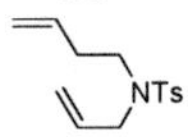 **21a**	 **21b**	1.5	95
5	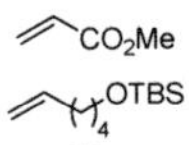 **22a**	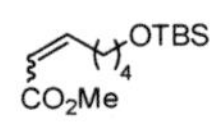 **22c**	24	80

Conditions: 5 mol% of **29a**, 0.5M, H_2O, rt.

9 Summary and Outlook

In summary we reported ruthenium catalyst **14** bearing carboxylate ligand which posses interesting properties. It is almost inactive at room temperature and can be activated by various acids or thermally. Unique structural motif of carboxylate allow to introduce new carboxylate derived ligand (free, tagged or supported) without use of any silver of thallium salt. Catalyst **14** is efficiently removed by single silica gel chromatography – products of metathesis reactions is not contaminated with ruthenium waste. Finally derivatives of **14** can be used in ionic liquids and aqueous emulsions with good results.

Acknowledgements Authors thanks the Foundation for Polish Science for "Mistrz" professorship.

References

[1] (a) Trnka TM, Grubbs RH (2001) Acc Chem Res 34:18–29; (b) Grubbs RH (ed.) (2003) Handbook of metathesis. Wiley-VCH, Weinheim, p. 1204; (c) Schrock RR, Hoveyda AH (2003) Angew Chem 115:4740–4782; Angew Chem Int Ed 42:4592–4633; (d) Connon SJ, Blechert S (2003) Angew Chem 115:1944–1968; Angew Chem Int Ed 42:1900–1923; (e) Astruc D (2005) New J Chem 29:42–56; (f) Dragutan V, Dragutan I, Verpoort F (2005) Platinum Metals Rev 49:33–40

[2] Clavier H, Grela K, Kirschning A, Mauduit M, Nolan SP (2007) Angew Chem Int Ed 46:6786–6801

[3] (a) Krause JO, Nuyken O, Wurst K, Buchmeiser MR (2004) Chem Eur J 10:777–784; (b) Halbach TS, Mix S, Fischer D, Maechling S, Krause JO, Sievers C, Blechert S, Nuyken O, Buchmeiser MR (2005) J Org Chem 70:4687–4694; (c) Mayr M, Wang D, Kröll R, Schuler N, Prühs S, Fürstner A, Buchmeiser MR (2005) Adv Synth Catal 347:484–492; (d) Krause JO, Lubbad SH, Nuyken O, Buchmeiser MR (2003) Macromol Rapid Commun 24:875–878; (e) Krause JO, Lubbad SH, Nuyken O, Buchmeiser MR (2003) Adv Synth Catal 345:996–1004

[4] Monfette S, Fogg DE (2006) Organometallics 25:1940–1944

[5] Zhang W, Liu P, Jin K, He R (2007) J Mol Catal A: Chem 275:194–199

[6] Samec JSM, Grubbs RH (2008) Chem Eur J 14:2686–2692

[7] Bieniek M, Bujok R, Cabaj M, Lugan N, Lavigne G, Arlt D, Grela K (2006) J Am Chem Soc 128:13652–13653

[8] Michrowska A, Mennecke K, Kunz U, Kirschning A, Grela K (2006) J Am Chem Soc 128:13261–13267

[9] Gawin R, Makal A, Woźniak K, Mauduit M, Grela K (2007) Angew Chem Int Ed 46:7206–7209

[10] Barbasiewicz M, Bieniek M, Michrowska A, Szadkowska A, Makal A, Woźniak K, Grela K (2007) Adv Synth Catal 349:193–203

[11] Goldup SM, Pilkington CJ, White AJP, Burton A, Barrett AGM (2006) J Org Chem 71:6185–6191

[12] (a) Michrowska A, Gułajski Ł, Grela K (2006) Chem Commun 841–843; (b) Grela K, Kim M (2003) Eur J Org Chem 963–966

[13] (a) Kirschning A, Gułajski Ł, Mennecke K, Meyer A, Busch T, Grela K (2008) Synlett: 2692–2696; (b) Gułajski Ł, Michrowska A, Narożnik J, Kaczmarska Z, Rupnicki L, Grela K (2008) ChemSusChem 1:103–109; (c) Michrowska A, Gułajski Ł, Kaczmarska Z, Mennecke K, Kirschning A, Grela K (2006) Green Chem 8:685–688

[14] (a) Śledź P, Mauduit M, Grela K (2008) Chem Soc Rev 37:2433–2442; (b) Rix D, Caïjo F, Laurent I, Gułajski Ł, Grela K, Mauduit M (2007) Chem Commun 3771–3773; (c) Audic N, Clavier H, Mauduit M, Guillemin J-C (2003) J Am Chem Soc 125:9248–9249; (d) Clavier H, Nolan SP, Mauduit M (2008) Organometallics 27:2287–2292

[15] Gawin R, Grela K, Mauduit M unpublished results

[16] (a) Burtscher D, Grela K (2008) Angew Chem Int Ed. 48:442–454 (b) Lipshutz BH, Aguinaldo GT, Ghorai S, Voigtritter K (2008) Org Lett 10:1325–1328; (c) Lipshutz BH, Ghorai S, Aguinaldo GT (2008) Adv Synth Catal 350:953–956; (d) Lipshutz BH, Ghorai S (2008) Aldrichim Acta 41:59–72

Ruthenium–Arene Complexes Derived from NHC·CO_2 and NHC·CS_2 Zwitterionic Adducts and Their Use in Olefin Metathesis

Lionel Delaude,* **Albert Demonceau**

Laboratory of Macromolecular Chemistry and Organic Catalysis, Institut de Chimie (B6a), University of Liege, Sart-Tilman, 4000 Liege, Belgium

*E-mail: l.delaude@ulg.ac.be

Abstract A range of imidazol(in)ium-2-carboxylates and -dithiocarboxylates bearing alkyl or aryl groups on their nitrogen atoms were prepared by reacting the corresponding N-heterocyclic carbenes (NHCs) with either carbon dioxide or carbon disulfide. All the zwitterionic products were characterized by various analytical techniques, including thermogravimetric analysis (TGA). Their ability to act as NHC ligand precursors for in situ catalytic applications was investigated in the ruthenium-promoted ring-opening metathesis polymerization (ROMP) of cyclooctene. Upon exposure to the $[RuCl_2(p\text{-cymene})]_2$ dimer, the NHC CO_2 adducts readily dissociated to generate $[RuCl_2(p\text{-cymene})(NHC)]$ complexes that were highly active catalyst precursors for olefin metathesis. Conversely, the NHC CS_2 betaines retained their zwitterionic nature and led to new cationic complexes of the $[RuCl(p\text{-cymene})(NHC\ CS_2)]^+PF_6^-$ type that were devoid of any significant catalytic activity in the reaction under consideration.

Keywords Arene · Betaines · Cyclooctene · Ring-opening metathesis polymerization · Ruthenium

1 Introduction

Since they were first isolated and characterized by Arduengo and coworkers in 1991 [1–4], stable N-heterocyclic carbenes (NHCs) have been extensively studied [5–7]. Over the past 2 decades, they have already afforded a whole new generation of nucleophilic reagents [8] and organometallic catalysts [9–16], including chiral ones [17–19], that have revolutionized key areas of organic synthesis and polymer chemistry. Currently, the NHCs most commonly encountered are based on the

V. Dragutan et al. (eds.), *Green Metathesis Chemistry: Great Challenges in Synthesis, Catalysis and Nanotechnology*,

imidazole ring system. This electron-rich heterocycle provides a suitable framework that stabilizes the carbene center located between two nitrogen atoms [20]. Depending on the presence or the absence of a double bond between C4 and C5, imidazolin-2-ylidene and imidazolidin-2-ylidene species are obtained. They are usually prepared by deprotonating the corresponding imidazol(in)ium salts with a strong base, such as potassium *tert*-butoxide or sodium hydride [4]. We have applied this procedure to synthesize a wide range of ruthenium–arene complexes bearing NHC ligands (Scheme 1). The catalytic activity of these species, either preformed or generated in situ, was investigated in a number of transformations [21]. Fine tuning the steric and electronic properties of the substituents on the nitrogen atoms afforded highly efficient catalytic systems for the ring-opening metathesis polymerization (ROMP) of strained and low-strain cycloolefins [22, 23], for olefin cyclopropanation with diazoesters [21], and for atom transfer radical addition (ATRA) or polymerization (ATRP) of vinyl monomers [24, 25].

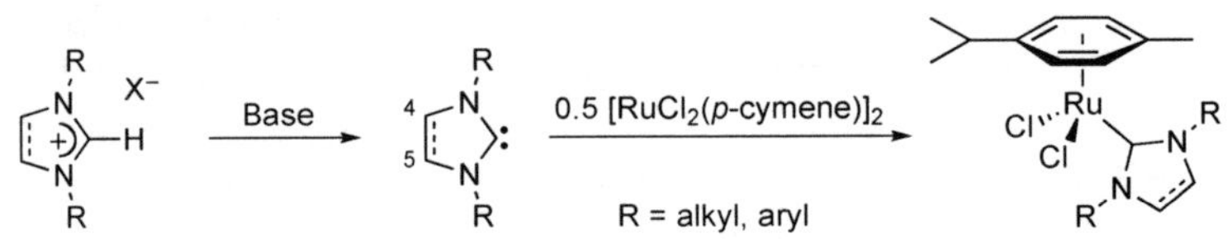

Scheme 1 Preparation of ruthenium–arene complexes bearing NHC ligands from $[RuCl_2(p\text{-cymene})]_2$, an imidazol(in)ium salt, and a base

As part of our continuous endeavor to develop convenient synthetic methods based on the association of carbene ligands and transition-metal catalysts, we became interested in alternative sources to imidazol(in)ium salts for generating active species in situ. Stable adducts resulting from the insertion of NHCs into acidic C–H bonds have already been successfully employed to generate various ruthenium–NHC complexes. For instance, Grubbs and coworkers used either chloroform or *tert*-butanol adducts of 1,3-dimesitylimidazolidin-2-ylidene (nicknamed SIMes) to prepare their second generation ruthenium–alkylidene metathesis catalyst $RuCl_2(=CHPh)(PCy_3)(SIMes)$ (Scheme 2) [26, 27]. Blechert et al. followed a similar pathway to substitute SIMes for a triphenylphosphine ligand in a ruthenium–indenylidene complex, starting from the *t*-BuOH adduct [28]. Although highly effective, these strategies were suitable only for introducing saturated imidazolidin-2-ylidene ligands, since the clean formation of insertion products was not achieved with unsaturated imidazolin-2-ylidene species [29]. Moreover, the experimental procedures required thermal activation to induce the decomposition of the NHC adducts. They also implied the release of a stoichiometric amount of chloroform or alcohol in the reaction mixtures. Hence, they have not been used so far for in situ catalytic applications with ruthenium complexes.

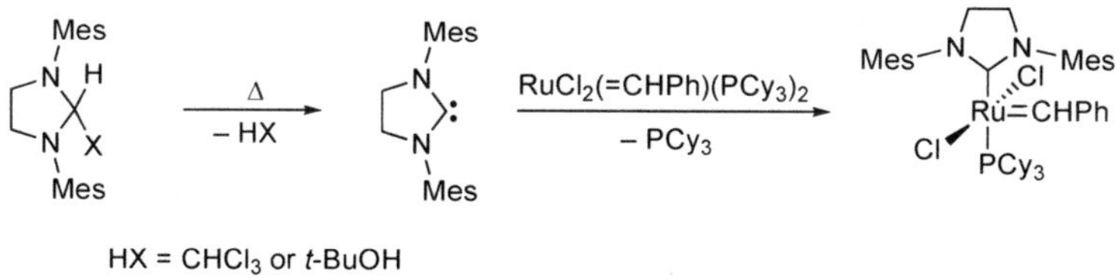

Scheme 2 Preparation of the Grubbs second generation catalyst using chloroform or *t*-BuOH adducts of 1,3-dimesitylimidazolidin-2-ylidene

We reasoned that the use of carbon dioxide to reversibly convert air- and moisture-sensitive carbenes into more stable adducts would alleviate most concerns of interference with catalytic systems. Indeed, Louie and coworkers reported in 2004 that the CO_2 adducts of 1,3-dimesitylimidazolin-2-ylidene (IMes · CO_2) and 1,3-bis(2,6-diisopropylphenyl)imidazolin-2-ylidene (IDip CO_2) were labile zwitterionic compounds that readily exchanged their carboxylate groups in solution [30]. These observations prompted us to investigate the recourse to imidazol (in)ium-2-carboxylates as NHC precursors in ruthenium–arene complexes for in situ catalytic applications. We felt comforted in our approach by a report from Crabtree et al. published in 2005 while our work was in progress. The Yale group showed that 1,3-dimethylimidazolium-2-carboxylate (IMe · CO_2) efficiently transferred its carbene fragment to various transition-metal complexes, including the [$RuCl_2$(*p*-cymene)]$_2$ dimer, to afford the corresponding NHC complexes in high yields (Scheme 3) [31]. Experimental observations and DFT calculations ruled out the dissociation of the betaine prior to its reaction with a metal source. Thus, IMe · CO_2 was believed to coordinate via two *cis* vacant sites on the metal, followed by C–C bond cleavage and transfer of the NHC moiety to the metal [32].

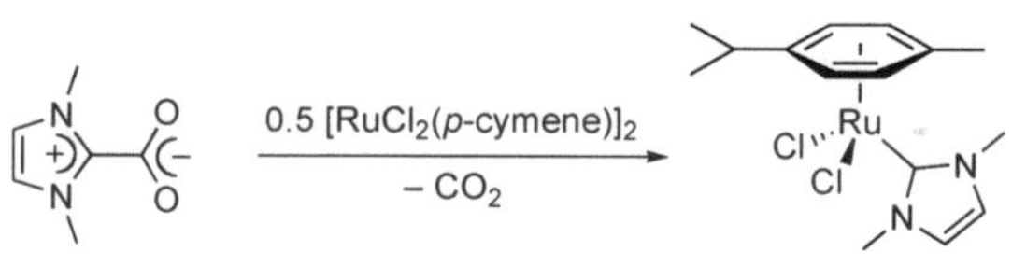

Scheme 3 Preparation of a ruthenium–arene complex from [$RuCl_2$(*p*-cymene)]$_2$ and an imidazolium-2-carboxylate

In this contribution, we briefly review the various experimental methods described in the literature for the preparation of NHC · CO_2 betaines and we detail our own strategy for accessing this class of inner salts. In parallel, we also discuss the synthesis of the related NHC · CS_2 adducts. The ability of both types of zwitterionic compounds to serve as catalyst modifiers was probed in the ruthenium-mediated ROMP of cyclooctene. To account for the differences of reactivity observed, we also compared the thermal stabilities of NHC · CO_2 and NHC · CS_2 betaines by thermogravimetric analysis. The final part of this work involved the isolation and characterization of ruthenium–arene complexes obtained from various NHC · CX_2 adducts (X = O, S) in order to validate the assumptions gathered from the in situ experiments.

2 Synthesis of NHC • CX_2 Zwitterionic Adducts (X = O, S)

2.1 *Synthesis of Imidazol(in)ium-2-carboxylates*

The first report describing the synthesis and characterization of a NHC · CO_2 adduct appeared in 1974 when Schössler and Regitz isolated 1,3-diphenylimidazolinium-2-carboxylate by hydrolysis of more complex zwitterionic carbamate derivatives [33]. In 1999, Kuhn et al. used gaseous CO_2 to trap 1,3-diisopropyl-4,5-dimethylimidazolin-2-ylidene as a stable crystalline adduct [34]. The free carbene that served as a starting material was obtained by reduction of the corresponding 2-thioketone with potassium in refluxing THF (Scheme 4) [35].

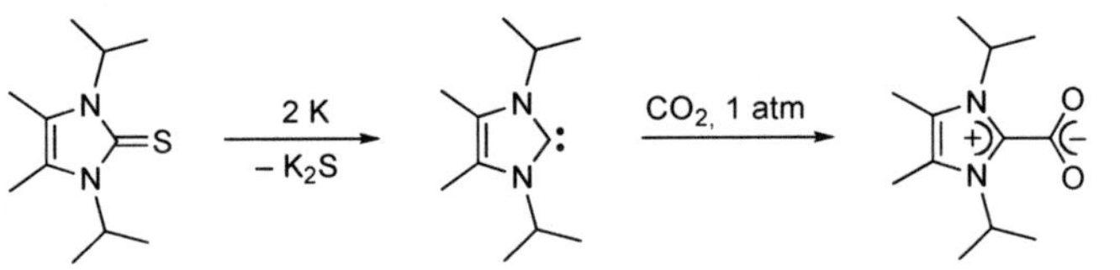

Scheme 4 Synthesis of 1,3-diisopropyl-4,5-dimethylimidazolium-2-carboxylate from a thioketone precursor

In 2002, the group of Ishiguro and Sawaki identified 1,3-di-*tert*-butylimidazolium-2-carboxylate as a transfer product in reactions of the parent carbene with singlet oxygen [36]. In 2003, Tkatchenko and coworkers reported the unexpected formation of 1,3-dimethylimidazolium-2-carboxylate from 1-methylimidazole and dimethyl carbonate [37]. Tommasi and Sorrentino subsequently extended the procedure to the preparation of 1-butyl-3-methylimidazolium-2-carboxylate [38]. Last but not least, Louie et al. disclosed in 2004 the direct carboxylation of IMes and IDip by deprotonation of the corresponding imidazolium chlorides with potassium *tert*-butoxide under a CO_2 atmosphere [30]. We have slightly modified this procedure to remove the potassium chloride by-product and any unreacted solid reagents prior to the precipitation of the organic carboxylates. We have also tested several strong bases to achieve the deprotonation step most efficiently. Potassium bis(trimethylsilyl)amide afforded reasonable reaction and filtration rates, yet potassium *tert*-butoxide or potassium hydride activated by a few drops of DMSO or *t*-BuOH were equally suitable in most cases [39]. The revised reaction sequence is summarized in Scheme 5.

In addition to IMes · CO_2 and IDip · CO_2, we have prepared their saturated heterocycle analogues SIMes · CO_2 and SIDip · CO_2 in 79% and 61% yield, respectively.

To further enlarge the scope of our catalytic studies with imidazol(in)ium-2-carboxylates, we also chose to include a representative adduct bearing alkyl groups instead of aryl substituents on its nitrogen atoms. Thus, we elected 1,3-dicyclohexylimidazolium-2-carboxylate (ICy · CO_2) to complement our set of NHC · CO_2 adducts. The 38% isolated yield obtained for the synthesis of this inner salt was the lowest of the series and the only one that did not exceed the 50% mark. A recalcitrant deprotonation step was the main cause for this poor result [39].

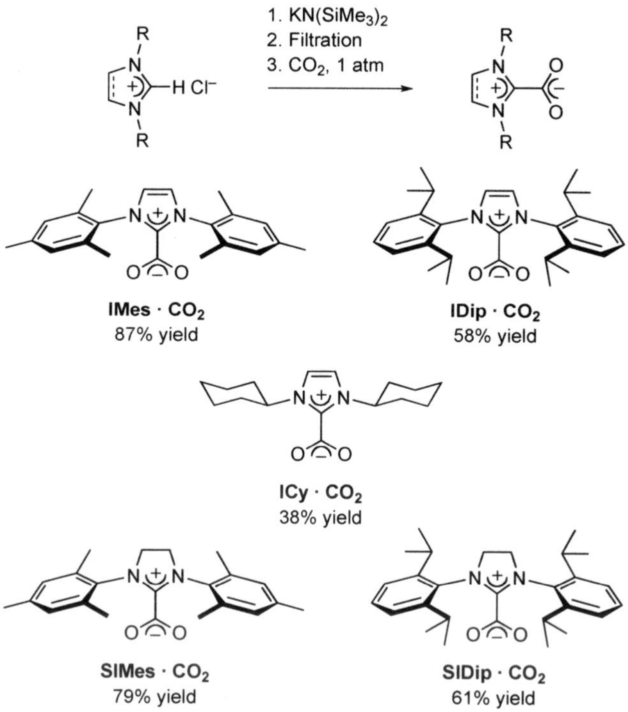

Scheme 5 Synthesis of imidazol(in)ium-2-carboxylates from imidazol(in)ium salts

2.2 *Synthesis of Imidazol(in)ium-2-dithiocarboxylates*

Although stable free carbenes were first isolated in 1991 [1–4], the chemistry of their formal enetetramine dimers has been under investigation since the 1960s [40–46]. It was soon recognized that these electron-rich olefins could be dissociated in exothermic reactions with various nucleophiles, including carbon disulfide, to afford stable zwitterionic adducts. Thus, initial reports describing the preparation of imidazol(in)ium-2-dithiocarboxylates were based on this approach (Scheme 6). The procedure was successfully applied to a range of N,N′-dialkyl or N,N′-diaryl bis(imidazolidin-2-ylidene) starting materials (R = Me [47], Et [47, 48], CH_2Ph [47], Ph [33, 47, 49], *p*-C_6H_4Cl [47]). It was also subsequently extended to benzimidazolidine derivatives with R = Me [50], Et [51], or CH_2CH_2Ph [51].

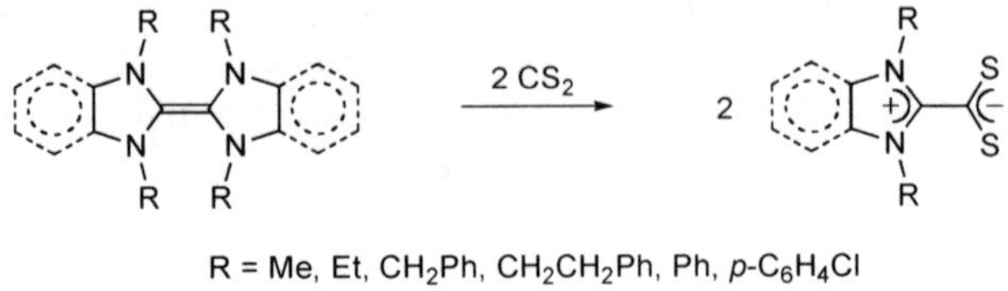

Scheme 6 Synthesis of (benz)imidazolium-2-dithiocarboxylates from enetetramine dimmers

Another less common strategy to obtain NHC · CS_2 betaines from stable NHC precursors involved the displacement of chloroform by carbon disulfide upon thermolysis of NHC · $CHCl_3$ adducts (Scheme 7). To the best of our knowledge, only two occurrences of this reaction were reported in the literature. They concerned, respectively, saturated imidazolidine heterocycles bearing *p*-anisyl [47] or mesityl [52] substituents on their nitrogen atoms.

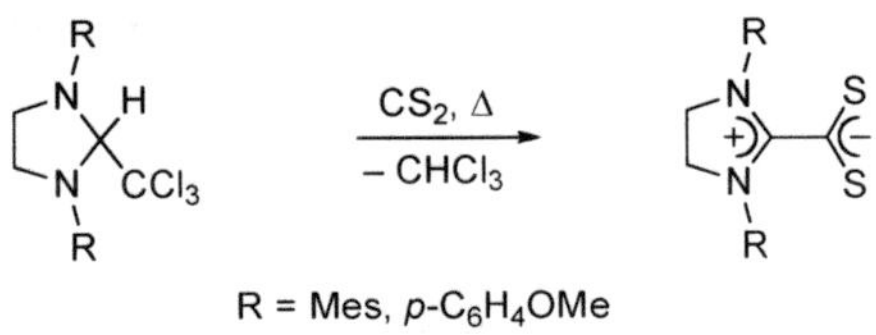

Scheme 7 Synthesis of imidazolinium-2-dithiocarboxylates from chloroform adducts

In 1993, Kuhn et al. isolated a series of 1,3-dialkyl-4,5-dimethylimidazolin-2-ylidenes by reduction of the corresponding thioketones with potassium in refluxing THF [35]. The free carbenes were reacted with a stoichiometric amount of carbon disulfide in THF at 0°C to afford the dithiocarboxylate inner salts in good yields (Scheme 8). A first report focused on the preparation and characterization of three adducts bearing simple alkyl groups on their nitrogen atoms (R = Me, Et, *i*-Pr) [53]. The authors later complemented this study by introducing alkoxy-terminated alkyl chains on the heterocycle 1,3-positions (R = $(CH_2)_2OMe$ or $(CH_2)_3OMe$) [54]. Synthesis of a 1,3-dimesitylimidazolium-2-dithiocarboxylate with silylated alkynyl functional groups on its 4,5-positions was also accomplished by Faust and Göbelt using a similar procedure [55].

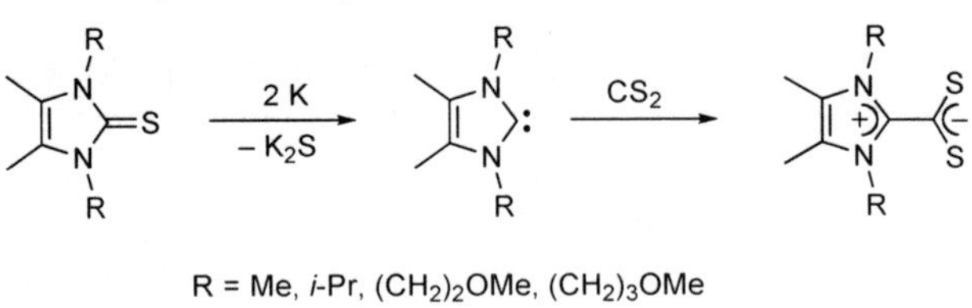

Scheme 8 Synthesis of 1,3-dialkyl-4,5-dimethylimidazolium-2-dithiocarboxylates from thioketone precursors

Currently, deprotonation of imidazol(in)ium chlorides or tetrafluoroborates with a strong base provides the most convenient and general access to NHCs, whether it is for preparative purposes [1–4, 56–58] or for in situ catalytic applications [23, 59–64]. This is due, in part, to the existence of efficient and flexible synthetic procedures that allow the straightforward preparation of a wide range of imidazol(in)ium salts from readily available acyclic precursors [65–67]. Thus, we chose this approach to release NHCs in solution prior to their trapping with carbon disulfide. Once the deprotonation step was completed, the suspensions were allowed to settle down and the inorganic by-product (NaCl) was filtered off, along with any unreacted starting materials. A small excess of carbon disulfide was added at once to the free carbene solutions. It led to an instantaneous color change and to the rapid formation of zwitterionic adducts (Scheme 9). 1,3-Dimesitylimidazolium-2-dithiocarboxylate (nicknamed IMes · CS_2) and its saturated heterocycle analogue (nicknamed SIMes · CS_2) precipitated from the reaction medium, whereas betaines bearing 2,6-diisopropylphenyl (IDip · CS_2 and SIDip · CS_2) or cyclohexyl substituents (ICy · CS_2) on their nitrogen atoms remained soluble in THF. The solvent was removed under reduced pressure to afford the zwitterionic products in solid form. All the compounds afforded beautiful ruby-like crystals suitable for X-ray diffraction analysis by slow evaporation of saturated solutions left in an open vessel at room temperature under a fume hood [68].

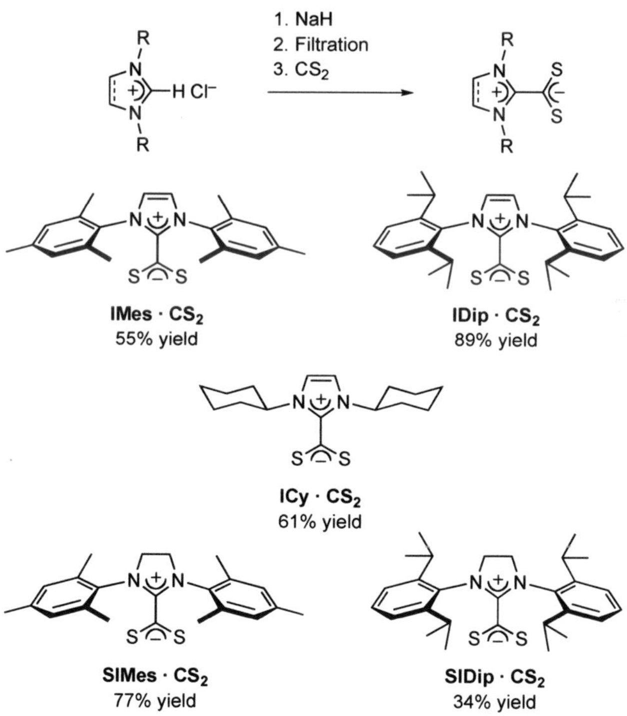

Scheme 9 Synthesis of imidazol(in)ium-2-dithiocarboxylates from imidazol(in)ium salts

Overall yields for the preparation of dithiocarboxylate adducts from imidazol(in)ium chlorides according to the method described above were usually good (cf. Scheme 9). The reaction between carbon disulfide and free carbenes occurred rapidly and quantitatively at room temperature. Due to the remarkable tendency of the zwitterionic adducts to form crystalline materials, only minor losses were encountered during the final purification. The main cause for an unsatisfactory mass balance was a recalcitrant deprotonation step. This was evidenced by a control experiment in which the 1,3-dimesitylimidazolin-2-ylidene free carbene (IMes) was isolated and recrystallized prior to its reaction with carbon disulfide in THF. Under these conditions, the corresponding dithiocarboxylate betaine was isolated in 91% yield, whereas the in situ deprotonation of IMes · HCl with NaH followed by filtration and direct addition of carbon disulfide afforded only a 55% isolated yield of IMes · CS_2 [68].

3 ROMP of Cyclooctene Using In Situ Generated Catalysts

The ability of imidazol(in)ium-2-carboxylates or -dithiocarboxylates to serve as NHC precursors for generating ruthenium–arene catalysts in situ was investigated in the ROMP of cyclooctene (Scheme 10). In our laboratory, this reaction serves as a benchmark test to assess the metathetical activity of new catalyst precursors [21, 23, 69, 70]. Polymerizations were carried out in chlorobenzene at 60°C or at room temperature using a stoichiometric mixture of $[RuCl_2(p\text{-cymene})]_2$ and NHC · CX_2 adducts (X = O, S). The monomer-to-ruthenium ratio was 250 and an ordinary neon tube placed 10 cm away from the Pyrex reaction flasks complemented the experimental setup. This device ensured a strong, reproducible visible illumination required to trigger the catalytic process. Indeed, earlier studies had shown that visible light induced the ROMP of cycloolefins catalyzed by ruthenium–arene complexes bearing NHC ligands. A photoactivation step was held responsible for the decoordination of the η^6-arene ligand, thereby affording highly unsaturated ruthenium centers that evolved into propagating alkylidene species, although the intimate details of the mechanism remain elusive [22].

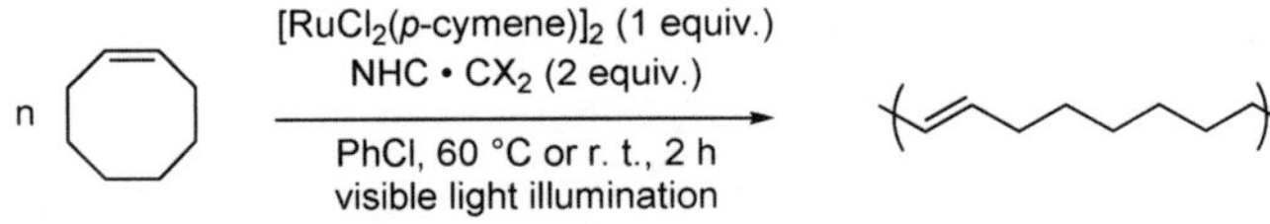

Scheme 10 ROMP of cyclooctene with catalyst precursors generated in situ from $[RuCl_2(p\text{-cymene})]_2$ and imidazol(in)ium-2-carboxylate (X = O) or -dithiocarboxylate (X = S) adducts

In a first series of experiments, the ROMP of cyclooctene was carried out at 60°C in the presence of $[RuCl_2(p\text{-cymene})]_2$ and either one of the five NHC · CO_2 zwitterions or their NHC · CS_2 counterparts. The two types of adducts displayed very different behaviors (Table 1). No polymer was isolated with the dithiocarboxylate betaines and only oligomers accounted for the low monomer consumption observed with these inner salts (entries 1–5). Conversely, almost quantitative conversion of the starting material occurred with imidazol(in)ium-2-carboxylates bearing mesityl- or 2,6-diisopropylphenyl groups on their nitrogen atoms and high yields of polyoctenamer were attained within 2 h (entries 6–9). The presence of alkyl substituents on ICy · CO_2 was less favorable to the reaction. It led to a 30% conversion and afforded only soluble oligomers and no high molecular weight polymer (entry 10). This result did not come as a surprise. The patent superiority of NHCs bearing aryl groups on their nitrogen atoms over related N,N′-dialkyl derivatives was already recognized when $RuCl_2$(*p*-cymene)(NHC) complexes, either preformed or generated in situ from imidazol(in)ium chlorides and a base, were investigated [22]. Earlier studies had also shown that the C4–C5 double bond in the imidazole ring was not crucial to achieve high catalytic efficiencies [23].

Table 1 ROMP of cyclooctene in chlorobenzene at 60°C for 2 h catalyzed by $[RuCl_2(p\text{-cymene})]_2$ and various NHC precursors

Entry	NHC precursor	Conv. (%)	Yield (%)	M_n (kg/mol)[a]	M_w/M_n[a]	σ_{cis}[b]	Ref.
1	IMes · CS_2	10	0	–	–	–	[71]
2	IDip · CS_2	4	0	–	–	–	[71]
3	SIMes · CS_2	14	0	–	–	–	[71]
4	SIDip · CS_2	0	0	–	–	–	[71]
5	ICy · CS_2	9	0	–	–	–	[71]
6	IMes · CO_2	100	91	632	2.32	0.21	[39]
7	IDip · CO_2	79	64	798	2.70	0.29	[39]
8	SIMes · CO_2	100	85	692	2.87	0.19	[39]
9	SIDip · CO_2	89	77	383	2.11	0.24	[39]
10	ICy · CO_2	30	0	–	–	–	[39]
11	IMes · HCl + KO-*t*-Bu	99	92	659	2.02	0.20	[23]
12	IDip · HCl + KO-*t*-Bu	99	60	398	3.09	0.20	[23]
13	SIMes · HCl + KO-*t*-Bu	99	93	512	2.19	0.23	[23]
14	SIDip · HCl + KO-*t*-Bu	57	44	838	2.13	0.51	[23]

[a]Determined by GPC in THF versus monodisperse polystyrene standards.
[b]Fraction of *cis* double bonds within the polymer chain, determined by ^{13}C NMR.

The similarity between the catalytic systems generated from NHC · CO_2 adducts or from NHC · HCl salts and a base at 60°C was further evidenced by comparing the physico-chemical parameters of the polymers obtained in both cases (see entries 6–9 and 11–14 in Table 1). No significant differences were observed in chain length distributions and *cis/trans* ratios. Most of the reactions led to high molecular weight polymers with a polydispersity index comprised between 2 and 3 and a sizable proportion (ca. 80%) of *trans* double bonds within the unsaturated backbones [39]. Fluctuations in the molecular weights should not be over-considered. Unlike related cationic ruthenium [72, 73] or osmium [74] complexes possessing an alkylidene, vinylidene, allenylidene, or indenylidene moiety in their coordination sphere, the ruthenium–(*p*-cymene) catalyst precursors examined in this study lack a suitable metal–carbene fragment to initiate olefin metathesis. Ill-defined mechanisms involving arene disengagement and monomer coordination were held responsible for their transformation into active species during the course of the reaction [75, 76]. The main drawback of this system is a poor control of the initiation step that results in high molecular weights and rather broad polydispersities. Because only a small number of propagating centers are present in solution, slight changes in the initiation efficiency from one experiment to another lead to significant variations in the molecular weights attained.

Yet, the good results obtained with 1,3-diarylimidazol(in)ium-2-carboxylates at 60°C prompted us to launch a second series of ROMP experiments at room temperature (Table 2) [39]. All the other experimental parameters were kept unchanged. Within the 2 h allowed for the polymerization to proceed, three out of the four adducts led to high conversions, now in the 80–90% range, while polyoctenamer was isolated in 70–80% yield (entries 1–4). The macromolecular chains still had a very high molecular weight but their polydispersity was significantly narrowed compared to the reactions performed at 60°C. The *cis*/*trans* distribution of the double bonds in the metathesis products was also noticeably altered by the temperature change. Working at a lower temperature gave rise to higher *cis* content. In other words, increasing the temperature favored the formation of thermodynamically more stable *trans* double bonds, as it could have been anticipated. Only SIDip · CO_2 displayed a considerable drop in activity when reacted at room temperature instead of 60°C. It also led to a much higher proportion of *cis* double bonds. Although we have no rational explanation for these anomalies, a similar trend was already observed for the corresponding hydrochloride salt at 60°C (see entry 14, Table 1).

Table 2 ROMP of cyclooctene in chlorobenzene at room temperature for 2 h catalyzed by $[RuCl_2(p\text{-cymene})]_2$ and various NHC precursors or $RuCl_2(p\text{-cymene})(IMes)$

Entry	NHC precursor	Conv. (%)	Yield (%)	M_n (kg/mol)[a]	M_w/M_n[a]	σ_{cis}[b]
1	IMes · CO_2	78	70	502	1.48	0.33
2	IDip · CO_2	80	72	1,017	1.58	0.31
3	SIMes · CO_2	88	78	706	1.51	0.33
4	SIDip · CO_2	33	28	699	1.53	0.55
5	$RuCl_2(p\text{-cymene})(IMes)$	85	78	899	1.51	0.30
6	IMes · HCl + KO-*t*-Bu	25	11	1,014	1.46	0.37
7	SIMes · HCl + KO-*t*-Bu	2	1	–	–	–

[a]Determined by GPC in THF versus monodisperse polystyrene standards.
[b]Fraction of *cis* double bonds within the polymer chain, determined by ^{13}C NMR.

To put the results obtained with the NHC · CO_2 adducts in perspective, we carried out the ROMP of cyclooctene at room temperature using preformed $RuCl_2(p\text{-cymene})(IMes)$ (Table 2, entry 5) or mesityl-based catalysts generated in situ from imidazol(in)ium salts and potassium *tert*-butoxide (entries 6 and 7). Recourse to the preformed complex did not further improve the polymerization outcome, while the multicomponent systems were almost completely devoid of catalytic activity. A polymer sample was only isolated in low yield with the combination of IMes · HCl and KO-*t*-Bu. A poor solubility of the ionic reagents in chlorobenzene at room temperature was invoked to account for this lack of initiation [22]. Thus, recourse to the labile carboxylates proved a successful strategy to maintain the high activity of the preformed ruthenium–NHC complex under mild conditions, while enabling the use of easily available, air-stable catalyst precursors [39].

4 Thermogravimetric Analysis

In order to rationalize the large discrepancy observed between NHC · CO_2 and NHC · CS_2 zwitterions in terms of catalytic activity, we probed their thermal stabilities in the solid state by thermogravimetric analysis (TGA). Decomposition profiles shown in Figure 1 for NHC · CO_2 and NHC · CS_2 adducts bearing 2,6-diisopropylphenyl groups on their nitrogen atoms provided graphic evidence for the greater lability of carboxylate inner salts compared to their dithiocarboxylate analogues. Thus, imidazol(in)ium-2-carboxylates are more likely to release carbene ligands in the presence of $[RuCl_2(p\text{-cymene})]_2$, a requisite for generating catalytically active species in situ. Similar TGA curves were recorded for the cyclohexyl- or mesityl-based adducts (curves not shown). As previously noted, the

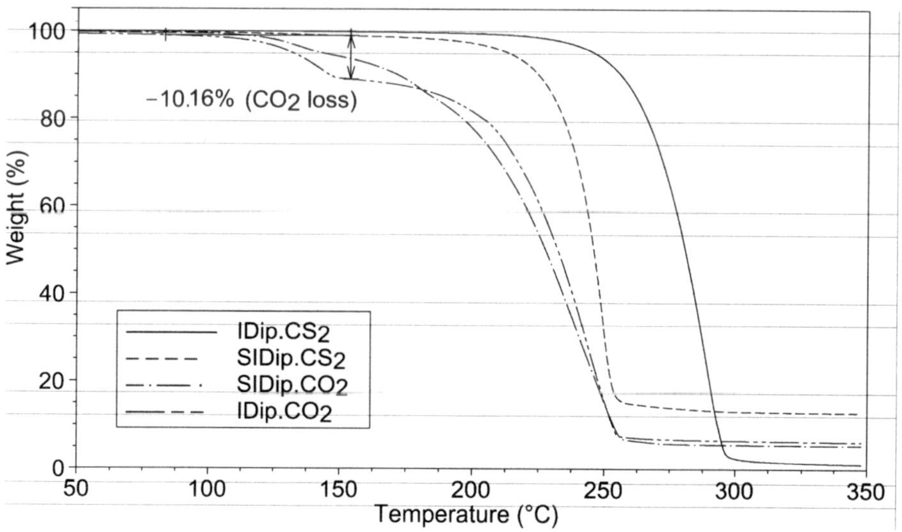

Figure 1 Comparison of the TGA curves of NHC · CO_2 and NHC · CS_2 adducts bearing 2,6-diisopropylphenyl groups on their nitrogen atoms

presence or the absence of a heterocyclic C4–C5 double bond did not significantly alter the onset of the degradation process [39]. Evidence for a clean release of the free carbene was detected only for IDip · CO_2 (theoretical weight loss for CO_2: 10.17%, found: 10.16%), as previously observed by Duong et al. [30].

5 Preparation of Ruthenium–Arene Complexes

5.1 Starting from Imidazol(in)ium-2-carboxylates

Because imidazol(in)ium-2-dithiocarboxylates obviously behaved very differently from the corresponding carboxylate adducts in catalytic systems generated in situ for the ROMP of cyclooctene, we have investigated in more detail the formation of preformed ruthenium–arene complexes from these two types of ligand precursors.

In a first series of experiments, [$RuCl_2$(*p*-cymene)]$_2$ was refluxed for 2 h in THF with 2 eq. of ICy · CO_2, IMes · CO_2, or IDip · CO_2. After work-up and purification by column chromatography, $RuCl_2$(*p*-cymene)(NHC) complexes were isolated in high yields (Scheme 11) [71]. Neutral 18-electron ruthenium–arene complexes bearing NHC ligands are well-known initiators for the ROMP of strained and low-strain cycloolefins [21]. Their synthesis from [$RuCl_2$(*p*-cymene)]$_2$ and free carbenes [22, 77–80] or a carboxylate adduct [31, 32] were already reported by several groups. Compared to the free carbenes, the zwitterionic adducts displayed a much greater stability toward air and moisture in the solid state. Thus, they could be easily handled with no particular precautions. Upon heating or dissolution, however, they readily lost the CO_2 moiety, thereby providing a convenient method to generate NHC ligands in situ. Contrary to the deprotonation of imidazol(in)ium salts, the decomposition of betaine adducts did not require the use of a strong base and was accompanied only by the evolution of carbon dioxide.

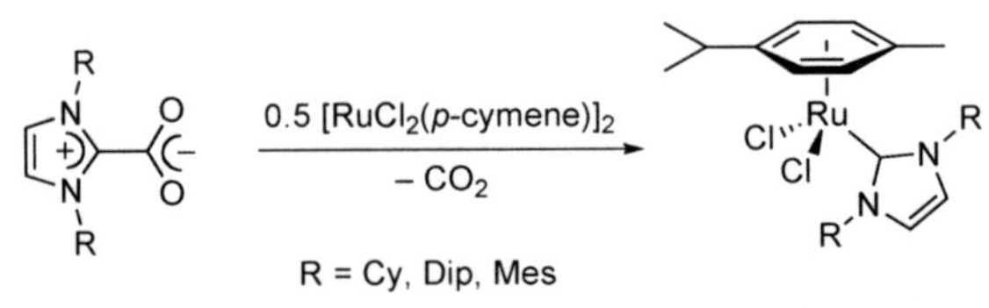

Scheme 11 Synthesis of ruthenium–arene complexes from [$RuCl_2$(*p*-cymene)]$_2$ and various imidazolium-2-carboxylate betaines

5.2 Starting from Imidazol(in)ium-2-dithiocarboxylates

Attempts to prepare the $RuCl_2$(*p*-cymene)(IMes) complex by refluxing [$RuCl_2$ (*p*-cymene)]$_2$ with 2 eq. of IMes · CS_2 instead of IMes · CO_2 remained unsuccessful. In all cases, ^{13}C NMR analysis of the crude reaction mixtures revealed the persistence of signals in the 195–225 ppm region, thereby showing that elimination of CS_2 had not occurred, even after prolonged heating under a slow stream of argon [71]. This observation did not come as a surprise taking into account the greater stability displayed by NHC · CS_2 adducts in the solid state compared to their NHC · CO_2 analogues (cf. Figure 1). Indeed, a review article published in 2005 confirmed that none of the various reactions of diaminocarbenium–dithiocarboxylate species reported that far involved the regeneration of diaminocarbenes [81]. Unlike the NHC · CO_2 adducts that readily dissociated and could be viewed as convenient surrogates for free carbenes, the NHC · CS_2 betaines retained their zwitterionic nature throughout chemical transformations. Thus, their reactivity is determined mainly by the presence of a Lewis acid center in the imidazol(in)ium ring, while the CS_2^- group acts as a Lewis or Brønsted base. To the best of our knowledge, this dual behavior has been exploited in a single catalytic application so far, namely the cyanosilylation of aldehydes with imidazolinium-2-dithiocarboxylates as organocatalysts [82].

In light of the above considerations, the reaction between $[RuCl_2(p\text{-cymene})]_2$ and NHC · CS_2 adducts was reinvestigated in the presence of excess carbon disulfide and potassium hexafluorophosphate (Scheme 12). The reaction partners were refluxed in ethanol for 3 h, followed by filtration and evaporation of the volatiles. Under these conditions, the dithiocarboxylate anions cleaved the μ-chloro bridges of the ruthenium dimer to afford new cationic complexes with the generic formula $[RuCl(p\text{-cymene})(NHC \cdot CS_2)]^+PF_6^-$ [71]. This experimental procedure was originally devised by Miguel et al. for the preparation of $[RuCl(arene)(PR_3 \cdot CS_2)]^+PF_6^-$ complexes (arene = benzene, *p*-cymene or hexamethylbenzene, PR_3 = P-*i*-Pr_3 or PCy_3) [83, 84]. In our hands, the reactions proceeded cleanly and afforded 16-electron ruthenium–arene complexes bearing imidazol(in)ium-2-dithiocarboxylate ligands in high yields. Crystals of $[RuCl(p\text{-cymene})(IMes\ CS_2)]^+PF_6^-$ suitable for X-ray diffraction analysis were obtained and their molecular structure determined (Figure 2).

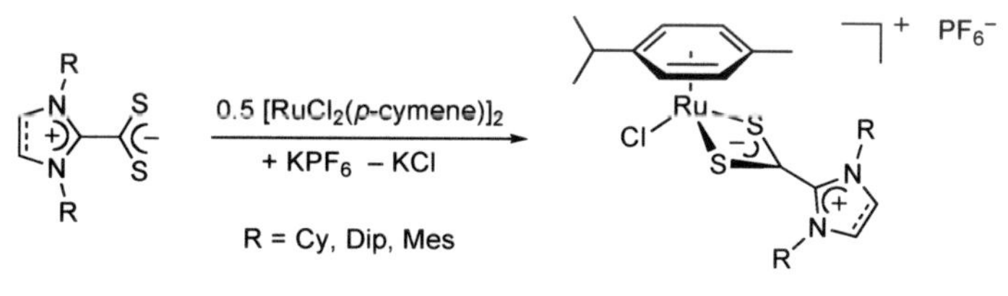

Scheme 12 Synthesis of cationic ruthenium–arene complexes from $[RuCl_2(p\text{-cymene})]_2$ and various imidazol(in)ium-2-dithiocarboxylate betaines

We have carried out control experiments to assess the catalytic activity of $[RuCl(p\text{-cymene})(NHC \cdot CS_2)]^+PF_6^-$ complexes in the ROMP of cyclooctene. Reactions were performed at 60°C for 2 h using the standard experimental protocol (cf. Scheme 10). In full agreement with the results obtained for active species generated in situ (cf. Table 1, entries 1–5), the preformed cationic complexes did not afford any polymer [71].

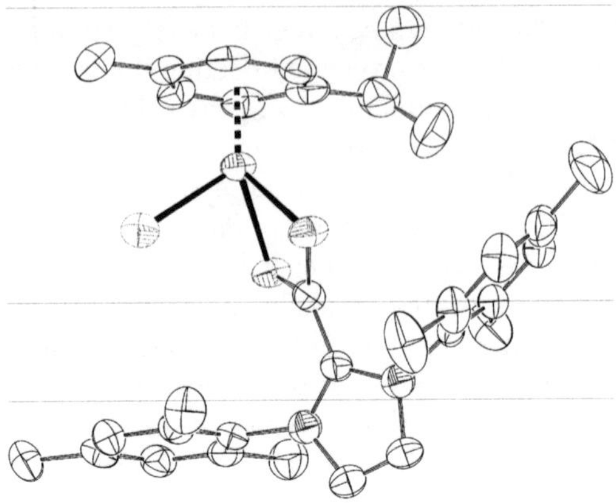

Figure 2 Molecular conformation (ORTEP diagram) of $[RuCl(p\text{-cymene})(IMes \cdot CS_2)]^+PF_6^-$ showing thermal displacement ellipsoids (50% probability). Hydrogen atoms and the PF_6^- counterion were omitted for clarity

6 Summary

A set of ten imidazol(in)ium-2-carboxylates and -dithiocarboxylates was prepared by deprotonating the corresponding imidazol(in)ium chlorides with a strong base, followed by reaction with carbon dioxide or carbon disulfide, respectively. The ability of the various zwitterionic adducts to act as NHC precursors was investigated in the ruthenium-promoted ring-opening metathesis polymerization of cyclooctene. Thermogravimetric analysis was employed to determine their thermal stability in the solid state.

Unlike the $NHC \cdot CO_2$ betaines that readily dissociated to afford highly active $RuCl_2(p\text{-cymene})(NHC)$ catalytic species when exposed to the $[RuCl_2(p\text{-cymene})]_2$ dimer, the $NHC \cdot CS_2$ inner salts retained their zwitterionic nature and led to cationic complexes of the $[RuCl(p\text{-cymene})(NHC \cdot CS_2)]^+PF_6^-$ type that were devoid of any significant catalytic activity in the reaction under consideration.

Acknowledgments We thank Prof. Johan Wouters (Facultés Universitaires N.-D. de la Paix, Namur) for the X-ray diffraction analysis of $[RuCl(p\text{-cymene})(IMes \cdot CS_2)]^+PF_6^-$. Financial support from the Romanian Academy and the Belgian C.G.R.I. is gratefully acknowledged.

References

[1] Arduengo AJ, III, Harlow RL, Kline M (1991) A stable crystalline carbene. J Am Chem Soc 113: 361–363

[2] Arduengo AJ, III, Rasika Dias HV, Harlow RL, Kline M (1992) Electronic stabilization of nucleophilic carbenes. J Am Chem Soc 114: 5530–5534

[3] Arduengo AJ, III, Goerlich JR, Krafczyk R, Marshall WJ (1998) 1,3,4,5-Tetraphenylimidazol-2-ylidene: the realization of Wanzlick's dream. Angew Chem Int Ed 37: 1963–1965
[4] Arduengo AJ, III (1999) Looking for stable carbenes: the difficulty in starting anew. Acc Chem Res 32: 913–921
[5] Bertrand G (ed.) (2002) Carbene chemistry: from fleeting intermediates to powerful reagents. Marcel Dekker, New York
[6] Nolan SP (ed.) (2006) N-heterocyclic carbenes in synthesis. Wiley-VCH, Weinheim
[7] Glorius F (ed.) (2007) N-heterocyclic carbenes in transition metal catalysis. Topics in Organometallic Chemistry, vol. 21. Springer, Berlin
[8] Nair V, Bindu S, Sreekumar V (2004) N-heterocyclic carbenes: reagents, not just ligands! Angew Chem Int Ed 43: 5130–5135
[9] Herrmann WA, Köcher C (1997) N-heterocyclic carbenes. Angew Chem Int Ed 36: 2162–2187
[10] Jafarpour L, Nolan SP (2001) Transition-metal systems bearing a nucleophilic carbene ancillary ligand: from thermochemistry to catalysis. Adv Organomet Chem 46: 181–222
[11] Herrmann WA (2002) N-heterocyclic carbenes: a new concept in organometallic catalysis. Angew Chem Int Ed 41: 1290–1309
[12] Herrmann WA, Weskamp T, Böhm VPW (2002) Metal complexes of stable carbenes. Adv Organomet Chem 48: 1–69
[13] Crudden CM, Allen DP (2004) Stability and reactivity of N-heterocyclic carbene complexes. Coord Chem Rev 248: 2247–2273
[14] Scott NM, Nolan SP (2005) Stabilization of organometallic species achieved by the use of N-heterocyclic carbene (NHC) ligands. Eur J Inorg Chem: 1815–1828
[15] Garrison JC, Youngs WJ (2005) Ag(I) N-heterocyclic carbene complexes: synthesis, structure, and application. Chem Rev 105: 3978–4008
[16] Dragutan V, Dragutan I, Delaude L, Demonceau A (2007) NHC–Ru complexes – friendly catalytic tools for manifold chemical transformations. Coord Chem Rev 251: 765–794
[17] Perry MC, Burgess K (2003) Chiral N-heterocyclic carbene-transition metal complexes in asymmetric catalysis. Tetrahedron: Asymmetry 14: 951–961
[18] Enders D, Balensiefer T (2004) Nucleophilic carbenes in asymmetric organocatalysis. Acc Chem Res 37: 534–541
[19] César V, Bellemin-Laponnaz S, Gade LH (2004) Chiral N-heterocyclic carbenes as stereodirecting ligands in asymmetric catalysis. Chem Soc Rev 33: 619–636
[20] Bourrisou D, Guerret O, Gabbaï FP, Bertrand G (2000) Stable carbenes. Chem Rev 100: 39–91
[21] Delaude L, Demonceau A, Noels AF (2006) Synthesis and application of new N-heterocyclic carbene ruthenium complexes in catalysis: a case study. Curr Org Chem 10: 203–215
[22] Delaude L, Demonceau A, Noels AF (2001) Visible light induced ring-opening metathesis polymerisation of cyclooctene. Chem Commun: 986–987
[23] Delaude L, Szypa M, Demonceau A, Noels AF (2002) New in situ generated ruthenium catalysts bearing N-heterocyclic carbene ligands for the ring-opening metathesis polymerization of cyclooctene. Adv Synth Catal 344: 749–756
[24] Delaude L, Delfosse S, Richel A, Demonceau A, Noels AF (2003) Tuning of ruthenium N-heterocyclic catalysts for ATRP. Chem Commun: 1526–1527
[25] Richel A, Delfosse S, Cremasco C, Delaude L, Demonceau A, Noels AF (2003) Ruthenium catalysts bearing N-heterocyclic carbene ligands in atom transfer radical reactions. Tetrahedron Lett 44: 6011–6015
[26] Scholl M, Ding S, Lee CW, Grubbs RH (1999) Synthesis and activity of a new generation of ruthenium-based olefin metathesis catalysts coordinated with 1,3-dimesityl-4,5-dihydroimidazol-2-ylidene ligands. Org Lett 1: 953–956
[27] Trnka TM, Morgan JP, Sanford MS, Wihelm TE, Scholl M, Choi T-L, Ding S, Day MW, Grubbs RH (2003) Synthesis and activity of ruthenium alkylidene complexes coordinated with phosphine and N-heterocyclic carbene ligands. J Am Chem Soc 125: 2546–2558

[28] Randl S, Gessler S, Wakamatsu H, Blechert S (2001) Highly selective cross metathesis with acrylonitrile using a phosphine free Ru-complex. Synlett: 430–432
[29] Arduengo AJ, III, Calabrese JC, Davidson F, Dias HVR, Goerlich JR, Krafczyk R, Marshall WJ, Tamm M, Schmutzler R (1999) C-H insertion reactions of nucleophilic carbenes. Helv Chim Acta 82: 2348–2364
[30] Duong HA, Tekavec TN, Arif AM, Louie J (2004) Reversible carboxylation of N-heterocyclic carbenes. Chem Commun: 112–113
[31] Voutchkova AM, Appelhans LN, Chianese AR, Crabtree RH (2005) Disubstituted imidazolium-2-carboxylates as efficient precursors to N-heterocyclic carbene complexes of Rh, Ru, Ir, and Pd. J Am Chem Soc 127: 17624–17625
[32] Voutchkova AM, Feliz M, Clot E, Eisenstein O, Crabtree RH (2007) Imidazolium carboxylates as versatile and selective N-heterocyclic carbene transfer agents: synthesis, mechanism, and applications. J Am Chem Soc 129: 12834–12846
[33] Schössler W, Regitz M (1974) Stable dipoles from 1,1′,3,3′-tetraphenyl-2,2′-biimidazolidinylidene and acyliso- or acylisothiocyanates. Chem Ber 107: 1931–1948
[34] Kuhn N, Steimann M, Weyers G (1999) Synthesis and properties of 1,3-diisopropyl-4,5-dimethylimidazolium-2-carboxylate, a stable carbene adduct of carbon dioxide. Z Naturforsch 54b: 427–433
[35] Kuhn N, Kratz T (1993) Synthesis of imidazol-2-ylidenes by reduction of imidazole-2(3*H*)-thiones. Synthesis: 561–562
[36] Ishiguro K, Hirabayashi K, Nojima T, Sawaki Y (2002) Nucleophilic-O transfer, cyclization, and decarboxylation of carbonyl oxide intermediate in the reaction of stable imidazolylidene and singlet oxygen. Chem Lett 31: 796–797
[37] Holbrey JD, Reichert WM, Tkatchenko I, Bouajila E, Walter O, Tommasi I, Rogers RD (2003) 1,3-Dimethylimidazolium-2-carboxylate: the unexpected synthesis of an ionic liquid precursor and carbene-CO_2 adduct. Chem Commun: 28–29
[38] Tommasi I, Sorrentino F (2005) Utilisation of 1,3-dialkylimidazolium-2-carboxylates as CO_2-carrier in the presence of Na^+ and K^+: application in the synthesis of carboxylates, monomethylcarbonate anions and halogen-free ionic liquids. Tetrahedron Lett 46: 2141–2145
[39] Tudose A, Demonceau A, Delaude L (2006) Imidazol(in)ium-2-carboxylates as N-heterocyclic carbene precursors in ruthenium-arene catalysts for olefin metathesis and cyclopropanation. J Organomet Chem 691: 5356–5365
[40] Wanzlick H-W, Kleiner H-J (1961) Nucleophile carbene chemistry: preparation of bis(1,3-diphenylimidazolidin-2-ylidene). Angew Chem 73: 493
[41] Wanzlick H-W (1962) Nucleophile carbene chemistry. Angew Chem 74: 129–134
[42] Wanzlick H-W, Esser F, Kleiner H-J (1963) New compounds of the type bis(1,3-diphenylimidazolidin-2-ylidene). Chem Ber 96: 1208–1212
[43] Wanzlick H-W, Lachmann B, Schikora E (1965) About the formation and reactivity of bis(1,3-diphenylimidazolidin-2-ylidene). Chem Ber 98: 3170–3177
[44] Lemal DM, Kawano KI (1962) Stepwise electron abstraction from a tetraaminoethylene. J Am Chem Soc 84: 1761–1762
[45] Lemal DM, Lovald RA, Kawano KI (1964) Tetraaminoethylenes. The question of dissociation. J Am Chem Soc 86: 2518–2519
[46] Winberg HE, Carnahan JE, Coffman DD, Brown M (1965) Tetraaminoethylenes. J Am Chem Soc 87: 2055–2056
[47] Krasuski W, Nikolaus D, Regitz M (1982) CS_2-dipoles from electron-rich olefins: preparation and cycloaddition reactions. Liebigs Ann Chem: 1451–1465
[48] Winberg HE, Coffman DD (1965) Chemistry of peraminoethylenes. J Am Chem Soc 87: 2776–2777
[49] Sheldrick WS, Schönberg A, Singer E, Eckert P (1980) About the reaction of 1,1′,3,3′-tetraphenyl-2,2′-bisimidazolidinylidene with carbon disulfide. Chem Ber 113: 3605–3609

[50] Küçükbay H, Çetinkaya E, Durmaz R (1995) Synthesis and antimicrobial activity of substituted benzimidazole, benzothiazole, and imidazole derivatives. Arzneim Forsch (Drug Res) 45: 1331–1334
[51] Küçükbay H, Durmaz R, Orhan E, Günal S (2003) Synthesis, antibacterial and antifungal activities of electron-rich olefins derived benzimidazole compounds. Farmaco 58: 431–437
[52] Nyce GW, Csihony S, Waymouth RM, Hedrick JL (2004) A general and versatile approach to thermally generated N-heterocyclic carbenes. Chem Eur J 20: 4073–4079
[53] Kuhn N, Bohnen H, Henkel G (1994) About the reaction of carbon disulfide carbene adducts with bromine and iodine. Z Naturforsch 49b: 1473–1480
[54] Kuhn N, Niquet E, Steimann M, Walker I (1999) Methoxyalkyl-functionalized 2,3-dihydroimidazol-2-ylidenes. Z Naturforsch 54b: 1181–1187
[55] Faust R, Göbelt B (2000) Persistent carbenes containing acetylenes: 4,5-dialkynylimidazol-2-ylidene. Chem Commun: 919–920
[56] Jafarpour L, Hillier AC, Nolan SP (2002) Improved one-pot synthesis of second-generation ruthenium olefin metathesis catalyst. Organometallics 21: 442–444
[57] Alcarazo M, J. RS, Alonso E, Fernandez R, Alvarez E, Lahoz FJ, Lassaletta JM (2004) 1,3-bis(N,N-dialkylamino)imidazolin-2-ylidenes: synthesis and reactivity of a new family of stable N-heterocyclic carbenes. J Am Chem Soc 126: 13242–13243
[58] Vehlow K, Maechling S, Blechert S (2006) Ruthenium metathesis catalysts with saturated unsymmetrical N-heterocyclic carbene ligands. Organometallics 25: 25–28
[59] Lee S, Hartwig JF (2001) Improved catalysts for the palladium-catalyzed synthesis of oxindoles by amide ⟨-arylation. Rate-acceleration, use of aryl chloride substrates, and a new carbene ligand for asymmetric transformation. J Org Chem 66: 3402–3415
[60] Viciu MS, Grasa GA, Nolan SP (2001) Catalytic dehalogenation of aryl halides mediated by a palladium/imidazolium salt system. Organometallics 20: 3607–3612
[61] Grasa GA, Viciu MS, Huang J, Nolan SP (2001) Amination reactions of aryl halides with nitrogen-containing reagents mediated by palladium/imidazolium salt systems. J Org Chem 66: 7729–7737
[62] Ma Y, Song C, Jiang W, Wu Q, Wang Y, Liu X, Andrus MB (2003) Sonogashira coupling using bulky palladium–phenanthryl imidazolium carbene catalysis. Org Lett 5: 3317–3319
[63] Navarro O, Kaur H, Mahjoor P, Nolan SP (2004) Cross-coupling and dehalogenation reactions catalyzed by (N-heterocyclic carbene)Pd(allyl)Cl complexes. J Org Chem 69: 3173–3180
[64] Altenhoff G, Goddard R, Lehmann CW, Glorius F (2004) Sterically demanding, bioxazoline-derived N-heterocyclic carbene ligands with restricted flexibility for catalysis. J Am Chem Soc 126: 15195–15201
[65] Arduengo AJ, III, Krafczyk R, Schmutzler R, Craig HA, Goerlich JR, Marshall WJ, Unverzagt M (1999) Imidazolylidenes, imidazolinylidenes and imidazolidines. Tetrahedron 55: 14523–14534
[66] Aidouni A, Demonceau A, Delaude L (2006) Microwave-assisted synthesis of N-heterocyclic carbene precursors. Synlett: 493–495
[67] Fürstner A, Alcarazo M, César V, Lehmann CW (2006) Convenient, scalable and flexible method for the preparation of imidazolium salts with previously inaccessible substitution patterns. Chem Commun: 2176–2178
[68] Delaude L, Demonceau A, Wouters J (2009) Assessing the potentials of zwitterionic NHC · CS_2 adducts for probing the stereoelectronic parameters of N-heterocyclic carbenes. Eur J Inorg Chem: accepted for publication
[69] Demonceau A, Stumpf AW, Saive E, Noels AF (1997) Novel ruthenium-based catalyst systems for the ring-opening metathesis polymerization of low-strain cyclic olefins. Macromolecules 30: 3127–3136
[70] Jan D, Delaude L, Simal F, Demonceau A, Noels AF (2000) Synthesis and evaluation of new $RuCl_2$(*p*-cymene)(ER_2R') and (η^1:η^6-phosphinoarene)$RuCl_2$ complexes as ring-opening metathesis polymerization catalysts. J Organomet Chem 606: 55–64

[71] Delaude L, Sauvage X, Demonceau A, Wouters J (2009) Synthesis and catalytic evaluation of ruthenium–arene complexes generated using imidazol(in)ium-2-carboxylates and -dithiocarboxylates. Eur J Inorg Chem: 1882–1891: 10.1002/ejic.200801110
[72] Castarlenas R, Vovard C, Fischmeister C, Dixneuf PH (2006) Allenylidene-to-indenylidene rearrangement in arene-ruthenium complexes: a key step to highly active catalysts for olefin metathesis reactions. J Am Chem Soc 128: 4079–4089
[73] Bruneau C, Dixneuf PH (2006) Metal vinylidenes and allenylidenes in catalysis: applications in anti-Markovnikov additions to terminal alkynes and alkene metathesis. Angew Chem Int Ed 45: 2176–2203
[74] Castarlenas R, Esteruelas MA, Oñate E (2006) N-heterocyclic carbene-osmium complexes for olefin metathesis reactions. Organometallics 24: 4343–4346
[75] Bencze L, Kraut-Vass A, Prókai L (1985) Mechanism of initiation of the metathesis of norbornene using $W(CO)_3Cl_2(AsPh_3)_2$ as catalyst. Chem Commun: 911–912
[76] Alvarez P, Gimeno J, Lastra P (2002) Catalytic synthesis of polynorbornene and polynorbornadiene of low polydispersity index by $[Ru(\eta^5\text{-}C_9H_7)Cl(COD)]$ (COD = 1,5-cyclooctadiene). Organometallics 21: 5678–5680
[77] Herrmann WA, Köcher C, Goossen LJ, Artus GRJ (1996) Heterocyclic carbenes: a high-yielding synthesis of novel, functionalized N-heterocyclic carbenes in liquid ammonia. Chem Eur J 2: 1627–1636
[78] Herrmann WA, Elison M, Fischer J, Köcher C, Artus GRJ (1996) N-heterocyclic carbenes: generation under mild conditions and formation of group 8–10 transition metal complexes relevant to catalysis. Chem Eur J 2: 772–780
[79] Jafarpour L, Huang J, Stevens ED, Nolan SP (1999) (*p*-cymene)$RuLCl_2$ (L = 1,3-bis(2,4,6-trimethylphenyl)imidazol-2-ylidene and 1,3-bis(2,6-diisopropylphenyl)imidazol-2-ylidene) and related complexes as ring closing metathesis catalysts. Organometallics 18: 3760–3763
[80] Lo C, Cariou R, Fischmeister C, Dixneuf PH (2007) Simple ruthenium precatalyst for the synthesis of stilbene derivatives and ring-closing metathesis in the presence of styrene initiators. Adv Synth Catal 349: 546–550
[81] Kirmse W (2005) Carbene complexes of nonmetals. Eur J Org Chem: 237–260
[82] Blanrue A, Wilhelm R (2004) Imidazolinium–carbodithioate zwitterions as organocatalysts for the cyanosilylation of aldehydes. Synlett: 2621–2623
[83] Cuyas J, Miguel D, Pérez-Martínez JA, Riera V (1992) The first cationic heterobinuclear complexes with bridging S_2CPR_3 ligands. X-ray structure of $[(\eta^6\text{-}C_6Me_6)Ru(\mu\text{-}Cl)(\mu\text{-}S_2CPCy_3)W(CO)_3]PF_6.CH_2Cl_2$. Polyhedron 11: 2713–2716
[84] Alvarez B, Miguel D, Pérez-Martínez JA, Riera V (1994) Arene-ruthenium complexes with S_2CPR_3 and trichlorostannate. J Organomet Chem 474: 143–147

Mono- and Bimetallic Ruthenium–Arene Catalysts for Olefin Metathesis: A Survey

Yannick Borguet, Xavier Sauvage, Albert Demonceau, Lionel Delaude*

Laboratory of Macromolecular Chemistry and Organic Catalysis, Institut de Chimie (B6a), University of Liege, Sart-Tilman, 4000 Liege, Belgium

*E-mail: l.delaude@ulg.ac.be

Abstract In this chapter, we summarize the main achievements of our group toward the development of easily accessible, highly efficient ruthenium–arene catalyst precursors for olefin metathesis. Major advances in this field are presented chronologically, with an emphasis on catalyst design and mechanistic details. The first part of this survey focuses on monometallic complexes with the general formula $RuCl_2$(*p*-cymene)(L), where L is a phosphine or N-heterocyclic carbene ancillary ligand. In the second part, we disclose recent developments in the synthesis and catalytic applications of homobimetallic ruthenium–arene complexes of generic formula (*p*-cymene)Ru(μ-Cl)$_3$RuCl(η^2-C_2H_4)(L) and their derivatives resulting from the substitution of the labile ethylene moiety with vinylidene, allenylidene, or indenylidene units.

Keywords Allenylidene ligand · Homogeneous catalysis · Indenylidene ligand· N-heterocyclic carbene · Phosphine ligand · Vinylidene ligand

1 Introduction

Thanks to the development of highly efficient, well-defined catalysts tolerant of many functional groups, olefin metathesis has emerged as a powerful tool for assembling unsaturated hydrocarbon backbones in organic synthesis and in polymer chemistry [1, 2]. Since Grubbs and coworkers first reported on the synthesis and metathetical activity of well-defined ruthenium–alkylidene complexes $RuCl_2(=CH–CH=CPh_2)(PCy_3)_2$ [3] and $RuCl_2(=CHPh)(PCy_3)_2$ (**1**) [4] in the mid-1990s, research groups around the world have engaged in designing further late transition-metal based catalysts with increased robustness and proficiency, in order to reduce catalyst loading and to perform metathesis of challenging substrates under milder conditions. Countless structural alterations have been made to the archetypal compound **1**, commonly referred to as the Grubbs first generation catalyst, in order to tailor its activity [5–9], stability [10–12], solubility [13–15],

V. Dragutan et al. (eds.), *Green Metathesis Chemistry: Great Challenges in Synthesis, Catalysis and Nanotechnology*,

recoverability [16–18], or latency [19–21] toward specific catalytic processes [22–26], sometimes in an asymmetric fashion [27–29]. Commercially available catalysts developed along these lines include the Grubbs second (**2**) [7] and third generation catalysts (**3**) [30], and the Hoveyda–Grubbs catalyst (**4**) [10] (Scheme 1).

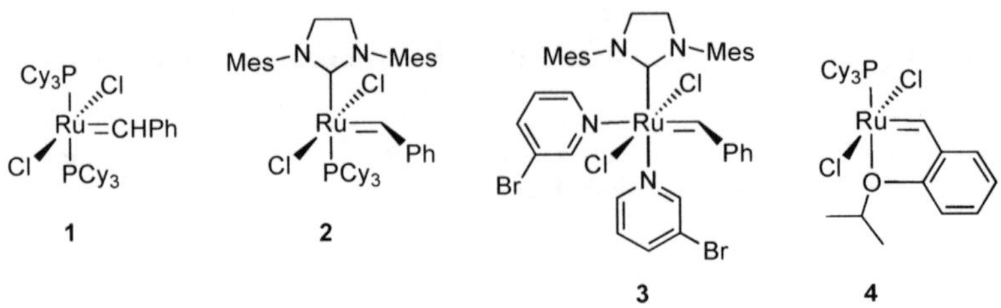

Scheme 1 Typical representatives of ruthenium–benzylidene catalysts for olefin metathesis

Because they are the closest analogues to the ruthenium–alkylidene active species believed to play a key role in the mechanism postulated for olefin metathesis [31], 16-electron ruthenium–benzylidene complexes have attracted a great deal of attention from the organometallic and catalysis communities. However, other related catalyst precursors yielding the same active species, such as ruthenium–vinylidene (**5**) [32–34], –allenylidene (**6**) [34], or –indenylidene complexes (**7**) [26, 35], have also proven efficient promoters for olefin metathesis (Scheme 2). In many cases, these precatalysts are more easily accessible than their benzylidene counterparts and they resist to harsher reaction conditions in terms of temperature and functional group tolerance. Thus, for some applications, they provide a valuable alternative to benzylidene-based precatalysts [36].

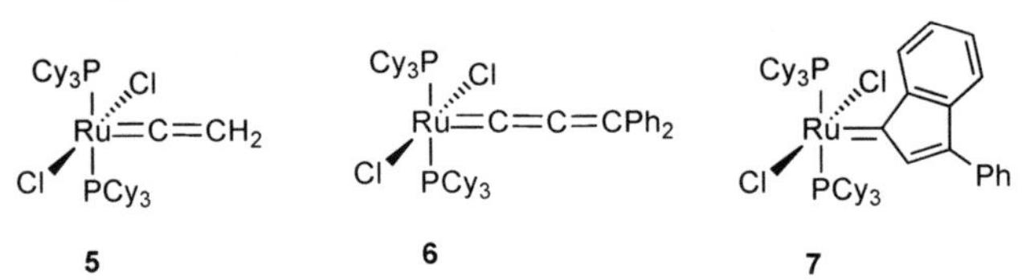

Scheme 2 Typical representatives of ruthenium–vinylidene (**5**), –allenylidene (**6**), and –indenylidene (**7**) catalysts for olefin metathesis

Ruthenium–arene complexes belong to a separate category of organometallic compounds. These 18-electron species with a distinctive piano-stool geometry have been known in the literature for more than 40 years [37–40]. Most representatives of this class of half-sandwich compounds were designed to include unfunctionalized arene ligands, such as benzene, *para*-cymene (1-methyl-4-isopropylbenzene), or hexamethylbenzene [41–44], but derivatives in which the aromatic hydrocarbon moiety is decorated with various chemical functions were also investigated [45]. Several neutral or cationic ruthenium–arene complexes are versatile and efficient catalyst precursors for various important organic transformations [46–51]. This is due in part to the lability of the η^6-arene ligand that can be easily removed upon thermal [52] or photochemical [53–55] activation to release highly active, coordinatively unsaturated species.

Our laboratory has been involved in the chemistry of ruthenium–arene complexes for more than 15 years. During the course of our studies, we showed that species of the $RuCl_2$(*p*-cymene) (L) type, where L is either a phosphine or a N-heterocyclic carbene (NHC) ligand, were highly efficient initiators for atom transfer radical additions and polymerizations [56–60]. Strikingly, Grubbs-type ruthenium–benzylidene catalysts were also found very apt at promoting controlled radical polymerizations [56, 61, 62], while ruthenium–arene compounds turned out to be effective catalyst precursors for olefin metathesis, thereby raising the question of a possible link between the two processes.

In this chapter, we summarize the main achievements of our group toward the development of easily accessible, highly efficient ruthenium–arene catalyst precursors for olefin metathesis. Major advances in this field are presented chronologically, with an emphasis on catalyst design and mechanistic details.

2 Results and Discussion

2.1 Monometallic Complexes Bearing Phosphine Ligands

During the 1990s, we demonstrated that $RuCl_2$(*p*-cymene)(PR_3) complexes bearing basic and bulky phosphine ligands, such as tricyclohexylphosphine (PCy_3), were highly effective catalyst precursors for the ring-opening metathesis polymerization (ROMP) of both strained and low-strain cycloolefins when activated by a suitable carbene precursor, such as ethyl diazoacetate [63] or trimethylsilyldiazomethane [64] (Scheme 3). Although we were not able to isolate the active species formed upon addition of the co-catalyst and monomer to the ruthenium–arene starting material, ^1H and ^{31}P NMR analyses of the reaction media supported the intermediacy of Grubbs type ruthenium–alkylidene complexes as actual ROMP initiators [65]. Indeed, an independent report from Thieuleux and Basset published in 2007 confirmed that the catalytically active species involved in the self-metathesis of ethyl oleate carried out in the presence of $RuCl_2$(*p*-cymene)(PCy_3) (**8**) and Me_3SiCHN_2 were probably the same as those resulting from the use of the single-component Grubbs first generation catalyst $RuCl_2(=CHPh)(PCy_3)_2$ (**1**). In the case of the multi-component system, de-coordination of the arene and redistribution of the PCy_3 ligands led to the in situ formation of ca. 5–10% of $RuCl_2(=CHR)(PCy_3)_2$ propagating species (R = $CH(CH_2)_6CH_3$ or $CH(CH_2)_6CO_2Et$), whereas complex **1** quantitatively afforded these species [66].

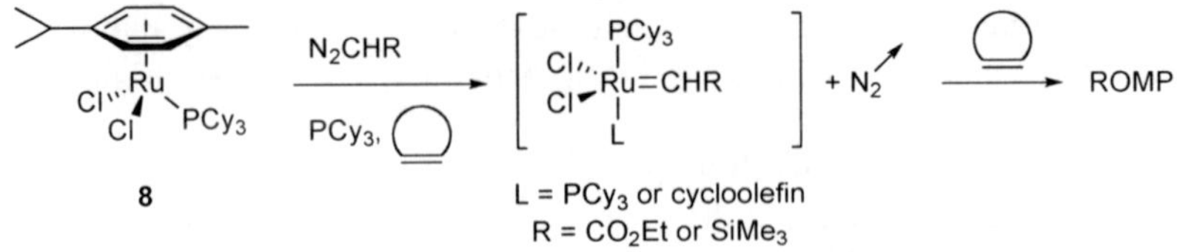

Scheme 3 Formation of metathetically active species from a ruthenium–arene complex and a diazo compound

As an alternative to the use of preformed $RuCl_2$(arene)(PR_3) complexes, highly active catalytic systems for the ROMP of norbornene and cyclooctene were also conveniently generated in situ by direct addition of a phosphine ligand to the stable [$RuCl_2$(arene)]$_2$ dimers, followed by activation with trimethylsilyldiazomethane. The influence of the arene and phosphine ligands and the role of the solvent on polynorbornene and polyoctenamer yields, molecular weight distributions, and microstructures were investigated [65]. The combination of durene (1,2,4, 5-tetramethylbenzene) or *para*-cymene as arene ligands, together with triisopropylphosphine or tricyclohexylphosphine, afforded the most active catalysts. The excellent functional group compatibility of these ruthenium initiators was illustrated by the synthesis of a variety of polyoctenamers bearing epoxide, acid, ether, ester, acetal, or bromine functionalities. In most cases, high yields of polymers were reached. Only, sulfide and azide functionalities in the monomers resulted in a deactivation of the catalyst. A striking positional influence of the functional group on the polymerization outcome was revealed by comparing two 4,5-substituted cyclooctenes with the corresponding allylic derivatives. Characterization of the polymers by IR and NMR spectroscopies indicated a lack of high regio- and stereospecificity in the propagation step [65].

2.2 Monometallic Complexes Bearing NHC Ligands

Replacement of the phosphine ligand in **8** by a N-heterocyclic carbene led to a second generation of monometallic ruthenium–arene complexes that were investigated at the dawn of the new millennium. Initial research efforts focused on preformed $RuCl_2$(*p*-cymene)(NHC) complexes, such as compound **9**. Compared to their phosphine analogues, these catalyst precursors did not require the addition of a diazo compounds co-catalyst to become active ROMP initiators. Instead, we found that visible light irradiation was required to trigger the polymerization of cyclooctene, most likely via a photo-induced decoordination of the arene ligand (Scheme 4) [67].

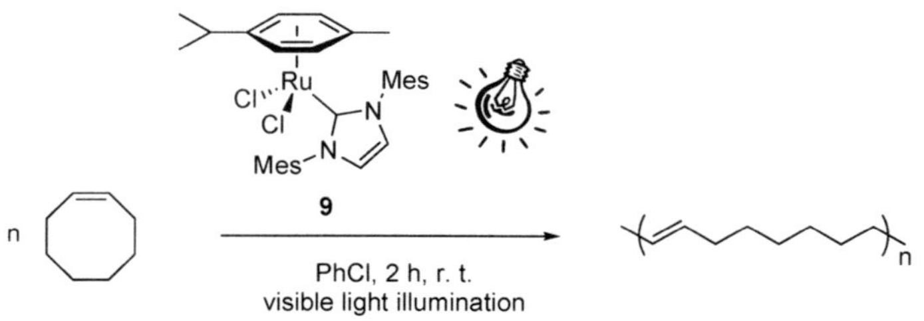

Scheme 4 Visible light induced ROMP of cyclooctene catalyzed by a ruthenium–arene complex bearing a NHC ligand

Next, a wide range of imidazolium and imidazolinium salts was synthesized and their ability to release NHC ligands upon deprotonation with a strong base was investigated in various catalytic processes. Thus, 1,3-diarylimidazol(in)ium chlorides bearing phenyl, 1-naphthyl, 4-biphenyl, 3,5-dimethylphenyl, 2-tolyl, 2, 6-dimethylphenyl, 2,4,6-trimethylphenyl (mesityl), and 2,6-diisopropylphenyl substituents on their nitrogen atoms were prepared. They were combined with the $[RuCl_2(p\text{-cymene})]_2$ dimer and potassium *tert*-butoxide or sodium hydride to generate the corresponding $RuCl_2(p\text{-cymene})(NHC)$ adducts in situ (Scheme 5, path A). The catalytic activity of all these species was investigated in the photoinduced ROMP of norbornene and cyclooctene. Results from this study showed that the C4–C5 double bond in the imidazole ring of the NHC ligands was not crucial to achieve high catalytic efficiencies. The presence or the absence of alkyl groups on the *ortho* positions of the phenyl rings had a more pronounced influence. Blocking all the *ortho* positions was a requisite for obtaining efficient catalysts. Failure to do so probably resulted in the *ortho*-metallation of the carbene ligand, thereby altering the coordination sphere of the ruthenium active centers [68].

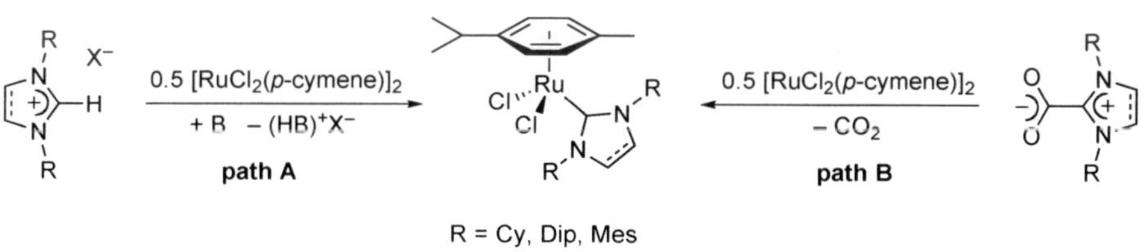

Scheme 5 Synthesis of ruthenium–arene complexes bearing NHC ligands from imidazol(in)ium salts and a base or from imidazol(in)ium-2-carboxylates

In 2006, we successfully demonstrated the aptitude of imidazol(in)ium-2-carboxylates to act as NHC precursors for various in situ catalytic applications based on ruthenium–arene complexes (Scheme 5, path B). When the visible light induced ROMP of cyclooctene was carried out at 60°C, the NHC · CO_2 adducts and their NHC · HX counterparts (X = Cl, BF_4) displayed similar high activities. When metathesis polymerizations were performed at room temperature, however, the carboxylates proved far superior to the corresponding imidazol(in)ium acid salts. Indeed, they displayed the same level of activity as the preformed $RuCl_2$(*p*-cymene)(IMes) complex (**9**), whereas the combinations of NHC · HX and KO-*t*-Bu were almost totally inactive [69].

2.3 Bimetallic Complexes Bearing NHC Ligands

In 2005, Severin and coworkers investigated the reaction of $[RuCl_2(p\text{-cymene})]_2$ with 1 eq. of PCy_3 under an ethylene atmosphere. Under these conditions, the ruthenium dimer afforded a new type of molecular scaffold, in which a $RuCl(\eta^2\text{-}C_2H_4)(PCy_3)$ fragment was connected via three μ-chloro bridges to a ruthenium–(*p*-cymene) moiety. The resulting complex (**10**) displayed an outstanding catalytic activity in atom transfer radical additions [70], but was not effective at promoting olefin metathesis [71]. In view of the enhancements brought by the replacement of phosphines by NHCs in monometallic ruthenium–arene catalyst precursors, we decided to investigate the effect of similar modifications in the bimetallic series. Thus, in 2007, we reported on the synthesis and catalytic activity of two new homobimetallic ruthenium–arene complexes of general formula (*p*-cymene) $Ru(\mu\text{-}Cl)_3RuCl(\eta^2\text{-}C_2H_4)$(NHC), where NHC = 1,3-bis(2,4,6-trimethylphenyl)imidazolin-2-ylidene (**11**) or 1,3-bis(2,4,6-trimethylphenyl)-4,5-dichloroimidazolin-2-ylidene (**12**) [71]. These compounds were isolated in high yields upon heating a toluene solution of $[RuCl_2(p\text{-cymene})]_2$ with 1 eq. of carbene ligand under an ethylene atmosphere (Scheme 6). Contrary to monometallic ruthenium–arene complexes of the $RuCl_2$(*p*-cymene)(L) type (L = phosphine or NHC ligand), the new homobimetallic species did not require the addition of a diazo compound nor visible light illumination to initiate the ring-opening metathesis of norbornene or cyclooctene. When α,ω–dienes were exposed to **11** or **12**, a mixture of cycloisomerization and ring-closing metathesis products was obtained in a non-selective way. Addition of a terminal alkyne co-catalyst enhanced the metathetical activity while completely repressing the cycloisomerization process. Thus, quantitative conversions of diethyl 2,2-diallylmalonate and N,N-diallyltosylamide were achieved within 2 h at room temperature using 2 mol% of catalyst precursor **11** and 6 mol% of phenylacetylene [71].

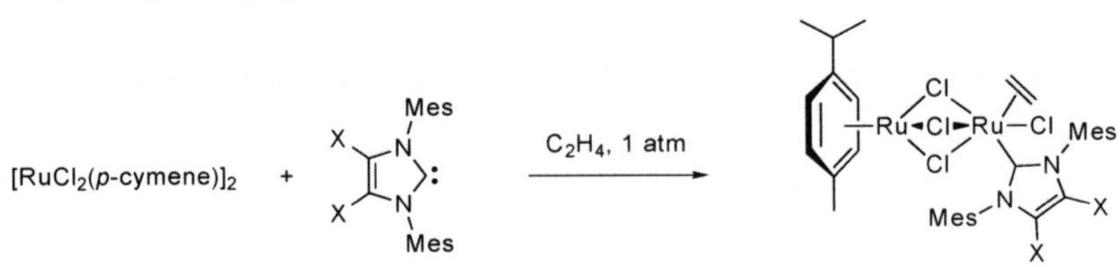

11 (X = H) **12** (X = Cl)

Scheme 6 Synthesis of homobimetallic ruthenium–arene complexes bearing NHC ligands

2.4 Bimetallic Complexes Bearing Phosphine Ligands

The positive influence of phenylacetylene on catalytic systems derived from complexes **11** and **12** prompted us to further investigate the reaction of alkynes with the labile ruthenium–ethylene complex **10** (Scheme 7) [72]. Hence, treatment of (*p*-cymene)Ru(μ-Cl)$_3$RuCl(η^2-C_2H_4)(PCy_3) with a slight excess of propargyl alcohol derivatives afforded quantitative yields of vinylidene complexes **13** and **14** within 2 h at room temperature. Although these products were stable enough to be fully characterized, they underwent a slow albeit irreversible transformation into ruthenium–allenylidene complex **15** in solution. Elimination of *n*-propanol from the γ-propoxyvinylidene unit in **14** proceeded cleanly and selectively without the need for any additive. Dehydration of the γ-hydroxyvinylidene ligand of **13** was better accomplished in the presence of 3 Å molecular sieves to suppress side-reactions. Although its structure was erroneously reported in 1999 [73–75], complex **15** had never been isolated so far. In the presence of an acidic promoter, it rearranged into the homobimetallic ruthenium–indenylidene compound **16**, whose molecular structure was unambiguously determined by X-ray diffraction analysis. A direct vinylidene-to-indenylidene interconversion was also successfully carried out in the presence of a drying agent and a strong acid, thereby affording complex **16** in three steps and 72% overall yield from [$RuCl_2$(*p*-cymene)]$_2$, 1 eq. of PCy_3, and 1,1-diphenylpropynol [72].

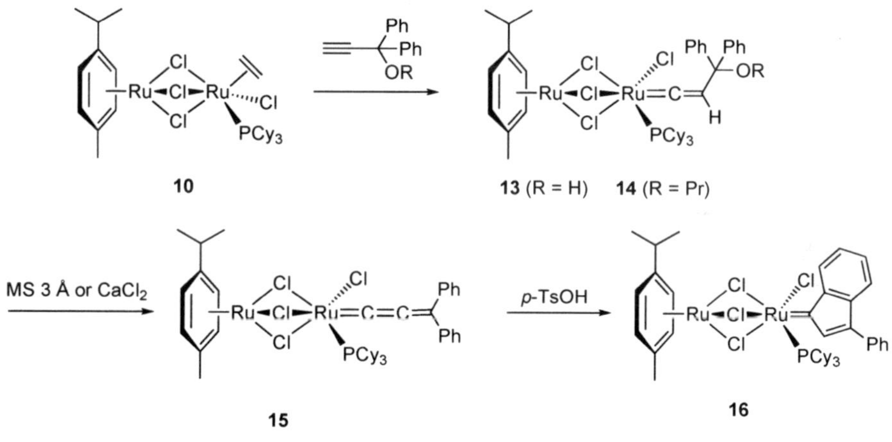

Scheme 7 Synthesis of homobimetallic ruthenium–arene complexes bearing a phosphine ligand and polyunsaturated carbon-rich fragments

The catalytic activity of complexes **13–16** was probed in various types of olefin metathesis reactions and compared with those of first (**1**), second (**2**), and third generation (**3**) monometallic Grubbs catalysts. For the ring-closing metathesis of diethyl diallylmalonate, ruthenium–indenylidene complex **16** outperformed all the ruthenium–benzylidene complexes under investigation and was only slightly less efficient than its monometallic parent **7**. Cross-metathesis experiments with ethylene showed that deactivation of bimetallic ruthenium–alkylidene or –indenylidene complexes was due to the rapid bimolecular decomposition of methylidene active species into ethylene complex **10**. Vinylidene and allenylidene complexes **13–15**, on the other hand, were far less efficient olefin metathesis initiators and remained inert under an ethylene atmosphere. Their catalytic activity was, however, substantially enhanced upon addition of an acidic co-catalyst that most likely promoted their in situ transformation into indenylidene species [72].

Due to its straightforward synthesis and high metathetical activity, homobimetallic ruthenium–indenylidene complex **16** was deemed an attractive intermediate to convert into alkoxyalkylidene species via stoichiometric cross-metathesis with 2-isopropoxystyrene. Thus, a convenient one-pot procedure was devised for the preparation of Hoveyda–Grubbs catalyst **4** from ethylene complex **10** via a vinylidene–allenylidene–indenylidene cascade pathway (Scheme 8). Taking into account the optimized synthesis of precursor **10** from $[RuCl_2(p\text{-cymene})]_2$ in a preliminary step, monometallic catalyst **4** was obtained in 85% overall yield. No large excess of organic reagents was required and the transition metal not incorporated into the final product could easily be recovered and recycled at the end of the process [72].

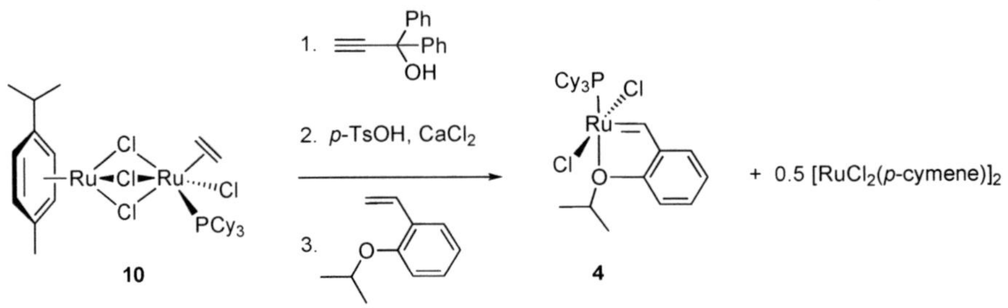

Scheme 8 One-pot synthesis of the Hoveyda–Grubbs catalyst (**4**) from homobimetallic ruthenium–arene complex **10**

3 Conclusion and Perspectives

Mono- and bimetallic ruthenium–arene complexes bearing phosphine or NHC ancillary ligands are versatile and efficient catalyst precursors for olefin metathesis. Their synthesis can usually be achieved in a single step starting from the widely available $[RuCl_2(p\text{-cymene})]_2$ dimer. Such a straightforward access sharply contrasts with the multiple steps required to obtain Grubbs' type ruthenium–benzylidene or indenylidene complexes of second and third generation. Indeed, the preparation of these compounds involves the intermediacy of first generation diphosphine complexes, whose transformation into more elaborate derivatives adds to the duration and cost of the overall process. Hence, ruthenium–arene complexes provide a convenient platform for catalytic engineering and ligand fine-tuning, as we have shown by carefully adjusting the nature of the ancillary ligands or by switching from a monometallic to a bimetallic molecular scaffold. Because all these 18-electron species lack a vacant coordination site and the mandatory alkylidene fragment needed for metathesis, their major drawback is a poor initiation efficiency. This may be a critical issue, for instance when controlled polymerizations are required. However, the ability to trigger the catalytic process upon addition of a co-catalyst or via photochemical irradiation is also a highly desirable feature for some practical applications. Combining the best of both worlds into a single, easily accessible catalytic system is our next challenge.

Acknowledgments Financial support from the Romanian Academy and the Belgian C.G.R.I. is gratefully acknowledged.

References

[1] Ivin KJ, Mol JC (1997) Olefin metathesis and metathesis polymerization. Academic Press, San Diego CA
[2] Grubbs RH (ed.) (2003) Handbook of metathesis. Wiley-VCH, Weinheim
[3] Nguyen ST, Grubbs RH (1993) Syntheses and activities of new single-component, ruthenium-based olefin metathesis catalysts. J Am Chem Soc 115: 9858–9859
[4] Schwab P, France MB, Ziller JW, Grubbs RH (1995) A series of well-defined metathesis catalysts – synthesis of [$RuCl_2(=CHR')(PR_3)_2$] and its reaction. Angew Chem Int Ed 34: 2039–2041
[5] Weskamp T, Schattenmann WC, Spiegler M, Herrmann WA (1998) A novel class of ruthenium catalysts for olefin metathesis. Angew Chem Int Ed 37: 2490–2493
[6] Huang J, Stevens ED, Nolan SP, Petersen JL (1999) Olefin metathesis-active ruthenium complexes bearing a nucleophilic carbene ligand. J Am Chem Soc 121: 2674–2678
[7] Scholl M, Ding S, Lee CW, Grubbs RH (1999) Synthesis and activity of a new generation of ruthenium-based olefin metathesis catalysts coordinated with 1,3-dimesityl-4,5-dihydroimidazol-2-ylidene ligands. Org Lett 1: 953–956
[8] Trnka TM, Morgan JP, Sanford MS, Wihelm TE, Scholl M, Choi T-L, Ding S, Day MW, Grubbs RH (2003) Synthesis and activity of ruthenium alkylidene complexes coordinated with phosphine and N-heterocyclic carbene ligands. J Am Chem Soc 125: 2546–2558
[9] Love JA, Sanford MS, Day MW, Grubbs RH (2003) Synthesis, structure, and activity of enhanced initiators for olefin metathesis. J Am Chem Soc 125: 10103–10109
[10] Kingsbury JS, Harrity JPA, Bonitatebus PJ, Jr., Hoveyda AH (1999) A recyclable Ru-based metathesis catalyst. J Am Chem Soc 121: 791–799
[11] Garber SB, Kingsbury JS, Gray BL, Hoveyda AH (2000) Efficient and recyclable monomeric and dendritic Ru-based metathesis catalysts. J Am Chem Soc 122: 8168–8179
[12] Wakamatsu H, Blechert S (2002) A highly active and air-stable ruthenium complex for olefin metathesis. Angew Chem Int Ed 41: 794–796
[13] Lynn DM, Mohr B, Grubbs RH, Henling LM, Day MW (2000) Water-soluble ruthenium alkylidenes: synthesis, characterization, and application to olefin metathesis in protic solvents. J Am Chem Soc 122: 6601–6609
[14] Hong SH, Grubbs RH (2006) Highly active water-soluble olefin metathesis catalyst. J Am Chem Soc 128: 3508–3509

[15] Michrowska A, Gułajski Ł, Kaczmarska Z, Mennecke K, Kirschning A, Grela K (2006) A green catalyst for green chemistry: synthesis and application of an olefin metathesis catalyst bearing a quaternary ammonium salt. Green Chem 8: 685–688

[16] Jafarpour L, Heck M-P, Baylon C, Lee HM, Mioskowski C, Nolan SP (2002) Preparation and activity of recyclable polymer-supported ruthenium olefin metathesis catalysts. Organometallics 21: 671–679

[17] Yang L, Mayr M, Wurst K, Buchmeiser MR (2004) Novel metathesis catalysts based on ruthenium 1,3-dimesityl-3,4,5,6-tetrahydropyrimidin-2-ylidenes: synthesis, structure, immobilization, and catalytic activity. Chem Eur J 10: 5761–5770

[18] Michrowska A, Mennecke K, Kunz U, Kirschning A, Grela K (2006) A new concept for the noncovalent binding of a ruthenium-based olefin metathesis catalyst to polymeric phases: preparation of a catalyst on Raschig rings. J Am Chem Soc 128: 13261–13267

[19] Ung T, Hejl A, Grubbs RH, Schrodi Y (2004) Latent ruthenium olefin metathesis catalysts that contain an N-heterocyclic carbene ligand. Organometallics 23: 5399–5401

[20] Hejl A, Day MW, Grubbs RH (2006) Latent olefin metathesis catalysts featuring chelating alkylidenes. Organometallics 25: 6149–6154

[21] Ledoux N, Drozdzak R, Allaert B, Linden A, Van der Voort P, Verpoort F (2007) Exploring new synthetic strategies in the development of a chemically activated Ru-based olefin metathesis catalyst. Dalton Trans: 5201–5210

[22] Trnka TM, Grubbs RH (2001) The development of L_2X_2Ru=CHR olefin metathesis catalysts: an organometallic success story. Acc Chem Res 34: 18–29

[23] Hoveyda AH, Gillingham DG, Van Veldhuizen JJ, Kataoka O, Garber SB, Kingsbury JS, Harrity JPA (2004) Ru complexes bearing bidentate carbenes: from innocent curiosity to uniquely effective catalysts for olefin metathesis. Org Biomol Chem 2: 8–23

[24] Deshmukh PH, Blechert S (2007) Alkene metathesis: the search for better catalysts. Dalton Trans 24: 2479–2491

[25] Bieniek M, Michrowska A, Usanov DL, Grela K (2008) In an attempt to provide a user's guide to the galaxy of benzylidene, alkoxybenzylidene, and indenylidene ruthenium olefin metathesis catalysts. Chem Eur J 14: 806–818

[26] Boeda F, Clavier H, Nolan SP (2008) Ruthenium–indenylidene complexes: powerful tools for metathesis transformations. Chem Commun: 2726–2740

[27] Seiders TJ, Ward DW, Grubbs RH (2001) Enantioselective ruthenium-catalyzed ring-closing metathesis. Org Lett 3: 3225–3228

[28] Van Veldhuizen JJ, Gillingham DG, Garber SB, Kataoka O, Hoveyda AH (2003) Chiral Ru-based complexes for asymmetric olefin metathesis: enhancement of catalyst activity through steric and electronic modifications. J Am Chem Soc 125: 12502–12508

[29] Funk TW, Berlin JM, Grubbs RH (2006) Highly active chiral ruthenium catalysts for asymmetric ring-closing olefin metathesis. J Am Chem Soc 128: 1840–1846

[30] Love JA, Morgan JP, Trnka TM, Grubbs RH (2002) A practical and highly active ruthenium-based catalyst that effects the cross metathesis of acrylonitrile. Angew Chem Int Ed 41: 4035–4037

[31] Hérisson J-L, Chauvin Y (1971) Catalysis of olefin transformation by tungsten complexes. II telomerization of cyclic olefins in the presence of acyclic olefins. Makromol Chem 141: 161–176

[32] Katayama H, Ozawa F (2004) Vinylideneruthenium complexes in catalysis. Coord Chem Rev 248: 1703–1715

[33] Dragutan V, Dragutan I (2004) Ruthenium vinylidene complexes. Synthesis and applications in metathesis catalysis. Platinum Met Rev 48: 148–153

[34] Bruneau C, Dixneuf PH (2006) Metal vinylidenes and allenylidenes in catalysis: applications in anti-Markovnikov additions to terminal alkynes and alkene metathesis. Angew Chem Int Ed 45: 2176–2203

[35] Dragutan V, Dragutan I, Verpoort F (2005) Ruthenium indenylidene complexes. Metathesis catalysts with enhanced activity. Platinum Met Rev 49: 33–40

[36] Clavier H, Nolan SP (2007) N-heterocyclic carbene and phosphine ruthenium indenylidene precatalysts: a comparative study in olefin metathesis. Chem Eur J 13: 8029–8036

[37] Jones D, Pratt L, Wilkinson G (1962) π-Cyclohexadienyl compounds of manganese, rhenium, iron, and ruthenium. J Chem Soc: 4458–4463

[38] Winkhaus G, Singer H (1967) Ruthenium(II) complexes with bidentate cycloheptatriene and benzene. J Organomet Chem 7: 487–491

[39] Zelonka RA, Baird MC (1972) Benzene complexes of ruthenium(II). Can J Chem 50: 3063–3072

[40] Bennett MA, Smith AK (1974) Arene ruthenium(II) complexes formed by dehydrogenation of cyclohexadienes with ruthenium(III) trichloride. J Chem Soc, Dalton Trans: 233–241

[41] Le Bozec H, Touchard D, Dixneuf PH (1989) Organometallic chemistry of arene ruthenium and osmium complexes. Adv Organomet Chem 29: 163–247

[42] Bennett MA (1997) Recent advances in the chemistry of arene complexes of ruthenium(0) and ruthenium(II). Coord Chem Rev 166: 225–254

[43] Pigge FC, Coniglio JJ (2001) Stoichiometric applications of η^6-arene ruthenium(II) complexes in organic chemistry. Curr Org Chem 5: 757–784

[44] Rigby JH, Kondratenko MA (2004) Arene complexes as catalysts. Top Organonomet Chem 7: 181–204

[45] Therrien B (2009) Functionalised η^6-arene ruthenium complexes. Coord Chem Rev 253: 493–519

[46] Küçükbay H, Çetinkaya B, Guesmi S, Dixneuf PH (1996) New (carbene)ruthenium-arene complexes: preparation and uses in catalytic synthesis of furans. Organometallics 15: 2434–2439

[47] Çetinkaya B, Özdemir I, Dixneuf PH (1997) Synthesis and catalytic properties of N-functionalized carbene complexes of rhodium(I) and ruthenium(II). J Organomet Chem 534: 153–158

[48] Fürstner A, Müller T (1999) Efficient total synthesis of resin glycosides and analogues by ring-closing olefin metathesis. J Am Chem Soc 121: 7814–7821

[49] Castarlenas R, Dixneuf PH (2003) Highly active catalysts in alkene metathesis: first observed transformation of allenylidene into indenylidene via alkenylcarbyne–ruthenium species. Angew Chem Int Ed 42: 4524–4527

[50] Castarlenas R, Eckert M, Dixneuf PH (2005) Alkenylcarbene ruthenium arene complexes as initiators of alkene metathesis: an enyne creates a catalyst that promotes its selective transformation. Angew Chem Int Ed 44: 2576–2579

[51] Yiğit M, Yiğit B, Özdemir I, Çetinkaya E, Çetinkaya B (2006) Active ruthenium-(N-heterocyclic carbene) complexes for hydrogenation of ketones. Appl Organomet Chem 20: 322–327

[52] Quebatte L, Haas M, Solari E, Scopelliti R, Nguyen QT, Severin K (2005) Atom-transfer radical reactions under mild conditions with $[\{RuCl_2(1,3,5\text{-}C_6H_3iPr_3)\}_2]$ and PCy_3 as the catalyst precursors. Angew Chem Int Ed 44: 1084–1088

[53] Hafner A, Mühlebach A, van der Schaaf PA (1997) One-component catalysts for thermal and photoinduced ring opening metathesis polymerization. Angew Chem Int Ed 36: 2121–2124

[54] Fürstner A, Ackermann L (1999) A most user-friendly protocol for ring closing metathesis reactions. Chem Commun: 95–96

[55] Fürstner A, Liebl M, Lehmann CW, Picquet M, Kunz R, Bruneau C, Touchard D, Dixneuf PH (2000) Cationic ruthenium allenylidene complexes as catalysts for ring closing olefin metathesis. Chem Eur J 6: 1847–1857

[56] Simal F, Demonceau A, Noels AF (1999) Highly efficient ruthenium-based catalytic systems for the controlled free-radical polymerization of vinyl monomers. Angew Chem Int Ed 38: 538–540

[57] Simal F, Jan D, Delaude L, Demonceau A, Spirlet M-R, Noels AF (2001) Evaluation of ruthenium-based complexes for the controlled radical polymerization of vinyl monomers. Can J Chem 79: 529–535

[58] Delaude L, Delfosse S, Richel A, Demonceau A, Noels AF (2003) Tuning of ruthenium N-heterocyclic catalysts for ATRP. Chem Commun: 1526–1527
[59] Richel A, Delfosse S, Cremasco C, Delaude L, Demonceau A, Noels AF (2003) Ruthenium catalysts bearing N-heterocyclic carbene ligands in atom transfer radical reactions. Tetrahedron Lett 44: 6011–6015
[60] Borguet Y, Richel A, Delfosse S, Leclerc A, Delaude L, Demonceau A (2007) Microwave-enhanced ruthenium-catalysed atom transfer radical additions. Tetrahedron Lett 48: 6334–6338
[61] Simal F, Demonceau A, Noels AF (1999) Kharasch addition and controlled atom transfer radical polymerisation (ATRP) of vinyl monomers catalysed by Grubbs' ruthenium–carbene complexes. Tetrahedron Lett 40: 5689–5693
[62] Simal F, Delfosse S, Demonceau A, Noels AF, Denk K, Kohl FJ, Weskamp T, Herrmann WA (2002) Ruthenium alkylidenes: modulation of a new clas of catalysts for controlled radical polymerization of vinyl monomers. Chem Eur J 8: 3047–3052
[63] Demonceau A, Noels AF, Saive E, Hubert AJ (1992) Ruthenium-catalysed ring-opening metathesis polymerization of cycloolefins initiated by diazoesters. J Mol Catal 76: 123–132
[64] Stumpf AW, Saive E, Demonceau A, Noels AF (1995) Ruthenium-based catalysts for the ring-opening metathesis polymerization of low-strain cyclic olefins and of functionalised derivatives of norbornene and cyclooctene. Chem Commun: 1127–1128
[65] Demonceau A, Stumpf AW, Saive E, Noels AF (1997) Novel ruthenium-based catalyst systems for the ring-opening metathesis polymerization of low-strain cyclic olefins. Macromolecules 30: 3127–3136
[66] Ahr M, Thieuleux C, Copéret C, Fenet B, Basset J-M (2007) Noels' vs. Grubbs' catalysts: evidence for one unique active species from two different systems! Adv Synth Catal 349: 1587–1591
[67] Delaude L, Demonceau A, Noels AF (2001) Visible light induced ring-opening metathesis polymerisation of cyclooctene. Chem Commun: 986–987
[68] Jan D, Delaude L, Simal F, Demonceau A, Noels AF (2000) Synthesis and evaluation of new $RuCl_2$(*p*-cymene)(ER_2R') and (η^1:η^6-phosphinoarene)$RuCl_2$ complexes as ring-opening metathesis polymerization catalysts. J Organomet Chem 606: 55–64
[69] Tudose A, Demonceau A, Delaude L (2006) Imidazol(in)ium-2-carboxylates as N-heterocyclic carbene precursors in ruthenium–arene catalysts for olefin metathesis and cyclopropanation. J Organomet Chem 691: 5356–5365
[70] Quebatte L, Solari E, Scopelliti R, Severin K (2005) A bimetallic ruthenium ethylene complex as a catalyst precursor for the Kharasch reaction. Organometallics 24: 1404–1406
[71] Sauvage X, Borguet Y, Noels AF, Delaude L, Demonceau A (2007) Homobimetallic ruthenium-N-heterocyclic carbene complexes: synthesis, characterization, and catalytic applications. Adv Synth Catal 349: 255–265
[72] Sauvage X, Borguet Y, Zaragoza G, Demonceau A, Delaude L (2009) Homobimetallic ruthenium vinylidene, allenylidene, and indenylidene complexes: synthesis, characterization, and catalytic studies. Adv Synth Catal 351: 441–455
[73] Harlow KJ, Hill AF, Wilton-Ely JDET (1999) The first co-ordinatively unsaturated Group 8 allenylidene complexes: insights into Grubbs' vs. Dixneuf-Fürstner olefin metathesis catalysts. J Chem Soc, Dalton Trans: 285–291
[74] Fürstner A, Hill AF, Liebl M, Wilton-Ely JDET (1999) Coordinatively unsaturated ruthenium–allenylidene complexes: highly effective, well defined catalysts for the ring-closure metathesis of α,ω-dienes and dienynes. Chem Commun: 601–602
[75] Fürstner A, Guth O, Düffels A, Seidel G, Liebl M, Gabor B, Mynott R (2001) Indenylidene complexes of ruthenium: optimized synthesis, structure elucidation, and performance as catalysts for olefin metathesis – application to the synthesis of the ADE-ring system of Nakadomarin A. Chem Eur J 7: 4811–4820

Mesoporous Molecular Sieves Based Catalysts for Olefin Metathesis and Metathesis Polymerization

Hynek Balcar,* Jiří Čejka

J. Heyrovský Institute of Physical Chemistry, v.v.i., Academy of Sciences of the Czech Republic, Dolejškova 3, 182 23 Prague 8, Czech Republic

*E-mail: balcar@jh-inst.cas.cz

Abstract Heterogeneous catalysts for olefin metathesis using different types of (i) siliceous mesoporous molecular sieves, and (ii) organized mesoporous alumina as supports are reported. The catalysts were prepared either by spreading of transition metal oxidic phase on the support surface or by immobilizing transition metal compounds (mostly organometallic) on the support. The activity of these catalysts in various types of metathesis reactions (i.e. alkene and diene metathesis, metathesis of unsaturated esters and ethers, RCM, ROMP and metathesis polymerization of alkynes) was described. The main advantages of these catalysts consist generally in their high activity and selectivity, easy separation of catalysts from reaction products and the preparation of products free of catalyst residue. The examples of pore size influence on the selectivity in metathesis reactions are also given.

Keywords Heterogeneous catalysts · Hybrid catalysts · Olefin metathesis · Metathesis polymerization

1 Introduction

Mesoporous molecular sieves of M41S family [1] (e.g. MCM-41, MCM-48, SBA-15) and several kinds of organized mesoporous alumina (OMA) [2] with unique characteristics like high surface area, large void volume and mesopores of narrow size distribution represent modern supports for catalytic systems active in various reactions (oxidation, hydrodesulfurization, C–C coupling reactions, etc.) [3, 4]. Recently, these supports have also been applied for preparation of olefin metathesis catalysts [5]. Transition metal oxides, chlorides and complexes were immobilized on the surface of siliceous mesoporous sieves (MMS) and/or OMA and catalysts prepared in this way were successfully used in various types of metathesis reaction. The aim of this paper is to summarize results achieved in this topic.

V. Dragutan et al. (eds.), *Green Metathesis Chemistry: Great Challenges in Synthesis, Catalysis and Nanotechnology*,

2 Catalysts Preparation

2.1 Supports

Table 1 gives a list of organized mesoporous supports used in the preparation of metathesis catalysts. All-siliceous MMS of two dimensional hexagonal architecture MCM-41 and SBA-15, three dimensional cubic structure MCM-48, several types of OMA with amorphous walls and "worm-hole like" structures of pores of different average sizes, and lath-like (LMA) and scaffold-like (SMA) aluminas with increased crystallinity of the framework walls are surveyed with their main texture characteristics, surface area S_{BET}, void volume V and average pore diameter D. Characteristics of conventional supports (silica Merck, γ-alumina Alcoa, mesoporous alumina Condea) are given for comparison.

Table 1 Survey of supports

Support	Architecture	S_{BET} (m²/g)	V (g/cm³)	D (nm)	Ref.
MCM-41	Hexagonal	1,070	0.871	3.1	[10]
SBA-15	Hexagonal	777	1.258	6.1	[10]
MCM-48	Cubic	1,334	0.927	2.8	[10]
SiO_2 Merck		559	0.473	4.5	[10]
OMA3	Wormhole-like	332	0.250	3.5	[6]
OMA5	Wormhole-like	267	0.529	4.9	[6, 7]
OMA6	Wormhole-like	314	0.745	6.5	[11]
OMA7	Wormhole-like	300	0.850	7.0	[11]
SMA7	Scaffold- like	308	0.780	7.0	[6, 8]
LMA4	Lath-like	342	0.620	3.9	[6, 8]
Alcoa		342	0.355	5.0[a]	[9, 11]
Condea		197	0.427	7.2[b]	[9]

[a]Broad distribution, 30% of V in micropores.
[b]Broad distribution.

2.2 Oxide Catalysts

Oxide catalysts include: (i) molybdenum oxide on siliceous MMS (MoO_x/MCM-41, MoO_x/MCM-48, MoO_x/SBA-15), and (ii) rhenium oxide on different types of organized mesoporous alumina, e.g. ReO_x/OMA.

(i) MoO_3, $(NH_4)_6Mo_7O_{24}$, $MoO_2(acac)_2$ can be used as a molybdenum source. Thermal spreading method consisting in heating physical mixture of selected Mo compound and sieves in pre-calculated amounts at 500°C was used for preparation of molybdenum phase on the sieve surface [10]. Wet impregnation can be also used (except MoO_3). Thermal spreading, however, seems to be more advantageous because of its simplicity, perfect preservation of supports architecture and quantitative transfer of Mo into sieves (up to 12 wt% of Mo). The pore diameter as well as pore size distribution did not change. The molybdenum phase contains different Mo oxide species covalently bound to the surface of the support channels, from isolated MoO_4 species (Figure 1a) through oligo- and polymolybdate species to MoO_3 microcrystals. Good Mo dispersion (i.e. high population of isolated Mo oxide species and absence of MoO_3 crystallites) depends on catalyst loading and on the way of catalyst preparation and activation. The optimum activity was found for loading about 6 wt% of Mo.

(ii) Rhenium oxide catalysts were prepared by thermal spreading of NH_4ReO_4 on OMA at 550°C [11]. The organized mesoporous structure of OMA as well as SMA and LMA was preserved and no losses of Re were observed during immobilization (up to 15 wt% of Re). However, the maximum catalytic activity was found for catalyst loading 9 wt% of Re. Re^{7+} and Re^{6+} oxide species (Figure 1b) are proposed to be present on the surface. Wet impregnation from aqueous solutions of NH_4ReO_4 or $HReO_4$ is also possible [12, 13].

Just before use, both type of catalysts need to be activated by heating at 500°C in a stream of dry air followed by stream of Ar. During this period, adsorbed water is removed and transition metals are partially reduced into the oxidation state suitable for creating catalytically active species.

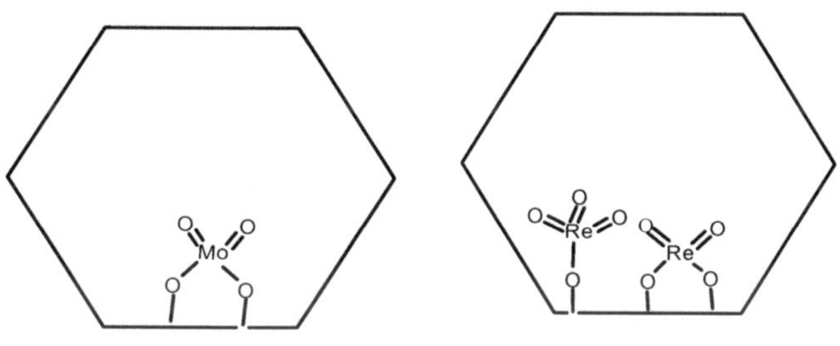

a b

Figure 1 Mo (**a**) and Re (**b**) species on MMS surface

2.3 *Grafted Catalysts*

By mixing solutions of selected Mo or Ru complexes with dried molecular sieves in inert atmosphere quantitative transfer of transition metal to the solid phase occurred in a short time. During this process, reactions of given complexes with surface OH groups are supposed to proceed according to Schemes 1, 2 and 3. After removing liquid phase, the catalysts were dried in vacuo and stored in inert atmosphere. In this way following hybrid catalysts were prepared: $MoCl_x$/MCM-41 (3 wt% of Mo), grafted Schrock carbenes Mo_F/MCM-41, Mo_F/MCM-48,

Mo_F/SBA-15, Mo_{Bu}/MCM-48 and Mo_{BPh}/MCM-41 (all 1 wt% of Mo) [14] and Ru grafted catalyst Ru(*p*-cymene)Cl_x/SBA-15 (1 wt% of Ru). In a similar way, Mo complex with bidentate aminophenolate ligand [15] was immobilized on SBA-15 surface modified with methylaluminoxane (MAO) giving $Mo_{N,O}Cl_x$/SBA-15/MAO hybrid catalyst (1 wt% of Mo) (Scheme 4). In all cases, negative filtration experiments, proving that catalytic activity is bound steadily to the solid phase, and negligible leaching observed in the course of the metathesis reaction confirmed the creation of strong bond between support surface and transition metal species.

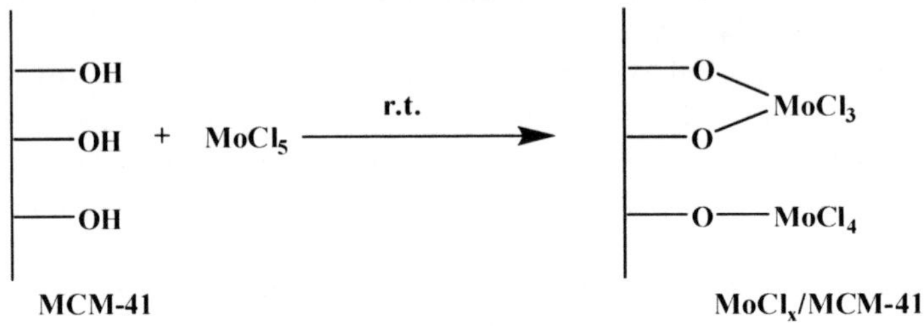

Scheme 1 Preparation of $MoCl_x$/ MCM-41 hybrid catalyst

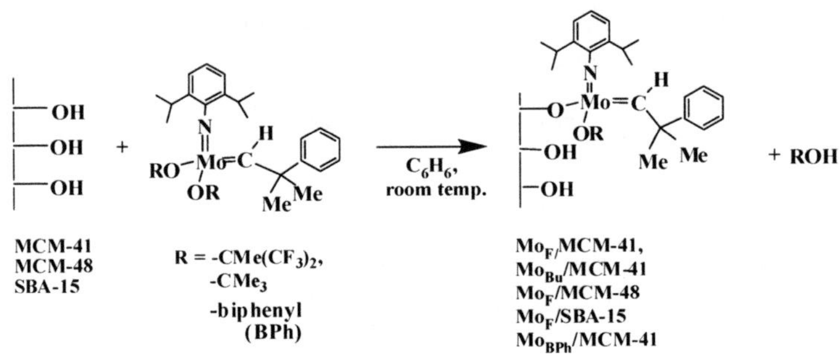

Scheme 2 Preparation of Schrock carbene grafted on mesoporous molecular sieves

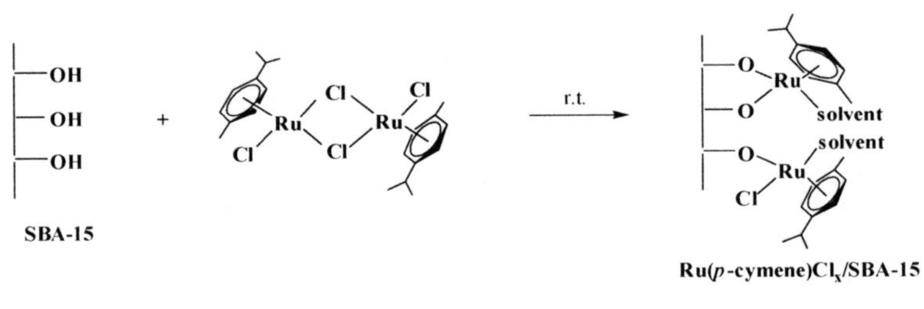

Scheme 3 Preparation of Ru(*p*-cymene)Cl_x/SBA-15

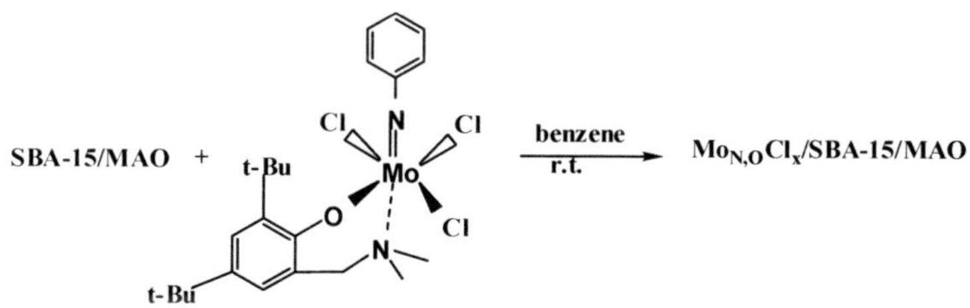

Scheme 4 Grafting of phenylimino-2,4-di-tert-butyl-6-((dimethylamino) methylphenolato)-trichloromolybdenum on SBA-15 modified with methyl aluminoxane

3 Application of Mesoporous Molecular Sieves Based Catalysts for Alkene and Diene Metathesis

3.1 Oxide Catalysts

In metathesis of liquid 1-alkenes at 40°C MoO_x/MCM-41 (6 wt% of Mo) exhibited about 1 order of magnitude higher activity (expressed as initial TOF) in metathesis of liquid 1-alkenes at 40°C MoO_x/MCM-41 (6 wt% of Mo) exhibited about 1 order of magnitude higher activity (expressed as initial TOF) in comparison with catalyst used conventional silica as a support MoO_x/SiO_2 (Figure 2) [10].

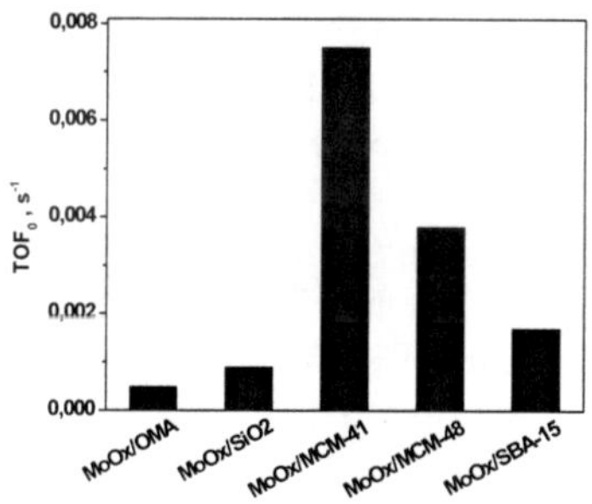

Figure 2 Specific activity of MoO_x heterogeneous catalysts in metathesis of neat 1-octene at 40°C, loading 8 wt% of Mo

The activity of MoO_x/MCM-48 and MoO_x/SBA-15 is lower but still higher than that of MoO_x/SiO_2. In all cases the selectivity to the main metathesis product was between 80% and 90%. On the other hand the activity of MoO_x on OMA5 support was very low and also the selectivity was about 20% only.

In the same type of reaction, the activity of rhenium oxide catalysts on all mesoporous organized alumina supports is higher than the activity of catalysts using conventional aluminas (Alcoa, Condea) as supports (Figure 3) [11]. In metathesis of 1-decene at 60°C the activity increases with increasing pore size in the range from 3 to 6 nm. The selectivity is in all cases about 95%.

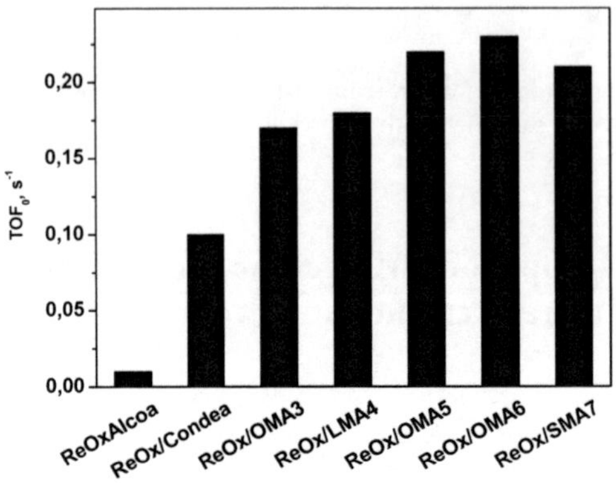

Figure 3 Specific activity of ReO_x heterogeneous catalysts in metathesis of neat 1-decene at 60°C, loading 9 wt% of Re

α,ω-Dienes undergo ADMET oligomerization except 1,7-octadiene, where RCM to cyclohexene strongly prevails. Similarly to alkene metathesis, the activity of ReO_x catalysts was found to increase in the order ReO_x/Alcoa << ReO_x/OMA3 < ReO_x/OMA6). Moreover, some differences in selectivity between ReO_x/OMA3 and ReO_x/OMA6 were found. In ADMET of 1,5-hexadiene and 1,9-decadiene, prevalence of higher oligomers was observed for ReO_x/OMA6 in comparison with ReO_x/OMA3 [7].

In alkene and diene metathesis, oxide catalysts on MMS can be used repeatedly and after it can be reactivated (by heating at 500°C) without any loss either in activity or selectivity. No leaching of Mo or Re into the liquid phase was observed.

3.2 Grafted Catalysts

Mo grafted catalysts were tested in metathesis of 1-alkenes and α,ω-dienes. Their activity was higher in comparison with oxide heterogeneous catalysts. In comparison with their homogeneous counterparts, however, the reaction rate of grafted catalysts was significantly lower as a result of slower diffusion in catalyst channels (for Schrock carbene catalyst see Figure 4a, b). On the other hand, with grafted catalysts, the reaction can be performed without any solvents (with neat substrates) which increases the reaction rate. Selectivity is very high (near 99% for 1-alkene) and fully comparable with corresponding homogeneous catalysts. The influence of pore size on catalytic activity was studied for grafted Schrock carbene complexes in 1-heptene metathesis and a slight decrease in the order Mo_F/SBA-15 > Mo_F/MCM-41 > Mo_F/MCM-48 was found. As concerns complexes used for grafting, the activity of hybrid catalysts decreases in the order Mo_F/MCM-41 ≈ Mo_{BPh}/MCM-41> Mo_{Bu}/MCM-41.

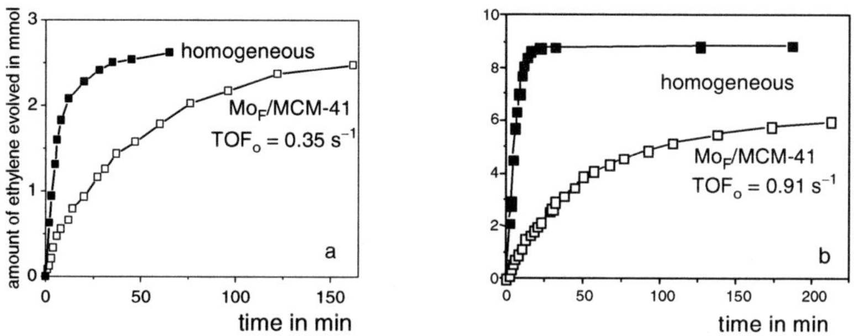

Mo_{BPh}/MCM-41> Mo_{Bu}/MCM-41.

Figure 4 Metathesis of 1-heptene - 1ml (**a**) and of 1,7-octadiene 1.5 ml (**b**) with Mo_F/MCM-41 and $Mo(=CHCMe_2Ph)(=N\text{-}2,6\text{-}iPr_2C_6H_3)[OCMe(CF_3)_2]_2$ used as a homogeneous catalyst. Benzene - 1ml, room temperature, c_{Mo} = 0.002 M. TOF expressed as mol amount of vinyl groups reacted per mol Mo and s

$MoCl_x$/MCM-41 needs to be activated by addition of co-catalysts. In accord with homogeneous systems Me_4Sn belongs to the best co-catalysts. Because of its toxicity, its replacement by different co-catalyst is desirable. Ph_2SiH_2 represents a good solution, although reaction rate lower than that with Me_4Sn was achieved (Figure 5).

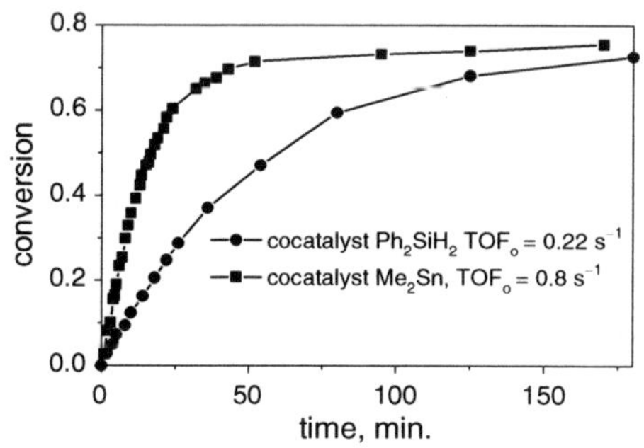

Figure 5 Metathesis of neat 1-heptene with $MoCl_x$/MCM-41 catalyst activated by Me_4Sn and Ph_2SiH_2, respectively. Room temp., molar ratio 1-heptene/Sn(Si)/Mo = 1,200/3/1; co-catalyst was added as a last component

Although filtration experiments proved catalytic activity being permanently bound to the solid phase and no Mo leaching was observed, the reusability of Mo grafted catalyst is rather poor. The reason may be in high sensitivity of these catalysts to traces of impurities. Their main advantage is easy catalyst separation and preparation of metal free products. Mo content in products was less than 5 ppb.

4 Application of Mesoporous Molecular Sieves Based Catalysts in Metathesis of Unsaturated Ethers and Esters

In comparison with alkenes and dienes, the activity of unsaturated ethers and esters in metathesis is often severely reduced due to the specific interaction of substrate oxygen atoms with catalysts. These interactions include blocking of catalytic active species and/or their irreversible deactivation. It leads to the decrease of the reaction rate and especially to the low TONs achieved with both homogeneous and heterogeneous catalysts. The extent of this undesirable interaction depends mainly on the transition metal used in the catalyst. It is usually accepted that the tolerance to the oxygen atoms in substrate molecules decreases in the order Ru >Re > Mo > W.

Table 2 gives TONs achieved in (i) metathesis of *p*-allylanisole (Equation 1), (ii) RCM of diethyl diallylmalonate (DEDAM) (Equation 2), and (iii) metathesis of 5-hexenyl acetate (HexAc) (Equation 3) with Mo and Re catalysts immobilized on organized mesoporous supports.

Table 2 TONs achieved in metathesis of *p*-allylanisole, DEDAM and HexAc with various catalytic systems

Substrate	MoO_x/MCM-41 + Me_4Sn[a] Ref. [16]	ReO_x/OMA5 + Me_4Sn[b] Ref. [6]	Mo_F/MCM-41[c]	Mo_F/SBA-15[c]
p-allylanisole	6	79	–	18
DEDAM	0.8	30	15	–
HexAc	–	8	23	–

[a]6 wt% of Mo, Sn/Mo molar ratio = 0.9, toluene, 40 °C.
[b]9 wt% of Re, Sn/Re molar ratio = 1, toluene, 25 °C.
[c]1 wt% of Mo, 25 °C, benzene.

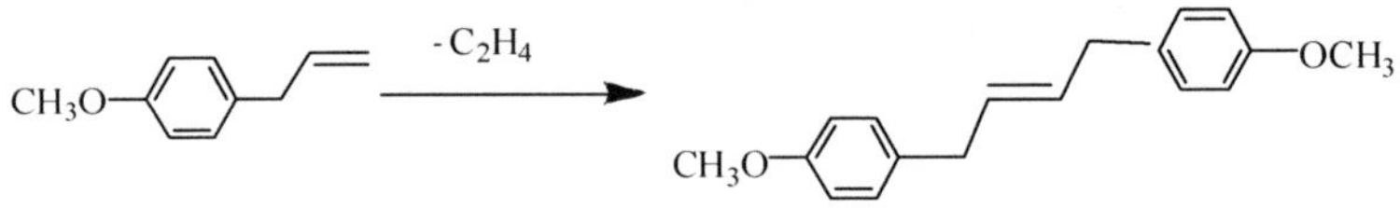

(1)

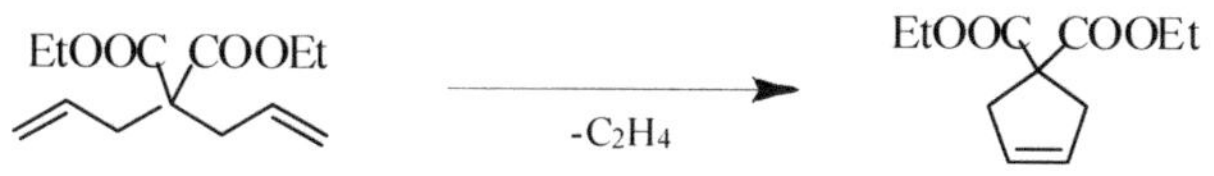

(2)

$$CH_2 = CH(CH_2)_4OOCCH_3 \xrightarrow[- C_2H_4]{} CH_3COO(CH_2)_4CH = CH(CH_2)_4OOCCH_3 \quad (3)$$

MoO_x/MCM-41 even with co-catalysts is not suitable for metathesis of given ethers and esters. ReO_x/OMA with Me_4Sn catalyst exhibits a satisfactory activity (TOF_o from 0.01 to 0.09 s^{-1}, according to the reaction conditions) and a high selectivity (near 100%). Although some positive effect of pore size on activity, indicating less rapid catalyst deactivation in the case of catalyst with larger pores, was observed [6], the TONs achieved were never higher than 100. Schrock carbenes on mesoporous molecular sieves are active without co-catalyst, they are highly selective (near 100%) but TONs achieved are about 20 only.

5 Application of Mesoporous Molecular Sieves Based Catalysts for Cycloalkene ROMP and Cycloalkene Cross-Metathesis

The catalysts on organized mesoporous supports were tested also in ROMP of cyclooctene (COE) and norbornene (NBE) (Equations 4, 5).

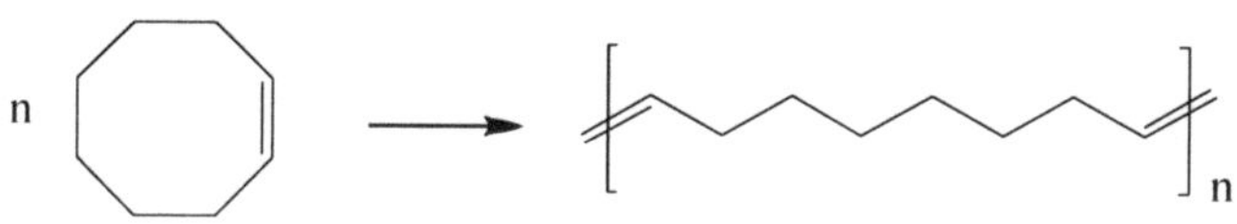

(4)

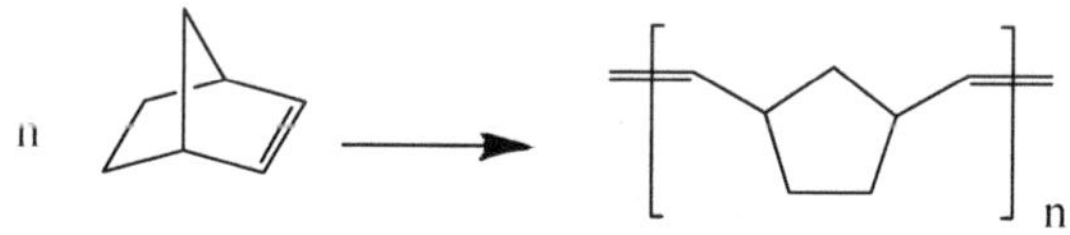

(5)

Molecular weights of the polymers were determined by SEC and values relative to polystyrene standards are given. It was found that these catalysts are effective in this type of reaction due to their pore size, which enables removal of polymeric products from catalysts.

With ReO_x/OMA6 (toluene, 15 mmol Re/l, mol ratio COE/Re = 400, 40°C,) ROMP of COE was accomplished producing in 30 min high molecular weight polymer (M_w = 850,000, M_n = 200,000, *E/Z* ratio = 4.26) in 65% yield [17]. Contrary to this, rhenium oxide on conventional alumina gives only oligomers under similar conditions [18, 19]. Results of ROMP of COE and NBE with Mo grafted catalysts are summarized in Table 3.

Table 3 ROMP of COE and NBE with Mo grafted catalysts

Catalyst	Monomer	Initial conc. c^o_{mon}/c^o_{Mo} (mol/l)	Polymer yield/reaction time	Molecular weight M_w (kDa)			Weight ratio $P_1/P_2/O$
				P_1	P_2	O	
$MoCl_x$/MCM-41 + 5 Me_4Sn	COE	2.5/ 0.003	69%/5 h	160	–	2.0	60/0/40
Mo_F/SBA-15	COE	0.5/ 0.001	87%/2 h	210	14	1.3	25/50/25
Mo_F/MCM-41	NBE	0.2/ 0.002	43%/2 h	1,100	41	2.3	50/25/25
Mo_F/SBA-15	NBE	0.1/ 0.001	37%/2 h	2,000	190	2.2	42/16/42
			45%/2 2h	1.900	190	2.1	40/18/42
			60%/22 h[a]	1,800	210	2.0	50/20/30
$Mo_{N,O}Cl_x$/ SBA-15/MAO	NBE	0.2/ 0.002	88%/1 h	Partially insoluble			–

Benzene, room temperature.
[a]After addition of pivalaldehyde, M_w = weight average molecular weight of polymer fraction P_1 and P_2, and of oligomer fraction O.

In all cases high molecular weight polymers were prepared in good yields. Especially M_w of poly(NBE)s prepared on grafted Schrock catalysts reached values higher than 1×10^6. Product molecular weight distribution, however, was multimodal. Large oligomer fraction (O) was present in all products; in many products two polymer fraction P_1 and P_2 were distinguished. During the polymerization, polymer formed was continuously released into the liquid phase due to the chain transfer by polymer. In cases where extremely high molecular weight product was formed, terminating agent was added at the end of the reactions, which led to the increase of polymer yield (see NBE polymerization with Mo_F/SBA-15 in Table 3). The content of Mo in polymer was 6 ppm. *E/Z* ratio of poly(NBE) prepared with Mo_F/MCM-41 was 2.57, similarly as in the polymer prepared with corresponding Schrock carbene used as homogeneous catalyst.

Molecular weight of polymers prepared can be effectively reduced by adding small amount of 1-alkene to the reaction mixture. Thus, using mixture of NBE with 1-heptene (in molar ratio NBE/1-heptene = 20) and Mo_F/SBA-15 catalyst, polymer of M_w = 105,000 was prepared in 80% yield under same condition as applied in Table 2. Using a higher amount of 1-alkenes, mixture of telomers according to Schemes 5 and 6 have been prepared either with ReO_x/OMA or with Mo grafted catalysts [7, 14].

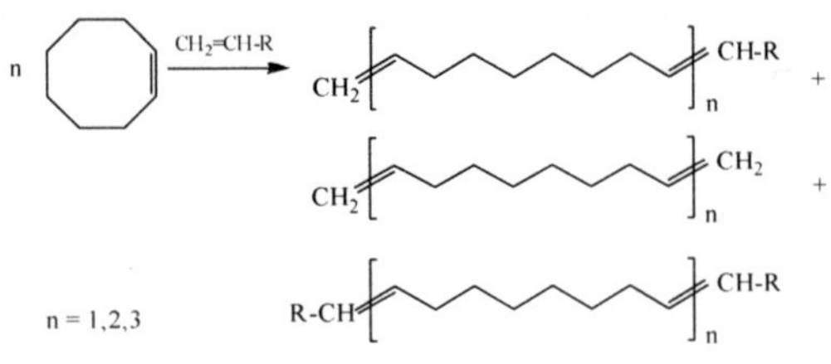

Scheme 5 Cross-metathesis of 1-alkene and COE

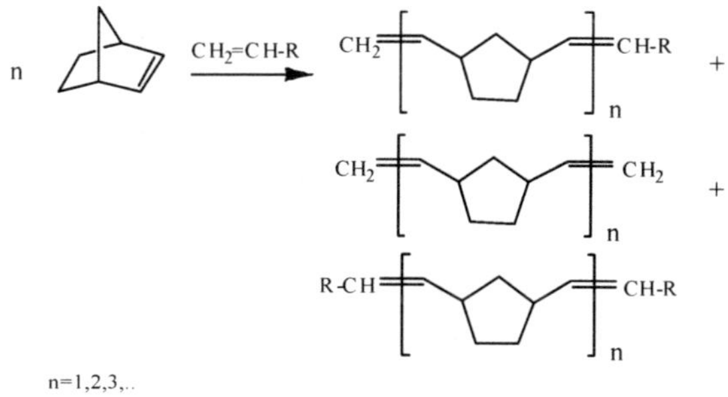

Scheme 6 Cross-metathesis of 1-alkene and NBE

The amount of different types of telomers depends on catalysts and reaction conditions. They need not follow the statistical proportions and some of them can strongly prevail in the reaction product [7, 14]. For example, in cross-metathesis of cyclooctene with 1-octene over ReO_x/OMA6 diene C_{22} and triene C_{30} strongly prevailed among telomers in the reaction mixture [7].

Ru(*p*-cymene)Cl_x/SBA-15 was used in ROMP of NBE. After catalyst activation with trimethylsilyldiazomethane before reaction, the yields of high molecular weight poly(NBE) (M_w = 2.8–7.4 × 10^5) up to 77% were obtained (1 h, 60°C). If conventional silica Merck was used instead of SBA-15 the polymer yield was significantly lower, which demonstrates the influence of supports on catalyst activity. The molecular weight distribution of poly(NBE) was unimodal and molecular weights were lower in comparison with those of polymers prepared on grafted Schrock catalysts. The *E/Z* ratio (1.12) was also lower than that of polymers prepared with Schrock catalysts.

6 Application of Mesoporous Molecular Sieves Based Catalysts for Metathesis Polymerization of Substituted Acetylenes

Molybdenum catalysts were also tested in metathesis polymerization of substituted acetylenes (Equation 6).

(6)

R= alkyl, aryl

As it is seen from Table 4, Mo heterogeneous catalysts are not efficient in polymerization of phenylacetylene (PhA). *t*-Butylacetylene (*t*-BuA) was polymerized into high molecular weight polymers by both $MoCl_x$/MCM-41 and MoO_x/MCM-41 catalyst with moderate yield. Surprisingly, MoO_x on mesoporous molecular sieves were found very effective catalysts for polymerization of linear aliphatic alkynes (1-hexyne, 1-decyne, 1-tetradecyne) [20]. For 1-hexyne the polymer yield as well as polymer molecular weight increased with increasing pore size in the order MoO_x/MCM-48 < MoO_x/MCM-41 < MoO_x/SBA-15. Only very low yield of polymer was obtained with MoO_x/SiO_2 demonstrating a beneficial effect of support mesoporous structure on polymer production. The activity of Mo_F/MCM-41 in 1-hexyne polymerization was similar as that of MoO_x/MCM-41, but the molecular weight of polymer was considerably lower (Table 4).

The separation of polymers formed from solid catalyst is easy. With all above given catalysts the polymer is released from catalysts without addition of terminating agent. We assume that during polymerization the polymer is disconnected from the catalysts by chain transfer by polymer (the growing polymer chain reacts with catalytic species located on the surface in their vicinity). In addition, the rest of polymer may be liberated from the surface by reaction with oxygen in the course of separation of liquid phase from catalyst performed under air by decantation and centrifugation. From the liquid phase polymers were isolated by solvent evaporation or by precipitation in methanol. The content of Mo in polymer was 8 ppm.

Table 4 Metathesis polymerization of substituted acetylenes with Mo heterogeneous catalysts

Catalyst	Monomer	Polymer yield (%)	Polymer M_w (kDa)
$MoCl_x$/MCM-41 + $3Me_4Sn$*	PhA**	6.5	42
$MoCl_x$/MCM-41+ $3Me_4Sn$*	*t*-BuA***	40	500
MoO_x/MCM-41	*t*-BuA	11	300
Mo_F/MCM-41	1-hexyne	38	23
MoO_x/MCM-48	1-hexyne	22	8.8
MoO_x/SBA-15	1-hexyne	64	30
MoO_x/SiO_2	1-hexyne	4	6.8
Mo_F/MCM-41	1-hexyne	34	7.5

Toluene, 40°C (*25°C), 3 h, initial Mo concentration = 0.01 M, mol ratio monomer/Mo = 70 (**100, ***300).

7 Conclusions

Heterogeneous catalysts for olefin metathesis and metathesis polymerization using siliceous mesoporous molecular sieves or organized mesoporous aluminas as supports were described. Activity and selectivity of these catalysts were tested in various type of metathesis reactions.

Catalysts consisting of molybdenum oxide on siliceous mesoporous sieves and of rhenium oxide on organized mesoporous alumina exhibited considerably higher activity in alkene and diene metathesis in comparison with catalysts supported on conventional silica or alumina. Rhenium oxide catalysts were also successfully used in metathesis of unsaturated esters and ethers.

Grafted molybdenum catalysts exhibited high activity and selectivity like their homogeneous counterparts. Their main advantage of their application in alkene and diene metathesis (as well as in metathesis of unsaturated esters and ethers) is easy separation of reaction products from catalysts and preparation of products free of catalyst residues.

In metathesis polymerization (ROMP and acetylene metathesis polymerization) catalysts based on mesoporous molecular sieves produce high molecular weight polymers in high yields. By easy separation of solid catalysts from polymer containing liquid phase, polymers of considerably reduced content of transition metal residues were prepared.

The additional advantage of catalysts based on mesoporous molecular sieves is a possibility to affect the selectivity by proper choice of pore size, as it was seen e.g. in metathesis of α,ω-dienes or cross-metathesis of COE with 1-alkenes.

Acknowledgements The authors thank the Grant Agency of the Academy of Sciences of the Czech Republic (IAA400400805) and the Academy of Sciences of the Czech Republic (KAN100400701) for the financial support.

References

[1] Beck JS, Vartuli JC, Roth WJ, Leonowicz ME, Kresge CT, Schmitt KD, Chu C-W, Olson DH, Sheppard EW, McCullen SB, Higgins JB, Schlenker JL (1992) J Am Chem Soc 114:10834

[2] Čejka J, Žilková N, Rathouský J, Zukal A (2001) Phys Chem Chem Phys 3:5076

[3] Thomas JM (1999) Angew Chem Int Ed 38:3588–3628

[4] Čejka J (2003) Appl Catal A: Gen 254:327

[5] Balcar H, Čejka J (2007) Mesoporous molecular sieves as supports for metathesis catalysts. In: Iammoglu Y, Dragutan V (eds.) Metathesis chemistry: from nanostructure design to synthesis of advanced materials. NATO Science Series II. Mathematics, Physics and Chemistry, vol. 243, pp. 151–166. Springer, Dordrecht, The Netherlands

[6] Balcar H, Hamtil R, Žilková N, Zhang Z, Pinnavaia TJ, Čejka J (2007) Appl Catal A: Gen 320:56–63

[7] Hamtil R, Žilková N, Balcar H, Čejka J (2006) Appl Catal A: Gen 302:193–200

[8] Zhang ZR, Pinnavaia TJ (2002) J Am Chem Soc 124:12294

[9] Balcar H, Žilková N, Bastl Z, Dědeček J, Hamtil R, Brabec L, Zukal A, Čejka J (2007). In Xu R, Gao Z, Chen J, Yan W (eds.) Stud Surf Sci Catal 170:1145–1152, Elsevier, Amsterdam

[10] Topka P, Balcar H, Rathouský J, Žilková N, Verpoort F, Čejka J (2006) Micropor Mesopor Mater 96:44–54

[11] Balcar H, Hamtil R, Žilková N, Čejka J (2004) Catal Lett 97:25–29

[12] Aguado J, Escola JM, Castro MC, Paredes B (2005) Appl Catal A: Gen 284:47

[13] Oikawa T, Ookoshi T, Tanaka T, Yamamoto T, Onaka M (2004) Micropor Mesopor Mater 74:93–103

[14] Balcar H, Žilková N, Sedláček J, Zedník J (2005) J Mol Catal A:Chem 232:53–58
[15] Lehtonen A, Balcar H, Sedláček J, Sillanpää R (2008) J Organomet Chem 693:1171–1176
[16] Topka P (2008) Ph.D. Thesis. Charles University, Prague
[17] Hamtil R (2008) Ph.D. Thesis. Technical University, Prague
[18] Warwel S, Katker H, Rauenbusch C (1987) Angew Chem Int Ed 26:702
[19] Saito K, Yamaguchi S, Tanabe K, Ogura T, Yagi M (1979) Bull Chem Soc Jpn 52:3192
[20] Balcar H, Topka P, Sedláček J, Zedník J, Čejka J (2008) J Polym Sci Part A Polym Chem 46:2593–2599

Binary and Ternary Catalytic Systems for Olefin Metathesis Based on $MoCl_5/SiO_2$

Victor I. Bykov, Boris A. Belyaev, Tamara A. Butenko, Eugene Sh. Finkelshtein*

Topchiev Institute of Petrochemical Synthesis, Russian Academy of Sciences, Leninsky prospect 29, Moscow 119991, Russia

*E-mail: fin@ips.ac.ru

Abstract Kinetics of α-olefin metathesis in the presence of binary ($MoCl_5/SiO_2$-Me4Sn) and ternary catalytic systems ($MoCl_5/SiO_2$–Me_4Sn–ECl_4, E = Si or Ge) was studied. Specifically, kinetics and reactivity of 1-decene, 1-octene, and 1-hexene in the metathesis reaction at 27°C and 50°C in the presence of $MoCl_5/SiO_2$–$SnMe_4$ were examined and evaluated in detail. It was shown that experimental data comply well with the simple kinetic equation for the rate of formation of symmetrical olefins with allowance for the reverse reaction and catalyst deactivation: $r = (k_1 \cdot c_\alpha - k_{-1} \cdot c_s) \cdot e^{-k_d \cdot \tilde{n}_{tot}}$. The coefficients for this equation were determined, and it was shown that these α-olefins had practically the same reactivity. It was found that reactivation in the course of metathesis took place due to the addition of a third component (silicon tetrachloride or germanium tetrachloride in combination with tetramethyltin) to a partially deactivated catalyst. The number of active centers was determined (5–6% of the amount of Mo) and the mechanisms of formation, deactivation, and reactivation were proposed for the binary and ternary catalytic systems. The role of individual components of the catalytic systems was revealed.

Keywords Olefin metathesis · Heterogeneous catalysts · Kinetics · Mechanisms of formation · Deactivation · Reactivation of active centers

1 Introduction

The catalytic metathesis of olefins is a very interesting and promising reaction in polymer chemistry, fine organic chemistry, and petroleum chemistry. In 1964, Banks and Bailey from the Phillips Petroleum Company passed propylene at 150°C over $LiAlH_4$-reduced $Mo(CO)_6/Al_2O_3$ and obtained ethylene and 2-butene [1]. Since then, a great number of studies have been dedicated to different versions

V. Dragutan et al. (eds.), *Green Metathesis Chemistry: Great Challenges in Synthesis, Catalysis and Nanotechnology*,

of this very promising reaction [2]. This fact was recognized by awarding the Nobel Prize in Chemistry in 2005 for the development of the metathesis method in organic synthesis. The contribution of the Nobel Prize Laureates to the development of the metathesis approach was primarily related to the discovery of a metal carbene mechanism (Y. Chauvin) and the development of efficient homogeneous catalysts containing 100% metal carbene active centers for metathesis: Ru carbene (R.H. Grubbs) and Mo (W) carbene (R.R. Schrock). These metal carbene complexes are capable of performing metathesis of complex unsaturated structures containing various functional groups [2].

A new flexible strategy was proposed at the Topchiev Institute of Petrochemical Synthesis, Russian Academy of Sciences (Moscow) for the synthesis of a wide range of pheromones (environmentally friendly insecticides) and other natural compounds in just a few stages. This strategy is based on the stereospecific cometathesis of petrochemical raw materials (α-olefins, cycloolefins, cyclooctadiene, and ethylene) [3–7] and so called ill-defined heterogeneous catalytic systems [8]. However, like other catalysts, these catalytic systems gradually lose their activity in the course of the reaction. This problem can be solved only on the basis of the knowledge of the mechanisms of formation, deactivation, and reactivation of the catalyst.

Here, we present experimental data on kinetics of α-olefin metathesis in the presence of binary ($MoCl_5/SiO_2$–Me_4Sn) and ternary catalytic systems ($MoCl_5/SiO_2$–Me_4Sn–ECl_4, E = Si or Ge). We determined the number of active centers and proposed possible mechanisms of formation, deactivation, and reactivation of these active centers. We first demonstrate that metal carbene active centers can be reactivated in the course of metathesis with the use of silicon tetrachloride or germanium tetrachloride as third components.

2 Experimental

The metathesis of α-olefins was performed in a thermostated glass reactor with a magnetic stirrer, which was equipped with a dropping funnel, a reflux condenser, and a gas burette for measuring the volume of ethylene released in the course of the reaction. A weighed portion of catalyst was loaded in the reactor, and specified amounts of an α-olefin, a cocatalyst (Me_4Sn), and a reactivator ($SiCl_4$ or $GeCl_4$) were placed in the dropping funnel. All of the operations were performed in a flow of dry argon of high purity grade. The kinetic data were obtained by volumetry and confirmed by determining the concentration of symmetrical olefin in the products of catalysis using GLC analysis. After completion of the reaction, the gas evolution practically ceased. The products of catalysis were removed, and a new α-olefin portion, which was equal to the preceding one, was added to the catalyst partially deactivated in the course of the reaction. The procedure was repeated five to eight times. In special experiments, we found that kinetic data were independent of the fact whether the cocatalyst was added with each new α-olefin portion or only with the first portion. Therefore, tetramethyltin was added only to the first reagent portion in all of the experiments. The purity of both starting and resulting compounds and the course of reactions were monitored using GLC analysis on an LKhM-8MD chromatograph with a flame-ionization detector (50 m × 0.2 mm quartz capillary column); stationary phases were SKTFP and SE-30; H_2 was a carrier gas. Analyses were performed under conditions of linear temperature programming (12°C/min) from 35°C to 100°C, lower than the boiling point of the corresponding compound. The ^{1}H and ^{13}C NMR spectra were measured on a Bruker MSL-300 spectrometer in $CDCl_3$ with reference to Me_4Si. The IR spectra in thin films were obtained on a Specord IR-75 instrument. The mass spectra were measured on a Finnigan MAT 95 XL 70 instrument (70 eV). All reactions, as well as the preparation of the starting compounds (α-olefins of chemically pure grade from Novocherkassk plant), were performed under an atmosphere of high purity argon using $LiAlH_4$ or Na as a drying agent. According to GLC data, the purity of the starting olefins was >99.9%.

3 Results and Discussion

3.1 *Binary Olefin Metathesis Catalytic Systems ($MoCl_5/SiO_2$–Me_4Sn)*

The metathesis of α-olefins occurs stoichiometrically, in accordance with Scheme 1, to yield 1 mol of ethylene and 1 mol of the symmetrical olefin from 2 mol of α-olefin.

$$2(R\text{–}CH{=}CH_2) \rightarrow R\text{–}CH{=}CH\text{–}R + CH_2{=}CH_2,$$

$$\text{where } R = C_8H_{17}, C_6H_{13}, \text{ and } C_4H_9.$$

Scheme 1 Metathesis of α-olefins

Figure 1 shows kinetic data for the metathesis of 1-decene at 27°C. Numerals written on the curves correspond to each new 1-decene portion, which was equal to the preceding one, added to the same catalyst sample. Analogous experimental data were obtained also for metathesis of 1-decene at 50°C and metathesis of 1-octene and 1-hexene at the same temperature. Although the evolved ethylene was removed from the reaction zone under experimental conditions, the conversion of 1-decene did not reach 100%, as can be seen in Figure 1.

This is evidently due to the fact that, when a certain symmetrical olefin concentration was reached, the following equilibrium between internal symmetrical olefins and α-olefins was established with the participation of secondary and primary metal carbene centers (Scheme 2):

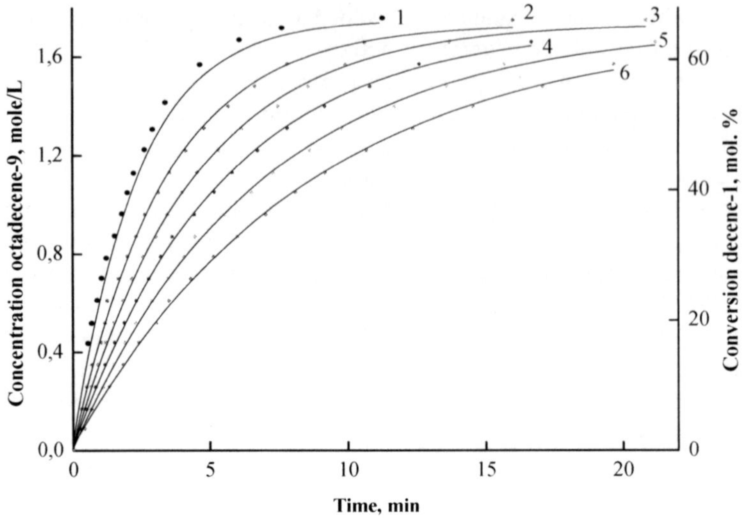

Figure 1 Kinetic curves of 1-decene metathesis at 27°C (molar ratio 1-decene/Mo = 250:1 for each of six consecutive portions). The numerals on the curves correspond to each new 1-decene portion, which was equal to the preceding one, added to the same catalyst

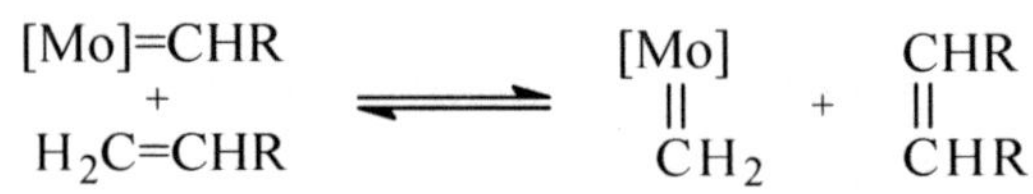

Scheme 2 Equilibrium between internal symmetrical olefins and α-olefins

The experimentally found first kinetic orders, with respect to the α-olefin and the catalyst, correspond to the stoichiometric coefficients for these reactants in equilibrium (Scheme 2). As can be seen in Figure 1, in each particular series of experiments, the next α-olefin portion added to the same catalyst sample reacted more slowly. This is evidently due to catalyst deactivation. It was found experimentally that the catalyst deactivation stops on completion of the metathesis reaction. Two sets of experiments were conducted. In the first set, each catalyzate portion was separated from the catalyst immediately after practical establishment of equilibrium. In the other set, catalyzate portions were held over the catalyst for a while after attaining equilibrium. In both cases, identical kinetic data were obtained. Blockage of the catalyst active sites by impurities from α-olefins cannot be the principal mechanism of deactivation, because additional purification of argon and α-olefins, as well as the use of unreacted α-olefins, led to the same results. This finding indicates that the mechanism of deactivation is inherent, i.e., related with the metathesis reaction itself.

We found that a decrease in the number of active centers ($n_{ac.}$) in the catalyst is adequately described by an exponential function (Equation 1).

$$n_{a.c.} = n_{\circ} \cdot e^{-k_d \cdot \bar{n}_{tot}} \tag{1}$$

where n_0 is the initial number of active centers, which is proportional to the amount of molybdenum (n_{Mo}, mol), k_d is the catalyst deactivation constant (mol Mo)/(mol symmetrical olefin)$_{tot}$, and $\widetilde{n}_{tot}$ is the total amount of the symmetrical olefin per mole of molybdenum atoms. The rate of reaction can be expressed as a differential equation (Equation 2).

$$r = (k_1 \cdot c_\alpha - k_{-1} \cdot c_s) \cdot e^{-k_d \cdot \widetilde{n}_{tot}} \tag{2}$$

where the rate of metathesis (r) is expressed in (mol symmetrical olefin) l^{-1} s^{-1} (mol Mo)$^{-1}$; k_1 and k_{-1} are the rate constants of the forward and reverse reactions, respectively; c_α and c_S are the current concentrations of α-olefin and symmetrical olefin, respectively.

The experimental data agree well with the calculated values obtained by numerical integration of (Equation 2) using the Runge–Kutta method. Note that in all runs the experimental points for the first portion of α-olefin lie above the calculated curve. The deviation is caused by the adsorption of α-olefins on a fresh catalyst; this effect is negligible for subsequent portions.

According to a material balance condition at a constant liquid phase volume, $c_\alpha = c_\alpha^\circ - 2c_S$, where c_α° is the initial α-olefin concentration. Inserting this relation into Equation 2 gives the following equation (Equation 3) for the rate of symmetrical olefin formation, which contains only one of the two current concentrations, namely the symmetrical olefin concentration c_S:

$$r = \{k_1 \cdot c_\alpha^\circ - (2 \cdot k_1 + k_{-1}) \cdot c_s\} \cdot e^{-k_d \cdot \widetilde{n}_{tot}} \tag{3}$$

Treatment of the kinetic data by the procedure described below allowed us to determine values for all the constants that appear into Equation 3. For example, four concentrations of symmetrical olefin (1) 0.48, (2) 0.83, (3) 1.18, and (4) 1.52 mol/l (i.e. at four different values of 9-octadecene concentration approaching the equilibrium value equal to 1.78 mol/l) were selected for the curves in Figure 1. Then, linear dependences were obtained for each of these values of concentration the c_S in the ln(r) $- \widetilde{n}_{tot}$ coordinates (where $\widetilde{n}_{tot}$ is the total amount of symmetrical olefin formed all together in the current and in all the previous experiments of this set) (Figure 2).

The slope of each of the straight lines thus obtained is equal to the catalyst deactivation constant k_d [mol Mo/(mol of symmetrical olefin)$_{tot}$] taken with the opposite sign. This value characterizes the catalyst stability and is the probability for active center decay, equal to $k_d = 3.1 \times 10^{-3}$ mol Mo/(mol of 9-octadecene)$_{tot}$ for this run. Extrapolation of straight lines in ln(r) $- \widetilde{n}_{tot}$ coordinates to the ordinate axis gives the values ln(r) at $\widetilde{n}_{tot} = 0$, i.e., values of the metathesis rate on the non-deactivated catalyst. If r at $\widetilde{n}_{tot} = 0$ is plotted versus the concentration c_s of the symmetrical olefin, the obtained straight line (Figure 3) enables us to find the rate constants of direct and reverse metathesis reactions. The intercept of this straight

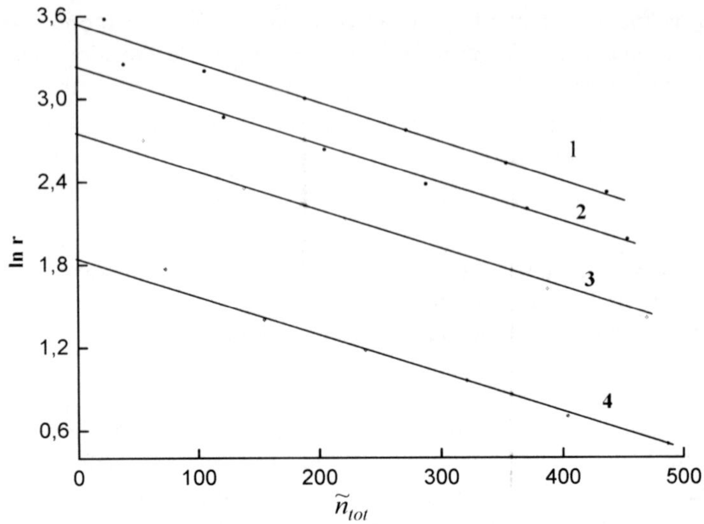

Figure 2 Plot of the logarithmic rate versus the total amount of product 9-octadecene for four different concentrations of the latter: (1) 0.48, (2) 0.83, (3) 1.18, and (4) 1.52 mol/l at 27°C

line with the ordinate gives the rate constant of the direct reaction k_1 multiplied by the initial α-olefin concentration, 1-decene in this case, and the slope of this line is the $(2k_1 + k_{-1})$ sum allowing the rate constant k_{-1} of the reverse reaction to be determined. The intersection point of the straight line with the abscissa determines the equilibrium concentration of the symmetrical olefin, coinciding with the experimental value of 1.78 mol/l.

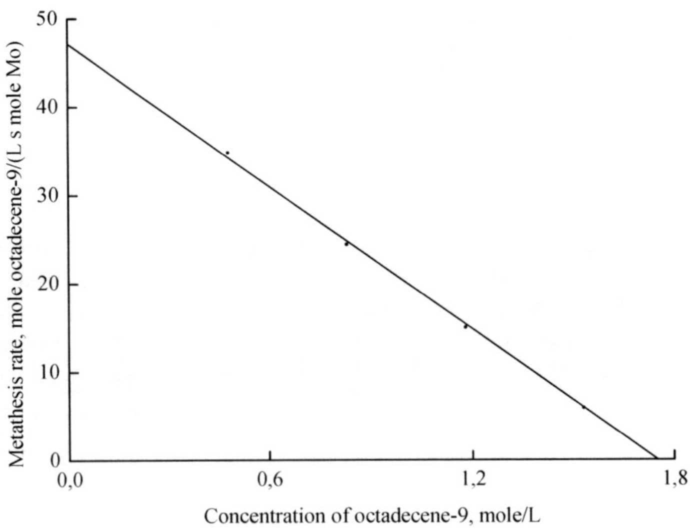

Figure 3 Plot of the rate of decene-1 metathesis, at 27°C, on the non-deactivated catalyst, versus the 9-octadecene concentration

Treatment of the kinetic data on metathesis of 1-decene, 1-octene, and 1-hexene according to the above procedure leads to the constants listed in Table 1.

Table 1 Rate constants for the α-olefin metathesis reaction and the deactivation constants

Olefin	**k_1 (s^{-1} mol^{-1})**		**k_{-1} (s^{-1} mol^{-1})**		**$k_d \times 10^{-3}$ (mol Mo/mol sym. Olefin)**	
	T = 27°C	T = 50°C	T = 27°C	T = 50°C	T = 27°C	T = 50°C
1-Decene	8.8	35	8.5	23	3.1	5.3
1-Octene	10	39	9.6	26	2.5	4.1
1-Hexene	14	50	13	33	2.0	3.2

Comparing the rate constants of the direct and reverse reactions given in Table 1 for various α-olefins, we may conclude that the constants increase with decreasing molecular mass of the olefin and, therefore, the α-olefins have unequal reactivity. However, expressing the metathesis rate (Equation 4) through the α-olefin conversion, substituting $c_s = \frac{\alpha}{2} \cdot c_\alpha^\circ$

$$\frac{d\alpha}{dt} = \left(k_1 \cdot \frac{2}{c_\alpha^\circ} \cdot c_\alpha - k_{-1} \cdot \frac{2}{c_\alpha^\circ} \cdot c_s \right) \cdot e^{-k_d \cdot \tilde{n}_{tot}} \tag{4}$$

where α is the conversion and recalling that the number of moles of α-olefins in 1 l is different, that is 5.28, 6.37, and 8.00 for 1-decene, 1-octene, and 1-hexene, respectively, we see that the values of $k_1^* = k_1 \cdot \frac{2}{c_\alpha^\circ}$ and $k_{-1}^* = k_{-1} \cdot \frac{2}{c_\alpha^\circ}$ are practically identical (Table 2). Thus, the apparent difference in the reactivity of α-olefins is due to the different concentrations of these substrates defined by the ratio of the density d_α to the molar mass M_α ($c_\alpha^\circ = d_\alpha/M_\alpha$), consequently, 1-decene, 1-octene and 1-hexene have practically the same reactivity in the homometathesis reaction.

Table 2 Reduced rate constants for the α-olefin metathesis reaction

Olefin	**k_1^* (l/(mol s mol Mo))**		**k_{-1}^* (l/(mol s mol Mo))**	
	T = 27°C	**T = 50°C**	**T = 27°C**	**T = 50°C**
1-Decene	3.3	13.3	3.2	8.7
1-Octene	3.1	12.2	3.0	8.2
1-Hexene	3.5	12.5	3.3	8.3

3.2 Ternary Olefin Metathesis Catalytic Systems ($MoCl_5/SiO_2$–Me_4Sn–ECl_4, E = Si or Ge)

The catalytic system was formed in the following manner: The interaction of molybdenum pentachloride with the surface hydroxyl groups of silica gel at 80°C occurred with the release of two HCl molecules per each $MoCl_5$ molecule (Scheme 3). In this way, $MoCl_5$ was completely bound to the surface of silica gel.

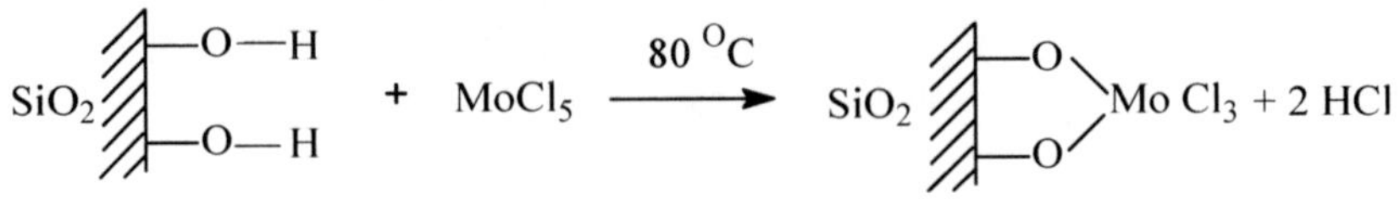

Scheme 3 Interaction of $MoCl_5$ with the surface hydroxyl groups of silica gel

It is believed that the fraction of molybdenum atoms bound to three oxygen atoms of silica gel is negligibly small. A temperature of 170°C is required for the release of three HCl equivalents. At this temperature for immobilization of $MoCl_5$ on the surface of silica gel, the latter is inactive in the metathesis reaction. It is of interest that an almost inactive catalyst was formed upon the immobilization of $MoOCl_4$ on the surface of silica gel with the release of 2 eq. of HCl.

According to mass-spectrometric data, only trimethylchlorostannane (Me_3SnCl) was formed during interaction of Me_4Sn with MoCl5 immobilized on the surface of silica gel. In this case, as determined by gas-chromatographic analysis with consideration of a calibration factor, the amount of trimethylchlorostannane was close to 2 eq. with respect to $MoCl_5$. Thus, the alkylation stage can be represented as in Scheme 4.

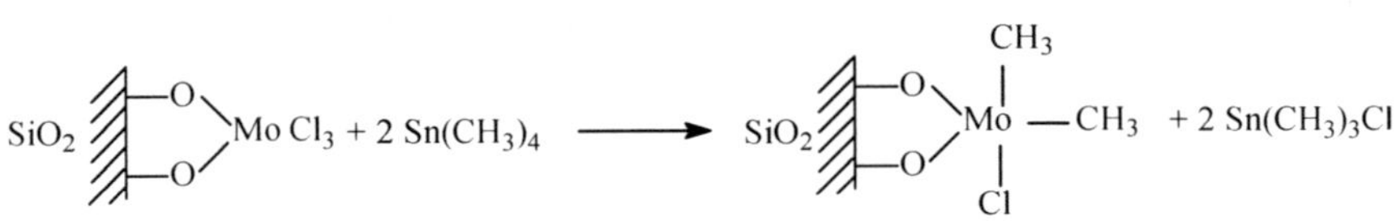

Scheme 4 Interaction of Me_4Sn with $MoCl_5$ immobilized on the surface of silica gel

The dimethyl derivative spontaneously decomposed by α-H elimination to form a primary molybdenum carbene active center and methane (Scheme 5).

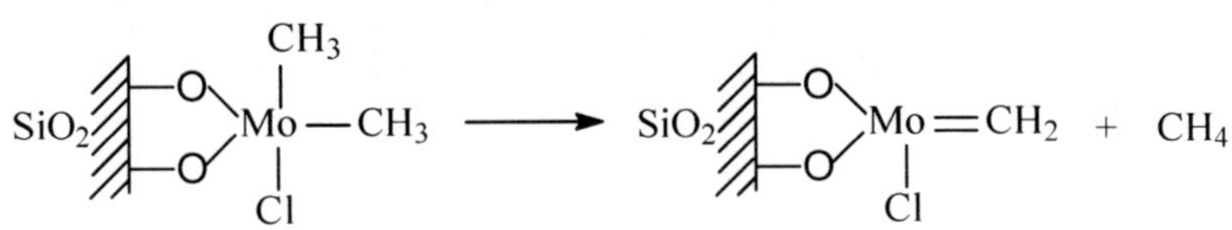

I

Scheme 5 Formation of molybdenum carbene active center

Determination of the number of active centers is based on the fact that the primary active center **I** reacts with a symmetrical olefin to form a secondary active center and the α-olefin. For example, 1-octene was formed by interaction with 7-tetradecene; the amount of 1-octene can be easily determined using octane as an internal standard (Scheme 6).

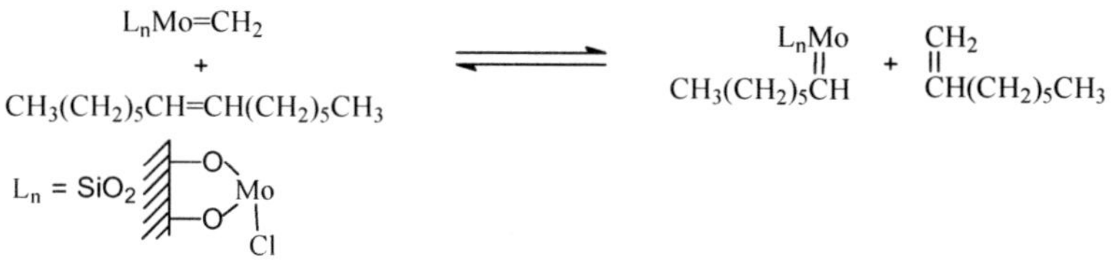

Scheme 6 Formation of 1-octene by interaction of 7-tetradecene with the molybdenum carbene

Because 7-tetradecene was taken in a large excess over $MoCl_5$, the chemical equilibrium of Scheme 6 was strongly shifted to the right, toward 1-octene formation. Thus, it is believed that the amount of 1-octene practically corresponds to the number of primary molybdenum carbene active centers. The GLC analysis demonstrated that the proportion of these centers in the total number of molybdenum atoms in $MoCl_5$ was 5.1% ± 0.3% at 27°C or 6.3% ± 0.3% at 50°C. Evidently, such a low number of active centers was related to the fact that the α-elimination process, which leads to the formation of a molybdenum carbene center and methane (Scheme 5), occurred to a lesser degree than the reductive elimination process giving ethane and inactive trivalent Mo (Scheme 7).

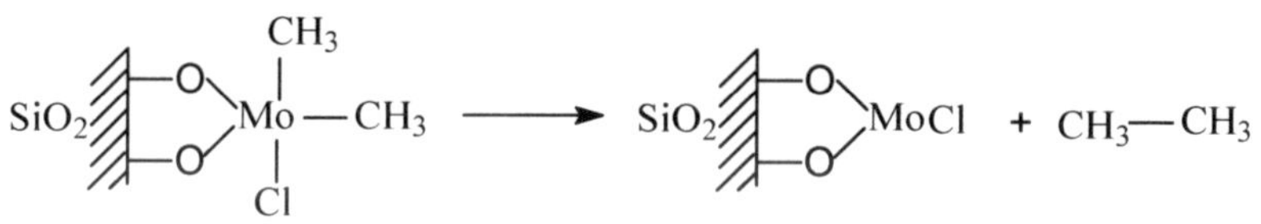

Scheme 7 Reductive elimination leading to inactive trivalent Mo

The formation of methane and ethane was determined using chromatography–mass spectrometry. The ratio between the amounts of methane and ethane was 0.07, which is consistent with Schemes 5 and 7 and the data concerning the determination of the number of active centers.

As noted above, the constant of deactivation characterizes the stability of the catalyst and equals the probability of decay of the active centers. We found the true values of deactivation constants as expressed in (mol active centers)/(mol symmetrical olefin)$_{tot}$. The reciprocal of k_d shows the total number of product (symmetrical olefin) molecules formed at an active center of a catalyst before its decay. Table 3 summarizes the true values and the reciprocals of the deactivation constants. It can be seen that the probability of decay of an active center increases with temperature and exhibits maximum and minimum values for 1-decene and 1-hexene, respectively. Taking into account that the mechanism of deactivation is related to the metathesis reaction, we assume that the mechanism of deactivation can be represented by Scheme 8. There is a probability of the asynchronous degradation of the metallacyclobutane intermediates with the formation of inactive hydride forms like **II** and **III**.

It is evident that the constant of deactivation increased on going from 1-hexene to 1-decene because the probability of the asynchronous degradation of metal cyclobutane increases with increasing α-olefin chain length. The longer the carbon chain, the greater the number of its vibrational degrees of freedom; therefore, it more strongly destabilizes the cyclic transition state. An increase in the temperature leads to the same result.

Table 3 Catalyst deactivation constants and reciprocal values

α-Olefin	T, °C	$k_d \times 10^4$ (mol active centers)/(mol symmetrical olefin)$_{tot}$	$1/k_d$ (mol symmetrical olefin)$_{tot}$/(mol active centers)
1-Decene	27	1.6	6,250
	50	3.3	3,000
1-Octene	27	1.3	7,700
	50	2.6	3,850
1-Hexene	27	1.0	10,000
	50	2.0	5,000

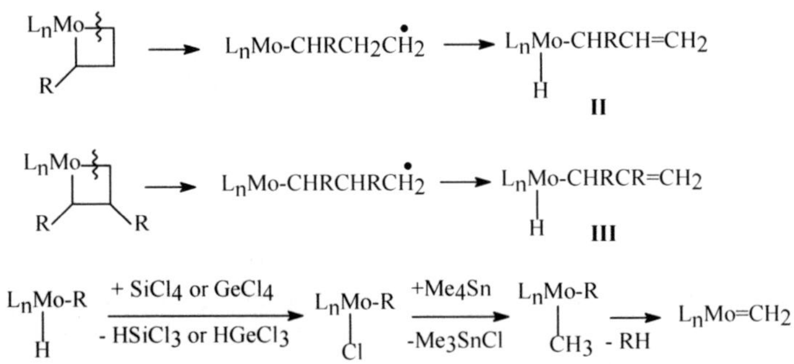

Scheme 8 Mechanism of deactivation-reactivation

Obviously, the reactivation mechanism (Scheme 8) is related to the fact that silicon chloride or germanium chloride reacts with an inactive molybdenum hydride species by substituting chlorine for a hydrogen atom. In this case, silicon chloride or germanium chloride was converted into hydride forms. The formation of trichlorosilane hydride was demonstrated by mass spectrometry. The mass spectrum of the products of the interaction of $SiCl_4$ with a deactivated catalyst exhibited peaks due to ions with m/z = 99 and 101, which are characteristic of

$HSiCl_3$. These ions were practically absent from the mass spectra of parent $SiCl_4$ and the products of the interaction of $SiCl_4$ with the undeactivated catalyst. The ions $^{28}SiH^{35}Cl_2^+$ and $^{28}SiH^{35}Cl^{37}Cl^+$ made the main contribution to the intensities of signals with $m/z = 99$ and 101. Trace amounts of ions with $m/z = 99$ and 101 in the parent $SiCl_4$ were due to the isotopic fragments $^{29}Si^{35}Cl_2^+$ and $^{29}Si^{35}Cl^{37}Cl^+$. Thus, mass-spectrometric data unambiguously demonstrate the presence of silane hydride in the products of the interaction of $SiCl_4$ with the deactivated catalyst.

Thus, we were the first to develop self-reactivating ternary catalytic systems for metathesis ($MoCl_5/SiO_2$– Me_4Sn–ECl_4; E = Si or Ge), in which the active center was reactivated in the course of metathesis. We determined the number of active centers in heterogeneous catalysts and proposed mechanisms for their formation, deactivation, and reactivation with consideration of the role of each particular component in the binary and ternary catalytic systems for metathesis.

Acknowledgments This work was supported by the Russian Foundation for Basic Research (project № 08-03-00007), the Presidium of the Russian Academy of Sciences (basic research program "Development of Methods for the Preparation of Chemicals and New Materials," subprogram "Development of the Methodology of Organic Synthesis and the Design of Compounds with Valuable Applied Properties," coordinator, Academician V.A. Tartakovskii), the Department of Chemistry and Materials Science of the Russian Academy of Sciences (basic research program "Theoretical and Experimental Studies of the Nature of Chemical Bonding and the Mechanisms of the Most Important Chemical Reactions and Processes," coordinator, Academician O.M. Nefedov).

References

[1] Banks RL, Bailey GC (1964) Ind Eng Chem Prod Res Dev 3:170
[2] Grubbs RH (ed.) (2003) Handbook of metathesis, Vol. 2. Wiley-VCH, Weinheim
[3] Bykov VI, Butenko TA, Finkelshtein ESh, Henderson PT (1994) J Mol Catal 90:111
[4] Bykov VI, Butenko TA, Kelbakiani LV, Finkelshtein ESh (1996) Dokl Akad Nauk 349(2):198; Dokl Chem Eng Ed 349(1–3):168
[5] Bykov VI, Butenko TA, Kelbakiani LV, Finkelshtein ESh (1996) Izv Akad Nauk Ser Khim 8:2127
[6] Bykov VI, Finkelshtein ESh (1998) J Mol Catal 133:17
[7] Bykov VI, Butenko TA, Petrova EB, Finkelshtein ESh (1999) Tetrahedron 55:8249
[8] Bykov VI et al. (1991) USSR Inventor's Certificate No. 1666177 Bull Izobret:28

PART II. CONCEPTS AND CHALLENGES IN SUSTAINABLE CHEMICALS SYNTHESIS

Ring-Closing Metathesis Synthesis of Medium and Large Rings: Challenges and Implications for Sustainable Synthesis

Sebastien Monfette, Deryn E. Fogg*

Center for Catalysis Research and Innovation, Department of Chemistry, University of Ottawa, Ottawa, ON, Canada K1N 6N5

*E-mail: D.Fogg@uotawa.ca

Abstract Synthesis of medium-sized and macrocyclic rings by ring closing metathesis (RCM) reactions of dienes involves challenges not found in synthesis of the more common five- and six-membered rings. This review discusses factors that determine the probability and efficiency of cyclization, and experimental methods that have been used to increase selectivity for RCM products, with specific reference to the concentrations at which RCM can be achieved. These issues have important implications for the environmental and economic sustainability of large-scale synthetic processes utilizing RCM for assembly of rings larger than six members.

Keywords Ring-closing metathesis · Cyclization · Oligomerization · Ring-chain equilibria · Gauche interactions · Non-covalent interactions · Metal complexes

1 Introduction

Ring-closing metathesis (RCM) of dienes has transformed synthetic approaches to the assembly of carbocyclic and heterocyclic molecules [1–4]. However, while high efficiencies can be attained in synthesis of the common five- and six-membered rings, this is often not the case for medium and large ring sizes [5]. Improving control over the selectivity for intramolecular, versus intermolecular, reaction is essential for improving process sustainability and economics in synthesis of such targets by RCM methods. Here we present an overview of current understanding. Issues discussed include the thermodynamic and kinetic factors that govern this selectivity, as well as experimental parameters that influence the probability of liberating RCM products through backbiting equilibria. Also discussed are experimental methods that have been used to improve selectivity for RCM products in synthesis of medium and large rings.

V. Dragutan et al. (eds.), *Green Metathesis Chemistry: Great Challenges in Synthesis, Catalysis and Nanotechnology*,

2 General Considerations in Cyclization Reactions

The competition between intramolecular cyclization and intermolecular oligomerization reactions, a central issue in the synthesis of medium and large rings, has been recognized since the pioneering work of Ruggli, Ruzicka and Ziegler in the early part of the last century [6]. The parameters involved have been studied for a wide range of bifunctional molecules and mechanisms: both thermodynamic and kinetic factors come into play, though a major distinction between the majority of the classical systems and those to be considered in this review lies in the fact that the former are typically irreversible, stoichiometric reactions. The yield and distribution of metathesis products can be dramatically affected by the operation of equilibria at certain stages, as discussed in the following sections.

2.1 Thermodynamic Considerations

The fundamental challenge in synthesis of medium and large rings can be analyzed in terms of the Gibbs–Helmholtz equation, $\Delta G = \Delta H - T\Delta S$. The enthalpic picture is straightforward: the major contributor is ring strain, arising principally from imperfect staggering and transannular strain between atoms forced into proximity from opposite sides of the ring. The positive enthalpic term reaches a maximum in the medium-ring regime, though the precise position of this maximum with respect to ring size depends on the nature and hybridization of the ring atoms. Because entropic gains must pay the enthalpic penalty associated with cyclization, the success of any attempted cyclization will normally be determined by the extent to which the entropy term $T\Delta S$ can be biased in its favour. Figure 1 summarizes key entropic contributions.

The majority of cyclization reactions involve loss of some leaving group, and thus conversion of n precursor molecules into 2n products. An entropic bias toward cyclization results from the corresponding increase in the number of translational and rotational degrees of freedom. Oligomerization, as shown, is "revenue-neutral" in converting n molecules into n products: this factor, as well as the greater mobility (i.e. translational and overall rotational entropy) associated with molecules of smaller size, favours cyclic products over oligomers. These advantages are offset by the lower conformational flexibility of atoms within a cyclic backbone, particularly for medium rings, relative to atoms present in the backbone of a chain. The relative weighting of these two entropic factors is sensitive to concentration, because translational mobility declines with increasing viscosity, while conformational motion is much less affected. Elevated temperatures can favour cyclization by reducing viscosity and maximizing thermal motion: they also serve to reinforce any entropic bias in the $T\Delta S$ term. Two experimental variables of particular importance in creating a thermodynamic bias toward cyclic products are therefore high temperatures and high dilutions. A further, more minor entropic effect is the probability of encounter between reactive endgroups: although high for common ring sizes, this declines as the "monomer" backbone length increases.

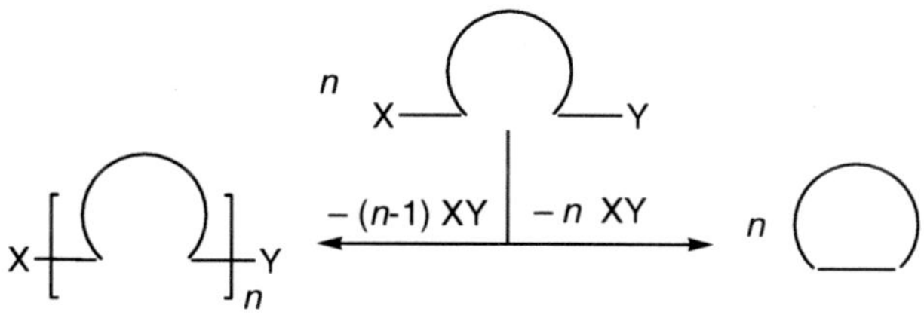

			Favours:
n	Number of molecules	$2n$	rings
high	Conformational motion	low	oligomers
low	Overall rotation	high	rings
	Translational motion:		
low	• in dilute solution	high	rings
low	• neat	low	neither

Figure 1 Entropic balance sheet for cyclization versus oligomerization. For simplicity, only linear oligomers are shown

2.2 *Kinetic Factors*

The importance of high dilution in creating a kinetic bias toward unimolecular cyclization, over bimolecular oligomerization, is central to the Ziegler concept of "infinite dilution", in which the substrate is infused into a relatively large volume of solvent at a rate equal to the rate of cyclization. This strategy, or less rigorous but more convenient variations thereon, is a standard tool for the synthesis of medium and macrocyclic rings. More intrinsic factors affecting rates of cyclization, versus oligomerization, emerge from the relative energies of activation for the two processes. The relevant factors parallel those discussed in the preceding section: in particular, maximizing translational and (overall) rotational motion is key to intramolecular rate accelerations [7]. For a bimolecular reaction, achieving the (mono-molecular) transition state reduces the number of independent species in the system, with a consequent loss of three translational and up to three rotational degrees of freedom. No such loss is incurred for the corresponding first-order, intramolecular reaction. Introduction of conformational constraints can be valuable in lowering the transition state energy for cyclization, via modulation of the enthalpy and/or entropy of activation. Specific examples of this strategy are discussed in Section 4, with an examination of their influence on the dilutions required for successful cyclization.

2.3 Assessing the Probability of Cyclization

The Jacobson-Stockmayer (JS) theory of macrocyclization equilibria describes the distribution of (strainless) cyclic and linear polymers at equilibrium in concentrated solutions [8]. Later modifications address the equilibrium yields and distributions of cyclic oligomers obtained under dilute conditions [9], as well as the effect of ring strain on the ease of cyclization [10–12]. A central concept in JS theory is the critical monomer concentration (or "cut-off point"), which in polymer chemistry defines the equilibrium concentration of monomer below which only cyclics are present, and above which the concentration of cyclooligomers remains constant, and linear chains emerge [13]. Below the critical monomer concentration, a distribution of cyclooligomers is obtained, the ratio of which depends on their respective free energies of formation. Because it does not distinguish between rings of different sizes, the critical monomer concentration has little predictive value in organic synthesis, where selectivity for a particular product (typically the smallest ring) is desired.

More generally useful in organic synthesis is the concept of effective molarity (EM) [6, 10, 14, 15]. The thermodynamic EM value, EM_T, is equivalent to the macrocyclization equilibrium constant of polymer chemistry [16], and corresponds to the concentration at which $K_{intra}/K_{inter} = 1$. Mandolini has noted the fundamental similarity between the concepts of effective molarity, the older term effective concentration, and macrocyclization equilibria [15]. EM values are valuable as an empirical predictor of the bias toward cyclization, but do not represent physically real concentrations [6]. In contrast, the effective concentration quantifies the bias toward intra- versus intermolecular reaction by evaluation of a physically real concentration of one reacting entity, as experienced by its partner. Because these values are not readily determined, however, the term effective concentration is less commonly used than the EM value. For that matter, few EM_T values have been reported, owing to the dominance of studies focusing on irreversible reactions: much more common are kinetic EM values, in which the efficiency of cyclization is assessed from the ratio of the rate constants for the intramolecular and intermolecular reactions. The kinetic EM value is the concentration at which $k_{intra}/k_{inter} = 1$, higher values indicating a greater bias toward intramolecular reaction (i.e. preference for the first-order cyclization process, over the second-order process of chain growth).

Comparison of EM and EM_T values for lactones reveals that while the trends tend to track together, kinetic EM values are consistently higher up to the limit of "strainless" large rings [6]. Where no strain energy is present in the transition states, the kinetic and thermodynamic EM values coincide [10]. Thus, while the difference between EM and EM_T diminishes as the difference in strain energy between the transition state and cyclic product decreases [9], higher dilutions are required for cyclization under thermodynamic conditions for any ring where strain is present. This is particularly acute in the medium-ring regime, in which EM_T

values are much lower than the corresponding kinetic values. Both, however, are much lower than the corresponding values for large rings, consistent with the need for higher dilutions in synthesis of medium rings [10].

Originally used to assess the ease of cyclization reactions [6, 10, 14, 15], EM values have been used more recently to evaluate the probability of self-assembly in other contexts [17, 18], including in dynamic combinatorial chemistry [19, 20] and multivalent interactions [21–23]. Limitations to EM_T values in cyclodepolymerization have been reported [24]. While useful as a general guide, reported EM values should not be expected to show great precision in RCM. Literature values predominantly describe saturated compounds (for the impact of incorporation of sp^2 centers on ring strain, see Section 4); as well, the large difference between kinetic and thermodynamic EM values means that very different behaviour may be seen for a given substrate using different catalysts or conditions, depending on whether the reaction operates under kinetic or thermodynamic control. This issue is considered further in the next section. An open question, moreover, remains the potential of the metal complex to perturb strain energies in the cyclic transition states. Although kinetic EM values have been shown to be largely independent of the nature of the reacting groups, reaction mechanism, and solvent for "strainless" rings, for which EM is entropically controlled [25], the strain-free condition is clearly violated in the medium and large-ring regime.

3 The Operation and Role of Equilibria in RCM

First described by Villemin nearly 3 decades ago [26], RCM has risen to astonishing prominence in organic synthesis over the last decade, owing largely to the development of relatively robust, well-defined catalysts that enable controlled reaction [27], RCM is now one of the most important synthetic methods in use for the construction of cyclic molecules [28, 29]. Examples abound in recent reviews of important macrocyclic [2, 3, 29–35] and medium-ring [28, 30, 36–39] targets prepared by RCM, including natural products [40–43], compounds of biological and medicinal relevance [37, 44, 45], and topologically interesting molecules, including "molecular machines" [32, 46–48]. As the discussion of Section 2 suggests, however, RCM competes with oligomerization reactions (often described as "acyclic diene metathesis", ADMET), and observation of oligomers (in macrolide synthesis) [26] or cyclodimers (in attempted synthesis of cyclooctanoids) [49] dates back to the earliest reports. A distinction from the majority of classical cyclization reaction emerges from the fact that olefin metathesis is, in principle, fully reversible. This has important implications for the synthesis of medium and large rings. The following sections examine the extent of reversibility as a function of the nature of the substrates, the catalysts, and reaction conditions. The catalysts discussed are shown in Figure 2.

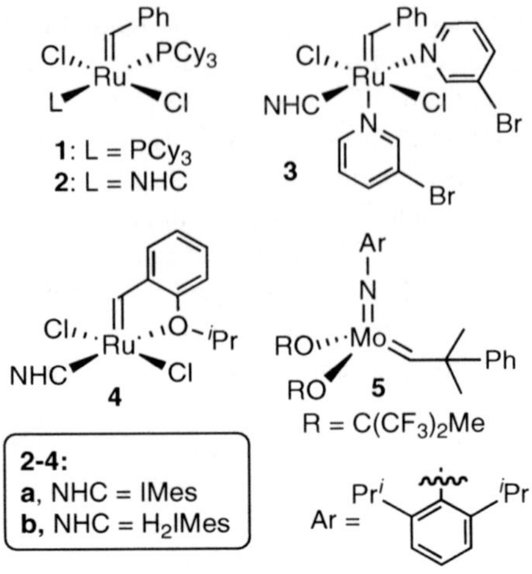

Figure 2 Common Ru and Mo metathesis catalysts. IMes = *N,N′*-bis(mesityl)imidazol-2-ylidene

3.1 Fully Reversible Metathesis

Fully reversible metathesis results in "living" metathesis products, related by a series of Chauvin [2+2] cycloaddition–cycloreversion equilibria [13, 50]. A thermodynamic distribution of olefinic products is therefore obtained. Factors that increase the likelihood of full reversibility – i.e. operation according to the standard representation of Figure 3 – include long catalyst lifetimes and/or high catalyst loadings, inhibition of non-metathetical pathways, reactivity sufficient to enable attack at 1,2-disubstituted double bonds (as required for backbiting at internal olefinic sites in the metal-terminated oligomers), and retention of the released olefin in solution. For vinylic α,ω-dienes, retention of the volatile coproduct

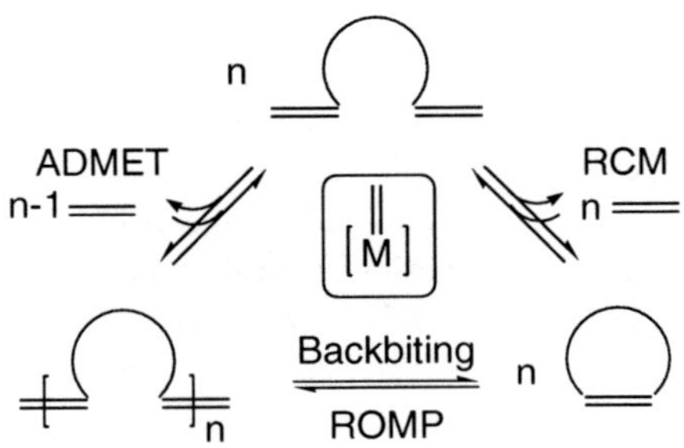

Figure 3 Conventional representation of olefin metathesis: equilibria relate all olefinic species

ethylene is maximized by use of sealed vessels with little headspace. In metathesis of 1,2-disubstituted, tri- or tetrasubstituted dienes, the lower volatility and increased solubility [51] of the olefinic coproducts increases the probability that they will participate in ring-opening–ring-closing equilibria (including reformation of diene) even under conventional synthetic conditions.

While the ring-chain equilibrium relating oligomeric and RCM products can be exploited to "correct" initial product distributions (see later), full reversibility has its drawbacks. Retention of diene in solution limits the driving force for reaction, and increases the proportion of catalyst lifetime spent in unproductive reformation and re-reaction of diene. (Relay RCM, illustrated in Figure 4 [52], represents a clever exception to this rule: here the olefin coproduct is cyclopentene, the resistance of which to metathesis enables irreversible reaction. Liberation of cyclopentene favours RCM both entropically and enthalpically, though the entropic driving force remains much lower than that attained by volatilization.) Where ethylene is retained in solution, productivity is also adversely affected by conversion of Ru-alkylidene species into shorter-lived and less reactive methylidene species [53–55]. (The Mori and Diver groups have noted, however, beneficial effects of ethylene pressure in enyne metathesis [56, 57].) Finally, retention of ethylene in solution for reactions carried out in sealed small vessels can lead to significant discrepancies in reaction rates and product distribution, relative to the identical reactions in open vessels.

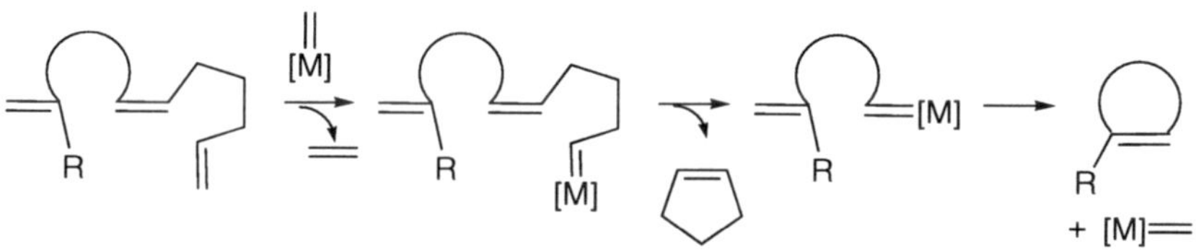

Figure 4 Relay RCM

3.2 Limitations on Reversibility in RCM Reactions

Completely unimpeded reversibility in metathesis of α,ω-dienes, as illustrated in Figure 3, is rare. Common restrictions include formation of by-products that cannot reenter the metathesis cycle by reason of efficient volatilization or low reactivity (see above), and inhibition of backbiting by rigidity, ring strain, catalyst deactivation prior to establishment of equilibrium, low catalyst reactivity toward internal olefins, or non-metathetical reaction pathways such as isomerization. These issues are discussed below. We explicitly neglect, however, the particular case of irreversible RCM associated with very high EM substrates such as diethyldiallyl malonate.

One of the most common limitations on full reversibility in olefin metathesis is the irreversible liberation of olefin coproduct. Despite growing examples of RCM of tri- and tetrasubstituted olefins, the majority of substrates studied contain unsubstituted α,ω-dienes, and therefore eliminate ethylene as co-product. Because RCM is typically carried out at elevated temperatures in a vessel open to an atmosphere of N_2 or Ar, the low solubility of ethylene in organic solvents [58] normally results in its rapid, irreversible loss from solution. Volatilization can be accelerated further by deliberately sparging reaction solutions: Reiser and coworkers have noted that this has a beneficial effect on yields [59]. The entropic gain adds greatly to the driving force for both oligomerization and RCM: because regeneration of the starting α,ω-diene is impossible once the ethylene has escaped, the metathesis equilibria simplify to the pathways shown in Figure 5. The only equilibrium then still operational is that between oligomeric and RCM products (provided that the catalyst is competent to effect both ROMP and backbiting), and its concentration-dependence can be exploited to manipulate the product ratios, as discussed below [60].

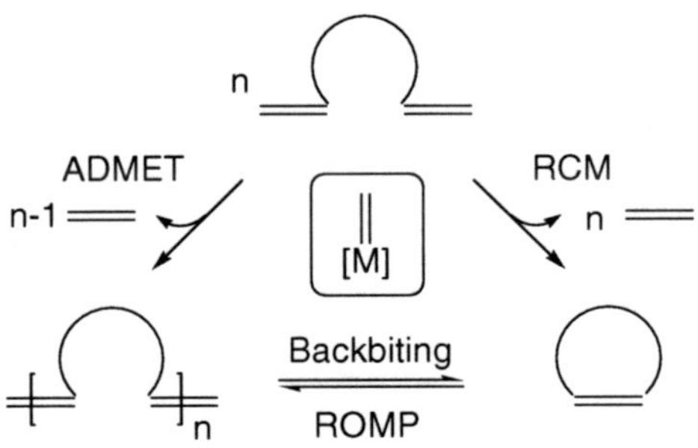

Figure 5 Olefin metathesis pathways invoking irreversible loss of ethylene

Decomposition of the metal alkylidene complex required to mediate cycloaddition will inhibit establishment of equilibrium product distributions. Catalyst decomposition may account for the somewhat surprisingly poor performance of the Schrock catalyst **5** to induce backbiting of ADMET oligomers, which their high reactivity should be competent to address [61]. The thermal sensitivity of these group 6 catalysts, as well as their susceptibility to decomposition by traces of oxygen or water, may be responsible. The Grubbs-class Ru catalysts, while less fragile, are also not very long-lived under typical RCM conditions (usually refluxing CH_2Cl_2 or toluene). The methylidene intermediate is particularly vulnerable [53, 62, 63]. If, furthermore, decomposition occurs via a unimolecular pathway, rates of catalyst depletion will be unaffected by the high dilutions required to synthesize medium and large rings, while rates of metathesis will suffer. Deactivation of the catalyst prior to complete reaction will yield a non-equilibrium mixture of the constituents shown in Figure 3. High catalyst loadings (>10 mol%) are often used to effect high conversions, particularly in demanding transformations in (e.g.) total synthesis [2, 3, 29, 31, 34]. While the catalyst precursor is not itself immune to decomposition, it is less vulnerable than the 14-electron active species, and a reservoir of unreacted catalyst can therefore be beneficial. In this regard, the low turnon efficiency of **4b** may confer an unexpected advantage.

Non-metathetical pathways can disrupt equilibrium RCM by depleting the concentration of the metathesis-active species, as discussed above, or by altering the product distribution. Among the wide range of non-metathesis pathways promoted by the Grubbs catalysts [64], olefin isomerization is particularly common as an unintended side-reaction [65]. Isomerization of the diene precursor can alter the product ring size, or disrupt RCM completely. Isomerization of ADMET polymers (i.e. double bond migration along the polymer backbone) can likewise result in ring contraction or expansion upon backbiting [66].

Limitations on reversibility can also, however, arise from low catalyst reactivity. High catalyst reactivity is essential to enable the backbiting reaction of a metal endgroup with the 1,2-olefinic sites present in ADMET oligomers (Figure 6). Limitations on the reactivity of 1 toward sterically more congested olefins have been described [67]. While efficient in RCM of vinylic α,ω-dienes, 1 effects cyclodepolymerization [68] (or indeed chain transfer reactions [69]) of ROMP polymers relatively slowly. The low reactivity of 1 toward internal olefins also limits its efficacy in RCM of unsymmetrically substituted dienes in which one end is deactivated by steric or electronic means. Preferential dimerization of such substrates is common even using more reactive catalysts [67, 70–75].

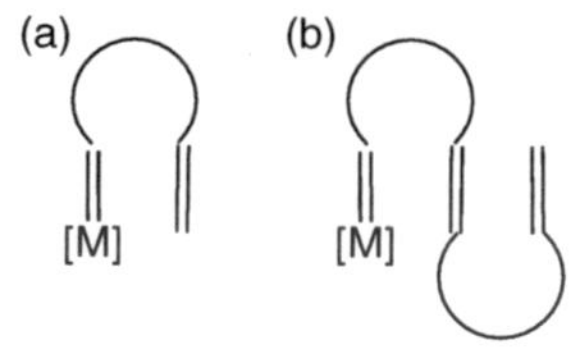

Figure 6 Intermediates involved in (**a**) direct RCM of vinylic α,ω-olefins, versus (**b**) backbiting

Chelation of substrate substituents to the metathesis catalyst can also inhibit reactivity. The Grubbs-class ruthenium metathesis catalysts, while less sensitive to oxygen or water than the group 6 systems, are nevertheless susceptible to deactivation through chelation of (e.g.) carbonyl functionalities [76–79]. Instances of sensitivity to alcohol or amine functionalities have also been reported [80, 81] (see also Section 4.1.2). Premature catalyst deactivation can result in a non-equilibrium mixture of products in RCM of such substrates. In some cases, catalyst productivity can be improved by capping or protecting amine or alcohol groups, or by Lewis acid "protection" of carbonyl functionalities: see Section 4.

3.3 Ring-Chain Equilibria in RCM

The operation of ring-chain equilibria in ROMP chemistry has been recognized since the pioneering work by Hocker 3 decades ago [82]. However, despite the common observation of oligomerization during intended RCM assembly of large or medium rings, and occasional reports of the conversion of oligomers to RCM products (vide infra), the generality and implications of these equilibria to RCM has only recently gained attention [60]. Indeed, because oligomerization is viewed as fundamentally inimical to RCM, considerable effort is often invested in limiting it. We recently showed that this effort may be unnecessary. In a detailed examination of the RCM of a range of conformationally flexible dienes (Figure 7) via catalysts **2a** and **3a**, we reported that ADMET was, unexpectedly, kinetically favoured even at high dilutions (5 mM) and elevated temperatures [60]. At short reaction times, the major products in metathesis of these substrates were oligomeric chains of up to 12 repeat units, as indicated by MALDI-MS analysis. Importantly, however, near-quantitative cyclization was effected on longer reaction time (see, e.g., Figure 8). Given the ease with which ethylene is lost from solution under the open-vessel conditions used, we proposed that the dominant mechanism for cyclization of these substrates involved not direct RCM, but a concentration-dependent ring-chain equilibrium, in which the key steps are reinstallation of a metal endgroup on the oligomer, and cyclodepolymerization via backbiting. The oligomers are thus key intermediates en route to the cyclic products. Synthetic protocols designed to promote initial oligomerization were shown to significantly reduce total reaction times, thereby limiting the potential for undesired side reactions, including those triggered by catalyst decomposition.

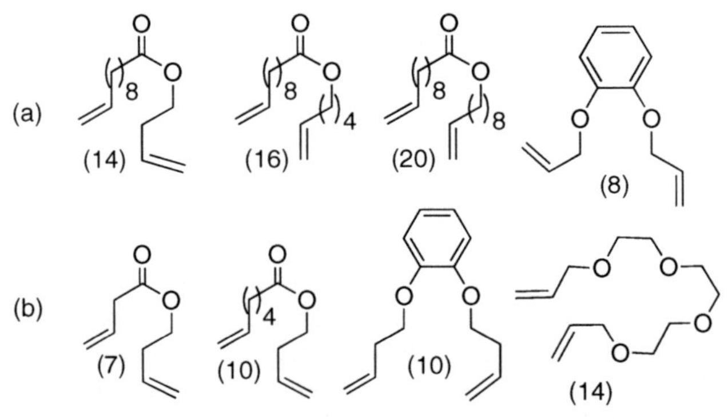

Figure 7 Representative substrates shown to undergo RCM via the concentration-dependent oligomerization-backbiting pathway at (**a**) 5 mM, (**b**) 0.5 mM concentration of diene. Numbers in brackets indicate the ring size of the RCM product obtained

The importance of such "oligomerization-cyclodepolymerization" behaviour in RCM of conformationally flexible dienes may have been masked in earlier work by the nature of the catalysts originally used. As noted above, establishment of these equilibria rests on the ability of the catalyst to effect both living ROMP of low-strain cyclic products, and backbiting of the oligomers. A dual requirement for high reactivity and good lifetimes is evident. While our study demonstrated that **2a**, in particular, was competent in both respects [60], the lower reactivity of the original Grubbs catalyst **1** limits it to essentially irreversible oligomerization or RCM, while more sensitive catalyst species such as 5 may be deactivated before equilibrium can be achieved. Slow rates of attack on internal, 1,2-disubstituted olefinic sites would account for reports of ADMET dimerization using a comparatively unreactive catalyst such as **1**, and subsequent production of macrocycles by treating the isolated dimer with a more reactive catalyst [67, 70, 71, 73]. Of interest, pseudohalide analogues of catalyst **3**, in which the chloride ligand(s) are replaced by one or more aryloxide ligands, do not share this strong kinetic bias toward oligomers, but enable rapid RCM [83]. Whether this represents improved selectivity for RCM over ADMET, or simply faster backbiting, is not yet clear.

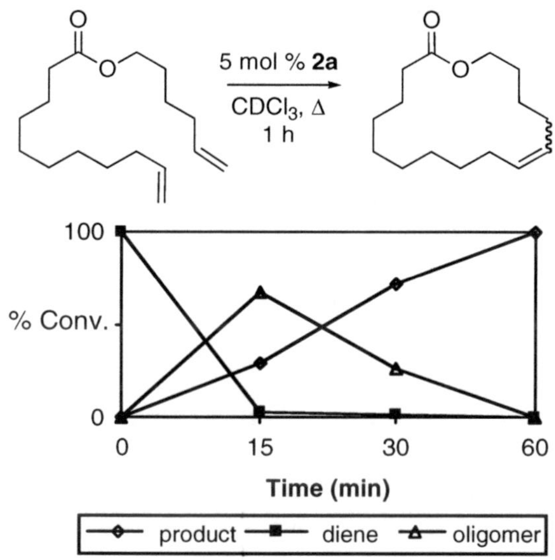

Figure 8 Product speciation in a representative RCM reaction as a function of time. Dropwise addition, [S] << 5 mM, GC-FID analysis

Where the ring-chain equilibrium is viable, the product distribution is governed by concentration. Thus, high concentration favours ADMET, and high dilutions favour RCM [60]. Suitable concentrations for the substrates shown in Figure 7 were 5 and 0.5 mM. Higher cyclooligomers were formed as coproducts even on doubling these concentrations. In the extreme of 700 mM, Hodge and Kamau demonstrated that macrocyclic olefins containing up to 84 ring atoms could be induced to undergo efficient, entropically-driven ROMP [84].

4 Strategies for Promoting RCM over Oligomerization in Synthesis of Medium and Large Rings

Cyclization of medium and large rings remains a challenge to sustainable synthesis, for which no general solution yet exists. Synthesis in dilute solution is almost invariably required, for the reasons discussed above [29]. The ring-chain equilibrium can offer an opportunity to correct RCM reactions initially carried out at too high a concentration: where backbiting is possible, dilution can increase yields of desired cyclic species. Equilibration to desired "cyclomonomer" species can be prohibited, however, by a number of factors, including low catalyst activity, catalyst deactivation, substrate rigidity, and excessive ring strain, as discussed above. Where the target rings can be obtained from low-EM substrates, the large volume of solvent required is a major impediment to sustainable and economical synthesis. The scale of the problem is highlighted by a recent GSK report estimating that solvents account for >85% of the mass of chemicals used in pharmaceutical manufacturing, 60% of the overall energy used to produce active pharmaceutical ingredients, and 50% of the total post-treatment greenhouse gas emissions [85, 86]. Indeed, solvent use is the major contributor to the notoriously high E factors ubiquitous in pharmaceutical and fine-chemicals manufacturing (E = ton of waste per ton of product) [87, 88]. A number of strategies have been developed in an attempt to promote cyclization over dimerization and/or oligomerization in RCM synthesis of medium and large rings. Below we summarize such approaches, with representative examples, and analyze their impact on the dilutions required.

4.1 Conformational Pre-organization Through Substrate Gearing

The value of conformational pre-organization in macrocyclization reactions has long been recognized, and the subject was recently reviewed [89]. A "gearing" element that exerts a bias toward RCM can be created by, inter alia, the gem-dialkyl effect, appropriately positioned protecting groups, introduction of trigonal heteroatom or unsaturated sites, or non-covalent interactions. In certain cases, these effects permit RCM at concentrations significantly higher than those typically required (0.5–10 mM) for cyclization of medium and macrocyclic rings; vide infra. Among the many studies aimed at incorporating gearing elements to facilitate RCM, however, few explicitly examine the impact on concentration. This probably reflects the "target-focused" approach common in small-scale syntheses, where the impact of solvent costs or E-factor constraints is minimal. We note, however, that the issue of concentration merits routine examination even under these circumstances, given the negative impact of high dilutions on both catalyst productivity (see Section 3.2) and RCM yields (i.e. conversions and isolated yields following removal of spent catalyst).

4.1.1 Substituent-induced conformational constraints

Acyclic conformational constraints are notoriously challenging to introduce. The ability of a proximate gem-dialkyl group to promote cyclization has received much attention, and the kinetic and thermodynamic contributions are examined in a recent review [90]. In a much-cited example of this effect in RCM, Forbes, Wagener, and coworkers reported near-quantitative cyclization of 6b, bearing two allylic gem-dimethyl groups, to afford the seven-membered ketone **7b** (Figure 9) [91]. In comparison, solely ADMET oligomer was observed for the unsubstituted analogue **6a**. While important for common ring sizes, the effects of geminal substitution drop off sharply as the acyclic chain length increases [6, 92], and this strategy has found little success in RCM synthesis of larger rings. Attempted RCM of 11-membered **8**, for example, yielded only oligomers (Figure 9) [91].

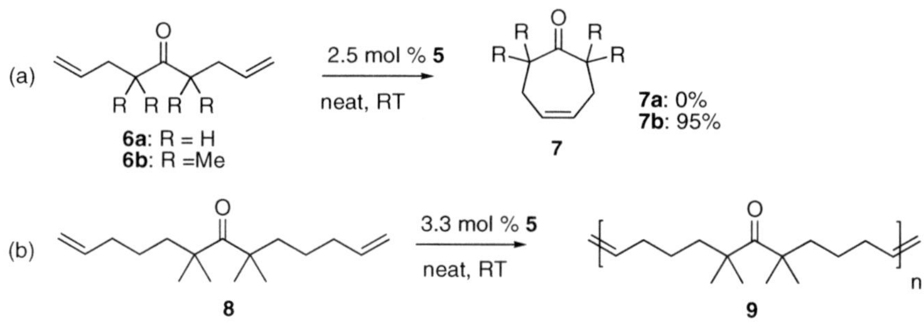

Figure 9 Influence of the gem-dimethyl effect on attempted cyclization to form (**a**) common, (**b**) large rings. Catalyst loading converted from 1 wt% used [91]

In a related observation by Crimmins and coworkers, a nine-membered cyclic ether was obtained in ca. 95% yield using **1** at a diene concentration of 3 mM (Figure 10) [93]. Successful cyclization was attributed to the presence of two gauche interactions between substituents on both sides of the central ether functionality, which was proposed to bring the olefins into close proximity (for other examples, see below). In larger rings, the conformational impact of functionalization is often unpredictable, particularly as the molecular complexity increases. Even the position of the double bonds relative to ancillary functional groups can have a profound impact [94, 95]. In RCM to form the gleosporone framework, for example, the macrolactone could be formed only after inserting an additional methylene group on the "ether" side of the ester unit: this outcome was attributed to gearing effects, but could not be predicted a priori [73]. A number of examples have also been reported in which remote substituents dramatically influence RCM yields [96–100].

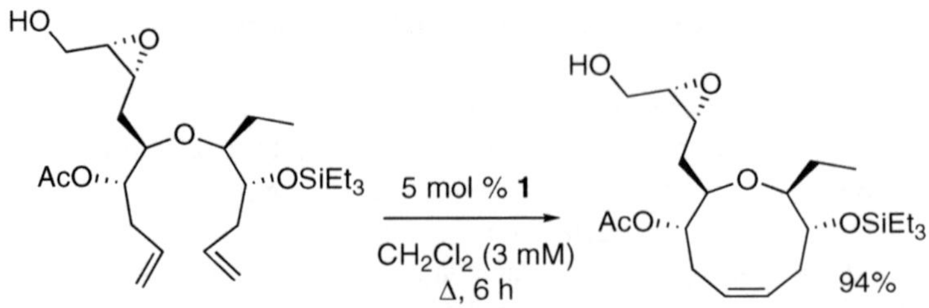

Figure 10 Gauche interactions facilitating cyclization

Small groups can be effective in exerting a gearing effect in conjunction with a rigid element in the diene backbone. During synthesis of the 13-membered macrocycle **11**, Murphy and coworkers found that installation of a methyl group increased RCM yields to 70%, from 15% for its unsubstituted counterpart **10a** (Figure 11; both reactions at 7 mM) [101]. A similar proportion of dimers was observed by GC-MS (30% and 22% for **10a** and **10b**, respectively), the mass balance presumably being accounted for by larger, involatile oligomers.

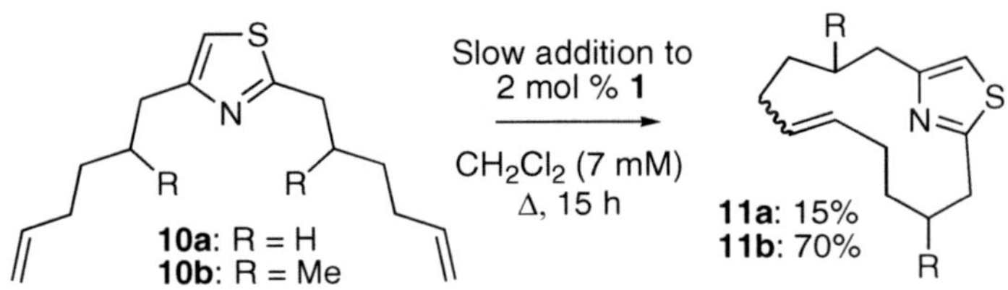

Figure 11 Effect of backbone methyl groups near a rigid element

Appropriately sited metal substituents can likewise be suitable gearing elements [80, 102, 103]. Licandro and coworkers reported the construction of seven-, eight-, and nine-membered heterocyclic rings (**13a–f**, respectively; Figure 12), using a chromium alkylidene substituent a to the acyclic nitrogen group [103]. While incorporation of a $Cr(CO)_5$ unit enabled synthesis of **13a** and **13b** in 74–100% yield (2–10 mol% 1) at an impressively high concentration of 280 mM, the nine-membered ring could not be obtained. Interestingly, use of a $Cr(CO)_4$ alkylidene substituent instead gave access to all three ring sizes (**13d–f**) in 40–74% yield, behaviour attributed to chelation of chromium by the pendant morpholine group. Also relevant, however, may be the higher dilutions used (14 mM), and the increase in catalyst loading (10–40 mol% **1**). Removal of the alkylidene group was not undertaken. Related work has utilized cobalt [80] or iron [102] carbonyl complexes as gearing elements complexed to alkenyl or alkynyl units. These were found to aid in synthesis of structurally more complex seven- to nine-membered rings. The Young group reported the removal of the p-bound $Co_2(CO)_6$ unit using ceric ammonium nitrate, albeit in low yields (24–31%) [80]. Paley and coworkers described decomplexation of $Fe(CO)_3$ from a 1,3-diene moiety in ca. 90% yield, *via* treatment with $FeCl_3$ in acetonitrile at –15°C [102].

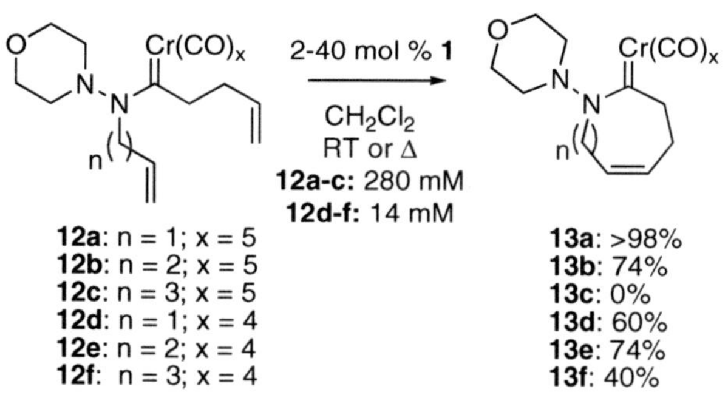

Figure 12 Metal-induced conformation bias toward RCM

4.1.2 Protecting groups as tunable gearing elements

A recent microreview highlights a range of synthetic reactions in which protecting groups are utilized to induce conformational pre-organization, thereby facilitating macrocyclization [89]. Protection of protic functionalities (e.g. hydroxyl and primary or secondary amines) can improve the performance of Ru metathesis catalysts [78, 104, 105], as noted above: this can also offer a highly modular means of introducing such conformational constraints. Specific examples of protecting group effects relevant to synthesis of epothilones have also been reviewed; beneficial effects in some cases associated with the unprotected alcohols were attributed to the stabilization of desired conformations by hydrogen-bonding interactions [106].

Protecting group-directed RCM was recently exploited by Mohaptra and coworkers in the total synthesis of an anti-malarial nonenolide [107], and by Boehringer–Ingleheim researchers in the synthesis of BILN 2061 (a protease inhibitor for the hepatitis virus containing a 15-membered macrocyclic core; Figure 13) [108, 109]. A key advance in the latter synthesis was the initial observation of some sensitivity of the RCM yield to the protecting group on the amide nitrogen within 14, this leading Shu and coworkers to screen a range of such groups, as well as solvents. Both affected the product distribution, and ultimately enabled RCM at the astonishingly high substrate concentration of 200 mM, 20 times higher than that originally required for >90% conversion to the cyclic product [109]. While these effects could not be predicted, the ease and modularity with which different protecting groups can be installed is a clear asset to screening programs.

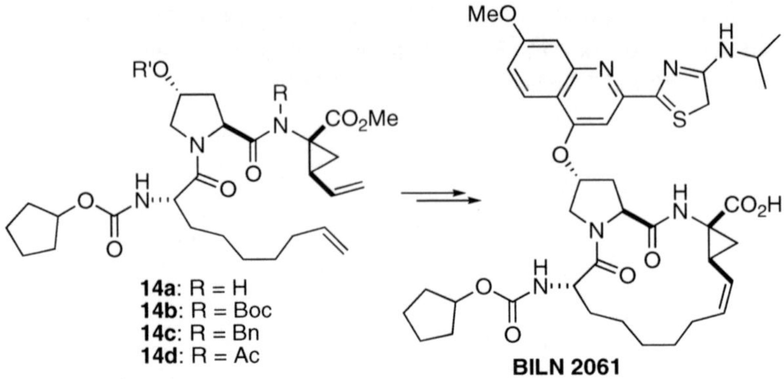

Figure 13 Protecting group effects in the RCM of BILN 2061 [109]

In another example, Martin and coworkers effected synthesis of seven- and nine-membered azabicyclic rings in high yields (82–91%; Figure 14) [110]. The success of these reactions was attributed to the influence of the N-acyl functionality in favouring axial positioning of the allyl substituents in 15. Notably, **16a** was obtained in >90% yield at 100 mM, but yields of the larger ring **16b** are reduced to 67% at this concentration. Details concerning side-product(s) were not discussed, but **16b** could be obtained in ca. 80% isolated yield at dilutions tenfold higher.

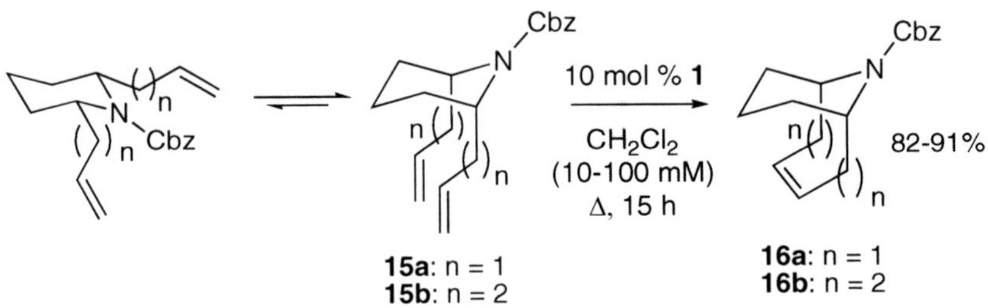

Figure 14 Conformational bias exerted by an N-protecting group

Unprotected alcohol groups, particularly those in allylic and homoallylic positions, can have beneficial or negative effects in RCM [104]. The potential benefits of hydrogen-bonding interactions in epothilone synthesis were noted above [106]. A recent macrocyclization study by Nicolaou and coworkers [111] described smooth RCM in the presence of an allylic alcohol, and decomposition or polymerization in its absence, even at a concentration of 5 mM. In other cases, however, protection of the alcohol functionalities improves RCM yields, possibly through amplified conformational effects. Bulky silyl protecting groups have been shown to aid in cyclization of ten-membered rings precursors in the synthesis of microtubule-stabilizing agents of the eleutheside family [112], although dilutions on the order of 5–10 mM were still necessary to limit oligomerization. A number of other examples have been reported in which hydroxyl protection improved RCM yields in synthesis of medium and large rings [2, 107, 108, 113, 114].

4.1.3 Use of planar centers to relieve ring strain

Installation of oxygen or nitrogen atoms, or sp^2-carbon centers, can relieve transannular strain in RCM substrates, while also increasing the probability of encounter between pendant olefinic sites. Deiters and Martin have reviewed the impact of the "gauche effect" exerted by 1,2-dioxygen substitution, which is thought to promote cyclization by stabilizing a conformation in which the pendant olefinic side chains are gauche [115] (see also Section 4.1.1 above) [93]. Several important examples also appear in a recent review of advances in the RCM assembly of non-annulated medium rings [104]. Cyclic conformational constraints associated with the presence of a preexisting ring structure have likewise been much used to promote RCM synthesis of medium-sized rings [28, 104]. A particularly notable example was the construction of the nine-membered core of the marine natural product ciguatoxin [116].

Grubbs and coworkers reported a dramatic change in product distribution on substituting methylene for oxygen in prolactones **17** (Figure 15) [73]. A mixture of cyclodimer and cyclotrimer products were formed for methylene derivative **17a**, but solely cyclodimer **18b** was observed for the ether derivative. These reactions were carried out at concentrations of 3-6 mM: the nine-membered "cyclomonomer" was not observed. Another example of substrate tailoring to maximize macrocyclization efficiency was reported by Schreiber [117]. By appropriately siting an ester or amide group within the molecule, the transannular strain associated with the 12-membered macrocycle was sharply curtailed, and RCM proceeded smoothly at a concentration of 3–8 mM.

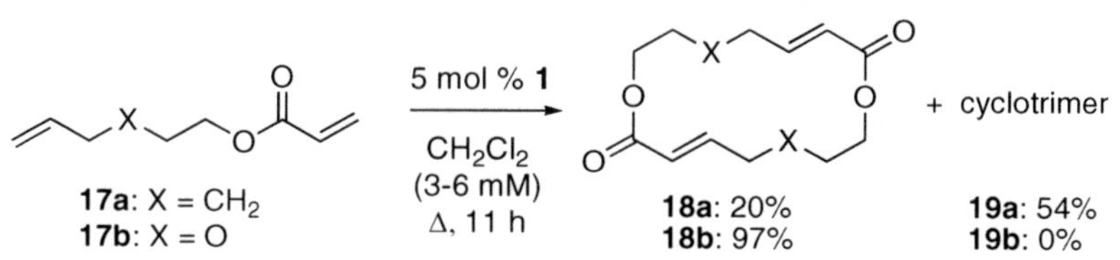

Figure 15 Effect of heteroatoms on cyclization

4.1.4 Non-covalent interactions

Less generally used as potential gearing elements are weak intramolecular forces, including hydrogen-bonding and π-system interactions. A number of such approaches are described in a recent review by Blankenstein and Zhu [89]; Danishefsky has reported examples of favourable H-bonding interactions in epothilone synthesis, as noted above [106]. A spectacular example in context of catenane synthesis was reported by Leigh [118] and coworkers, in chemistry that also emphasizes the role of reversible RCM. In this contribution, 30-membered macrocycles were found to undergo ring opening-ring closing metathesis equilibria with the corresponding [2]catenane in the presence of catalyst **1** (1–5 mol%). Hydrogen bonds between the amide hydrogens and carbonyl groups resulted in near-quantitative formation of catenanes as the thermodynamic products at 200 mM. Above 200 mM, higher cyclooligomers emerged, and the yield of catenanes dropped. Importantly, the catenanes reverted to the macrocyle product upon diluting to 0.2 mM in the presence of **1**, presumably reflecting the higher translational entropy of the non-catenated rings. Macrocycles could also be reformed on breaking the hydrogen-bonding network in the catenanes by incorporation of $(CF_3CO)_2O$ in the presence of catalyst **1**. A similar influence of noncovalent interactions in rotaxane [119] and catenane [120] formation was subsequently described by the Grubbs group.

Weak perfluorophenyl-phenyl interactions were utilized by Collins and coworkers in the RCM synthesis of paracyclophanes (Figure 16) [121] The perfluorinated ring present in **20b** was proposed to "shield" one side of the cyclophane, orienting both allyl chains in close proximity [122], and enabling RCM in moderate yields (41%). In the absence of this moiety, as in **20a**, solely cyclodimer was obtained. Even for **20b**, however, dilutions of 0.4 mM was required.

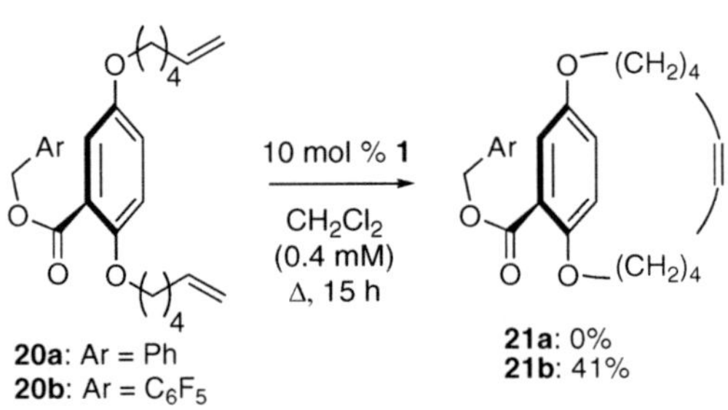

Figure 16 Exploiting non-covalent interaction in synthesis of [12]paracyclophanes

Hydrogen bonding has been proposed to aid in cyclization of vinyl-functionalized cyclic peptides [123]. Initial cross-metathesis, followed by RCM of the dimer, was accomplished by use of the Grubbs catalyst $RuCl_2(PCy_3)_2(=CHCH=CPh_2)$ (26 mol%), furnishing the cyclodimer product in 65% yield after 2 days at a concentration of 5 mM.

4.2 Use of Lewis Acid Additives

Covalent modification of substrates to induce a conformational bias toward cyclization is a hit-and-miss strategy, and is especially high-risk where the RCM step comes near the end of a total synthesis. A potentially more general solution, for substrates bearing appropriate polar functionalities, is the use of Lewis acid additives to create a conformational bias through dative interactions with Lewis base donor sites. While additive residues must then be removed following reaction, the impact on yields can be minimal if this can be achieved by the same means as removal of spent catalyst.

Two distinct strategies have been adopted in use of Lewis acids to promote RCM. Inorganic templating agents, ranging from simple alkali or alkaline earth metals to transition metal complexes, can bring together dienes separated by two or more Lewis basic donor sites. A complementary approach utilizes Lewis acids to block the conformational flexibility of substrates bearing a single polar site. Concentration suitable for ring-closing are again generally in the order of 0.1–20 mM even in the presence of template. In some cases, however, concentrations of up to 100 mM could be employed without crippling yields.

4.2.1 Metal ions as templates

Substrates containing polar functionalities can chelate an appropriately size-matched metal ion through Lewis acid–Lewis base interactions. An increased probability of encounter between subtended olefinic groups can result, if the length of these groups is sufficient. Maynard and Grubbs reported the use of lithium ion as a templating agent for the high-yield, cis-selective synthesis of 12- and 15-membered crown ether-like rings via catalyst **1** at a concentration of 20 mM [124]. In the absence of Li, only 20% of the desired product was obtained, the remainder consisting of low molecular weight polymer (MW ~ 10,000). Other templating agents used with good success, often in conjunction with coordinating pyridine (phenanthroline, terpyridine, etc.) donors, include potassium, copper, iron, zinc, or calcium ions [31, 125].

The strategy is not invariably successful, and in some cases it is not required (particularly in RCM via **1**, which we have found to exhibit a lower kinetic bias toward oligomerization than catalysts of type **2** and **3**) [126]. Thus, Nabeshima and coworkers reported higher yields of the non-templated reaction, albeit without the high cis selectivity associated with use of a metal template, at concentrations of 1–5 mM (Figure 17) [127]. Elevated concentrations were not explored, however. In a related approach, Rychnovsky and coworkers described the failure of a Ca-templated approach in the attempted synthesis of a dimeric macrolide (Figure 18) [128]. A mixture of unreacted diene and the cyclomonomer (i.e. the conventional RCM product) was obtained. Details regarding concentration were not described, but an intriguing inference is the potential use of excessive dilution in this case.

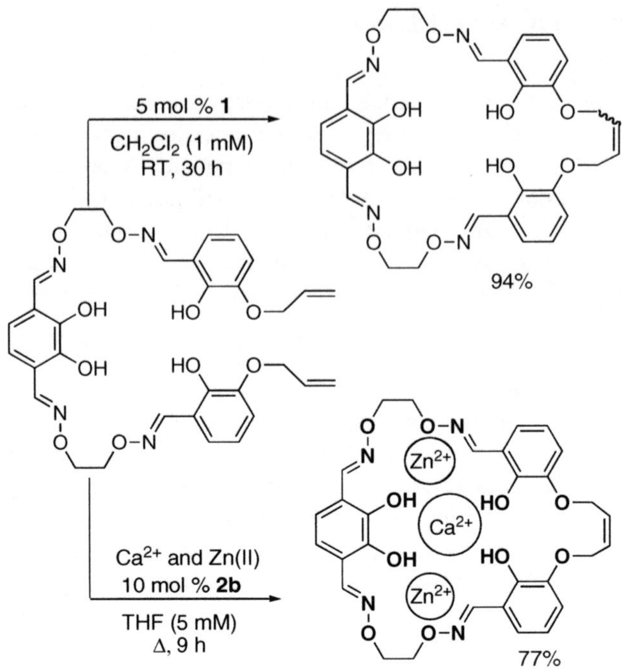

Figure 17 Metal templating in construction of macrocycles

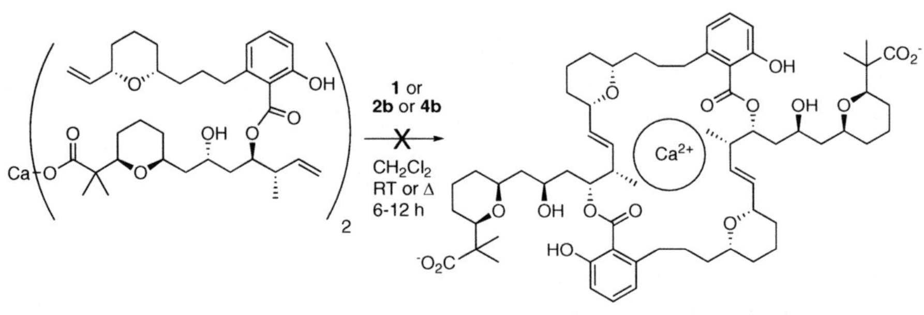

Figure 18 Failed synthesis of a macrocyclodimer using Ca^{2+} as template

4.2.2 Organic molecules as templates

While organic templates are less commonly used than inorganic templates, a number of instances of their use in RCM have been reported. Grubbs and coworkers have described a "magic ring" synthesis of a crown ether analogue via RCM, using **1** in the presence of a dumbbell-shaped secondary ammonium ion [119]. Yields of a hydrogen-bonded [2]rotaxane were improved to ca. 70% yield at 100 mM, from ca. 50% at 5 mM for the untemplated reaction [119]. The ring-opening–ring-closing equilibrium was also deliberately exploited in an alternative route involving equilibration of a preformed macrocyclic olefin with **2b**.

More recently, the same group reported cyclotrimerization of a dibenzo [24]crown-8-diene using a template containing three dialkylammonium ions [129]. The enforced proximity of the three sets of dienes yielded a mixture of cyclotrimer and cyclodimer (22% and 69%, respectively) within 4 h on reaction with 20 mol% **2b** in refluxing CH_2Cl_2, at a concentration of 1 mM. In the absence of template, the cyclodimer was obtained as the major product (56%), in preference to the cyclotrimer (0%). Similar behavior was reported by Nguyen and coworkers in a synthesis of hollow porphyrin prisms [130]. Use of tris((4-pyridyl)ethynyl)benzene as template afforded the desired trimer in over 70% yield; in its absence, solely dimers were obtained. Again, high dilutions (0.3 mM) were essential to limit formation of ADMET products.

Neutral amines can also be used as templates to increase RCM yields, where the RCM substrate itself contains Lewis acid sites. Thus, Takeuchi and coworkers reported the diamine-promoted cyclization of Zn porphyrin tetramers [131]. In the absence of the diamine template, only one pair of olefins reacted; addition of 2.4 eq. of the diamine afforded the desired product in 70% yield at a concentration of 0.25 mM (40 mol% **4b**). This strategy was recently applied to the synthesis of calix[4]arenes [132] and porphyrins [133].

4.2.3 Metal complexes as templates

Use of transition metal complexes as templates has enabled the construction of many topologically challenging molecules, via RCM of olefinic groups on the periphery of large, typically metalloorganic, substrates. Gladysz and coworkers have described the synthesis of molecular gyroscopes [134, 135], insulated "wires" [136], and parachute-like structures [137], while the Sauvage group has reported numerous examples of molecular knots, rotaxanes, and catenanes [46, 138, 139]. van Koten and coworkers described a general approach to the RCM synthesis of macroheterocycles [140, 141] utilizing a rigid tris-platinum pincer complex for pre-organization of the olefinic "tails". High dilutions (~1 mM) are generally essential to minimize oligomerization. In a recent showcase example, Ko and coworkers reported the RCM synthesis of a 90-membered hexa(pyridyl)macrocycle using a $Pt(PEt_3)_2$ template, at a dilution of 0.1 mM (Figure 19) [142]. Competing decomposition of the Ru catalyst imposed a requirement for very high loadings of **1** (30 mol%), but the desired product could be isolated by preparative thin-layer chromatography in 80% yield after 3 days at room temperature. The template was removed by treatment with NaI to afford the free macrocycle product isolated in 74% yield. For a comprehensive overview, and a fascinating discussion of the relevance of these structures to the assembly of molecular machines, readers are referred to two excellent reviews [47, 48].

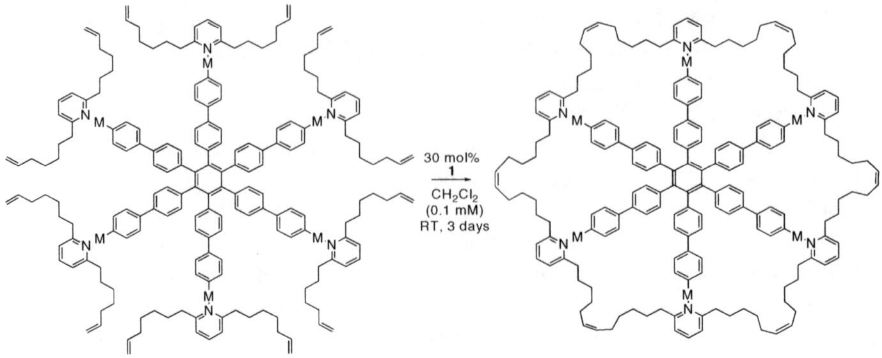

Figure 19 Template-directed synthesis of a 90-membered macroheterocycle (M = *trans*-$Pt(PEt_3)_2$)

4.2.4 Lewis acid-assisted metathesis

Lewis acids have been widely used as co-catalysts in RCM. The predominant interest has lain in inhibiting chelation of polar functionalities that can poison the catalyst (e.g. alcohols, esters, amides, etc.; see Section 4.1.2) [77, 78, 143–145]. However, Lewis acids can also favour cyclization by blocking the conformational mobility of the substrate. The Percy group recently showed the combined effect of a Lewis acid and hydroxyl protection in the synthesis of an eight-membered lactone at 10 mM concentrations (Figure 20) [146]. In the presence of 5 mol% 2b and 30 mol% of Ti(O-^{i}Pr)$_4$ as co-catalyst, dienes **22** underwent RCM without apparent formation of dimers or other oligomers in refluxing CH_2Cl_2 over 18 h. The EM of this substrate, in the presence of the Ti(IV) additive, could be increased from a value of 8 mM for **22a** to 240 mM for **22b**, by installation of a benzyl protecting group. Importantly, the EM of **22b** decreased sharply (to 48 mM) in the absence of the Ti(IV) additive, indicating that both the protecting group and the Lewis acid contribute to the measured EM values.

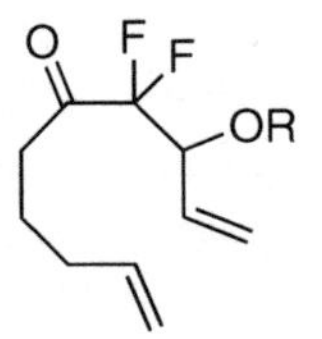

22

R = **a**: H; **b**: Bz

Substrate	Additive	EM
22a	Ti(O-^{i}Pr)$_4$	0.0082
22b	Ti(O-^{i}Pr)$_4$	0.24
22b	none	0.048

Figure 20 Increasing effective molarities (EM; values in M) through use of Lewis acid additives

4.3 RCM in Supercritical CO_2

Despite the advances described above, high dilutions are still routinely required to maximize yields of medium-sized and macrocyclic RCM products. Of great interest as an environmentally benign, relatively low-cost alternative diluent is supercritical CO_2 ($scCO_2$). Increasing the density of $scCO_2$ mimics the effect of dilution by decreasing the ratio of reactants relative to inert molecules. This technology offers the added advantage of enabling dilution at constant volume, and hence at a standard reactor size, simply by increasing $scCO_2$ density. Few uses of $scCO_2$ in RCM have been reported to date, however. The first example of this approach was described by Leitner, Furstner and coworkers in 1997, in macrolactonization via catalyst **1** (Figure 21) [147]. At low CO_2 densities of 0.55 g/ml, ADMET oligomers predominated, but the macrolactone was obtained in ca. 90% yield by increasing the density to 0.83 g/ml. More recently, RCM in $scCO_2$ was reported using **1** and **2b**, and an immobilized Hoveyda-type Ru catalyst [148]. Near-quantitative conversions were found for all catalysts and a range of different substrates. However, because this study was restricted to common ring sizes, the effect of the CO_2 density on selecting between cyclization and oligomerization could not be examined. An unrelated paper recently reported extraction of Ru residues using $scCO_2$ [149]. While this is of potential interest as a means of alleviating the difficulties in post-RCM purification, it points toward potential difficulties in ensuring the mutual solubility of substrate and catalyst in the reaction medium.

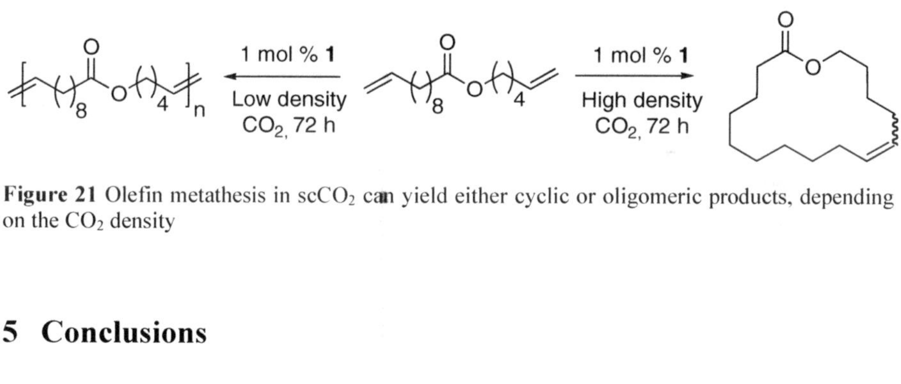

Figure 21 Olefin metathesis in $scCO_2$ can yield either cyclic or oligomeric products, depending on the CO_2 density

5 Conclusions

The foregoing describes some of the challenges specific to construction of medium and large rings by olefin metathesis. An important difference between the metathesis chemistry and more conventional routes to cyclic species is the potential for reversibility in the former. While many classic, stoichiometric approaches to cyclic products terminate in a kinetic distribution of cyclic and oligomeric products, unfavourable product distributions can potentially be "corrected" in metathesis reactions by appropriate choice of catalysts and reaction conditions. Of particular importance is the ability to shift the ring-chain (or ring-ring) equilibria in favour of the desired cyclic products by diluting the reaction. Where equilibrium metathesis is precluded by (e.g.) catalyst lifetime or low reactivity, however, a kinetic distribution of products is obtained. The initial concentration of substrate is of paramount importance under such conditions, as this "correction by dilution" strategy is no longer possible.

A number of approaches have been successfully used to increase the effective molarity of RCM substrates, including use of Lewis acid and other templates, or substituents that can create a conformational bias toward cyclization. Such effects are unpredictable, however, and must be identified and optimized by trial and error. Little explored to date is the possibility of modifying the EM values associated with a given catalyst-substrate pair by modulation of catalyst properties to exert some conformation bias on the bound substrate. Also of interest are alternative means of dilution: $scCO_2$ strategies are of considerable interest, as are supported metathesis strategies, should current limitations of catalyst lifetime in the latter be successfully addressed. Advances in these areas, as well as in development of methods for amplifying the EM values of RCM substrates, have the potential for major impact on the environmental and economic sustainability of metathesis routes to medium and macrocyclic rings.

Acknowledgements This work was supported by the Natural Sciences and Engineering Research Council of Canada and the Canada Foundation for Innovation.

References

[1] Fürstner A (2000) Angew Chem Int Ed 39:3012–3043
[2] Prunet J (2003) Angew Chem Int Ed 42:2826–2830; Angew Chem Int Ed 42:3322
[3] Nicolaou KC, Bulger PG, Sarlah D (2005) Angew Chem Int Ed 44:4490–4527
[4] Grubbs RH (ed.) (2003) Handbook of metathesis. Wiley-VCH, Weinheim
[5] Throughout this discussion, we adopt the following terminology for different ring sizes: small ring are those containing 3–4 members; common rings, 5–7 members; medium rings, 8–11 members; and large or macrocyclic rings, 12 or more
[6] Illuminati G, Mandolini L (1981) Acc Chem Res 14:95–102
[7] Page MI, Jencks WP (1971) Proc Natl Acad Sci USA 68:1678–1683
[8] Jacobson H, Stockmayer WH (1950) J Chem Phys 18:1600–1606
[9] Ercolani G, Mandolini L, Mencareli P, Roelens S (1993) J Am Chem Soc 115:3901–3908
[10] Galli C, Mandolini L (2000) Eur J Org Chem:3117–3125
[11] Chen Z-R, Claverie JP, Grubbs RH, Kornfield JA (1995) Macromolecules 28:2147–2154
[12] The Jacobson-Stockmayer theory of ring-chain equilibria is based on the following assumptions: (1) all rings are strainless and there is no heat of cyclization, (2) the end-to-end distances of linear chains obey Gaussian statistics, (3) the probability of ring formation is governed by the fraction of all configurations for which the ends coincide, and (4) the reactivity of the chain ends is independent of chain length. It follows from the first assumption that only the entropic term contributes to the equilibrium constant. There are two contributions to the entropy change: a positive one due to the dissociation of one molecule into two and a negative one due to the decreased number of configurations on going from a linear chain to a linear and a cyclic product

[13] Ivin KJ, Mol JC (1997) Olefin metathesis and metathesis polymerization. Academic Press, New York
[14] Kirby AJ (1980) Adv Phys Org Chem 17:183–278
[15] Mandolini L (1986) Adv Phys Org Chem 22:1–111
[16] Flory PJ (1969) Statistical mechanics of chain molecules. Wiley Interscience, New York
[17] Ercolani G (2006) Struct Bonding 121:167–215
[18] Ercolani G (2003) J Phys Chem B 107:5052–5057
[19] Corbett PT, Leclaire J, Vial L, West KR, Wietor J-L, Sanders JKM, Otto S (2006) Chem Rev 106:3652–3711
[20] Cacciapaglia R, Di Stefano S, Mandolini L (2005) J Am Chem Soc 127:13666–13671
[21] Mulder A, Huskens J, Reinhoudt DN (2004) Org Biomol Chem 2:3409–3424
[22] Li X, Liu DR (2004) Angew Chem Int Ed 43:4848–4870
[23] Huskens J, Mulder A, Auletta T, Nijhuis CA, Ludden MJW, Reinhoudt DN (2004) J Am Chem Soc 126:6784–6797
[24] Hubbard PA, Brittain WJ, Mattice WL, Brunelle DJ (1998) Macromolecules 31:1518–1522
[25] Cacciapaglia R, Di Stefano S, Mandolini L (2004) Acc Chem Res 37:113–122
[26] Villemin D (1980) Tetrahedron Lett 21:1715–1718
[27] (a) Schrock RR (2006) Angew Chem Int Ed 45:3748–3759; (b) Grubbs RH (2006) Angew Chem Int Ed 45:3760–3765
[28] Maier ME (2000) Angew Chem Int Ed 39:2073–2077
[29] Gradillas A, Perez-Castells J (2006) Angew Chem Int Ed 45:6086–6101
[30] Nishida A, Nagata T, Nakagawa M (2006) Top Heterocycl Chem 5:255–280
[31] Fürstner A (2004) Eur J Org Chem :943–958
[32] Ibrahim YA (2006) J Mol Catal A 254:43–52
[33] Majumdar KC, Rahaman H, Roy B (2007) Curr Org Chem 11:1339–1365
[34] Gaich T, Mulzer J (2005) Curr Top Med Chem 5:1473–1494
[35] Brenneman JB, Martin SF (2005) Curr Org Chem 9:1535–1549
[36] Chattopadhyay SK, Karmakar S, Biswas T, Majumdar KC, Rahaman H, Roy B (2007) Tetrahedron 63:3919–3952
[37] Van de Weghe P, Eustache J (2005) Curr Top Med Chem 5:1495–1519
[38] Kaliappan KP, Kumar N (2005) Tetrahedron 61:7461–7469
[39] Rassu G, Auzzas L, Battistini L, Casiraghi G (2004) Mini-Rev Org Chem 1:343–357
[40] Arisawa M, Nishida A, Nakagawa M (2006) J Organomet Chem 691:5109–5121
[41] Clark JS (2006) Chem Commun:3571–3581
[42] Kotha S, Lahiri K (2007) Synlett:2767–2784
[43] Schmidt B, Hermanns J (2006) Curr Org Chem 10:1363–1396
[44] Brik A (2008) Adv Synth Catal 350:1661–1675
[45] Martin WHC, Blechert S (2005) Curr Top Med Chem 5:1521–1540
[46] Dietrich-Buchecker C, Jimenez-Molero MC, Sartor V, Sauvage J-P (2003) Pure Appl Chem 75:1383–1393
[47] Collin JP, Dietrich-Buchecker C, Hamann C, Jouvenot D, Kern JM, Mobian P, Sauvage JP (2004) Compr Coord Chem II 7:303–326
[48] Champin B, Mobian P, Sauvage J-P (2007) Chem Soc Rev 36:358–366
[49] Miller SJ, Kim S-H, Chen Z-R, Grubbs RH (1995) J Am Chem Soc 117:2108–2109
[50] Hérisson J-L, Chauvin Y (1971) Makromol Chem 141:161–176
[51] The limiting solubility of ethylene in toluene is ca. 20% that of propylene: 77 mM, vs. 376 mM, under 1 atm of the corresponding gas. See: Atiqullah M, Hammawa M, Hamid H (1998) Eur Polym J 34:1511–1520

[52] Hoye TR, Jeffrey CS, Tennakoon MA, Wang JZ, Zhao HY (2004) J Am Chem Soc 126:10210–10211
[53] Ulman M, Grubbs RH (1999) J Org Chem 64:7202–7207
[54] Sanford MS, Love JA, Grubbs RH (2001) J Am Chem Soc 123:6543–6554
[55] Lysenko Z, Maughon BR, Mokhtar-Zadeh T, Tulchinsky ML (2006) J Organomet Chem 691:5197–5203
[56] Mori M, Sakakibara N, Kinoshita A (1998) J Org Chem 63:6082–6083
[57] Giessert AJ, Diver ST (2004) Chem Rev 104:1317–1382
[58] Lee L-S, Ou H-J, Hsu H-L (2005) Fluid Phase Equilib 231:221–230
[59] Nosse B, Schall A, Jeong WB, Reiser O (2005) Adv Synth Catal 347:1869–1874
[60] Conrad JC, Eelman MD, Duarte Silva JA, Monfette S, Parnas HH, Snelgrove JL, Fogg DE (2007) J Am Chem Soc 129:1024–1025
[61] Craig SW, Manzer JA, Coughlin EB (2001) Macromolecules 34:7929–7931
[62] Hong SH, Wenzel AG, Salguero TT, Day MW, Grubbs RH (2007) J Am Chem Soc 129:7961–7968
[63] Hong SH, Day MW, Grubbs RH (2004) J Am Chem Soc 126:7414–7415
[64] Alcaide B, Almendros P (2003) Chem Eur J 9:1259–1262
[65] Fürstner A, Thiel OR, Ackermann L, Schanz HJ, Nolan SP (2000) J Org Chem 65:2204–2207
[66] Petkovska VI, Hopkins TE, Powell DH, Wagener KB (2005) Macromolecules 38:5878–5885
[67] Kirkland TA, Grubbs RH (1997) J Org Chem 62:7310–7318
[68] Kamau SD, Hodge P, Hall AJ, Dad S, Ben-Haida A (2007) Polymer 48:6808–6822
[69] Bielawski CW, Benitez D, Morita T, Grubbs RH (2001) Macromolecules 34:8610–8618
[70] Fürstner A, Thiel OR, Ackermann L (2001) Org Lett 3:449–451
[71] Xu Z, Johannes CW, Houri AF, La DS, Cogan DA, Hofilena GE, Hoveyda AH (1997) J Am Chem Soc 119:10302–10316
[72] Yamamoto K, Biswas K, Gaul C, Danishefsky SJ (2003) Tetrahedron Lett 44:3297–3299
[73] Lee CW, Grubbs RH (2001) J Org Chem 66:7155–7158
[74] Lemarchand A, Bach T (2004) Tetrahedron 60:9659–9673
[75] Rivkin A, Biswas K, Chou T-C, Danishefsky SJ (2002) Org Lett 4:4081–4084
[76] Haigh DM, Kenwright AM, Khosravi E (2004) Tetrahedron 60:7217–7224
[77] Fürstner A, Langemann K (1997) J Am Chem Soc 119:9130–9136
[78] Yang Q, Xiao W-J, Yu Z (2005) Org Lett 7 871–874
[79] McNaughton BR, Bucholtz KM, Camaano-Moure A, Miller BL (2005) Org Lett 7:733–736
[80] Young DGJ, Burlison JA, Peters U (2003) J Org Chem 68:3494–3497
[81] Sheddan NA, Arion VB, Mulzer J (2006) Tetrahedron Lett 47:6689–6693
[82] (a) Höcker H, Reimann W, Riebel K, Szentivanyi Z (1976) Makromol Chem 177:1707–1715; (b) Höcker H, Reimann W, Reif L, Riebel K (1980) J Mol Catal 8:191–202; (c) Reif L, Höcker H (1984) Macromolecules 17:952–956
[83] Conrad JC, Parnas HH, Snelgrove JL, Fogg DE (2005) J Am Chem Soc 127:11882–11883
[84] Hodge P, Kamau SD (2003) Angew Chem Int E 42:2412–2414
[85] Jimenez-Gonzalez C, Curzons AD, Constable DJC, Cunningham VL (2005) Clean Technol Environ Policy 7:42–50
[86] Jimenez-Gonzalez C, Curzons AD, Constable DJC, Cunningham VL (2004) Int J Life Cycle Assess 9:115–121

[87] Sheldon RA (2007) Green Chem 9:1273–1283
[88] Sheldon RA (2007) Green chemistry and catalysis. Wiley-VCH, Chichester
[89] Blankenstein J, Zhu JP (2005) Eur J Org Chem :1949–1964
[90] Jung ME, Piizzi G (2005) Chem Rev 105:1735–1766
[91] Forbes MDE, Patton JT, Myers TL, Maynard HD, Smith DW Jr, Schulz GR, Wagener KB (1992) J Am Chem Soc 114:10978–10980
[92] Galli C, Giovannelli G, Illuminati G, Mandolini L (1979) J Org Chem 44:1258–1261
[93] Crimmins MT, Emmitte KA (2001) J Am Chem Soc 123:1533–1534
[94] Goldring WPD, Hodder AS, Weiler L (1998) Tetrahedron Lett 39:4955–4958
[95] Fürstner A, Langemann K (1997) Synthesis :792–803
[96] Vassilikogiannakis G, Margaros I, Tofi M (2004) Org Lett 6:205–208
[97] Fürstner A, Thiel OR, Blanda G (2000) Org Lett 2:3731–3734
[98] Yang K, Blackman B, Diederich W, Flaherty PT, Mossman CJ, Roy S, Ahn YM, Georg GI (2003) J Org Chem 68:10030–10039
[99] Aissa C, Riveiros R, Ragot J, Fürstner A (2003) J Am Chem Soc 125:15512–15520
[100] Fürstner A, Dierkes T, Thiel OR, Blanda G (2001) Chem Eur J 7:5286–5298
[101] Commeureuc AGJ, Murphy JA, Dewis ML (2003) Org Lett 5:2785–2788
[102] Paley RS, Estroff LA, Gauguet J-M, Hunt DK, Newlin RC (2000) Org Lett 2:365–368
[103] Licandro E, Maiorana S, Vandoni B, Perdicchia D, Paravidino P, Baldoli C (2001) Synlett:757–760
[104] Michaut A, Rodriguez J (2006) Angew Chem Int Ed 45:5740–5750
[105] Fu GC, Nguyen ST, Grubbs RH (1993) J Am Chem Soc 115:9856–9857
[106] Rivkin A, Cho YS, Gabarda AE, Yoshimura F, Danishefsky SJ (2004) J Nat Prod 67:139–143
[107] Mohapatra DK, Ramesh DK, Giardello MA, Chorghade MS, Gurjar MK, Grubbs RH (2007) Tetrahedron Lett 48:2621–2625
[108] Shu C, Zeng X, Hao M-H, Wei X, Yee NK, Busacca CA, Han Z, Farina V, Senanayake CH (2008) Org Lett 10:1303–1306
[109] Yee NK, Farina V, Houpis IN, Haddad N, Frutos RP, Gallou F, Wang X-J, Wei X, Simpson RD, Feng X, Fuchs V, Xu Y, Tan J, Zhang L, Xu J, Smith-Keenan LL, Vitous J, Ridges MD, Spinelli EM, Johnson M, Donsbach K, Nicola T, Brenner M, Winter E, Kreye P, Samstag W (2006) J Org Chem 71:7133–7145
[110] Neipp CE, Martin SF (2003) J Org Chem 68:8867–8878
[111] Nicolaou KC, Leung GYC, Dethe DH, Guduru R, Sun Y-P, Lim CS, Chen DYK (2008) J Am Chem Soc 130:10019–10023
[112] Caggiano L, Castoldi D, Beumer R, Bayon P, Telser J, Gennari C (2003) Tetrahedron Lett 44:7913–7919
[113] Ma C, Schiltz S, Le Goff XF, Prunet J (2008) Chem Eur J 14:7314–7323
[114] Nicolaou KC, Liu JJ, Yang Z, Ueno H, Sorensen EJ, Claiborne CF, Guy RK, Hwang CK, Nakada M, Nantermet PG (1995) J Am Chem Soc 117:634–644
[115] Deiters A, Martin SF (2004) Chem Rev 104:2199–2238
[116] Hirama M, Oishi T, Uehara H, Inoue M, Maruyama M, Oguri H, Satake M (2001) Science 294:1904–1907
[117] Lee D, Sello JK, Schreiber SL (1999) J Am Chem Soc 121:10648–10649
[118] Kidd TJ, Leigh DA, Wilson AJ (1999) J Am Chem Soc 121:1599–1600
[119] Kilbinger AFM, Cantrill SJ, Waltman AW, Day MW, Grubbs RH (2003) Angew Chem Int Ed 42:3281–3285
[120] Guidry EN, Cantrill SJ, Stoddart JF, Grubbs RH (2005) Org Lett 7:2129–2132
[121] El-Azizi Y, Schmitzer A, Collins SK (2006) Angew Chem Int Ed 45:968–973

[122] Zakarian JE, El-Azizi Y, Collins SK (2008) Org Lett 10:2927–2930
[123] Clark TD, Kobayashi K, Ghadiri MR (1999) Chem Eur J 5:782–792
[124] Marsella MJ, Maynard HD, Grubbs RH (1997) Angew Chem Int Ed 36:1101–1103
[125] Rapenne G, Dietrich-Buchecker C, Sauvage J-P (1999) J Am Chem Soc 121:994–1001
[126] Monfette S, Fogg DE, unpublished results
[127] Akine S, Kagiyama S, Nabeshima T (2007) Inorg Chem 46:9525–9527
[128] Cheung LL, Marumoto S, Anderson CD, Rychnovsky SD (2008) Org Lett 10:3101–3104
[129] Hou H, Leung KCF, Lanari D, Nelson A, Stoddart JF, Grubbs RH (2006) J Am Chem Soc 128:15358–15359
[130] Youm K-T, Nguyen SBT, Hupp JT (2008) Chem Commun:3375–3377
[131] Wakabayashi R, Kubo Y, Hirata O, Takeuchi M, Shinkai S (2005) Chem. Commun:5742–5744
[132] Rudzevich Y, Cao Y, Rudzevich V, Bohmer V (2008) Chem Eur J 14:3346–3354
[133] Van Gerven PCM, Elemans JAAW, Gerritsen JW, Speller S, Nolte RJM, Rowan AE (2005) Chem Commun:3535–3537
[134] Shima T, Hampel F, Gladysz JA (2004) Angew Chem Int Ed 43:5537–5540
[135] Nawara AJ, Shima T, Hampel F, Gladysz JA (2006) J Am Chem Soc 128:4962–4963
[136] De Quadras L, Hampel F, Gladysz JA (2006) Dalton Trans:2929–2933
[137] Skopek K, Barbasiewicz M, Hampel F, Gladysz JA (2008) Inorg Chem 47:3474–3476
[138] Frey J, Kraus T, Heitz V, Sauvage J-P (2005) Chem Commun:5310–5312
[139] Frey J, Kraus T, Heitz V, Sauvage J-P (2007) Chem Eur J 13:7584–7594
[140] Chuchuryukin AV, Chase PA, Dijkstra HP, Suijkerbuijk B, Mills AM, Spek AL, van Klink GPM, van Koten G (2005) Adv Syn Catal 347:447–462
[141] Chuchuryukin AV, Dijkstra HP, Suijkerbuijk BMJM, Klein Gebbink RJM, van Klink GPM, Mills AM, Spek AL, van Koten G (2003) Angew Chem Int Ed 42:228–230
[142] Song KH, Kang SO, Ko J (2007) Chem Eur J 13:5129–5134
[143] Li Y, Zhang T, Li Y-L (2007) Tetrahedron Lett 48:1503–1505
[144] Selvakumar N, Kumar PK, Reddy KCS, Chary BC (2007) Tetrahedron Lett 48:2021–2024
[145] Chen X, Wiemer DF (2003) J Org Chem 68:6597–6604
[146] Mitchell L, Parkinson JA, Percy JM, Singh K (2008) J Org Chem 73:2389–2395
[147] Furstner A, Koch D, Langemann K, Leitner W, Six C (1997) Angew Chem Int Ed 36:2466–2469
[148] Michalek F, Maedge D, Ruehe J, Bannwarth W (2006) Eur J Org Chem :577–581
[149] Gallou F, Saim S, Koenig KJ, Bochniak D, Horhota ST, Yee NK, Senanayake CH (2006) Org Process Res Dev 10:937–940

Functionalisation of Vinylsubstituted (Poly)Siloxanes and Silsesquioxanes via Cross-Metathesis and Silylative Coupling Transformations

Bogdan Marciniec,* **Cezary Pietraszuk**

Adam Mickiewicz University, Faculty of Chemistry, Grunwaldzka 6, 60-780 Poznań, Poland
*E-mail: bogdan.marciniec@amu.edu.pl

Abstract Applications of the catalytic transformations of vinyl group at silicon, i.e., cross-metathesis, silylative coupling with olefins and silylative coupling with acetylenes for functionalisation of vinylsubstituted (poly)organosiloxanes were overviewed. Cross-metathesis and silylative coupling of olefins with vinylsilicon compounds catalysed by ruthenium complexes were shown to constitute two valuable, complementary synthetic routes leading to functionalised (poly)siloxanes, cyclosiloxanes, silsesquioxanes and spherosilicates of great practical importance. Moreover, first examples of selective synthesis of variety of siloxanes with acetylene functionality via silylative coupling of acetylenes with vinylsiloxanes were described.

Keywords Cross-metathesis · Silylative coupling · Vinylsiloxanes · Acetylenes, Grubbs catalysts · Ruthenium complexes

1 Introduction

Vinyl-substituted silanes and siloxanes (both cyclic and linear) constitute a class of unsaturated organosilicon compounds of prospective wide applicability in organic synthesis [1, 2]. Functionalised cyclosiloxanes are used as the basic substrates for anionic and cationic ring opening polymerisation (copolymerisation) to yield respective functional silicon polymers [3]. Cubic vinylsubstituted silsesquioxanes containing nanosized inorganic core $(ViSiO_{1.5})_8$ similarly to other octasilsesquioxanes offer considerable potential for producing hybrid material as the vertices of the cube can be functionalised with a variety of functional groups, e.g., [4]. Functionalisation of poly(vinyl)siloxanes addresses need for polymers of specific properties, combining e.g. high thermal and oxidative stability, moisture resistance and a variety of other advantages of siloxane framework with properties, e.g. optoelectronic arisen from organic component [3].

In the last 2 decades two universal, effective and synthetically attractive methods for synthesis of well-defined molecular compounds with vinylsilicon functionality were developed in our group i.e. cross-metathesis (Equation 1) and silylative coupling (Equation 2). Both methods are based on catalytic transformations of vinylsilicon compounds with olefins and lead to synthesis of respective functionalised vinylsilicon derivatives (for reviews see [5]).

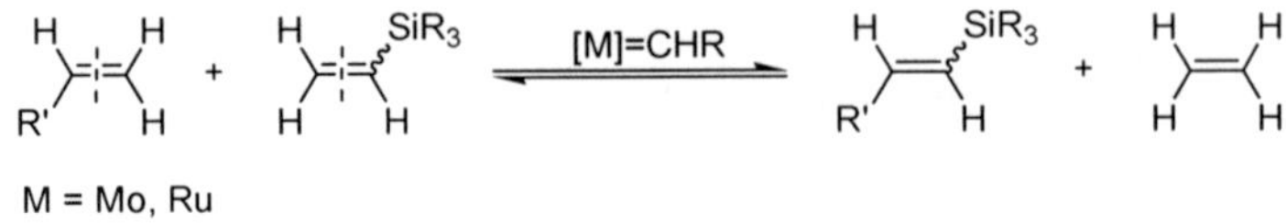

(1)

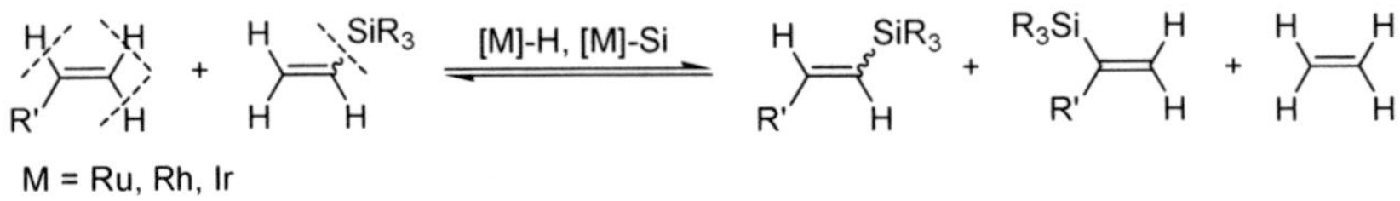

(2)

Cross-metathesis proceeds via the carbene mechanism and is catalysed by well-defined alkylidene complexes of W, Mo or Ru [6]. The family of ruthenium-based catalysts (e.g. **I**, **II** or **III**, Figure 1), tolerant of normal organic and polymer processing conditions and preserving their catalytic properties in the presence of the majority of functional groups has allowed a great number of new applications which made olefin metathesis an important synthetic tool in organic and polymer chemistry [6]. We have previously shown that Grubbs type catalysts effectively catalyse the cross-metathesis of trialkoxy-, trichloro- and generally electron withdrawing substituted vinylsilicon compounds [7]. It was also proved that vinylsilanes (and vinylsiloxanes) bearing at least one methyl substituent at silicon undergo efficient reaction with Grubbs catalyst leading to its deactivation [8].

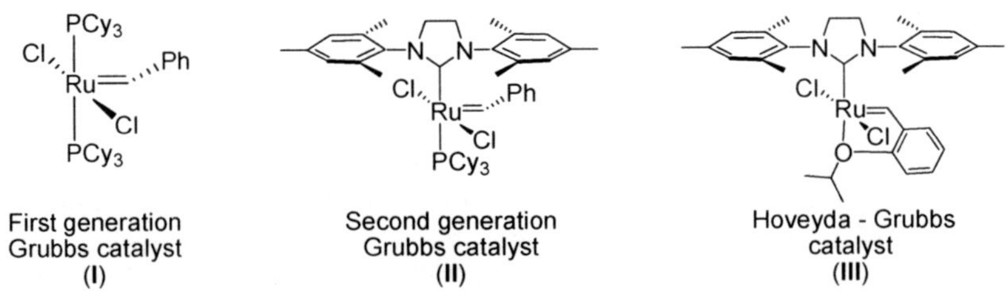

Figure 1 Grubbs type ruthenium alkylidene complexes

The silylative coupling of olefins with vinylsilanes is catalysed by complexes containing or generating hydride ([M]-H) or silyl ([M]-Si) ligands (silicometallics) (where M = Ru, Rh, Ir, Co) and proceeds according to a mechanism involving the activation of =C–H and Si–C= bonds (Equation 1) [9]. Ruthenium hydride [$RuHCl(CO)(PPh_3)_3$] (**IV**), [$RuHCl(CO)(PCy_3)_2$] (**V**) and silyl [$Ru(SiMe_3)Cl(CO)(PPh_3)_2$] (**VI**) complexes (also in the presence of copper(I) chloride as cocatalyst) were shown to be among the most active and convenient catalysts for silylative coupling of olefins with vinylsilanes.

Moreover, another catalytic transformation of vinylsilanes i.e. their coupling with alkynes has recently been discovered in our group (Equation 3) [10a, 10b, 10c] to extend the role of vinylsilicon compounds over catalytic activation of ≡C–H bonds [10d].

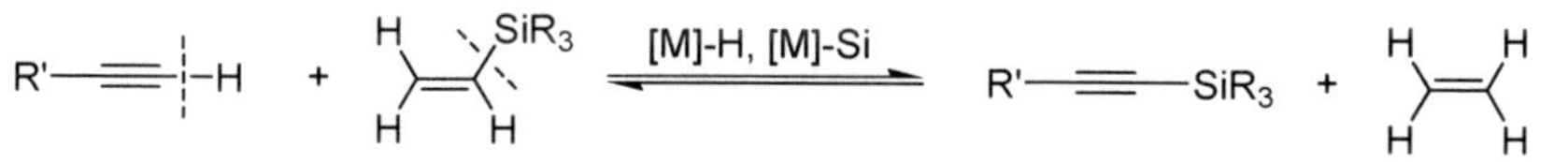

(3)

M = Ru, Rh

The aim of this review is to present the application of cross-metathesis and silylative coupling transformations for modification of the vinyl group at silicon in the variety of compounds bearing siloxy framework. Most procedures use highly active catalytic system, require mild conditions and result in efficient and selective functionalisation. Advantages and drawbacks of the reactions being general synthetic routes for the synthesis of functionalised vinylsiloxanes are discussed.

2 Model Study for Modification of Poly(vinyl)siloxanes via Metathesis and Silylative Coupling

3-Vinyltrisiloxanes of the type illustrated in Figure 2 were chosen as model compounds for the study of catalytic functionalisation of oligo- and polyvinyl-siloxanes. Styrene was used as a reaction partner because it is highly reactive and does not isomerize in the reaction conditions [5].

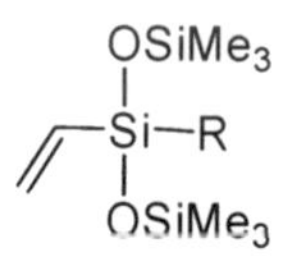

R = $OSiMe_3$, Ph, C_6H_4OMe-4, C_6H_4Me-4

Figure 2 Vinylsiloxanes tested in cross-metathesis with olefins

Treatment of a mixture of 3-vinyltrisiloxane and styrene in the presence of Grubbs' catalyst (**I**) or (**II**) in methylene chloride at 40°C gives rise to evolution of ethene and formation of respective 3-styryltrisiloxane (Equation 4) [7a, 11]. Under optimised conditions the products were formed in moderate to high yields. The reactions with styrenes proceed stereoselectively giving exclusively isomer *E*. Cross-metathesis with olefins proceeds with high yield but slightly lower stereoselectivity [7b, 11]. Noteworthy is that the reaction is not accompanied by the migration of double bond. Cross-metathesis is sensitive to the electronic properties of silicon at the vinyl group. When phenyl-substituted vinylsiloxane was used as a reaction partner, the reaction proceeded with modest yields. However, introduction of the electron-withdrawing group permits nearly quantitative conversion and high yield of the cross-metathesis products.

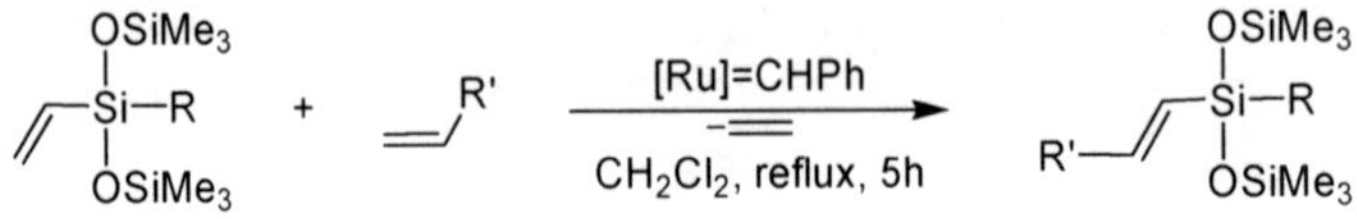

R	R'	cat.	yield [%]	*E*/*Z*
$OSiMe_3$	Ph	I	97	E
	C_8H_{17}	I	89	10/1
Ph	Ph	I	28	E
	C_8H_{17}	I	21	10/1
C_6H_4OMe-4	Ph	II	93	E
	C_8H_{17}	II	93	10/1
$C_6H_4(CF_3)$-4	Ph	I	97	E
	C_8H_{17}	I	95	10/1
	C_6H_4OMe-4	II	95	E
	CH_2SiMe_3	II	100	25/1
	CH_2OBu	II	95	7/1

(4)

Very modest conversion was observed, when 3-vinylheptamethyltrisiloxane was tested in the cross-metathesis with styrene in the presence **I** or **II**. It was proved earlier, that vinylsiloxanes similar to other vinylsilicon compounds bearing at least one methyl substituent at silicon undergo equimolar reaction with Grubbs catalyst leading to its deactivation [8].

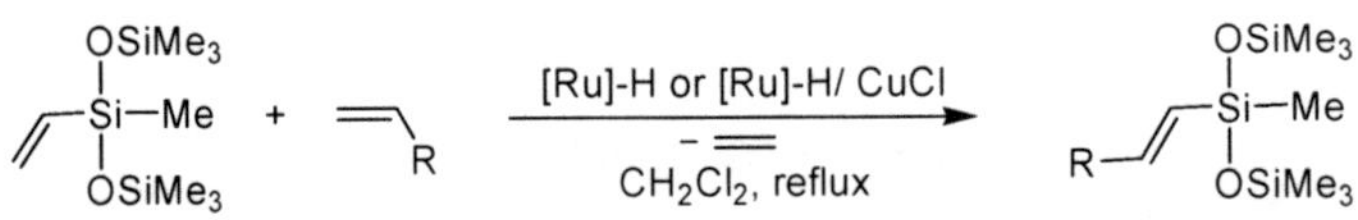

R	cat.	CuCl	time [h]	yield [%]
Ph	IV	-	24	88[a]
	IV	-	24	0
	IV	+	18	61
	V	-	18	69
	V	+	3	99
C_6H_4Me-4	V	+	3	99
C_6H_4OMe-4	V	+	2	99

[a]toluene, 100°C

(5)

3-Vinylheptamethyltrisiloxane in the presence of ruthenium hydride complex [RuHCl(CO)(PPh$_3$)$_3$] (**III**) and in particular [RuHCl(CO)(PCy$_3$)$_2$] (**V**) undergoes efficient silylative coupling with styrene leading to efficient formation of ethene and 3-styrylheptamethyltrisiloxane (Equation 5) [12, 13]. In the presence of (**V**) efficient conversion was observed already at 45°C. An addition of CuCl as a co-catalyst leads to a significant increase in the catalytic activity, irrespective of the hydride complex used [14]. The reaction is highly stereoselective. Only *E*-isomer of silylstyrene was observed by the ^{1}H NMR spectroscopy. The reaction was found to proceed efficiently also in the presence of substituted styrenes.

Vinylheptamethyltrisiloxane undergoes effective coupling with 1-ethynyl-1-(trimethylsiloxy)cyclohexane [10a]. In the fourfold excess of vinylsilanes, the reaction leads to efficient formation of substituted silylacetylene (Equation 6). No reaction was observed when phenylacetylene was used as an acetylenic substrate.

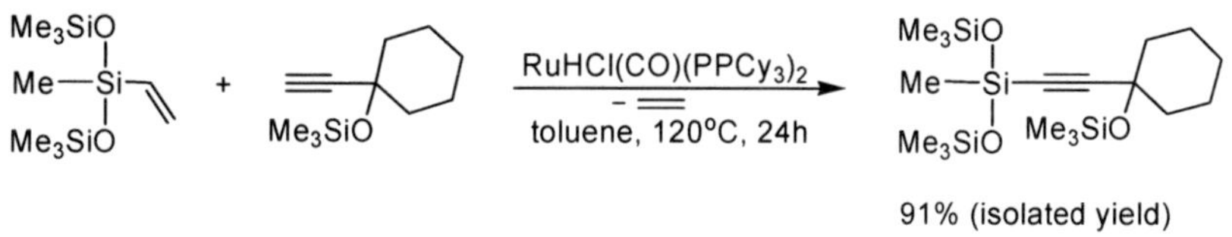

(6)

Analogously, silylative coupling of 1-vinylpentamethyldisiloxane with 1-ethynyl-1-(trimethylsiloxy)cyclohexane gives respective product with 90% of isolated yield [10a].

3 Divinyldisiloxanes

A series divinyldisiloxanes were tested with respect to their reactivity in olefin metathesis [15]. Cross-metathesis of divinyl-substituted disiloxanes with olefins in the presence of 5–10 mol% of catalyst **I**, **II** or **III** in boiling CH_2Cl_2 produces a mixture of mono- and disubstituted vinylsilanes and ethene (Equation 7). To achieve a high conversion of divinylsiloxanes and satisfactory selectivity of disubstituted products a tenfold excess of olefin had to be used. The reaction is accompanied by competitive olefin homo-metathesis that could not be avoided. Under the optimum conditions the products were obtained with moderate to high yields and selective formation of *E*-isomer (when styrenes were used as olefins) or a mixture of isomers with high excess of *E* or *E,E* (when 1-decene was used). Because of the presence of methyl substituents at silicon no activity in metathesis was observed for divinyltetramethyldisiloxane. Tetraphenylsubstituted analogue, does not influence the longevity of catalyst. However, also for this reagent almost no conversion was observed in cross-metathesis with styrene under the conditions used. Introduction of the electron-withdrawing group into a phenyl ring results in a substantial increase in reactivity. High conversions were observed in the reaction of $ViSi[C_6H_4(CF_3)\text{-}4]_2OSi[C_6H_4(CF_3)\text{-}4]_2Vi$ with 1-octene or styrenes. Significant amounts of monosubstituted product were obtained when divinyltetraethoxydisiloxane was treated with moderate (fourfold) excess of 1-decene in the presence of **I** (Equation 7).

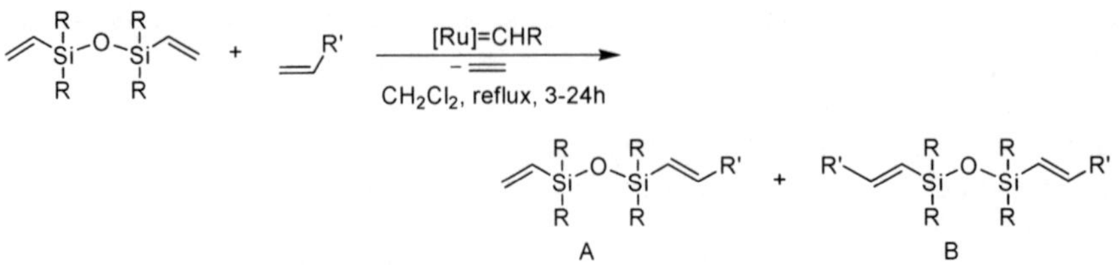

(7)

R	R'	cat.	yield of A [%]	*E/Z*	yield of B [%]	*E,E/E,Z*
$C_6H_4(CF_3)$-4	C_8H_{17}	II	8	*E*	92	15/1
	C_6H_4Cl-4	III	12	*E*	87	*E,E*
OEt	C_8H_{17}	I	54	*E*	5	*E,E*
	C_6H_4Cl-4	II	2	*E*	95	*E,E*

Divinyltetramethyldisiloxane was successfully tested as a monomer in silylative coupling polycondensation and copolycondensation with dienes [5]. Relatively few information concerns its reactivity with simple olefins. High yield of disubstituted product was observed in the silylative coupling of divinyltetramethyldisiloxane with styrene in the presence of $[\{(cod)Rh(\mu\text{-}OSiMe_3)\}_2]$ (Equation 8) [16].

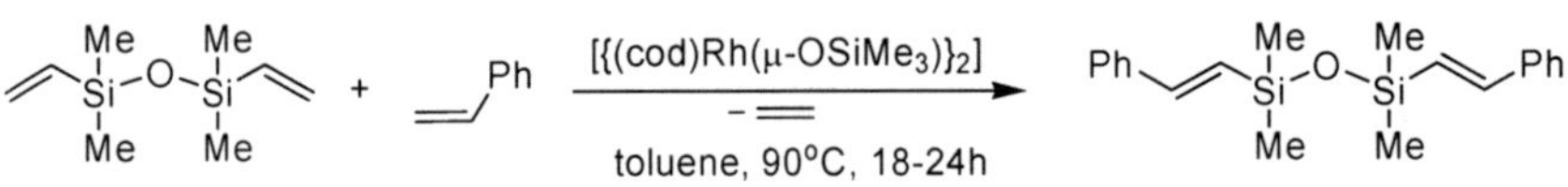

(8)

Recently, 9-vinylcarbazole was demonstrated to undergo quantitative and selective coupling with divinyltetramethyldisiloxane in the presence of $[RuHCl(CO)(PCy_3)_2]$ (**V**) (Equation 9) [17].

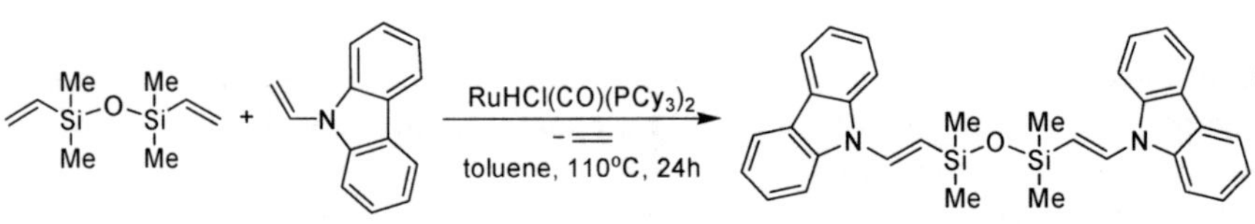

(9)

yield of *E,E*-isomer = 99%

Silylative coupling of divinyltetramethyldisiloxane with selected acetylenes yields monoalkynylvinyldisiloxane with high yields and selectivity (Equation 10). The reaction requires divinylsiloxane to be used in high excess (up to tenfold molar excess) and is accompanied by the product of homocoupling of the vinylsilicon substrate [10a].

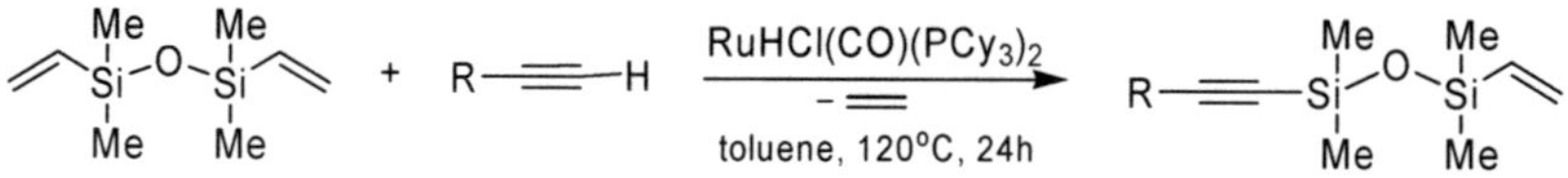

≡—R	yield [%]
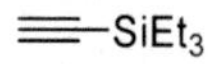	100
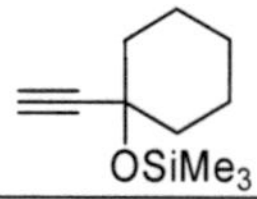	100

(10)

4 Vinylcyclosiloxanes

Effective metathesis transformation is observed when trivinylsubstituted cyclotrisiloxanes are treated with olefins in the presence of metathesis catalyst **I**, **II** or **III** (Equation 11) [11].

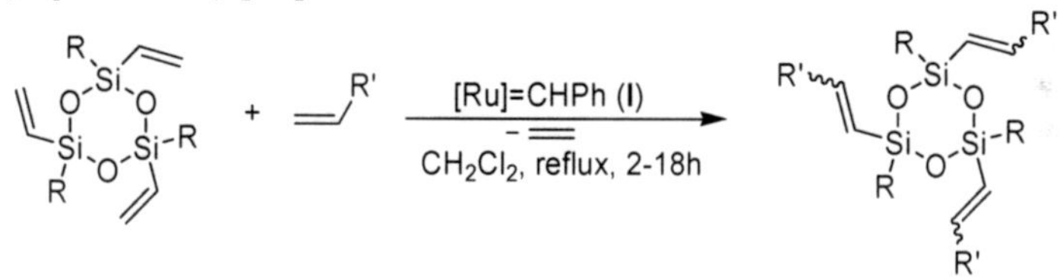

R	R'	yield [%]	*E/Z*
$OSiMe_3$	Ph	95	*E*
	C_6H_4Cl-4	96	*E*
	C_6H_4OMe-4	96	*E*
	C_8H_{17}	78	8/1
Ph	Ph	96	*E*
	C_8H_{17}	90	8/1
C_6H_4OMe-4	Ph	96	*E*
	C_8H_{17}	91	10/1
Me	Bu	70	55/45
	t-Bu	60	69/31

(11)

The reaction requires an excess of olefin. Therefore, formation of olefin homo-metathesis products could not be avoided. However, the proper choice of substrates permits getting high yields and selectivities of the substituted cyclosiloxanes at least for selected reaction systems. The optimum catalyst of the reactions conducted in the presence of excessive amount of olefin is complex **I**, which is a consequence of its low activity in the process of homo-metathesis of styrenes, under the conditions used. In the optimum conditions of the process, the products of cross-metathesis were obtained effectively in the yields up to 96% and the reaction is accompanied by formation of only trace amounts of the products of olefin homo-metathesis. In the presence of **II** the reaction takes place with high yields but the products of the cross-metathesis are accompanied by considerable amounts of those of olefin homo-metathesis. High catalytic activity of Grubbs catalyst (**I**) was observed for the reaction of vinylsilanes with vinyl ethers as we have proved previously [18]. However, the results of stoichiometric and catalytic experiments provided convincing evidence that instead of cross-metathesis, the silylative coupling occurs, via an insertion-elimination mechanism initiated by Ru-H complex generated in situ from complexes **I** or **II** [18].

Trivinyltrimethylcyclotrisiloxane undergoes efficient silylative coupling with styrene and some other olefins to form trisubstituted products with high yield (Equation 12) [19]. Reaction with styrene proceeds stereoselectively with exclusive formation of *E,E,E*-isomer. High yield of products but moderate selectivity was obtained when vinyl ethers were used as olefins.

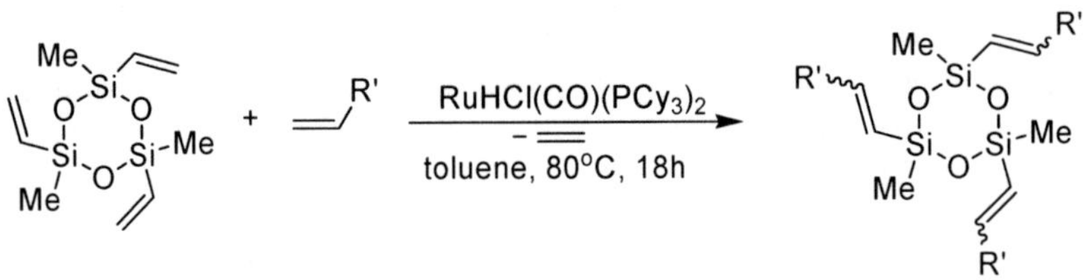

(12)

R'	cat.	yield [%]	*E/Z*
Ph	V	95	*E*
OBu	V	99	58/42
O(t-Bu)	V	99	72/28

Tetravinylcyclotetrasiloxane was successfully functionalised with styrenes and various heteroatom-substituted olefins in the presence of $[RuHCl(CO)(PCy_3)_2]$ (**V**) (Equation 13) [20].

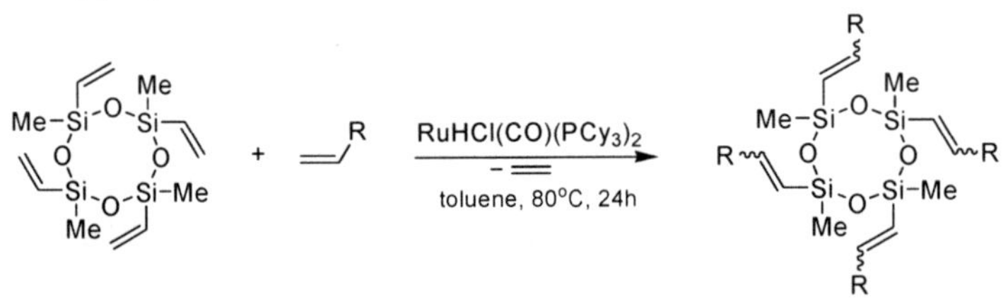

R	yield [%]	*E/Z*
Ph	92[a]	E
C_6H_4Me-4	92	E
C_6H_4OMe-4	93	E
C_6H_4Br-4	95	E
OBu	90	95/5
O(*t*-Bu)	93	92/8
$SiMe_3$	94	90/10[b]
N-pyrrolidinone	87[c]	99/1
9-carbazole	86[c]	98/2

[a]18h; [b]E/gem; [c]110°C, 24h

(13)

Exclusive formation of *E*-isomers was observed for the reaction with styrenes. Selectivities obtained for the heteroatom-substituted olefins were slightly lower. Nevertheless, most of the reactions proceeded with strong preference for the formation of the (*E*)-isomers.

Treatment of trivinyltrimethylcyclotrisiloxane with selected terminal alkynes in the presence of **V** leads to highly regio- and chemoselective formation of monoalkynyl-substituted cyclotrisiloxane (Equation 14) [21].

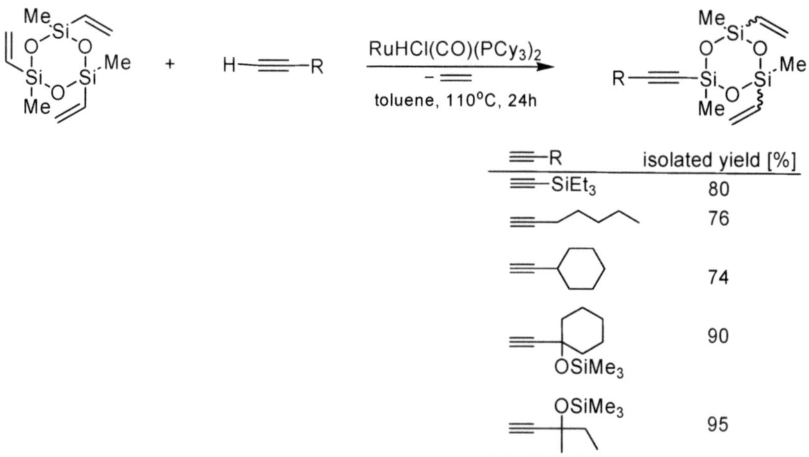

(14)

Analogously, tetravinyltetramethylcyclotetrasiloxane can be efficiently converted into monoalkynyl-substituted trivinylcyclotetrasiloxanes via silylative coupling with substituted acetylenes in the presence of **V** (Equation 15) [21]. The reactions proceed regio- and chemoselectively and permit getting the respective products with high yields.

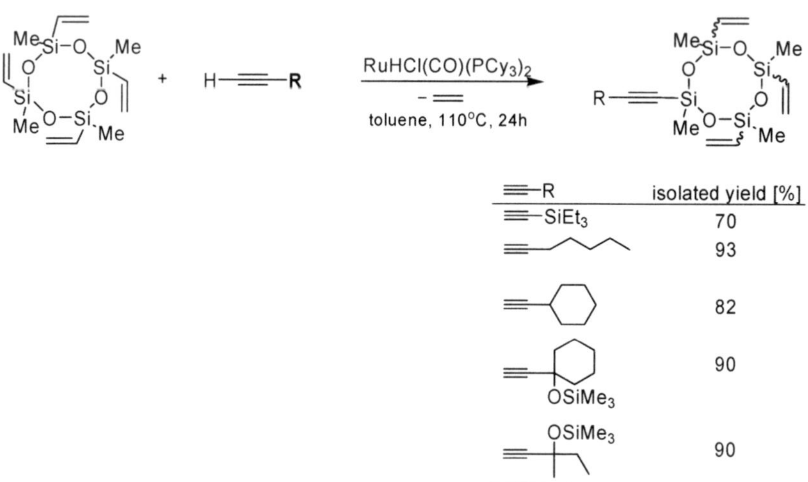

(15)

Mono-alkynyl-substituted cyclotrisiloxanes (Equation 16) and cyclotetrasiloxanes underwent functionalisation of the vinyl groups attached to silicon atoms with styrene in the presence of ruthenium catalyst (**V**) [21] to form very attractive monomers for ring opening (co)polymerisation of cyclosiloxanes.

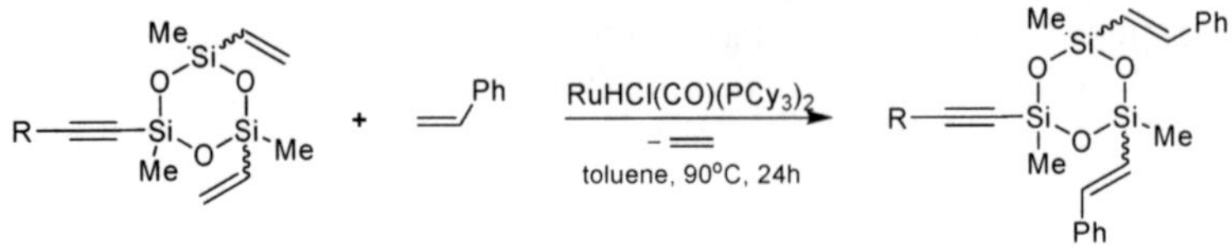

(16)

R	isolated yield [%]
$GeEt_3$	90
$OSiMe_3$ 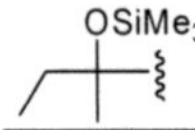	92

5 Vinylsilsesquioxanes and Spherosilicates

Cubic octavinylsilsesquioxane can be modified with pent-4-en-1-on and 5-brompent-1-ene in the presence of Schrock catalyst [22]. High catalytic activity of complex **I** in cross-metathesis of octavinylsilsesquioxanes with selected olefins has also been found in our laboratory (Equation 17) [23]. The reaction constitutes the basis for synthesis of highly functionalised silsequioxane framework. Treatment of this substrate with small excess of styrene (1.5 eq. in relation to vinyl group) leads to nearly quantitative and stereoselective formation of octastyrylsilsesquioxane. Effective cross-metathesis with 1-hexene requires the use of threefold excess of olefin (in relation to the vinyl group). In such conditions the reaction is accompanied by olefin homo-metathesis.

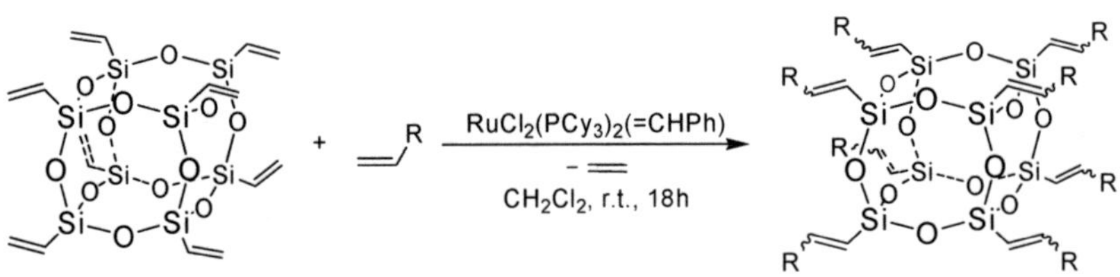

(17)

R	yield [%]	*E*/*Z*
Ph	96	*E*
Bu	72	94/6
CH_2SiMe_3	99	>99/1
OBu	70[a]	55/45[a]
O(*t*-Bu)	72[a]	68/32[a]
$OSiMe_3$	61[a]	37/63[a]
S(*t*-Bu)	95[b]	97/3[b]

[a]toluene, 80°C; [b]toluene, 60°C, (**II**)

High conversion of vinyl ethers was observed in the presence of Grubbs' catalysts. However, as described above, this reaction occurs according to the insertion-elimination mechanism characteristic for silylative coupling [18].

Treatment with styrene (1.5 eq. in relation to vinyl group) with octavinylsilsesquioxane in the presence of **V** leads to efficient and selective formation of octastyrylsilsesquioxane (Equation 18) [23].

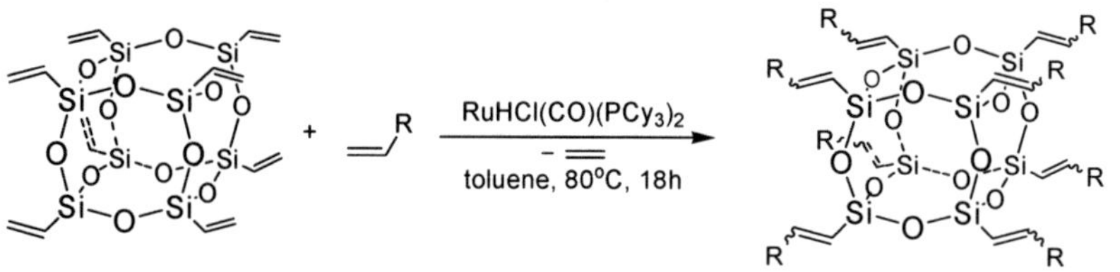

R	yield [%]	*E/Z*
Ph	92	E
$SiMe_3$	95	76/24*
OBu	99	57/43
O(*t*-Bu)	99	79/21
$OSiMe_3$	99	45/55
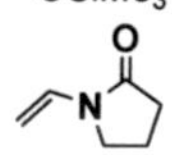	69	95/5

* E/(gem)

(18)

Reaction of this substrate with vinyltrimethylsilane results in formation of octasubstituted product in high yield but moderate regioselectivity, that is characteristic of vinylsilane homocoupling reaction [5]. Functionalisation via silylative coupling with vinyl ethers proceeds efficiently, however the fully substituted product formed is actually a mixture of stereoisomeric components.

Monovinylsiloxyheptaisobutylsilsesquioxane and octavinylspherosilicate of the formula $(ViMe_2SiO)(i\text{-}Bu)_7Si_8O_{12}$ and $(ViMe_2SiO)_8Si_8O_{12}$, respectively were tested in cross-metathesis and silylative coupling with a variety of olefins [24]. Cross-metathesis performed in the presence of (**I**) or (**II**) did not result in formation of respective product because of the catalyst decomposition occurring in the presence of methylvinylsilyl unit as discussed previously. On the other hand, the above mentioned substrates undergo efficient conversion with styrene and selected olefins in the presence of the $[Ru(H)(Cl)(CO)(PCy_3)_2]$ (**V**) under silylative coupling conditions (Equation 19). The reaction with styrene permits quantitative conversion of the substrate and takes place highly stereo- and regioselectively. The use of N- or O-vinyl derivatives results in formation of respected products in slightly lower yield and stereoselectivity.

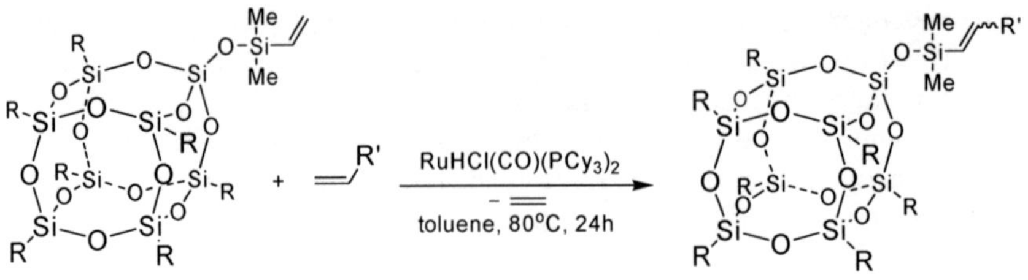

R	R'	yield [%]	E/Z
i-Bu	Ph	94	*E*
	OBu	93	62/38
	O(t-Bu)	95	55/45
	N-pyrrolidinone	92[a]	98/2[a]
	N-carbazole	95[a]	93/7[a]
$OSiMe_2Vi$	Ph	89	*E*
	OBu	85	64/36
	O(t-Bu)	83	57/43
	N-pyrrolidinone	80[a]	98/2[a]
	N-carbazole	81[a]	94/6[a]

[a]48h

(19)

6 Poly(vinyl)siloxanes

The reactivity of vinylsiloxanes provided important clues concerning the functionalisation and cross-linking of poly(vinyl)siloxanes. Therefore, selected poly (aryl,vinyl)siloxanes were tested in cross-metathesis with styrenes (Equation 20) [11].

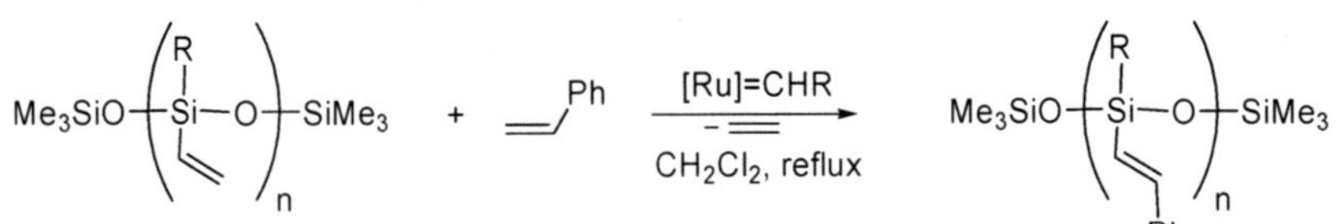

$R = C_6H_5, C_6H_4Me\text{-}4, C_6H_4OMe\text{-}4, C_6H_4(CF_3)\text{-}4$

(20)

R	yield [%]
C_6H_5	0
C_6H_4Me-4	0
C_6H_4OMe-4	32
$C_6H_4(CF_3)$-4	99

The experiments have indicated very high sensitivity of the reaction to the electronic properties of the double bond. No reaction was observed for phenyl-substituted siloxanes. Introduction of methoxy group to the phenyl ring results in an increase in the observed substitution of the vinyl groups. Finally, introduction of the strongly electron-withdrawing trifluoromethyl group ensured quantitative transformation.

Successful silylative coupling of 3-vinyltrisiloxane suggested the possibility of application of the reaction for functionalisation of vinylsiloxanes of higher molecular mass of styrene towards vinylsubstituted polysiloxanes. Therefore, we decided to test the reactivity of the commercially available poly (methylvinyl) siloxane (molecular weight 1,000–1,500) containing an average number of 13 vinyl side groups in the siloxane chain (as calculated on the basis of ^{1}H NMR spectroscopic analysis) in the silylative coupling with styrene. In the presence of complex **V** combined with copper(I) chloride, the reaction proceeds efficiently and stereoselectively according to Equation 21 [13]. To decrease the reaction time and increase its yield, styrene was used in threefold excess (in relation to the calculated content of vinyl groups). Under the conditions used the reaction proceeds chemoselectively and is not accompanied by unwanted polymerisation of styrene or homo-coupling of vinylsiloxane groups.

Trimethylsiloxy-terminated poly(dimethylsiloxane-*co*-methylvinyl-siloxane) containing 4.5 mol% of vinylmethylsiloxane unit (as calculated on the basis of ^{1}H NMR spectrum) undergoes silylative coupling with styrene in the presence of $[RuHCl(CO)(PCy_3)_2]$/CuCl to give fully substituted product (Equation 22) with complete stereoselectivity [13].

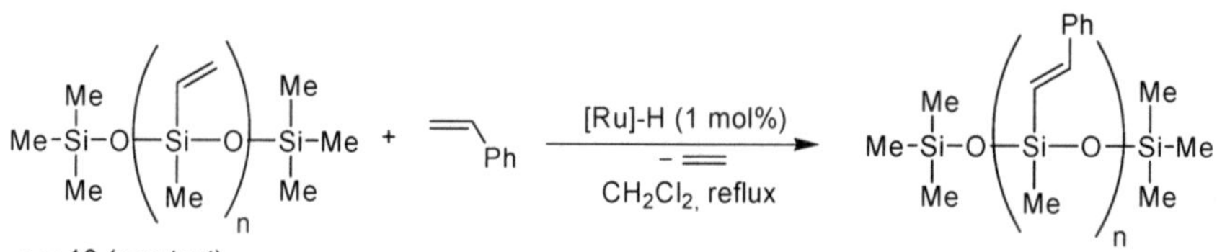

n = 13 (see text)

cat.	CuCl	time [h]	yield [%]
IV	-	24	60[a]
IV	+	24	0
V	-	8	60
V	+	8	100
V	+	3	100[b]

[a]benzene, 80°C; [b]3 mol% of **V**

(21)

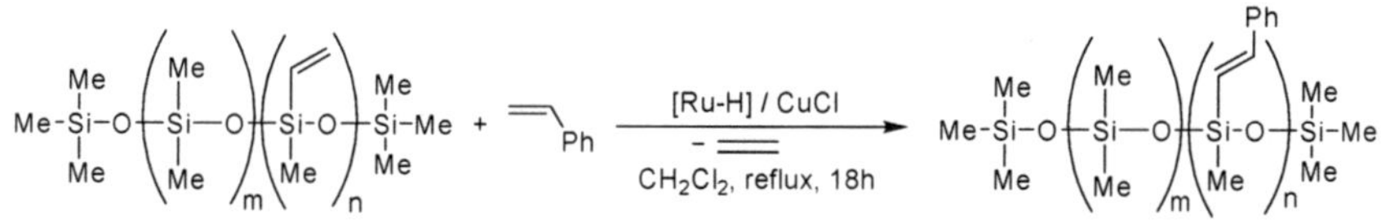

(22)

[Ru-H] = $[RuHCl(CO)(PCy_3)_2]$

(isolated yield = 80%)

The co-polymer exhibits lower reactivity in the reaction than oligovinylsiloxanes and its satisfactory conversion requires prolonged reaction time and increased catalyst concentration.

In contrast to cross-metathesis, silylative coupling can be successfully used for functionalisation of (poly)siloxanes bearing terminal vinyl groups [13]. 1,5-divinylhexamethyltrisiloxane and vinyldimethylsilyl-terminated poly(dimethylsiloxane) (Mw ≈ 6,000) containing an average number of 78 dimethylsiloxy units in the siloxane chain (as calculated on the basis of ^{1}H NMR spectrum) were treated with styrene in the presence of [RuHCl(CO)(PCy$_3$)$_2$] (**V**) combined with copper(I) chloride. A complete conversion of vinyl groups in 1,5-divinyltrisiloxane and quantitative formation of silylative coupling product (Equation 23) was observed already after 3 h. Efficient functionalisation of vinyl-terminated polysiloxane required the reaction to be performed for 18 h. Under such conditions full conversion and selective formation functionalised product was observed (Equation 23). In both cases reactions result in exclusive formation of the *E*,*E*-isomers.

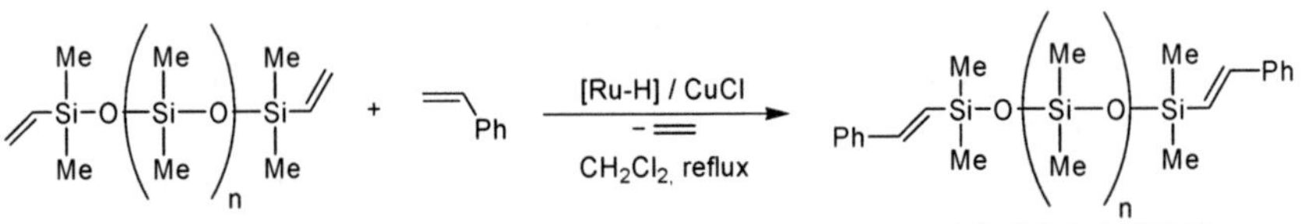

(23)

n = 1, 78 (see text)
[Ru-H] = [RuHCl(CO)(PCy$_3$)$_2$]

n = 1 (isolated yield 92%)
n = 78 (isolated yield 90%)

In all tests performed, particular attention was paid to detect the possible formation of the product of the undesirable styrene polymerisation. All described procedures ensure the efficient progress of silylative coupling and permit avoidance of the unwanted styrene polymerisation without the need of addition of inhibitors.

Although great progress has been made in investigation of the catalytic transformations of vinylsiloxanes, commercially important organosilicon products, many areas related to this problem still require exploration, in particular the problems of modification and cross-linking of poly(vinyl)siloxanes of high molecular mass.

7 Conclusion

Cross-metathesis and silylative coupling of olefins with vinylsilicon compounds catalysed by ruthenium complexes constitute two valuable, complementary synthetic routes leading to functionalised (poly)siloxanes, cyclosiloxanes, silsesquioxanes and spherosilicates of great practical importance as precursors and/or components of nanomaterials as well as organometallic reagents for organic synthesis. On the other hand, silylative coupling of acetylenes with vinylsilicon derivatives is a new efficient method for selective synthesis of variety of compounds with acetylene functionality.

Acknowledgement We thank the Ministry of Science and Higher Education (Project numbers PBZ-KBN-118/T09/17 and NR 05-0005 04/2008) for the financial support of this investigation.

References

[1] (a) Chan TH, Fleming I (1979) Synthesis 761; (b) Weber WP (1983) Silicon reagents for organic synthesis, Ch 7. Springer, Berlin; (c) Colvin EW (1988) Silicon reagents in organic synthesis Ch 3. Academic Press, London; (d) Luh T-Y, Liu S-T (1998) In: Rappoport Z, Apeloig Y (eds.) The chemistry of organosilicon compounds. Wiley, Chichester

[2] (a) Denmark SE, Sweis RF (2004) Organosilicon compounds in cross-coupling reaction. In: de Meijere A, Diederich F (eds.) Metal-catalysed cross-coupling reactions, Wiley-VCH, Weinheim; (b) Denmark SE, Ober MH (2003) Aldrichim Acta 36: 75–85

[3] Boutevin F, Guida-Pietrasanta F, Ratsimihety A (2000) Side group modified polysiloxanes. In: Jones RG, Ando W, Chojnowski J (eds.) Silicon containing polymers, Kluwer, Dordrecht

[4] Baney RH, Itoh M, Sakakibara A, Suzuki T (1995) Chem Rev 95: 1409–1430

[5] (a) Marciniec B, Pietraszuk C (2003) Metathesis of silicon-containing olefins. In: Grubbs RH (ed.) Handbook of metathesis, Wiley-VCH, Weinheim; (b) Marciniec B, Pietraszuk C (2000) Curr Org Chem 7: 691–743; (c) Marciniec B (2005) Coord Chem Rev 249: 2374–2390

[6] Grubbs RH (ed.) (2003) Handbook of metathesis, Wiley-VCH, Weinheim

[7] (a) Pietraszuk C, Marciniec B, Fischer H (2000) Organometallics 29: 913–917; (b) Pietraszuk C, Fischer H, Kujawa M, Marciniec B (2001) Tetrahedron Lett 42: 1175–1178; (c) Pietraszuk C, Marciniec B, Fischer H (2003) Tetrahedron Lett 44: 7121–7124

[8] Pietraszuk C, Fischer H (2000) Chem Commun 2463–2464

[9] (a) Wakatsuki Y, Yamazaki H, Nakano M, Yamamoto Y (1991) J Chem Soc Chem Commun 703–704; (b) Marciniec B, Pietraszuk C (1995) J Chem Soc Chem Commun 2003–2004; (c) Marciniec B, Pietraszuk C (1997) Organometallics 16: 4320–4326

[10] (a) Marciniec B, Dudziec B, Kownacki I (2006) Angew Chem Int Ed 45: 8180–8184; (b) Pol. Pat. Appl. 380422; (c) WO 2008020774; (d) Marciniec B (2007) Acc Chem Res 40: 943–952

[11] Zak P, Pietraszuk C, Marciniec B (2008) J Mol Catal A: Chem 285: 101–108

[12] Marciniec B, Pietraszuk C, Kujawa M (1998) J Mol Catal A: Chem. 133: 41–49

[13] Zak P, Skrobanska M, Pietraszuk C, Marciniec B (2008) Organometallics (submitted)

[14] Pat PL 195453

[15] Pietraszuk C, Rogalski S, Majchrzak M, Marciniec B (2006) J Organomet Chem 691: 5476–5481

[16] Marciniec B, Walczuk-Gusciora E, Pietraszuk C (2001) Organometallics 20: 3423–3428

[17] Prukała W, Marciniec B, Majchrzak M, Kubicki M (2007) Tetrahedron 63: 1107–1115

[18] Marciniec B, Kujawa M, Pietraszuk C (2000) Organometallics 19: 1677–1681

[19] Itami Y, Marciniec B, Kubicki M (2003) Organometallics 22: 3717–3722

[20] Marciniec B, Waehner J, Pawluć P, Kubicki M (2007) J Mol Catal A: Chem 265: 25–31

[21] Dudziec B, Marciniec B (2008) Organometallics 27: 5598–5604

[22] Feher FJ, Soulivong D, Eklund AG, Wydham KD (1997) Chem Commun 1185–1186

[23] Itami Y, Marciniec B, Kubicki M (2004) Chem Eur J 10: 1239–1248

[24] Waehner J, Marciniec B, Pawluć P (2007) Eur J Inorg Chem 2975–2980

The Olefin Metathesis Reactions in Dendrimers

Didier Astruc

Nanosciences and Catalysis Group, ISM, UMR CNRS N°5255, Université Bordeaux I, 33405 Talence Cedex, France

E-mail: d.astruc@ism.u-bordeaux1.fr

Abstract Dendrimers containing terminal olefins or ruthenium-benzylidene terminal groups undergo olefin metathesis reactions (RCM and ROMP types), and essentially results from our group are reviewed here. Dendrimers have been loaded at their periphery with ruthenium-chelating bis-phosphines, which leads to the formation of dendrimer-cored stars by ring-opening-metathesis polymerization (ROMP). $CpFe^{+}$-induced perallylation of polymethylaromatics ($Cp = \eta^{5}$-C_5H_5) followed by ring-closing metathesis (RCM) and/or cross metathesis (CM) leads to poly-ring, cage, oligomeric and polymeric architectures. In the presence of acrylic acid or methacrylate, stereospecific CM inhibits oligomerization, and dendritic olefins yield polyacid dendrimers. Finally, cros-metahesis reactions with dendronic acrylate allow dendritic construction and growth.

Keywords Metathesis · Polymerization · Dendrimer · Arene · Iron · Olefin

1 Introduction

The olefin metathesis reactions stand among the most powerful organometallic strategies to synthesize pharmaceutically useful compounds and a variety of polymers [1–4]. The alkene and alkyne metathesis reactions usually involves low-molecular weight unsaturated substrates, however. Here, we are dealing with dendrimers for alkene metathesis of various types. In the first section, we discuss dendrimer loaded with ruthenium-benzylidene complexes containing chelating *cis*-diphosphine ligands and their use in the catalysis of ROMP reactions. In the second section, we review olefin metathesis of olefin-terminated dendrimer including olefin-dendrimer functionalization and solubilization in water, and finally the selective cross-olefin-metathesis strategy to grow dendrimers.

V. Dragutan et al. (eds.), *Green Metathesis Chemistry: Great Challenges in Synthesis, Catalysis and Nanotechnology*,

2 Ruthenium Benzylidene-Terminated Dendrimers That Metathesize Norbornene to Dendrimer-Cored Stars by ROMP

Only very few examples of metallodendritic carbene complexes were known, and with only four branches before our study and metathesis activity [5], but good recyclability remains a challenge. The difficulty resided in the need to sustain both metathesis activity and stability of the metallodendrimer. Thus, we selected the ruthenium family of catalysts, and designed metallodendrimers containing ruthenium-benzylidene fragments [6] located at the dendrimer periphery and chelating phosphine ligands on the branch termini. The choice of chelating phosphines may seem counter-intuitive, because the activity of Grubbs' catalysts involves the decoordination of a phosphine from these *trans*-bis-phosphine complexes [6]. Studies by the groups of Hofmann, Fogg and Leitner, however, had shown the metathesis activity of *cis*-bis-phosphine ruthenium benzylidene catalysts [7]. We therefore used Reetz's bis-phosphines derived from the commercial polyamine DSM dendrimers [8]. These dendritic bis-phosphines are useful and versatile in metallodendritic catalysis and provided the first recyclable metallodendritic catalysts [8]. They also very cleanly yielded, with two phenyl groups on each phosphorus atom, the first dendrimers decorated with clusters at the periphery by an efficient electron-transfer-chain reaction using $[Ru_3(CO)_{12}]$ catalyzed by $[Fe^{I}Cp(\eta^6\text{-}C_6Me_6)]$, **1**, leading to the substitution of a carbonyl by a dendritic phosphine on each tether [9]. Related dendritic bis-phosphines with two cyclohexyl groups on each phosphorus were decorated with ruthenium benzylidene metathesis using Hoveyda's ruthenium benzylidene metathesis catalyst [10] as a starting point. These reactions provided the four generations of new, stable metallodendrimers containing ruthenium-benzylidene fragments at the periphery (Scheme 1) [11–13]. The fourth-generation metallodendrimer containing 32 ruthenium-benzylidene fragments, however, was found to have a rather low solubility in common organic solvents, unlike the three first-generation complexes that respectively contained 4, 8 and 16 ruthenium-benzylidene moieties. This weak solubility of the 32-Ru dendrimer is presumably due to steric congestion at its periphery. Such steric congestion is also responsible for the decrease of the catalytic activity of Ru and Pd high-generation dendritic catalysts shown in Scheme 3, even when these metallodendritic catalysts are soluble [14]. The X-ray crystal structure of the model mononuclear complex in which the dendritic branch was replaced by a benzyl group showed the distorted square pyramidal geometry and the classic geometric features of a Ru=C double bond. The oxygen atom of the isopropyl aryl ether group is not coordinated unlike in Hoveyda's complex **1**. The fundamental organometallic chemistry of this monomeric model complex was also original and was reported elsewhere [11, 12].

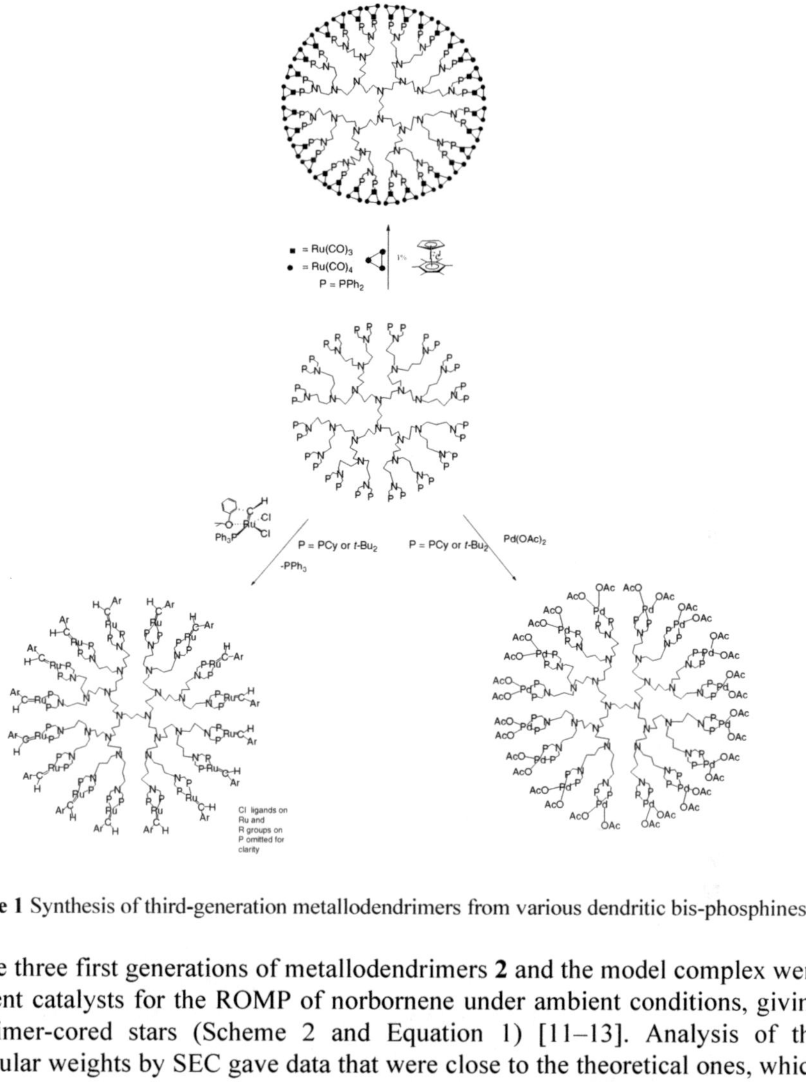

Scheme 1 Synthesis of third-generation metallodendrimers from various dendritic bis-phosphines

The three first generations of metallodendrimers **2** and the model complex were efficient catalysts for the ROMP of norbornene under ambient conditions, giving dendrimer-cored stars (Scheme 2 and Equation 1) [11–13]. Analysis of the molecular weights by SEC gave data that were close to the theoretical ones, which indicated that all the branches were efficiently polymerized. Dendritic-cored stars with an average of about 100 norbornene units on each dendritic branch were synthesized with the three first generations of ruthenium-carbene dendrimers containing respectively 4, 8 and 16 Ru=C bonds.

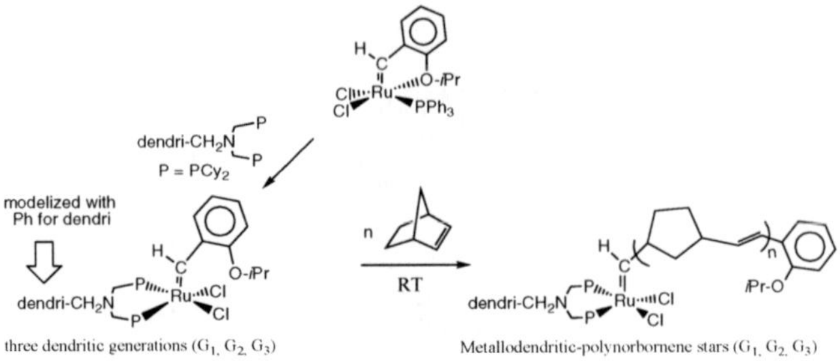

2

Scheme 2 Strategy for the ROMP of norbornene by Ru-benzylidene dendrimers to form dendrimer-cored stars

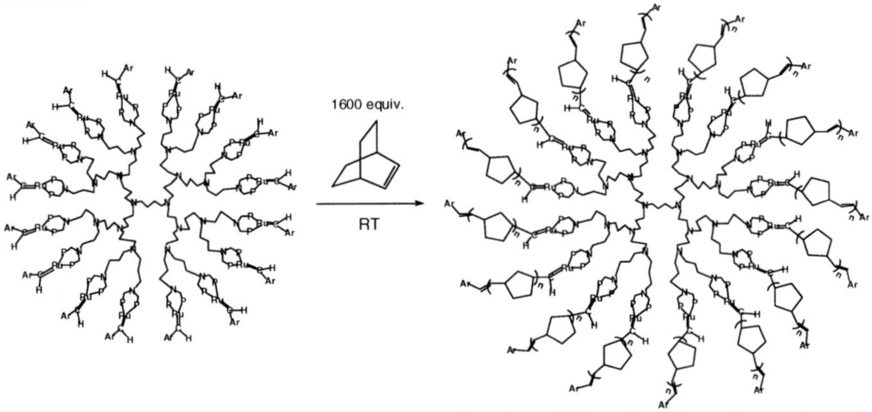

Equation 1 Third-generation (16 Ru atoms) ruthenium-benzylidene dendrimer that catalyzes the ROMP of norbornene at 25°C to form dendrimer-cored stars

Two kinds of dendritic effect were found upon analysis of the kinetic data:

(i) The dendrimers were more efficient catalysts than the monomeric model complex. This could possibly be due to labilization of metal-phosphine bonds that is facilitated in dendrimers as compared to the monomer for entropic reasons. Indeed, DFT calculation showed that the catalytic process must involve decoordination of a phosphorus atom, since the interaction of the olefin with the diphosphine complex is non-bonding. The dendritic ruthenium-benzylidene dendrimers were air-sensitive contrary to the monomer model complex, consistent with more rapid dissociation of the alkyl phosphine in the dendrimers than in the monomer.

(ii) The efficiency of catalysis decreased upon increasing the dendrimer generation. This second dendritic effect is thus a negative one, and it is probably related to the more difficult access to the metal center due to the increasing steric effect at the dendrimer periphery when the generation increases.

The analogous ruthenium benzylidene dendrimers were very recently synthesized with two tert-butyl groups on each phosphorus, and they were slightly more reactive ROMP catalysts for the polymerization of norbornene than those involving the cyclohexyl substituents [13]. These new dendritic ligands, in particular those of low generation (with up to eight branches) proved very efficient in palladium catalysis.

3 Metathesis of Dendrimers Containing Peripheral Olefin Termini: Coupling $CpFe^+$-Induced Arene Perfunctionalization with RCM and Cross Metathesis

Temporary coordination of arenes to the strongly electron-withdrawing cationic 12-electron group $CpFe^+$ largely increases the acidity of its benzylic protons (the *p*Ka's of the arenes in DMSO are lowered upon complexation with $CpFe^+$ by approximately 15 units, for instance from 43 to 28 for C_6Me_6) [15–17]. Therefore, deprotonation of the $CpFe(arene)^+$ complexes is feasible under mild conditions with KOH. Deprotonated $CpFe(arene)^+$ complexes are good nucleophiles, and reactions with electrophiles such as the alkyl halides lead to the formation of new C–C bonds. Coupling the deprotonation and the nucleophilic reactions *in situ* in the presence of excess substrates leads to perfunctionalization in cascade multi-step reactions. When the electrophile is allyl bromide, polyolefin compounds are produced, and these compounds are ideal substrates for ring closing metathesis (RCM) and cross metathesis (CM). For instance, below are shown new structures that were obtained using this strategy with durene (Scheme 3), *p*-xylene (Scheme 4), mesitylene (Scheme 7), and pentamethyl-cobaltocenium (Scheme 5) [18–20].

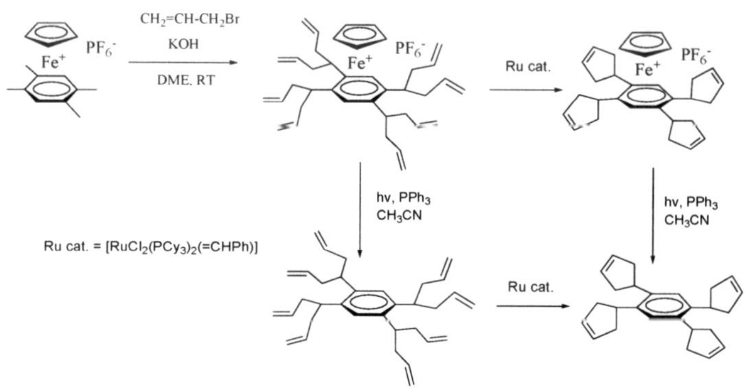

Scheme 3 $CpFe^+$-induced octa-allylation of durene followed by RCM

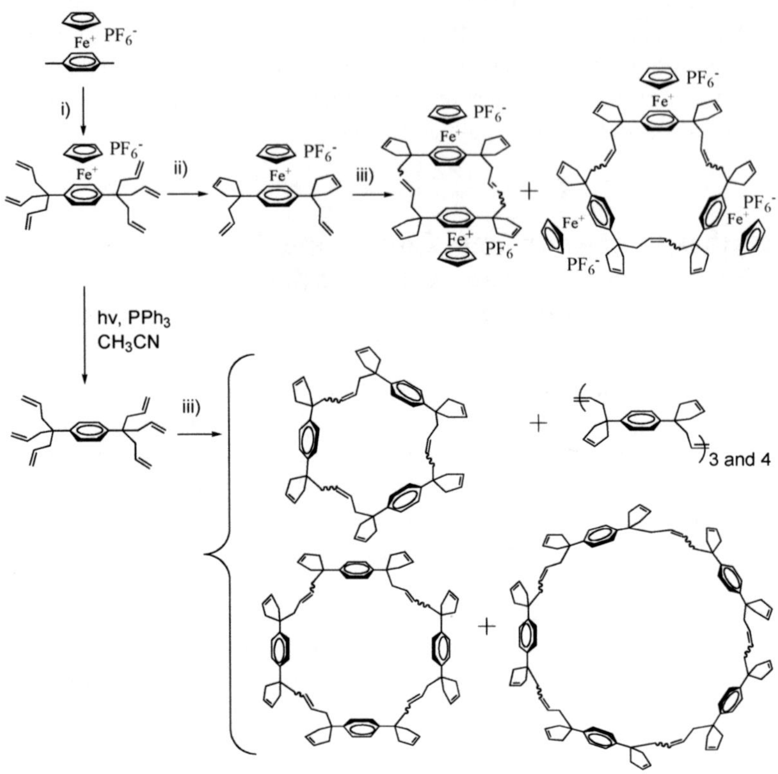

(i) Allyl bromide, KOH, DME

(ii) $[Ru(PCy_3)_2Cl_2(=CHPh)]$, CH_2Cl_2, RT

(iii) $[Ru(PCy_3)\{C(N(mesityl)CH)_2\}Cl_2(=CHPh)]$, **2**, $C_2H_4Cl_2$, 60°C

Scheme 4 Variety of structures obtained from p-xylene upon $CpFe^+$-induced perallylation followed by olefin metathesis catalyzed by Grubb's catalyst

Pentamethylcobalticinium was perallylated to yield a deca-allylated cobalticinium, and then RCM of the organometallic complex proceeded to afford a pentacyclopentenylcyclopentadienyl Co sandwich **3** complex using the catalyst $[Ru(PCy_3)_2Cl_2(=CHPh)]$ [6].

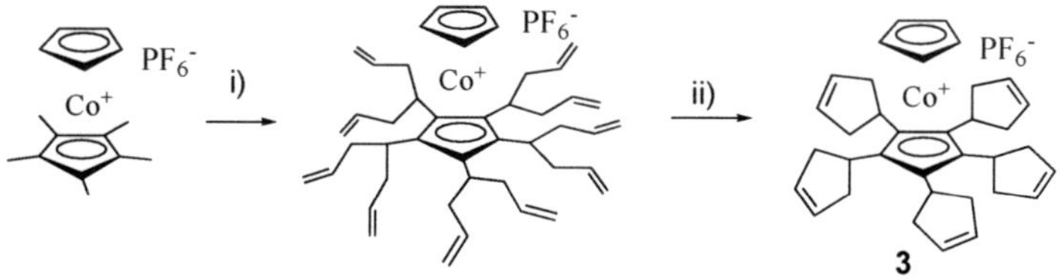

(i) Allyl bromide, KOH, DME; (ii) [Ru(PCy$_3$)$_2$Cl$_2$(=CHPh)], CH$_2$Cl$_2$, RT

Scheme 5 Deca-allylation of pentametylcobalticinium followed by quintuple RCM

The complex [CpFe(*p*-isopropylbenzenze)$^+$][PF$_6^-$] (Scheme 6) **47** is ideal for building a family of *p*-dialkenylaryl derivatives. In order to obtain organometallic [*n*] paracyclophanes, dialkenyl complex were synthesised with longer chains. Dialkenylation of the *p*-diisopropylbenzene complex **47** with a ω-alkenyl halide allowed the preparation of a large variety of *para* disubstituted substrates and intramolecular metathesis led to the desired paracyclophanes. Dialkenylation with 5-bromopentene or 6-bromohexene gave **48** and **49** respectively. Using **1** as catalyst in chloroform at room temperature, the substrate **48** led to a mixture of linear oligomers (two to six units) and mono-, bi- and trimetallic paracyclophanes identified by their molecular peaks in the MALDI TOF mass spectrum. On the other hand, **49**, containing alkenyl chains that are one methylene unit longerer than in **48,** selectively led to the cyclophane product **50.** In the ^{1}H NMR spectra of **19**, the signals of the β-hydrogens of the cyclophanes are shifted at 0.51 ppm because of the aromatic anisotropy [20].

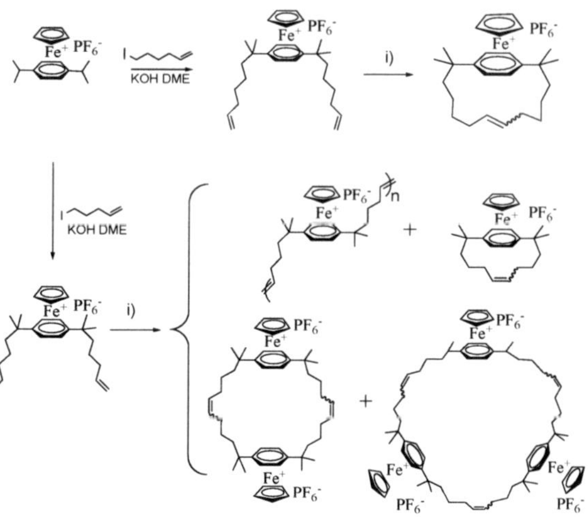

(i) [Ru(PCy$_3$)$_2$Cl$_2$(=CHPh)], CH$_3$Cl, RT

Scheme 6 Synthesis of p cyclophane Fe complexes by RCM

The activation of mesitylene by the $CpFe^+$ moiety followed by a one-pot perallylation yielded $[CpFe(nonaallylmesitylene)^+][PF_6^-]$, **4** (Scheme 9) [16, 17]. First, triple RCM reaction catalyzed by **1** proceeds in 10 min under ambient condition, and a tetracyclic intermediate iron arene complex was isolated. Furthermore and interestingly, when the metathesis reaction was carried out in refluxing dichloroethane and upon adding catalyst **2**, the di-iron cage compound **5** was formed. Similarly, the iron-free nonaallylated compound **6** gave, by metathesis catalyzed by **2**, the organic cage **7**. After hydrogenation with $H_2/Pd/C$ in CH_2Cl_2 of this tripled-bridged cage **7**, a single hydrogenated product is isolated. Another very useful feature is that the organic cage formation can be totally inhibited in the presence of acrylic acid to produce the triacid compound **8** by more rapid stereoselective cross metathesis (Scheme 7) [19, 20].

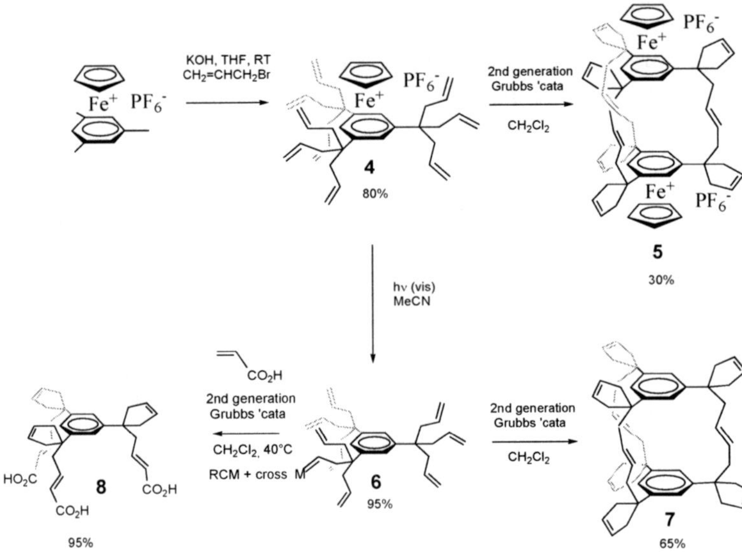

Scheme 7 $CpFe^+$-induced nona-allylation of mesitylene followed by rapid RCM catalyzed by $[RuCl_2(PCy_3)_2(=CHPh)]$ and slow cage formation by cross metathesis catalyzed by the second generation of Grubbs' catalyst vs. cross metathesis in the presence of acrylic acid (*bottom, left*)

Since successful cross metathesis with acrylic acid gives water-soluble compounds, this reaction was exploited to synthesize water-soluble dendrimers with carboxylate termini. Dendritic precursors were prepared with long tethers containing olefin termini so that no competitive RCM occurs unlike in the preceding example. Indeed, CM of these long-chain polyolefin dendrimers catalyzed by the second generation Grubbs metathesis catalyst proceeds selectively to produce dendrimers whose tethers are now terminated by carboxylic acid group (Schemes 8 and 9). The corresponding carboxylates are soluble in water. Higher-generation dendrimers with carboxylic acid termini were synthesized in this way [21, 22].

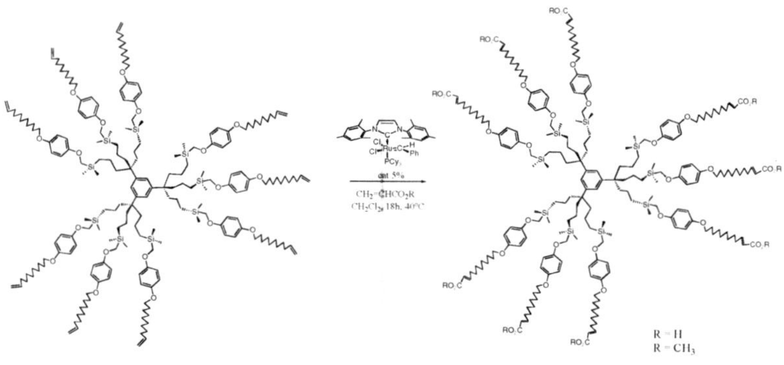

Scheme 8 Example of chemio-, regio- and steroselective cross metathesis of polyolefin dendrimers catalyzed by the second generation Grubbs metathesis catalyst producing water-soluble dendrimers

Other attempts have been reported in the literature for metathesis of polyolefin dendrimers or star compounds, and ring-closing metathesis products were formed. For instance, a third generation Fréchet-type dendrimer containing 24 allyl ether end-groups was synthesized by the Zimmerman group, cross-linked using the ring-closing metathesis (RCM) reaction, and the core was removed hydrolytically without significant fragmentation [23–25]. The results are analogous to those previously reported for homoallyl ether dendrimers suggesting that the less readily available homoallyl ether dendrimers can be replaced by their allyl ethers analogues. The strategy consisting in performing RCM of branches and then to remove the core has also been applied by the Peng group to nanoparticle-cored dendrimers [26, 27].

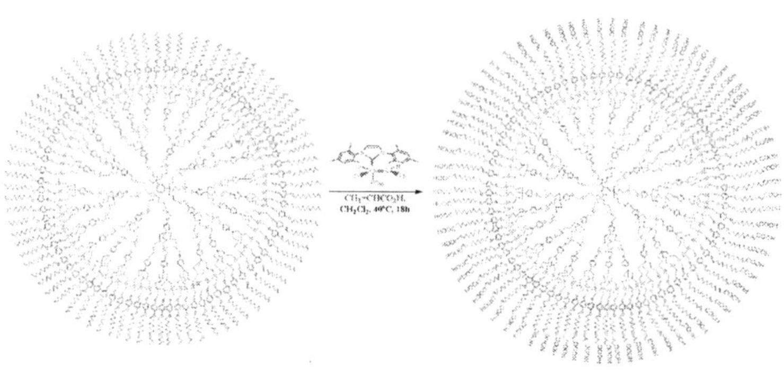

Scheme 9 Example of chemio-, regio- and steroselective cross metathesis of polyolefin dendrimers catalyzed by the second generation Grubbs metathesis catalyst: 81-terthered dendrimers

Finally, the cross metathesis of olefin-terminated dendrimer with acrylates has recently been extended to acrylates that contain a dendronic group [28, 29]. This strategy allows constructing dendrimers from one generation to the next. Thus, iteration allows synthesizing a dendrimer of second generation with 81 olefin termini from a dendritic core containing 9 allyl termini after two iterative metathesis-hydrosilylation reactions (Scheme 10). This principle was also extended to polymers and gold nanoparticles [22].

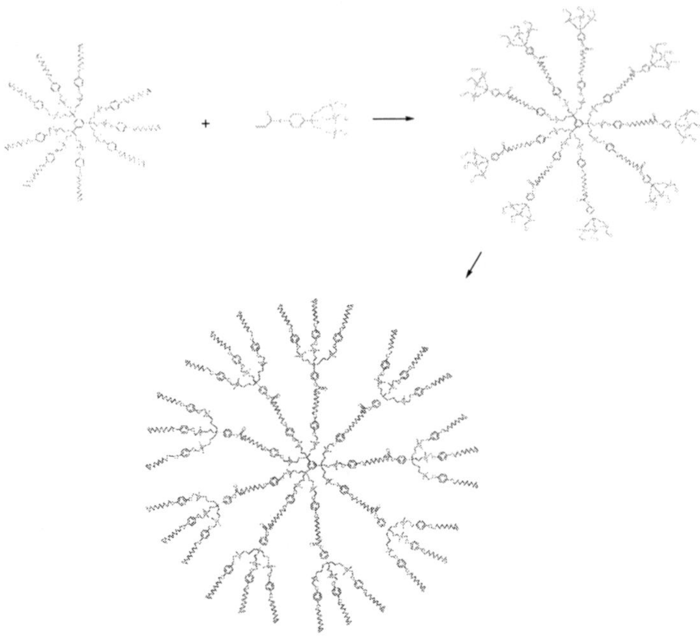

Scheme 10 Dendrimer construction scheme from a 9-olefinic dendrimer to a 27-olefinic dendrimer by regio-and stereoselective cross metathesis using Grubbs' second generation catalyst in CH_2Cl_2 at 40°C, followed by a Wiliamson reaction with p-HOC_6H_4-$O(CH_2)_8CH{=}CH_2$ in DMF a 80°C. The next iteration of identical reaction sequence yields the 81-olefinic dendrimer

4 Concluding Remarks

Introduction of ruthenium metathesis catalysts on the periphery of dendrimers is possible using a chelating phosphine, which lowers the metathesis activity but insures excellent framework stability. Such dendritic metathesis catalysts do not achieve RCM reactions, but ROMP of norbornene proceeds smoothly, yielding dendrimer-cored star polymers with several generations of dendritic catalysts (up to 16 Ru branches). With olefin-terminated dendrimers, RCM reactions are carried out using the first generation of Grubb's catalyst, $[RuCl_2(PCy_3)_2({=}CHPh)]$, whereas more difficult cage formation and cross metathesis reaction require the use of the second generation of Grubbs' ruthenium catalyst. In conclusion, alkene metathesis are very useful selective reactions for the assembly of nanoarchitectures including dendritic ones. Of remarkable interest is the novel strategy involving dendrimer construction using the olefin cross metathesis reaction between olefin-terminated dendrimers and dendronic acrylates.

Acknowledgement I should like to warmly thank all the students and colleagues who have greatly contributed to our studies on dendrimer metathesis and whose names are cited in the references. Financial support from the Institut Universitaire de France, the Université Bordeaux 1, the CNRS, the Ministère de la Recherche et de la Technologie, the Agence Nationale pour la Recherche (Project ANR-07-CP2D-05-03) is gratefully acknowledged.

References

[1] (a) Chauvin Y (2006) Angew Chem Int Ed 45:3740; (b) Schrock RR (2006) Angew Chem Int Ed 45:3748; (c) Grubbs RH (2006) Angew Chem Int Ed 45:3748; (d) Grubbs RH (ed) (2003) Handbook of metathesis, vols 1–3. Wiley-VCH, Weinheim

[2] (a) Astruc D (2007) Organometallic chemistry and catalysis, Ch 9, 15, 21. Springer, Berlin; (b) for a historical perspective and new developments on the metathesis reactions, see: Astruc D (2005) New J Chem 29:42

[3] For applications in total synthesis, see ref. 1d, vol. 2 and Nicolaou KC, Bulger PG, Sarlah D (2005) Angew Chem Int Ed 44:4490

[4] For applications in polymer chemistry, see ref. 1d, vol. 3 and Buchmeiser MR (2000), Chem Rev 100:1565

[5] (a) Wijkens P, Jastrzebski JTBH, van der Schaaf PA, Kolly R, Hafner A, van Koten G (2000) Org Lett 2:1621; (b) Garber SB, Kingsbury JS, Gray BL, Hoveyda AH (2000) J Am Chem Soc 122:8168; (c) Beerens H, Verpoort F, Verdonck L (2000) J Mol Catal 151:279; (d) Beerens H, Verpoort F, Verdonck L (2000) J Mol Catal 159:197

[6] (a) Sanford MS, Ulman M, Grubbs RH (2001) J Am Chem Soc 123:749; (b) Trnka TM, Grubbs RH (2001) Acc Chem Res 34:18

[7] (a) Hansen SM, Volland MAO, Rominger F, Eisenträger F, Hofmann P (1999) Angew Chem Int Ed 38:1273; (b) Hansen S, Rominger F, Metz M, Hofmann P (1999) Chem Eur J 5:557; (c) Adlhart C, Volland MAO, Hofmann P, Chen P (2000) Helv Chim Acta 83:3306; (d) Amoroso D, Fogg DE (2000) Macromolecules 33:2815; (e) Six C, Beck K, Wegner A, Leitner W (2000) Organometallics 19:4639

[8] Reetz MT, Lohmer G, Schwickardi R (1997) Angew Chem Int Ed 36:1526

[9] Alonso E, Astruc D (2000) J Am Chem Soc 122:3222

[10] Kingsbury J, Harrity JPA, Bonitatebus PJ, Hoveyda AH (1999) J Am Chem Soc 121:791

[11] Gatard S, Nlate S, Cloutet E, Bravic G, Blais J-C, Astruc D (2003) Angew Chem Int Ed 42:452

[12] Gatard S, Kahlal S, Méry D, Nlate S, Cloutet E, Saillard J-Y, Astruc D (2004) Organometallics 23:1313

[13] Méry D, Astruc D (2005) J Mol Catal A 227:1

[14] (a) Lemo J, Heuzé K, Astruc D (2005) Org Lett 7:2253; (b) Astruc D, Heuze K, Gatard S, Méry D, Nlate S, Plault L (2005) Adv Syn Catal 347:329; (c) Heuzé K, Méry D, Gauss D, Astruc D (2003) Chem Commun:2274; (d) Heuzé K, Méry D, Gauss D, Blais J-C, Astruc D (2004) Chem Eur J 10:3936; (e) Lemo J, Heuzé K, Astruc D (2007) Chem Commun:4351

[15] (a) Trujillo HA, Casado CM, Ruiz J, Astruc D (1999) J Am Chem Soc 121:5674; (b) Trujillo HA, Casado CM, Astruc D (1995) J Chem Soc Chem Commun:7; (c) Moulines F, Astruc D (1988) Angew Chem Int Ed 27:1347; (d) Moulines F, Djakovitch L, Boese R, Gloaguen B, Thiel W, Fillaut J-L, Delville M-H, Astruc D (1993) Angew Chem Int Ed

32:1075; (e) Catheline D, Astruc D (1983) J Organomet Chem 248:C9; (f) Astruc D, Blais J-C, Cloutet E, Djakovitch L, Rigaut S, Ruiz J, Sartor V, Valério C (2000) Top Curr Chem 120:229
[16] Astruc D, Nlate S, Ruiz J (2002). In: Astruc D (ed) Modern arene chemistry p 400. Wiley-VCH, Weinheim
[17] Sartor V, Djakovitch L, Fillaut J-L, Moulines F, Neveu F, Marvaud V, Guittard J, Blais J-C, Astruc D (1999) J Am Chem Soc 121:2929
[18] Martinez V, Blais J-C, Astruc D (2002) Org Lett 4:651
[19] Martinez V, Blais J-C (2003) Angew Chem Int Ed 42:4366
[20] (a) Martinez V, Blais J-C, Bravic G, Astruc D (2004) Organometallics 23:861; (b) Lacoste M, Rabaa H, Astruc D, Le Beuze A, Saillard J-Y, Précigoux G (1989) Organometallics 8:2233
[21] Ornelas C, Méry D, Ruiz J, Blais J-C, Cloutet E, Astruc D (2005) Angew Chem Int Ed 44:7399
[22] Ornelas C, Méry D, Clouet E, Ruiz J, Astruc D (2008) J Am Chem Soc 130:1495
[23] Beil JB, Lemcoff NG, Zimmerman SC (2004) J Am Chem Soc 126:13576
[24] Lemcoff NG, Spurlin TA, Gewirth AA, Zimmerman SC, Beil JB, Elmer SL, Vandeveer G (2004) J Am Chem Soc 126:11420
[25] Wendland MS, Zimmerman SC (1999) J Am Chem Soc 121:1389
[26] Aldama J, Wang Y, Peng X (2001) J Am Chem Soc 123:8844
[27] Wang YA, Li JJ, Chen H, Peng X (2002) J Am Chem Soc 124:2293
[28] Guo W, Li JJ, Wang X (2003) J Am Chem Soc 125:3901
[29] Guo W, Peng X (2003). In: Astruc D (ed) Dendrimers and nanosciences. C R Chimie 8–10 pp 989–997. Elsevier, Paris

Alkene Metathesis and Renewable Materials: Selective Transformations of Plant Oils

Raluca Malacea, Pierre H. Dixneuf*

Laboratory "Catalyse et Organométalliques", Institut Sciences Chimiques de Rennes, UMR 6226: CNRS-University of Rennes, Campus de Beaulieu, Av. Général Leclerc, 35042 RENNES, France

*E-mail: Pierre.Dixneuf@univ-rennes1.fr

Abstract The olefin metathesis of natural oils and fats and their derivatives is the basis of clean catalytic reactions relevant to green chemistry processes and the production of generate useful chemicals from renewable raw materials. Three variants of alkene metathesis: self-metathesis, ethenolysis and cross-metathesis applied to plant oil derivatives will show new routes to fine chemicals, bifunctional products, polymer precursours and industry intermediates.

Keywords Alkene metathesis · Vegetal oil · Ruthenium catalysts · Green chemistry · Self-metathesis · Ethenolysis · Cross-metathesis

1 Introduction

In recent years, sustainability has become increasingly important for the chemical industry and interest has developed for industrial applications from renewable resources. Plant oils have been established as useful materials for the chemical industry, and the current focus on the use of renewable materials has stimulated the search for new applications. Among the plant materials as renewable sources for the production of energy and chemicals, vegetal oils have received renewed attention because of their biodegradability, safety, price, and potential alternative or competitiveness with petrochemicals for the development of new value-added products. They contain a large paraffinic fragment and most of them are not chemically very different from some petroleum fractions. Therefore, in the last years, efforts have been devoted to the conversion of fats into fuels. In addition the vegetal and animal fats can also be used for the production of chemicals that have a larger added value like lubricants, surface coatings, polymers, pharmaceuticals and cosmetics. Fats and oils are obtained from plant and animal sources, mainly formed by triglycerides having mixed fatty acid moieties. A large proportion of vegetal oils, such as coconut, palm, and palm kernel oils, come from countries

V. Dragutan et al. (eds.), *Green Metathesis Chemistry: Great Challenges in Synthesis, Catalysis and Nanotechnology*,

with tropical climates. Soybean, rapeseed, and sunflower oils come from moderate climate countries. Although, in principle other parts of plants may yield oil, in practice the seeds are the almost exclusive sources. Seed oils are considered more healthful than animal fats because they contain less saturated fats, thus contain more unstaurations allowing functionalization. The composition of vegetal oils is variable, and so is their use as renewable feedstock for chemical production (Table 1) [1].

Table 1 Composition of vegetal oils

	Oil									
Fatty acid	**Soy bean**	**Palm**	**Rape seed**	**Sunflower**	**Cotton seed**	**Castor**	**Cori-ander**	**Olive**	**Coconut**	**Peanut**
Capric 10:0			0.6	0.2				7.3	6.4	
Lauric 12:0	0.1				0.5				48.5	
Myristic 14:0	0.3	2.5	0.1		0.9				17.6	0.1
Palmitic 16:0	10.9	40.8	5.1	6.5	20	2	3	11	8.4	11.6
Stearic 18:0	3.2	3.6	2.1	4.5	3	1		2.2	2.5	3.1
Oleic 18:1	24	45.2	57.9	21	25.9	6	31	77	6.5	46.5
Linoleic 18:2	54.5	7.9	24.7	68	48.8	3	13	8.9	1.5	31.4
Linolenic 18:3	6.8		7.9		0.3			0.6		
Arachidic 20:0	0.1		0.2							1.5
Ricinoleic						87				
Petroselinic							52			
Saturated	14.7	46.9	8.3	11	25	4	4	13.2	91.9	19.3
Unsaturated	85.3	53.1	91.7	89	75	96	96	86.8	8	78.1

Fatty acid monoesters are usually obtained from the transesterification of natural plant oils and fats with a lower alcohol, e.g., methanol with release of glycerol (Scheme 1). Glycerol is an intermediate for the synthesis of a large number of compounds used in industry like lubricants, solvents, pharmaceuticals, cosmetics, polymers, additives, plasticizers, adhesives, etc. Unsaturated fatty acids have alkyl chains that contain one or more double bonds. Mono-unsaturated fat is found mostly in vegetal oils such as soybean, olive, canola, and peanut oils. Poly-unsaturated fatty acids are found in nuts and oils such as soybean and sunflower oils and in fatty fish oils. Soybean oil is considered to be one of the most well-balanced oil, with a low saturated fat content of 15%, 24% of monounsaturated fatty acids, and around 61% of poly-unsaturated fatty acids.

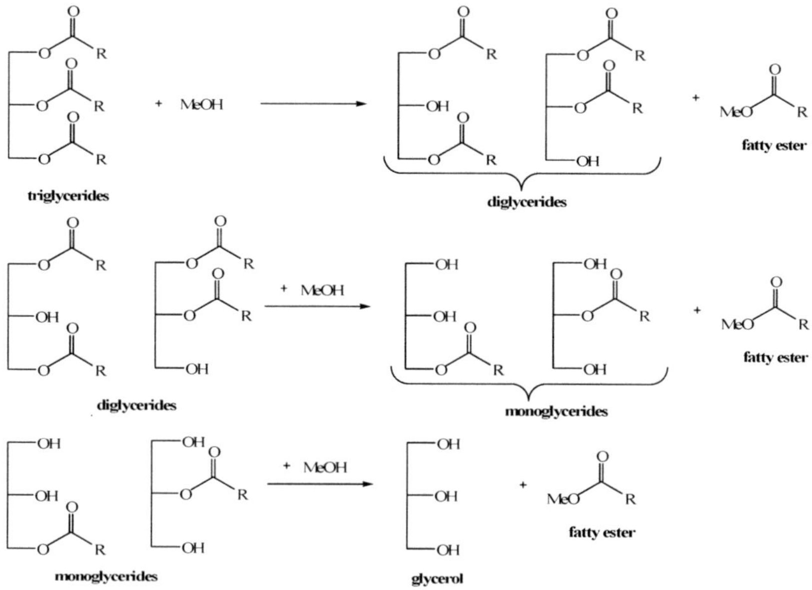

Scheme 1 Transesterification of triglyceride

More than 90% of the chemical reactions of fatty acid esters are carried out at the carboxyl function [2]. Basic chemicals, like fatty alcohols and amines can be produced and further used to generate emulsifiers, lubricants, surfactants, plasticizers, additives, bactericides, fungicides, etc. [1]. However, transformations by reactions involving the carbon–carbon double bond, such as hydrogenation, hydroformylation, hydrocarboxylation, epoxidation and epoxy ring-opening, ozonolysis and dimerization are becoming increasingly important industrially. This review will focus on another increasingly useful reaction of the C=C bond, the olefin metathesis reaction. Olefin metathesis is an important catalytic reaction in organic synthesis, in which olefins are converted into new products via the rupture and reformation of C=C double bonds. This process has been catalyzed by transition metal carbene complexes of Mo, Pd, Pt, Ru, Rh, Ir, Os, etc. (Scheme 2).

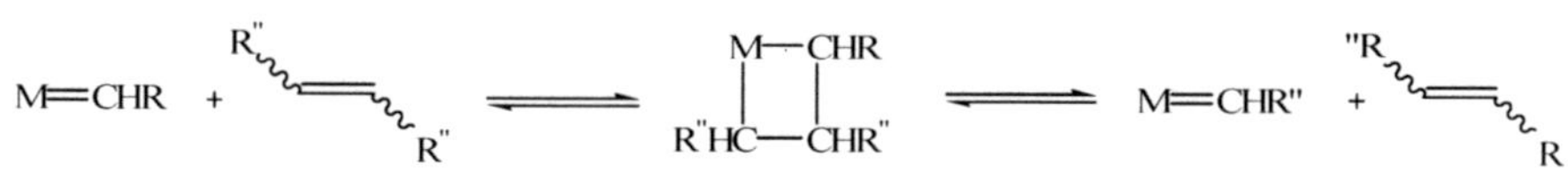

Scheme 2 Metathesis mechanism

The alkene metathesis of natural oils and fats and their derivatives is a clean catalytic reaction that can be considered as an example of green chemistry. A variety of oleo chemicals can be produced by the metathesis of natural oils containing unsaturated fatty acids on the conditions that isomerisation of double bonds does not take place [3].

The metathesis of natural oils can be differentiated in three categories, ***the self-metathesis***, when the reaction takes place between two identical molecules, ***the ethenolysis***, when the reaction takes place between an unsaturated chain and ethylene, and ***the cross-metathesis***, when the reaction takes place between a fatty chain and a different unsaturated molecule.

2 Self-Metathesis

The self-metathesis of fatty methyl esters is a simple way to synthesize alkenes and unsaturated diesters (Scheme 3). The olefins can be further on functionalised to give polyolefins, alcohols, lubricants, surfactants, etc. The diesters are often used for the synthesis of various macrocyclic compounds and as starting material for polyesters or polyamides.

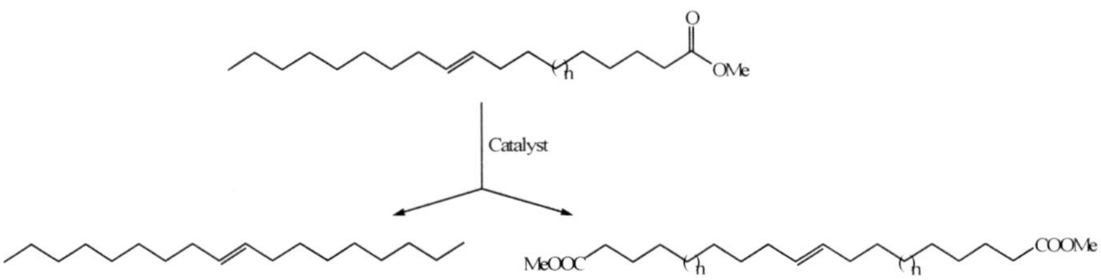

Scheme 3 Self-metathesis of fatty methyl esters

The first example of an active catalyst for metathesis of methyl oleate was described in 1972 by Boelhouwer [4]. The catalyst was a tungsten halide associated with a tetraalkyltin derivative as cocatalyst ($WCl_6/(CH_3)_4Sn$). Heterogeneous systems based on W, Sn, Re complexes have been developed in the meantime, but in the most of the cases a catalyst deactivation was observed [5].

Unsaturated diesters can be used for the production of macrocyclic compounds. For example, in 2002 Mol presented an application of olefin metathesis of natural oils for the synthesis of civetone which is an ingredient in musk perfumes [6]. The metathesis of ethyl oleate gave a diester which has been subjected to a Dieckmann condensation, followed by hydrolysis and decarboxylation, in order to give a *cis–trans* mixture of the unsaturated macrocyclic ketone 9-cycloheptadecen-1-one (**1**) (Scheme 4). The *cis* isomer form, *civetone*, is the active molecule in musk perfumes. If the synthesis starts with the metathesis of methyl oleate, a very pure substrate is required. The methyl oleate is first converted *via* a Claisen condensation to the unsaturated ketone oleon (9,26-pentatriacontadien-18-one). Oleon is then converted into 9-octadecene and a *cis–trans* mixture of 9-cycloheptadecen-1-one (**1**) *via* a ring-closing metathesis reaction. The reaction was carried out at room temperature in the presence of a heterogeneous Re_2O_7-based catalyst.

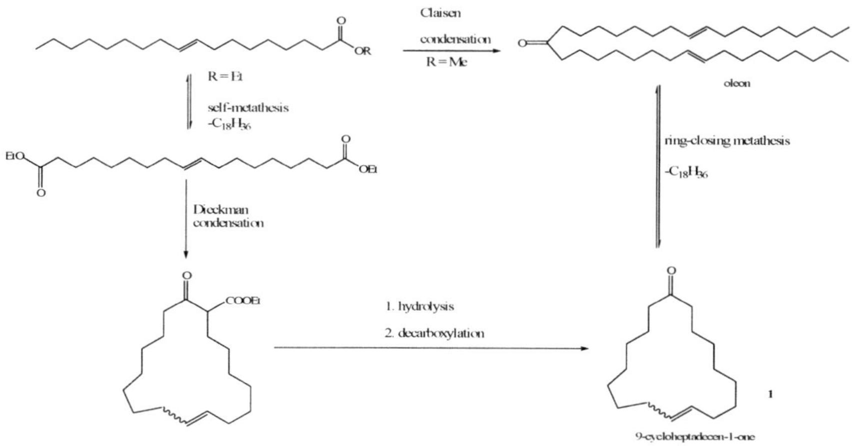

Scheme 4 Synthesis of cyclic ketone from alkyl oleate

Warwel showed that unsaturated polyesters, containing terminal double bond can be synthesized from oleochemical resources and petrochemical diols via two different synthetic pathways (Scheme 5) [7].

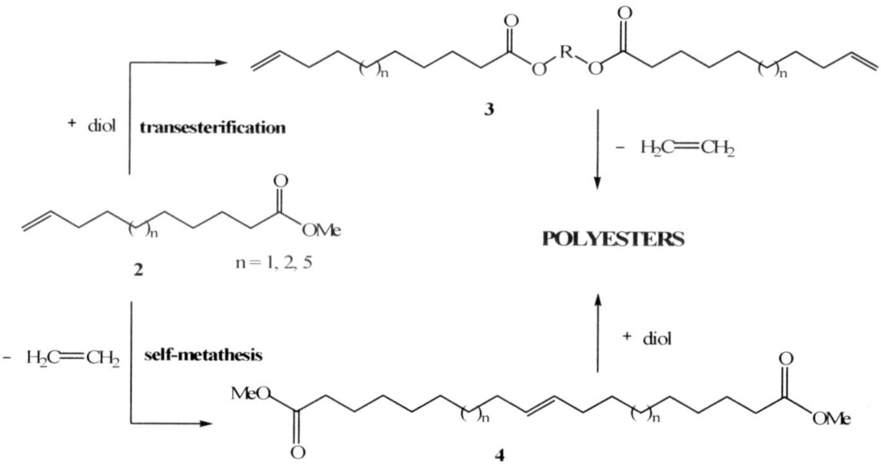

Scheme 5 Synthesis of polyesters from ω-unsaturated esters

9-Decenoic and 10-undecenoic acid methyl ester **2** were transesterified with petrochemical diols to produce the α,ω-alkylene di(un)decenoates **3**, which were subsequently converted into unsaturated polyesters by acyclic diene metathesis polymerization (ADMET). Different homogeneous and heterogeneous metathesis catalysts were evaluated and the best results were obtained with B_2O_3-Re_2O_7 /Al_2O_3-SiO_2-$SnBu_4$. Molecular weights (Mw) up to 100,000 g/mol were obtained.

Generally, the metathetical polymerization needed relatively high catalyst concentrations and the molecular weights were strongly dependent to the catalyst type and concentration. On the other hand, C10, C11 and C14-ω-unsaturated fatty acid methyl esters **2** were metathetically dimerized with a high efficiency using the homogeneous ruthenium $RuCl_2(=CHPh)(PCy_3)$, the Grubbs first generation catalyst, at an extremely low catalyst concentration. The resulting symmetrical, internally unsaturated, dicarboxylic acid dimethyl esters **4** were polycondensated with diol in bulk with conventional transesterification catalysts achieving molecular weights (Mw) up to 110,000 g/mol. Apart from the end groups, structurally identical polymers could be obtained with different thermal behaviour, which was dependent on the *cis/trans* ratio of the C=C double bonds in the polymer.

Heterogeneous Re_2O_7/Al_2O_3 catalyst system is able to efficiently metathesize methyl-10-undecenoate and the catalyst activity increased with the surface area of the Al_2O_3 support [8]. The self-metathesis of methyl-10-undecenoate is an interesting reaction, since the equilibrium can be shifted towards the diester formation if the formed ethylene is removed from the reaction (Scheme 6). The resulting diester can thus be produced in high yields and conversions and is, for instance, a valuable starting material for preparation of polyesters and polyamides.

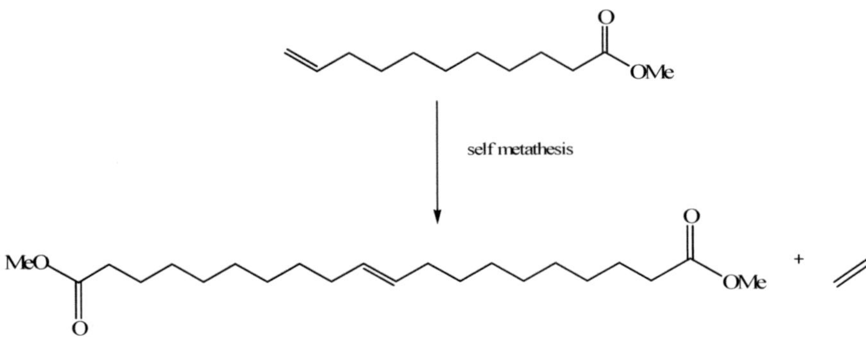

Scheme 6 Self metathesis of methyl-10-undecenoate

The well-defined ruthenium benzylidene complexes developed by Grubbs and coworkers, are excellent functional-group-tolerant metathesis catalysts [9]. One of the most recent homogeneous ruthenium catalyst used in metathesis of unsaturated fatty acid esters was developed in 2006 by Forman [10]. The phoban-indenylidene complex **II** is a robust catalyst for self-metathesis and ethenolysis reactions of methyl oleate (Scheme 7).

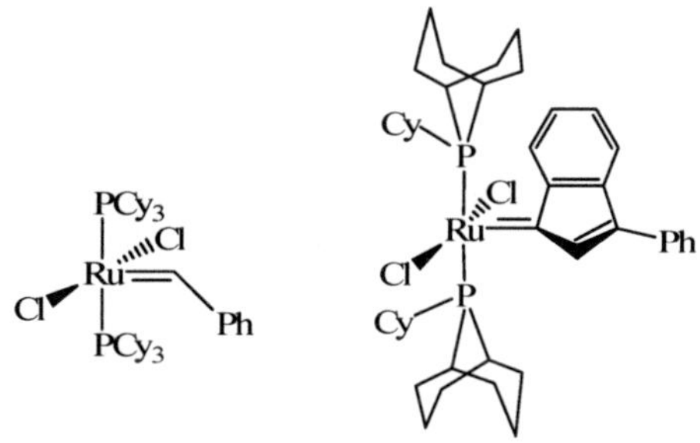

Scheme 7 Alkene metathesis ruthenium catalysts

The phoban-indenylidene catalyst **II** was used in various self-metathesis of methyl oleate, giving rise to significantly higher conversions compared to Grubbs first generation catalyst (Scheme 8, Table 2). These complexes are more stable and active than commercial Grubbs catalyst **I** and it can be accessed from simple precursors.

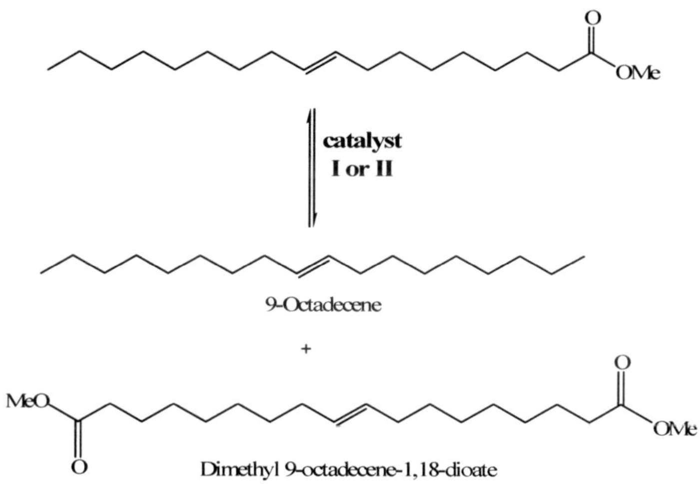

Scheme 8 Self-metathesis of methyl oleate

Table 2 Self-metathesis reactions of methyl oleate to 9-octadecene and dimethyl 9-octadecene-1,18-dioate via Ru-alkylidene catalysts

Substrate to catalyst ratio	Catalyst	Conv. (%)	Selectivity to products (%)	Productivity (t)
20,000:1	I	10.5	94.5	1,985
	II	49.6	97.8	9,702
40,000:1	I	6.2	92.1	2,284
	II	49.0	97.3	19,071
200,000:1	I	2.5	93.7	4,685
	II	37.1	96.5	71,603

3 Ethenolysis

The cross-metathesis of low-molecular-weight simple alkenes with oils or their derivatives, such as ethylene (ethenolysis), has received special attention due to the low cost and abundant supply of ethylene. Ethenolysis gives shorter alkenes or esters which can be further functionalized and used as intermediates in organic synthesis for polyolefins, lubricants, and surfactants [11].

Forman showed also that ethenolysis reaction of methyl oleate catalysed by phoban-indenylidene catalyst **II** gave rise to higher conversions to 1-decene and methyl 9-decenoate compared to Grubbs 1st generation catalyst **I** (Scheme 9, Table 3).

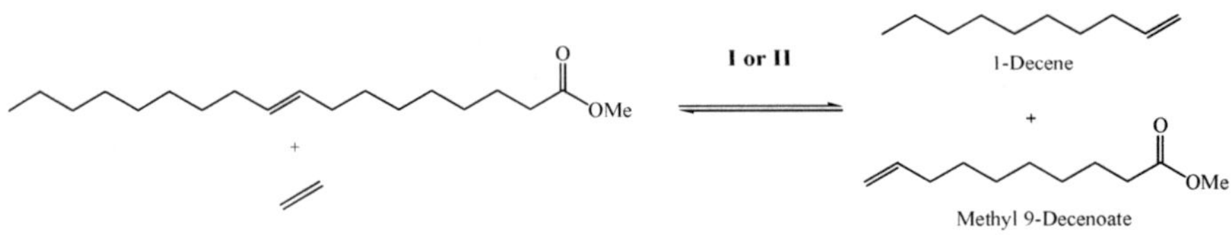

Scheme 9 Ethenolysis of methyl oleate

Table 3 Ethenolysis reactions of methyl oleate to 1-decene and methyl 9-decenoate promoted by Ru-alkylidene catalysts

Substrate to catalyst ratio	Catalyst	Conv. (%)	Selectivity to products (%)	Productive TON
5,000:1	**I**	53.0	98.9	2621
	II	60.0	99.2	3229
10,000:1	**I**	49.6	99.0	4918
	II	69.8	99.1	6917
200,000:1	**I**	43.0	98.5	8542
	II	63.9	97.4	12450

Olefin isomerisation/migration is one of the side reactions in olefin metathesis that can significantly alter the product distribution and decrease the yield of the desired product. Additionally, the side products resulting from unwanted isomerisation and subsequent metathesis are difficult to remove via standard purification techniques. The occurrence of olefin isomerisation during ethenolysis of fatty acid esters has limited its industrial application [12]. In 2005 Grubbs made a screening of different additives and found that electron-deficient benzoquinones can prevent olefin isomerisation of a number of allylic ethers and long-chain aliphatic alkenes during olefin metathesis reactions with ruthenium catalysts [13]. For the ethenolysis of meadow foam oil methyl ester **5** and 11-eicosenyl acetate **6** (Scheme 10) the 1,4-benzoquinone proved to be a good additive in order to suppress olefin isomerisation and it could be easily separated from the desired products. This inexpensive and effective method to block olefin isomerisation increases the synthetic utility of ethenolysis and olefin metathesis in general by improving product yield and purity.

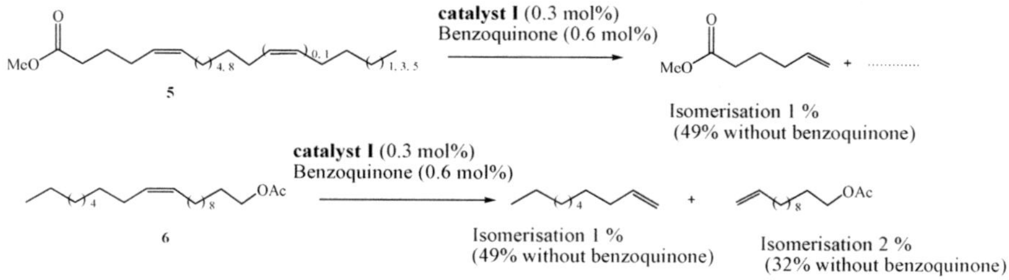

Scheme 10 Ethenolysis of foam oil methyl esters

The ethenolysis reaction can be made directly on triolein 7 at 100 bar of ethylene in the presence of the catalyst **I** and the corresponding unsaturated glycerol derivative can be further converted to a triol and used for the preparation of polyurethane (Scheme 11) [14].

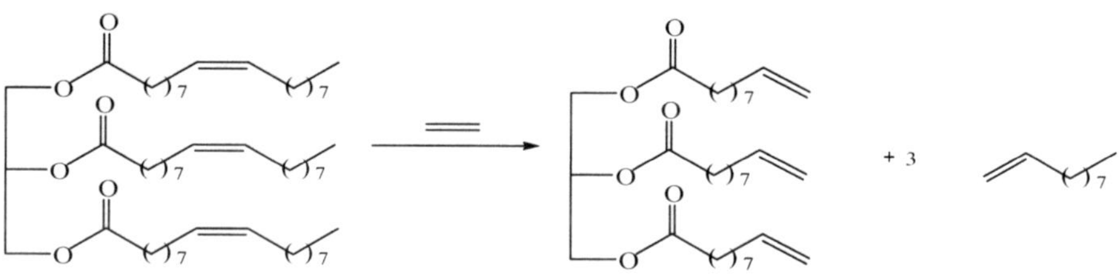

7

Scheme 11 Ethenolysis of triolein 7

In 2008 the Rennes team reported the efficient ethenolysis of methyl oleate under mild conditions with several ruthenium alkylidene catalysts in organic solvents and, for the first time, in imidazolium-type ionic liquids [15]. Alkene metathesis in room-temperature ionic liquids (RTILs) appeared in 1995 and has quickly developed during the last decade [16]. As shown previously, the main products of the methyl oleate ethenolysis are 1-decene and methyl 9-decenoate, but in parallel the self-metathesis reaction could take place giving the 9-octadecene and dimethyl-9-octadecene-1,18-dioate (Scheme 12).

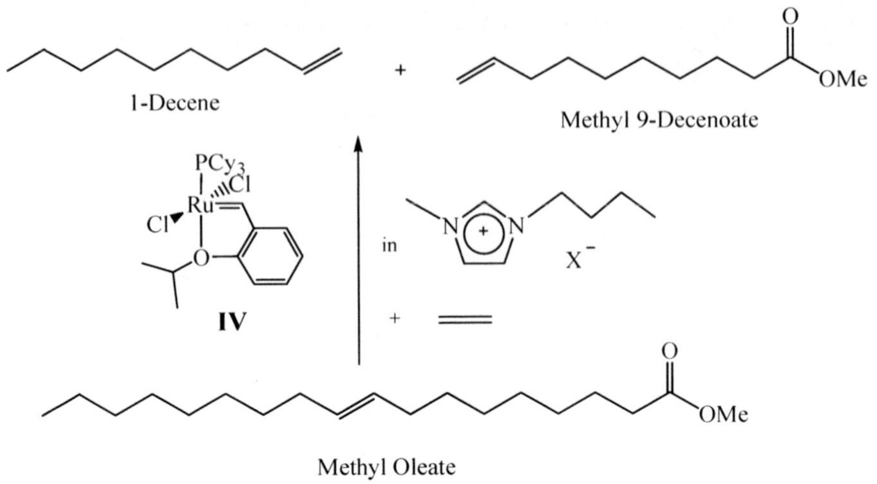

Scheme 12 Ethenolysis of methyl oleate in RT ionic liquids

The action of several ruthenium catalysts **I-VII** (Scheme 13) in toluene was investigated and the catalyst **IV** showed a better efficiency and selectivity than the other catalysts.

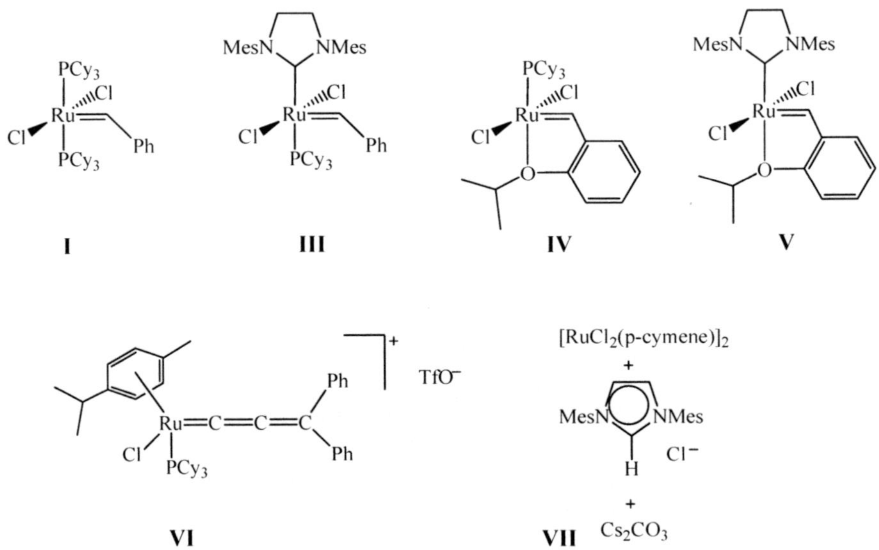

Scheme 13 Alkene metathesis ruthenium catalysts

Three imidazolium-based ionic liquids bearing different counter anions ([bmim][OTf], ([bmim]=1-butyl-3-methylimidazolium); [bmim][NTf_2] and [bdmim] [NTf_2]) were initially evaluated at room temperature in the presence of complex **I** as catalyst and a conversion of 83% of methyl oleate was reached using [bdmim][NTf_2].

Then, two Hoveyda-type catalysts **VIII** and **IX** bearing an ionic tag (Scheme 14) [17] were evaluated for the ethenolysis of methyl oleate in [bdmim][NTf_2]. Catalyst **IX** showed a better catalytic activity for the first run (89% conversion), but it was found to be poorly recyclable under ethenolysis conditions.

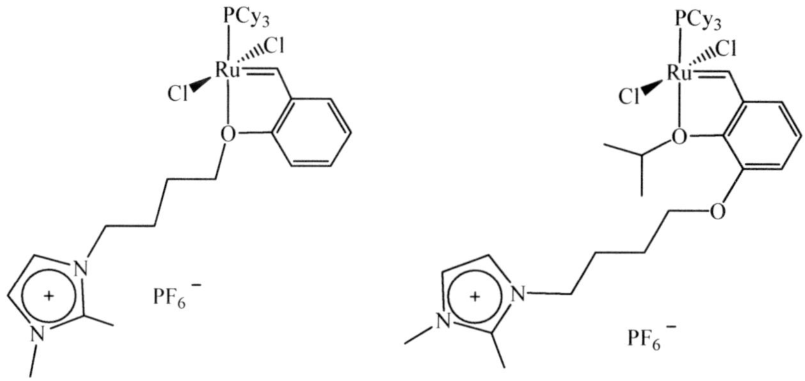

IX **X**

Scheme 14 Alkene ruthenium catalysts bearing an ionic tag

4 Cross-Metathesis

From a synthetic point of view, cross-metathesis reactions are very useful for the introduction of a new functionality and the production of a variety of chemicals with a precise number of carbons atom, which often are difficult to obtain by other means, and especially from petrochemicals. One of the major application of the cross metathesis of vegetal oils was presented by Patel et al. in 2005. They showed that the cross-metathesis of synthetic and natural triglycerides containing unsaturated fatty acids with 2-butene can be achieved with high conversion and excellent productive turnovers [18]. These reactions are catalysed by second-generation ruthenium-based olefin metathesis catalysts and can be conducted at -5°C in liquid 2-butene. The addition of catalysts **III**, **V** or **XI** (Scheme 15) to solutions of triolein (**7**) in liquid cis-2-butene resulted in conversion to the cross-metathesis products **8** and 2-undecene (**9**) (Scheme 16). Catalyst **V** was found to be the most active reaching maximum conversion of oleate chains to 9-undecenoate chains (95%) in less than 4 min with a 0.6 mol% loading of catalyst.

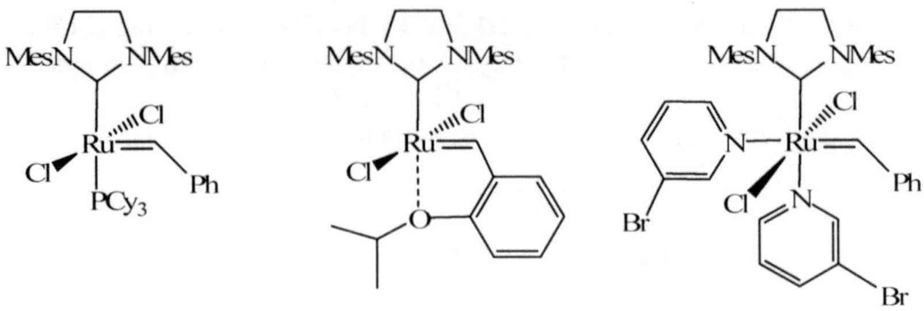

III V XI

Scheme 15 Ruthenium catalysts

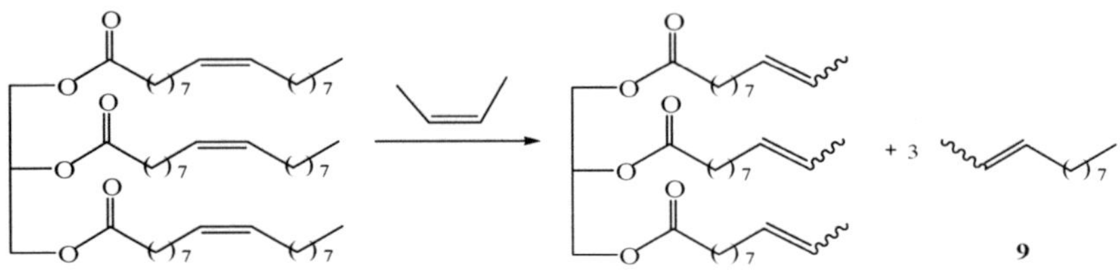

7 8 9

Scheme 16 Cross-metathesis of triolein with 2-butene

Various vegetal oils can be subjected to butenolysis in presence of ruthenium catalyst **III** using 2-cis-butene in excess [19]. Each oil examined was composed of different proportions of oleic acid (C18: 1), linoleic acid (C18: 2) and linolenic acid (C18: 3), and thus in addition to **8**, a range of linear alkenes and dienes was produced from the cross-metathesis with 2-butene (Scheme 17).

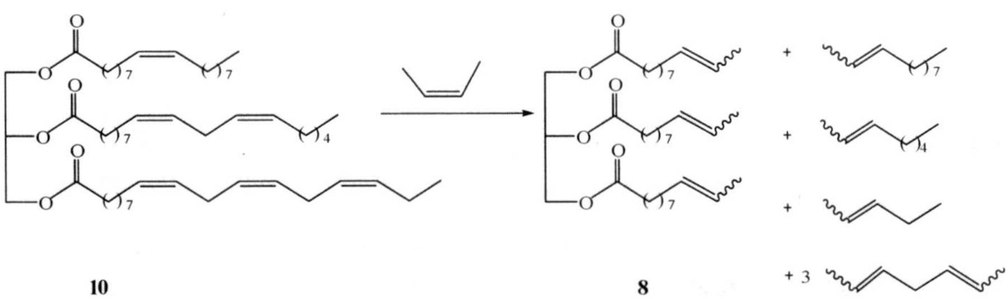

10 8

Scheme 17 Cross-metathesis of triglycerides with 2-butene

Each of the natural oils gave excellent conversion to butenolysis products with a low amount of ruthenium catalyst. The butenolysis products, possessing internal double bonds, could be further on ideal substrates for further metathesis driven transformations.

The same team developed a new process which involved the preparation of terminal oxygenates from renewable natural oils by a one-pot metathesis–isomerisation–methoxycarbonylation–transesterification reaction sequence [20]. The butenolysis was carried out in an autoclave under the previous conditions in order to evaluate the potential of vegetal oil. After elimination of the unreacted butene, the products were transformed by a sequence of reactions: isomerisation into terminal olefin, methoxycarbonylation and transesterification, in saturated mono and diesters (Scheme 18). Application of the one-pot reaction sequence to high oleic sunflower seed oil gave a product mixture very similar in composition to that obtained from the reaction of methyl oleate.

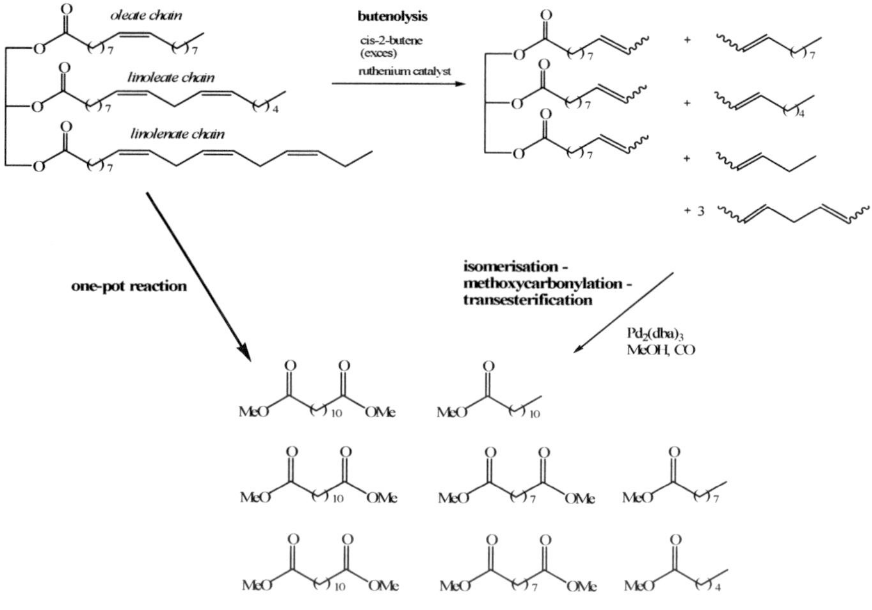

Scheme 18 One-pot metathesis isomerisation methoxycarbonylation transesterification reaction sequence

Another example is the synthesis of 1-triacontanol, $CH_3(CH_2)_{28}CH_2OH$, a plant growth stimulant. This synthesis was performed in a relatively simple two-step process by cross metathesis of methyl erucate with 1-octadecene followed by hydrogenation (Scheme 19) [21]. The metathesis reaction takes place in presence of a WCl_6/Me_4Sn catalyst, and the hydrogenation of the methyl 13-triacontenoate over a Cu/Zn catalyst at 280°C and 250 bar. In addition to the desired metathesis products, the self-metathesis products of the starting materials were also formed.

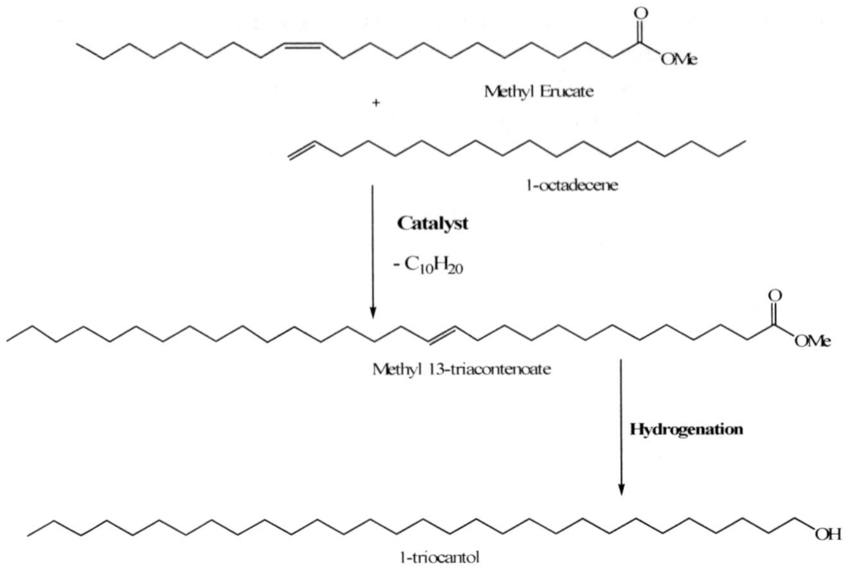

Scheme 19 Synthesis of 1-triacontanol

In 2002 Pederson described the synthesis of OLR (Omnovorous Leafroller) pheromone which is a particularly attractive target for metathesis because the metathesis reaction produces the pheromone with the desired isomeric (E to Z) ratio (Scheme 20) [22]. The two starting materials can be obtained from usual materials: 3-hexene was produced by homocoupling 1-butene and 11-eicosenyl acetate was produced from Jojoba oil. The Jojoba oil is a seed oil used in the cosmetic industry and which contains about 60% of 11-eicosenyl moiety. Cross metathesis of 11-eicosenyl acetate with 3-hexene produced the desired pheromone in 84% yield in the presence of the Grubbs second generation catalyst (**III**).

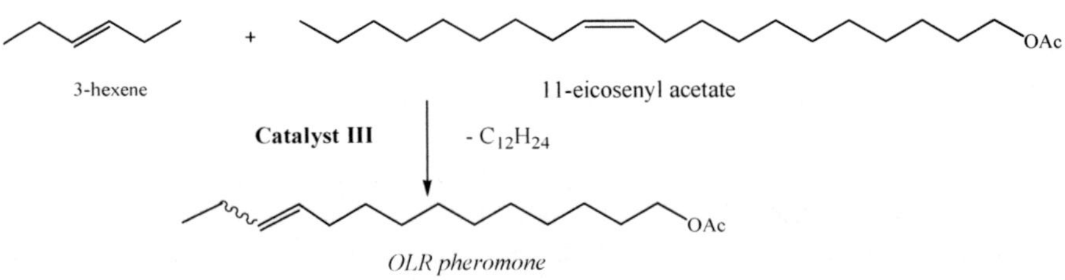

Scheme 20 Synthesis of Omnovorous Leafroller pheromone

Alkene metathesis of plant oils that contain triglycerides of unsaturated long-chain fatty acids proceeds intramolecularly as well as intermolecularly, the latter reaction strongly predominating. Thus, in the presence of the catalyst system

WCl_6/Me_4Sn, olive oil, which consists mainly of glyceryl trioleate (triolein), yields 9-octadecene and polymeric triglycerides, principally dimers and trimers (Scheme 21) [23, 24].

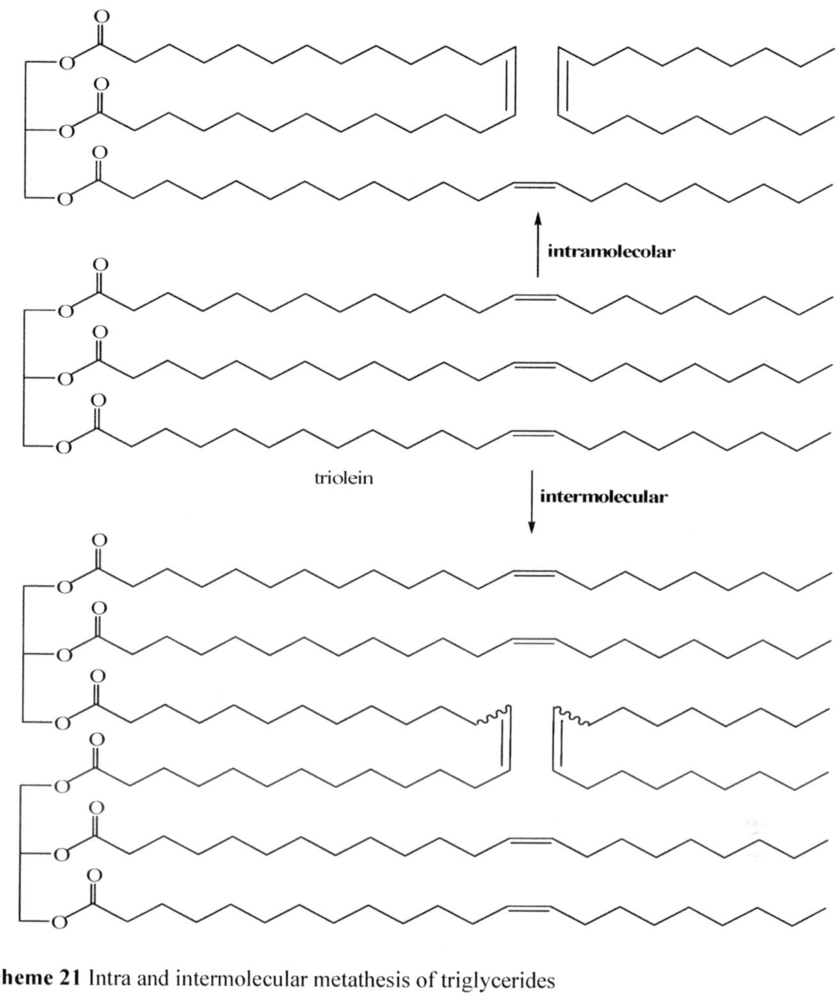

Scheme 21 Intra and intermolecular metathesis of triglycerides

Linseed oil and soybean oil, containing triglycerides of oleic, linoleic and linolenic acid, respectively, were also tested as substrates for metathesis reaction in order to obtain products of increased viscosity used for the manufacture of oil-based paint, printing ink, synthetic resins [25].

It was showed by Meier in 2007 that the cross-metathesis of electron deficient substrates, such as methyl acrylate with fatty acid derivatives was possible with second generation ruthenium metathesis catalysts **III** and **V** [26]. First the methyl oleate was used as a model substrate for the cross metathesis with methyl acrylate and 99% conversion was obtained with the catalyst **V** (Scheme 22). In the presence of other catalysts such as catalyst **IV**, the self metathesis products were observed in large amounts. The advantage of the cross metathesis with catalyst **V** is that the reaction can be performed without solvent, showing that the acrylate does not inhibit the catalyst.

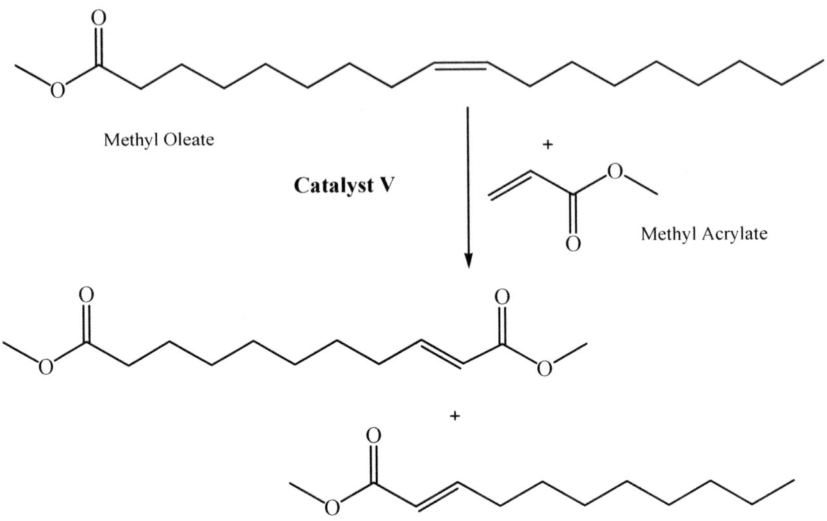

Scheme 22 Solvent free cross metathesis of methyl oleate with methyl acrylate

The cross-metathesis of methyl erucate and methyl petroselinate with methyl acrylate was also studied in presence of ruthenium catalysts **I**, **III** and **V** and in both cases the Hoveyda–Grubbs catalyst **V** showed the best activity, with the presence of self-metathesis products observed as minor products (Scheme 23) [27].

Recently, the same team studied the cross metathesis reaction of oleyl alcohol with methyl acrylate in solvent free conditions (Scheme 24). The reaction conditions were optimized for high conversions in combination with high cross-metathesis selectivity. The introduction of a protecting group for the alcohol functionality of oleyl alcohol was found to be a necessary step, in order to significantly reduce the amount of metathesis catalyst required to obtain fully conversions and good selectivities. For example, in presence of catalyst **V** the protected alcohol substrate allows the reduction of the catalyst amount of at least fivefold, while the conversion remains the same and the selectivity is slightly improved [28].

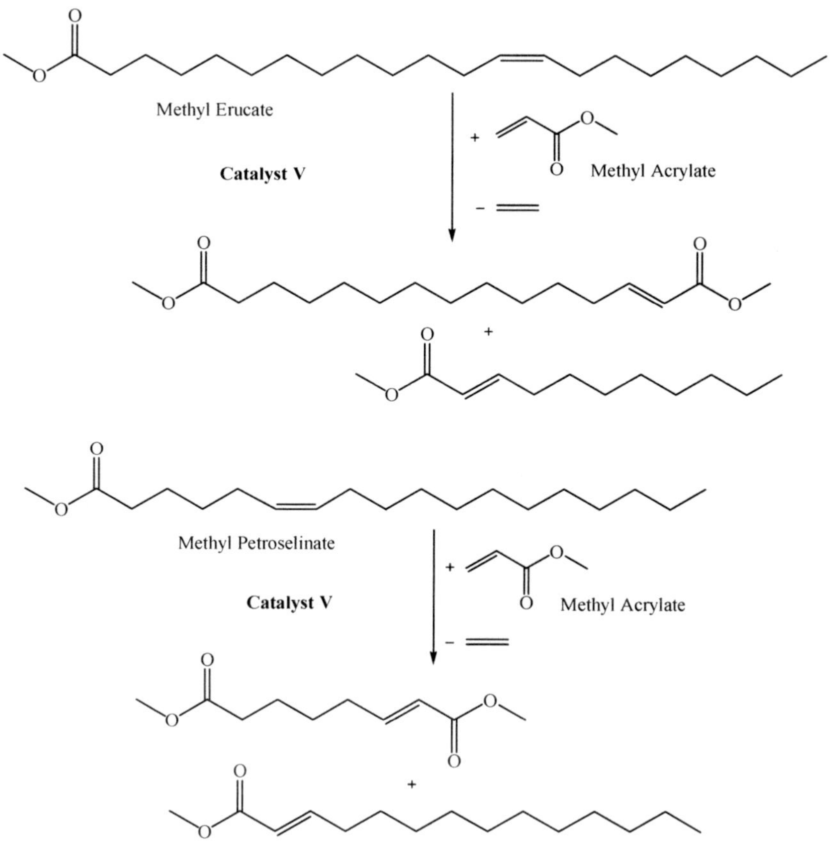

Scheme 23 Cross-metathesis of methyl erucate and methyl petroselinate with methyl acrylate

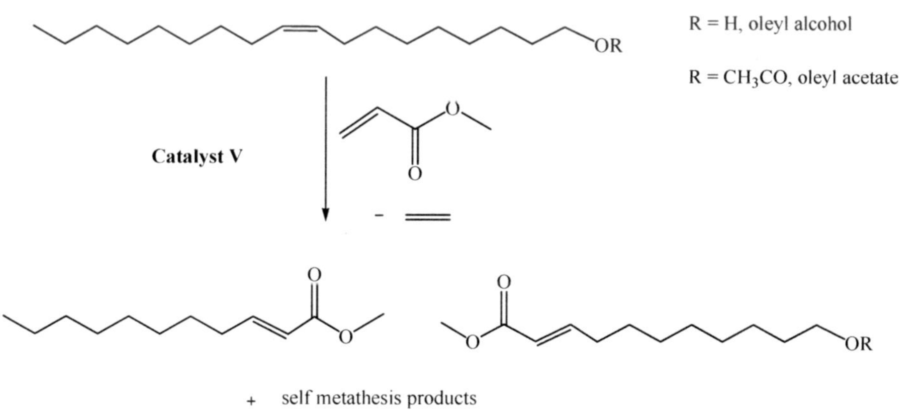

Scheme 24 Cross metathesis of oleyl derivatives with methyl acrylate

Recently, the Rennes team has investigated the cross-metathesis of acrylonitrile and fumaronitrile with unsaturated acids or esters arising from plant oils. Associated with the hydrogenation of both the C=C and CN bonds, this reaction constitutes a strategic way to generate linear amino acid monomers from renewable resources, industrially useful in particular for the production of polyamides [29]. We have thus considered the transformation *via* cross-metathesis with acrylonitrile of two unsaturated acids and related esters arising from renewable resources: the C11 ester **11** obtained from methanolysis of castor oil triglyceride and its related carboxylic acid **12**, and the C18 diester **13** and its diacid **14** obtained via bio-transformation or self-metathesis of oleic acid. We showed that the ruthenium catalyzed cross metathesis of acrylonitrile and the unsaturated C11 ester **11**, or even directly its acid **12**, easily leads to the corresponding nitrile ester and acid **15** and **16**, and that it can be applied to the internal double bond containing C18 diester **13** and its diacid **14**, without isomerisation of the substrates, to give the nitrile ester and acid **17** and **18**, the precursors of C12 and C11 linear aminoacids (Scheme 25). Different ruthenium catalysts were used and the best conversions in cross metathesis products were obtained with the ruthenium Hoveyda catalyst **V**.

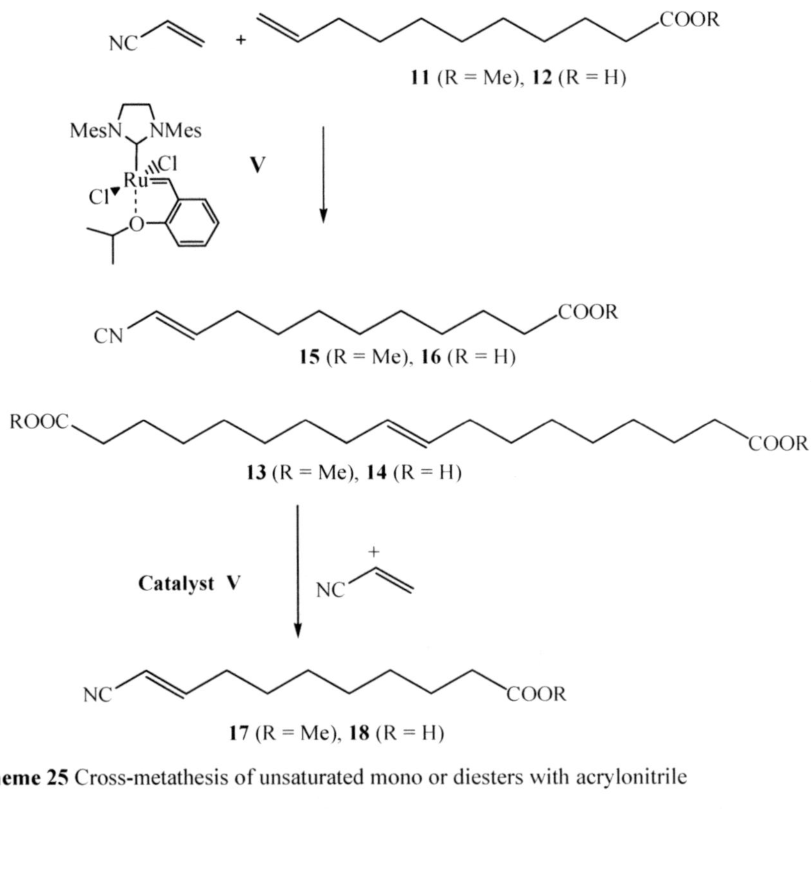

Scheme 25 Cross-metathesis of unsaturated mono or diesters with acrylonitrile

For the cross-metathesis of **13** and **14**, 5% molar of **V** are necessary to obtain total conversion in the cross metathesis products **17** and **18** but for the transformation of **11** or **12**, only 1% molar of **V** is necessary to obtain total conversion. It is noteworthy that the reaction is inhibited in acrylonitrile as solvent, but with only 2 eq. of acrylonitrile an excellent yield (90%) of nitrile esters can be obtained. The cross metathesis between acrylonitrile and an internal double bond containing olefin was previously described only in one example by Blechert, using the Grubbs second generation catalyst and copper chloride as additive [30].

Of special interest, the residual ruthenium catalyst obtained after cross-metathesis can be used as catalyst for the hydrogenation of the unsaturated nitrile esters double bond **16–18**. Two consecutive catalytic reactions are possible using only one starting ruthenium alkene metathesis catalyst that is in situ transformed into a hydrogenation catalyst.

The cross-metathesis of unsaturated esters can be performed directly using fumaronitrile as a source of unsaturated nitrile moiety (Scheme 26). Using 5% molar of ruthenium complex **V** and 2 eq. of fumaronitrile the nitrile ester **17** was obtained in 92% isolated yield [29].

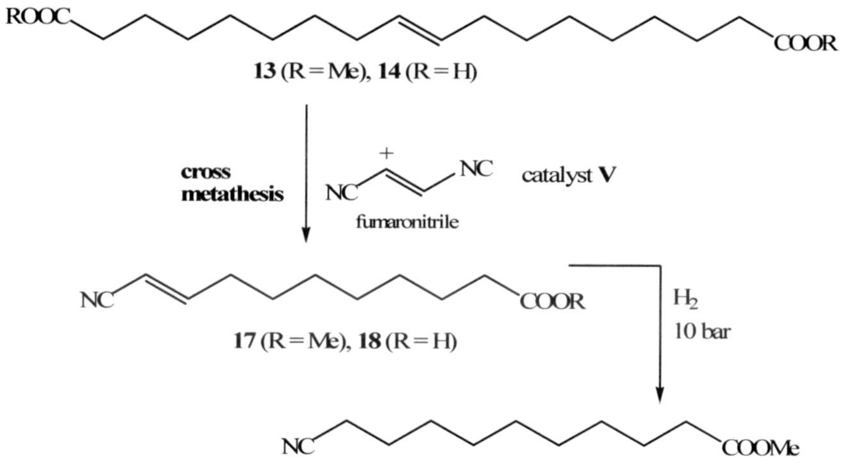

Scheme 26 Cross metathesis of unsaturated diester with fumaronitrile

A new and important method that allows the linkage of a suitably functionalised fatty acid or sugar to the side chains of lysine and cysteine by cross-metathesis was presented by Abell in 2004 [31] (Scheme 27). *N*-Boc-L-lysine **19** was transformed into derivative **20** that bears an olefin arm coming from a fatty acid. This substrate give a cross-metathesis reaction with a new molecule of a fatty acid in presence of the Grubbs catalyst **III** in order to obtain the product **21** in 65%.

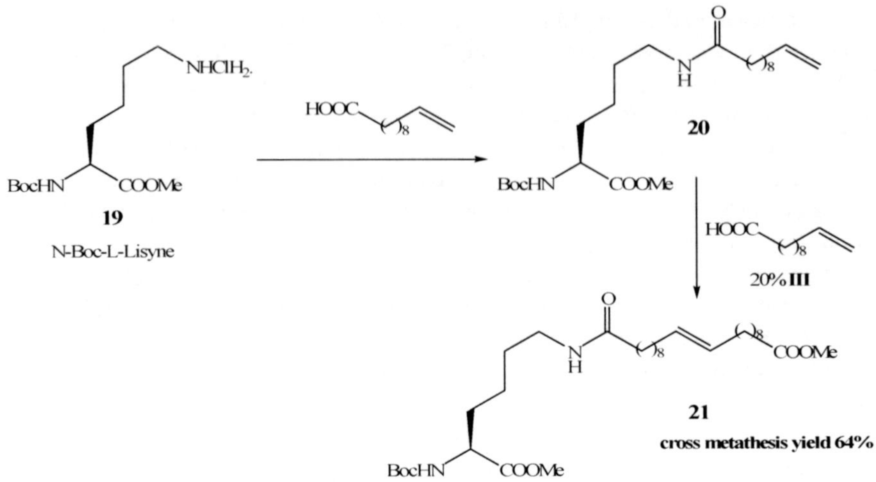

Scheme 27 Cross metathesis of aminoacids with fatty acids

Starting from the *N*-acylated derivative **20**, the cross-metathesis with the sugar derivative **22** give the sugar derivative **23** in 65% yield, in presence of the Grubbs catalyst **III** which can be hydrogenated further on Pd/C catalyst to give the fully deprotected and side chain reduced analogue **24** (Scheme 28).

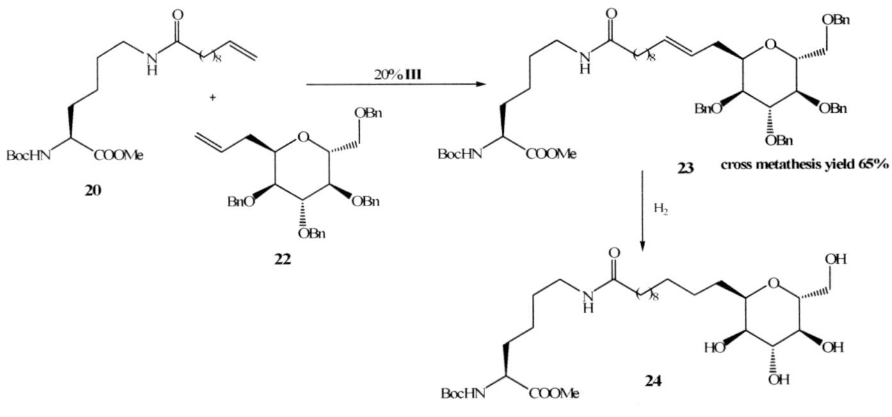

Scheme 28 Conjugation of an amino acid with a sugar by cross metathesis

Cross-metathesis of methyl oleate with allyl chloride has recently been performed and leads to α,ω-chloroester long chain derivatives, that are potential precursors by chloride substitution to linear amino acids and then polyamides. The reaction is performed without solvent and the best catalysts are the Hoveyda catalyst **V** and especially its substituted derivative **XII** (Scheme 29) [32].

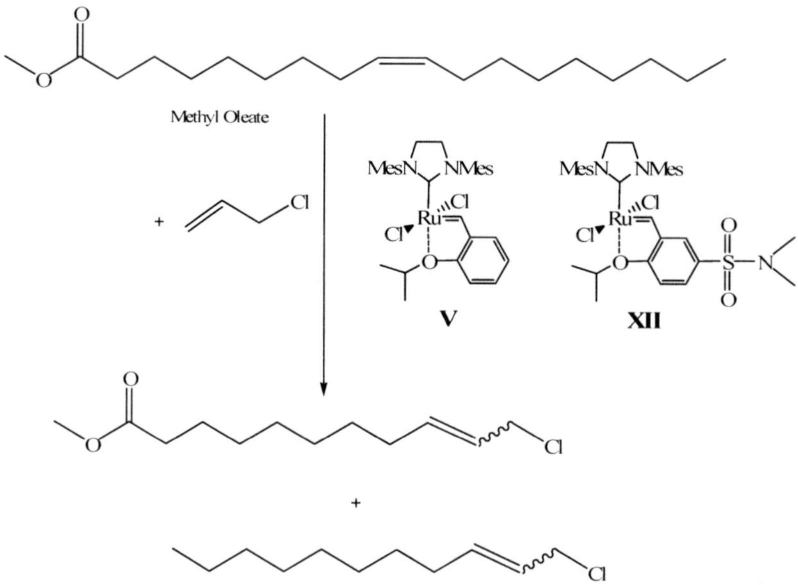

Scheme 29 Cross metathesis of methyl oleate with allyl chloride

The above results on alkene metathesis of the renewable plant oils to afford useful chemicals has attracted the interest of chemical industry, which has recently protected some plant oil metathesis catalysts and processes as well as the functionalisation of resulting simple olefins or unsaturated bifunctional products [33, 34].

5 Conclusion

The above presented results show that a variety of long chain unsaturated compounds derived from plant oils can be selectively transformed via alkene metathesis reactions: self-metathesis, ethenolysis and cross-metathesis, in the presence of different homogeneous of heterogeneous catalysts in high yields. In general, this reaction can be made under mild reaction conditions, and a good choice of the catalytic system could avoid isomerisation of the starting materials or of the generated products. A large variety of mono, bi- and multifunctional compounds could be obtained like esters, nitriles, amino acids that are precursors of various fine chemicals or polymers. Thus the metathesis of natural fats and oils and their derivatives constitutes a catalytic reaction which allows the synthesis of valuable chemicals in few steps from renewable raw materials.

References

[1] Corma A, Iborra S, Velty A (2007) Chem. Rev 107:2411
[2] Ivin KJ, Mol JC (1997) Olefin metathesis and metathesis polymerization. Academic, London
[3] Jackson WR et al. (2006) Green Chem 8:450
[4] Vandam PB, Boelhouwer C, Mittelme MC (1972) J Chem Soc, Chem Commun 1221
[5] Verkuijlen E, Kapteijn F, Mol JC, Boelhouwer C (1977) J Chem Soc, Chem Commun 198
[6] Mol JC (2002) Green Chem 4:5
[7] Warwel S, Tillack J, Demes C, Kunz M (2001) Macromol Chem Phys 202:1114
[8] Ellison A, Coverdale AK, Dearing PF (1983) Appl Catal 8:109
[9] (a) Nguyen ST, Grubbs RH, Ziller JW (1993) J Am Chem Soc 115:9858; (b) Schwab P, France MB, Ziller JW, Grubbs RH (1995) Angew Chem Int Ed 34:2039
[10] Forman GS et al. (2006) J Organomet Chem 691:5513
[11] Yadav GD, Doshi NS (2002) Green Chem 4:528
[12] Burdett KA, Harris LD, Margl P, Maughon BR, Mokhtar-Zadeh T, Saucier PC, Wasserman EP (2004) Organometallics 23:2027
[13] Grubbs RH et al. (2005) J Am Chem Soc 127:17160
[14] Zlatanic A, Petrovic ZS, Dusek K (2002) Biomacromolecules 3:1048
[15] Thurier C, Fischmeister C, Bruneau C, Olivier-Bourbigou H, Dixneuf PH (2008) ChemSusChem 1:118
[16] (a) Chauvin Y, Olivier-Bourbigou H (1995) Chemtech 9:26; (b) Gurtler C, Jautelat M, EP1035093A2; (c) Buijsman RC, Vuuren E, Sterrenburg JG (2001) Org Lett 3:3785; (d) Mayo KG, Nearhoof EH, Kiddle JJ (2002) Org Lett 4:1567; (e) Ding X, Lv X, Hui B, Chen Z, Xiao M, Guo B, Tang W (2006) Tetrahedron Lett 47:2921; (f) Williams DBG, Ajam M, Ranwell A (2006) Organometallics 25:3088; (g) Semeril D, Olivier-Bourbigou H, Bruneau C, Dixneuf PH (2002) Chem Commun 146; (h) Csihony S, Fischmeister C, Bruneau C, Horvath IT, Dixneuf PH (2002) New J Chem 26:1667
[17] Thurier C, Fischmeister C, Bruneau C, Olivier-Bourbigou H, Dixneuf PH (2007) J Mol Catal A 268:127
[18] Patel J, Elaridi J, Jackson WR, Robinson AJ, Serelisb AK, Suchb C (2005) Chem Commun 5546
[19] Patel J, Mujcinovic S, Jackson WR, Robinson AJ, Serelisb AK, Suchb C (2006) Green Chem 8:450
[20] Zhu Y, Patel J, Mujcinovic S, Jackson WR, Robinson AJ (2006) Green Chem 8:746
[21] Penninger J, Biermann M, Krouse HJ (1989) Fette Seifen Anstrichm 85:239
[22] Pederson RL, Fellows IM, Ung TA, Ishihara H, Hajela SP (2002) Adv Synth Catal 344:6
[23] Boelhouwer C, Mol JC (1985) Prog Lipid Res 242:43
[24] Van Dam PB, Mittelmeijer MC, Boelhouwer C (1974) J Am Oil Chem Soc 51:389
[25] Hellbardt S, Patzschke HP (1987) Ullmann's encyclopedia of industrial chemistry, 5th edn. VCH, Weinheim, A9:55
[26] Rybak A, Meier MAR (2007) Green Chem 9:1356
[27] Rybak A, Fokou PA, Meier MAR (2008) Eur J Lipid Sci Technol 110:797
[28] Rybak A, Meier MAR (2008) Green Chem 10:1099
[29] Malacea R, Fischmeister C, Bruneau C, Dubois JL, Couturier JL, Dixneuf PH (2009) Green Chem 11:152
[30] Rivard M, Blechert S (2003) Eur J Org Chem 2225
[31] Vernal AJ, Abell AD (2004) Org Biomol Chem 2:2555
[32] Jacobs A, Rybak A, Meier MAR (2008) Appl Catal A General 353:32–35
[33] Newman TH et al., WO 02/076920 A1
[34] Abraham T et al., WO 2008/048522 A1, WO 2008/060383 A2, WO 2008/048520 A2

Recent Applications of Alkene Metathesis in Fine Chemical Synthesis

Dario Bicchielli,[1] Yannick Borguet,[1] Lionel Delaude,[1] Albert Demonceau,[1*] Ileana Dragutan,[2] Valerian Dragutan,[2] Christo Jossifov,[3] Radostina Kalinova,[3] François Nicks,[1] Xavier Sauvage[1]

[1]Laboratory of Macromolecular Chemistry and Organic Catalysis, University of Liège, Sart-Tilman (B.6a), B-4000 Liège, Belgium

[2]Institute of Organic Chemistry of the Romanian Academy, 202B Spl. Independentei, 060023 Bucharest, P.O. Box 35-108, Romania

[3]Institute of Polymers, Bulgarian Academy of Sciences, Acad. Georgy Bonchev Str., bl. 103-A, 1113 Sofia, Bulgaria

[*]E-mail: A.Demonceau@ulg.ac.be

Abstract During the last decade or so, the emergence of the metathesis reaction in organic synthesis has revolutionised the strategies used for the construction of complex molecular structures. Olefin metathesis is indeed particularly suited for the construction of small open-chain molecules and macrocycles using cross-metathesis and ring-closing metathesis, respectively. These reactions serve, *inter alia*, as key steps in the synthesis of various agrochemicals and pharmaceuticals such as macrocyclic peptides, cyclic sulfonamides, novel macrolides, or insect pheromones. The present chapter is aiming at illustrating the great synthetic potential of metathesis reactions. Shortcomings, such as the control of olefin geometry and the unpredictable effect of substituents on the reacting olefins, will also be addressed. Examples to be presented include epothilones, amphidinolides, spirofungin A, and archazolid. Synthetic approaches involving silicon-tethered ring-closing metathesis, relay ring-closing metathesis, sequential reactions, domino as well as tandem metathesis reactions will also be illustrated.

Keywords Alkene metathesis · Cross-metathesis · Macrocycles · Ring-closing metathesis · Ring-opening metathesis · Ruthenium catalysts

1 Introduction

Natural products provide an immeasurable pool of lead structures for drug discovery, and more than 50% of today's prescription drugs are ultimately derived from compounds first obtained from natural sources. The screening of large

V. Dragutan et al. (eds.), *Green Metathesis Chemistry: Great Challenges in Synthesis, Catalysis and Nanotechnology*,

collections of natural products is nowadays – and will continue to be – an important strategy for the identification of new drug leads or compounds of biochemical and pharmacological interest [1]. Quite often, these new compounds are present in tiny amounts in the nature, which necessitates the handling of a – relatively – huge amount of the raw material. In some cases, as well, their structures were proposed on the basis of the spectroscopic data of a sample of less than 1 mg! and evaluation of their biological properties was hampered by their very low availability. Nature is also a remarkable source of inspiration for synthetic chemists, who are continuously developing new biologically active scaffolds through either simple peripheral derivatisation or extensive structural modification of existing natural product leads. Of utmost importance, in this context, are total syntheses that would provide meaningful amounts of scarce natural products for biological testing, and allow for systematic modification of their molecular structures.

Modern organic synthesis makes an increasing appeal to organometallic chemistry and homogeneous catalysis by transition metal complexes. In particular, during the last 2 decades or so the emergence of the metathesis reaction in chemical synthesis has revolutionised the way in which chemists approach the construction of molecular structures in the laboratory [2–4]. Over the past 10 years technology has developed to promote ring-closing, ring-opening, and cross-metathesis reactions. In particular, ring-closing reactions using the air-stable, commercially available ruthenium alkylidene catalysts, such as the first- and second-generation Grubbs' catalysts (Grubbs 1 (**1**) and Grubbs 2 (**2**), respectively) and the Hoveyda–Grubbs' catalyst (**3**) (Scheme 1), have achieved widespread use to promote a range of cyclisations. Alkenes, alkynes, and enynes have been used as substrates and the selective formation of one or even multiple rings has been reported in numerous reviews [5]. Despite the many successful applications, in some instances reactions can still be less than satisfactory, due to low reactivity and/or low selectivity, hence the recourse to the Schrock's complex (**4**), which is exceedingly efficient, to the detriment, however, of its tolerance of functional groups.

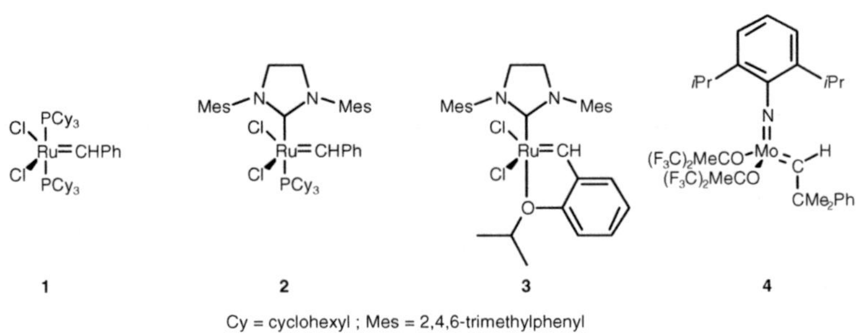

Scheme 1 Ruthenium and molybdenum catalysts for olefin metathesis

It is, of course, impossible to cover comprehensively in a single chapter the applications of olefin metathesis in fine organic synthesis. Therefore, a choice had to be made as to what subjects to include, and this is merely a matter of personal taste. Accordingly, the present chapter aims at illustrating the power of olefin metathesis, either cross-metathesis (CM) or ring-closing metathesis (RCM), by some recent syntheses of bioactive compounds.

2 Cross-Metathesis

Alkene cross-metathesis can provide a convenient access to a wide range of variously (poly)substituted olefins. The reaction has attracted much attention recently [6], although it has not yet attained the same level of synthetic usefulness as ring-closing metathesis or ring-opening metathesis polymerisation (ROMP). This is largely due to inherent difficulties in controlling selectivities toward cross-coupling instead of homodimerisation in a process that is not favoured either by a strong enthalpic driving force (like ROMP) or by entropic factors (like RCM).

Cross-metathesis has been used in the synthesis of a wide variety of biologically active compounds, such as siamenol (an anti-HIV agent) [7], decarestrictine D (a strong inhibitor of cholesterol biosynthesis in HEP-G2 liver cells) [9], isoprostanes (lipid oxidation metabolites) [11], and amphidinolide T1 (an antibiotic) [10] (Scheme 2). The potential of cross-metathesis is also nicely exemplified by the total synthesis of the potent antitumour agent apoptolidin A (**7**, Scheme 3) [13], which involved in a key step the challenging coupling of two sophisticated olefins, **8** and **9** (Scheme 4). The desired compound **10** was formed in 60% yield as a >20:1 *E*:*Z* mixture. Unfortunately, attempted cross-metatheses failed when the reactions were performed in the presence of glycosylated olefin **11**.

Cross-metathesis has also been employed in the total synthesis of platensimycin (**5**) [14], carbaplatensimycin (**6**) [15], adamantaplatensimycin [16], and platencin [17]. Platensimycin and its analogues attract nowadays considerable interest from both biological and chemical circles due to their unique pharmacological profile. Platensimycin is indeed the first antibiotic discovered in over 40 years that exerts its antibacterial effect through a novel mechanism of action, manifested by its impressive activity against a variety of drug-resistant bacteria, including methicillin- and vancomycin-resistant strains [1c, 18]. A variety of platensimycin analogues with varying degrees of complexity have been synthesised, in which cross-metathesis between vinyl boronate **12** and the tetracyclic motif **13** (Scheme 5) is common to most of their total syntheses.

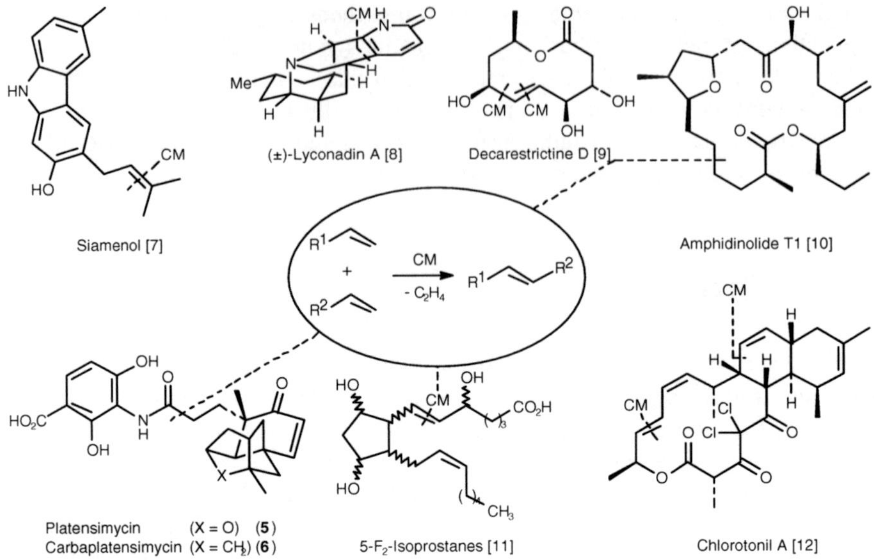

Scheme 2 Selected applications of cross-metathesis

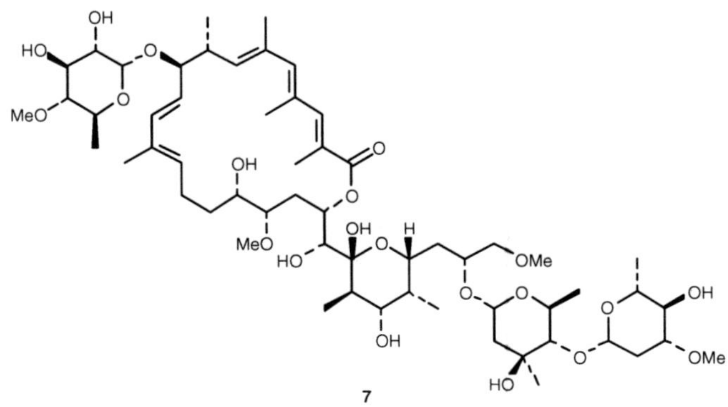

7

Scheme 3 Structure of apoptolidin A

Noteworthily, in the synthesis of the surrogate **15**, a ring-closing metathesis and cross-metathesis were employed successfully (Scheme 6) [19].

Another noteworthy application of cross-metathesis is in the synthesis of variously substituted *E*-stilbenes [20], which are derivatives or precursors of biologically important compounds such as resveratrol (**17**), piceatannol (**18**), and pinostilbene (**19**) (Scheme 7). Polyhydroxylated stilbenes are potent agents for the prevention and therapy of cancer. They also exhibit a variety of other interesting biological properties including cardiovascular protecting effects.

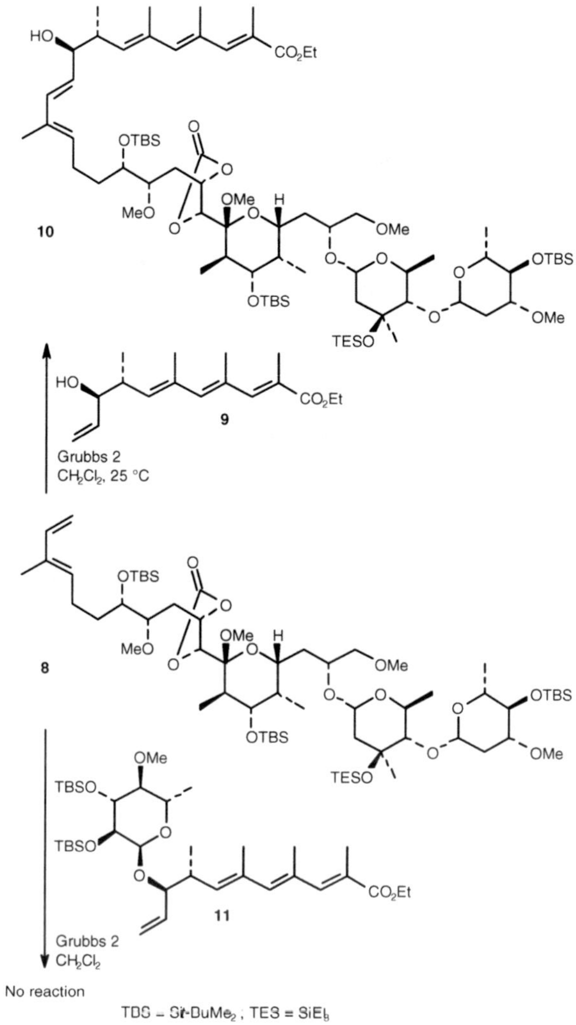

Scheme 4 Cross-metathesis in the synthesis of apoptolidin A

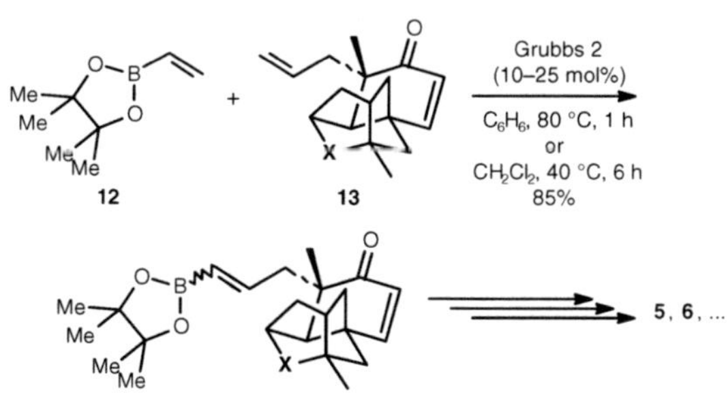

Scheme 5 Cross-metathesis in the synthesis of platensimycin analogues

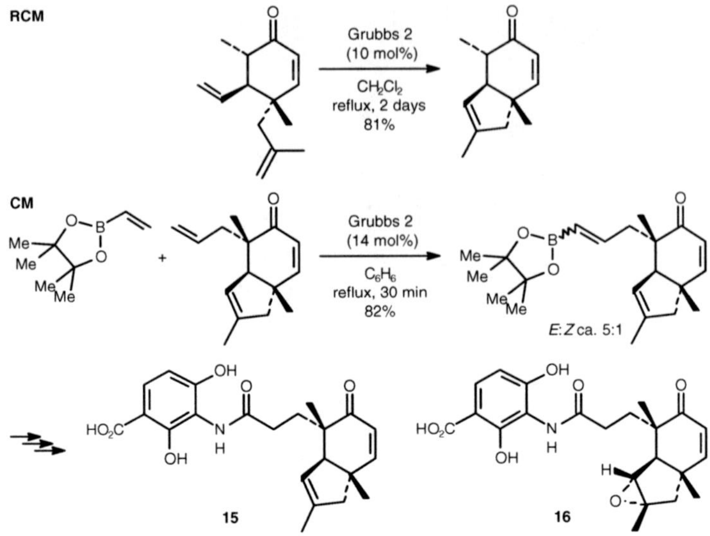

Scheme 6 Ring-closing metathesis and cross-metathesis in the synthesis of platensimycin analogues

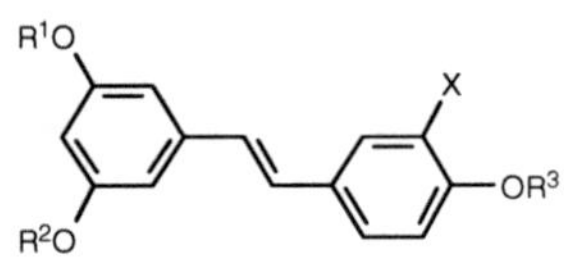

(**17**) Resveratrol (R^1 = R^2 = R^3 = X = H)
(**18**) Piceatannol (R^1 = R^2 = R^3 = H; X = OH)
(**19**) Pinostilbene (R^1 = R^3 = H; R^2 = Me; X = H)
(**20**) Piceid (R^1 = R^3 = X = H; R^2 = β-D-Glu)
(**21**) Pterostilbene (R^1 = R^2 = Me; R^3 = X = H)

Scheme 7 Structure of selected polyhydroxystilbenes

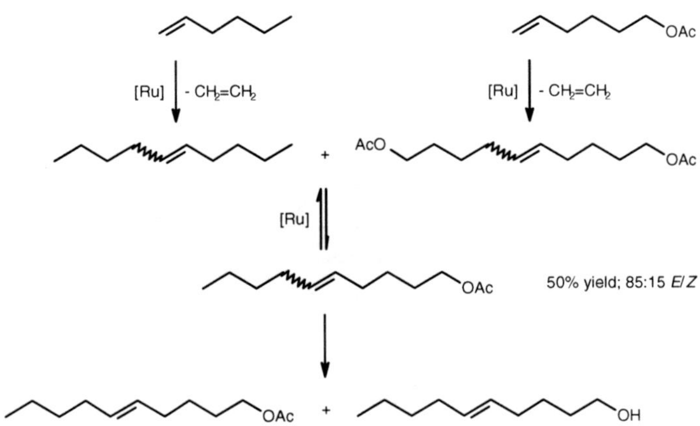

22 (83%) **23** (17%)

Scheme 8 Commercial application of cross-metathesis in the production of pheromones

According to the concept of "green chemistry", the solvent-free CM of 1-hexene with hexenyl acetate (Scheme 8) represents an example of the use of metathesis to generate environmentally friendly products. The product of this reaction can be readily converted into a mixture of compounds **22** and **23**, a pheromone of the peach twig borer moth, which can be used as an environmentally friendly means of insect control in lieu of the use of broad-spectrum pesticides. CM is currently under investigation for the commercial preparation of other pheromones, as this route represents an efficient means to generate these compounds starting from inexpensive starting materials [2c].

3 Ring-Closing Metathesis

Although the strategic advantages of ring-closing metathesis (RCM) have been illustrated many times and need no further confirmation [5], recent examples are particularly instructive, and highlight the superb application profile of Grubbs-type catalysts in general, for instance for the preparation of highly functionalised pharmaceutical agents, many of which are in the advanced stages of testing. For example, Boehringer Ingelheim recently disclosed the synthesis *via* RCM of **24**, a macrocyclic hepatitis C virus (HCV) NS3 protease inhibitor labelled BILN 2061 (Ciluprevir^{TM}) (Schemes 9 and 10).

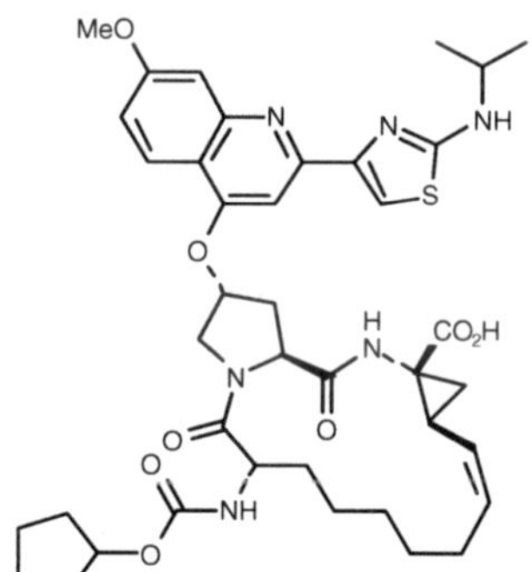

24

Scheme 9 Structure of BILN 2061

Among several ruthenium catalysts used in this reaction, Grela's catalyst **28** proved to be the most active one, generating macrocycle **26** with the desired *Z* selectivity [21]. Recently, the closely related catalyst **29** was used in the same transformation, giving similar results [22]. The process was successfully scaled-up to produce more than 400 kg of the 15-membered cycle **26** using Hoveyda–Grubbs type catalysts [23].

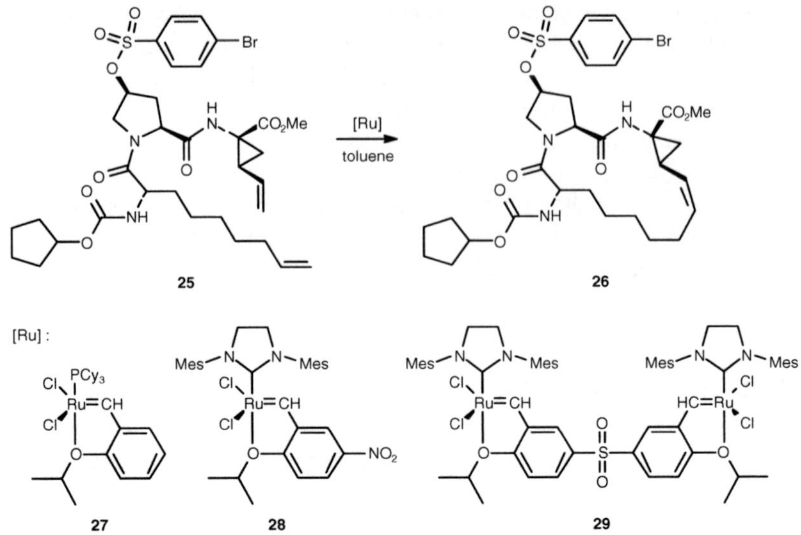

Scheme 10 Commercial application of ring-closing metathesis in the production of BILN 2061

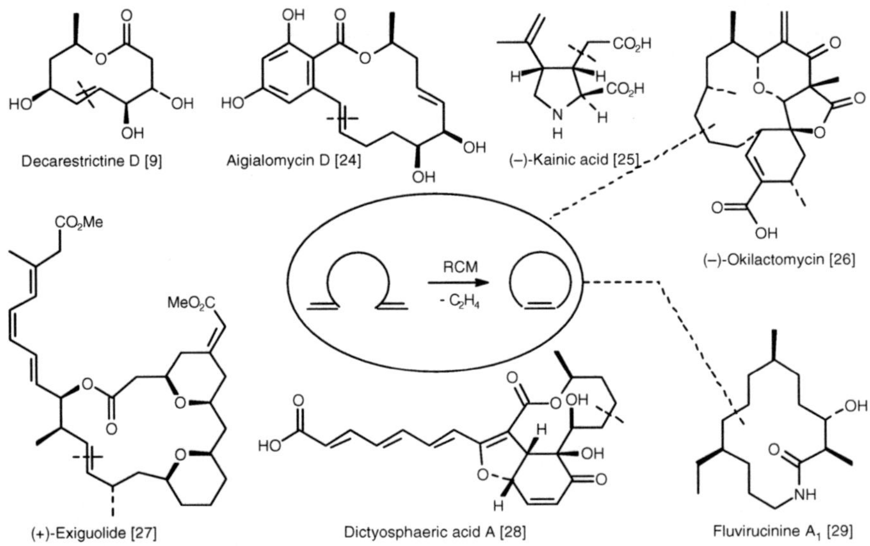

Scheme 11 Selected applications of ring-closing metathesis

Demonstration of the power of the RCM reaction for macrocyclisation can be found in the synthesis of a plethora of compounds, including antibiotics such as okilactomycin [26] and dictyosphaeric acid A [28] (Scheme 11). It is also worth noting that among the different synthetic strategies for oseltamivir phosphate (**32**, Gilead's Tamiflu, marketed by Roche), an important orally active anti-influenza drug, Yao's route is based on RCM of diene **30** as a key step (Scheme 12) [30].

Thus, using Grubbs 2 catalyst in CH_2Cl_2, the corresponding cyclohexene **31** was obtained in almost quantitative yield.

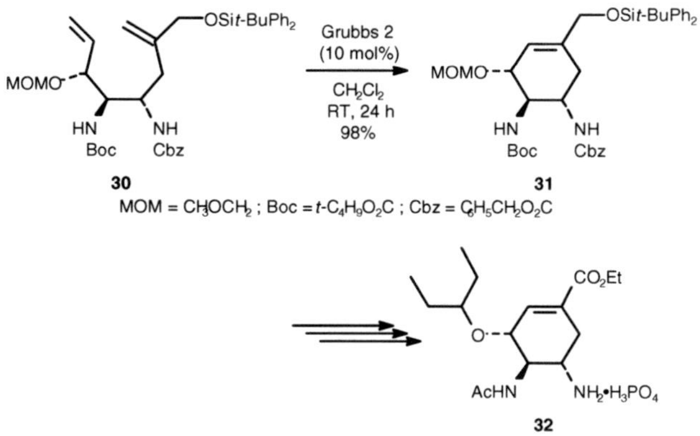

Scheme 12 Ring-closing metathesis in the Yao's total synthesis of Tamiflu

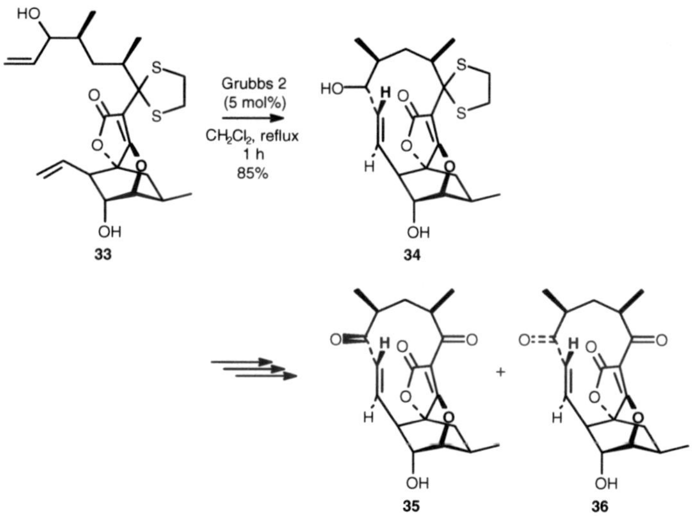

Scheme 13 Ring-closing metathesis in the synthesis of abyssomicin C

Abyssomicin C (**35**) is a recently discovered antibiotic with a novel molecular architecture and unique mechanism of action against drug-resistant bacteria [1c]. Because of its unique features, abyssomicin C and analogues thereof are nowadays targets of prime importance. Despite intense efforts, only two total syntheses have been reported so far [31]. As expected, the construction of the oxabicyclo [2.2.2] octane core and the macrocyclisation are key steps in the synthesis. Interestingly, in the Nicolaou's approach, the strained 11-membered macrocycle in **34** was forged by a RCM reaction (Scheme 13) [32]. In addition, this approach led not only to the total synthesis of abyssomicin C (**35**) but also to the formation of atrop-abyssomicin C (**36**), a conformer of the naturally occurring product that exhibits even more potent antibiotic activity than its originally isolated twin.

3.1 Synthesis of Epothilones

Since their isolation and characterisation by Reichenbach, Höfle, and coworkers in 1996 [33], epothilones have generated considerable interest in the scientific and medical community because of their remarkable cytotoxic activity, in particular against ovarian, breast, and colon tumour cell lines, which is superior to that of paclitaxel. Soon after this discovery, the first total syntheses of epothilone A (**37**) were reported in 1996/1997 by the research groups of Danishefsky [34], Nicolaou [35], and Schinzer et al. [36].

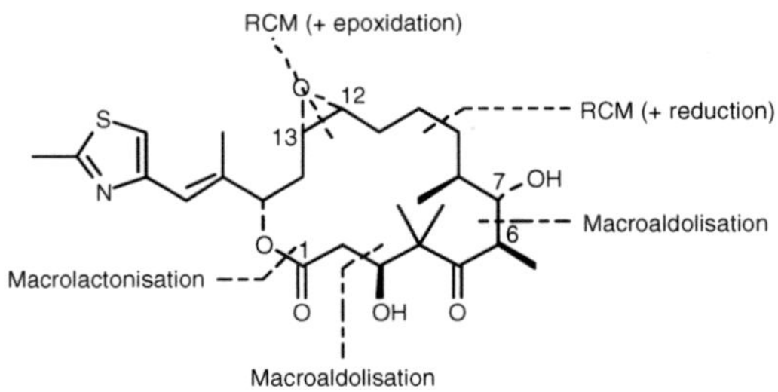

37

Scheme 14 Structure of epothilone A

From that time, several other epothilones have been isolated and fully characterised, and a plethora of epothilone analogues has been prepared [37]. Toward this end, three main strategies were developed, in which the macrocyclic skeleton was constructed using either a RCM reaction, a lactonisation, or an intramolecular aldol reaction (Scheme 14) [38]. Of significant interest is the recent development of 12-aza-epothilones **38** and **39** (Scheme 15) (termed "azathilones"), which are characterised by the replacement of a backbone carbon atom by nitrogen in the epothilone macrocycle. Their total syntheses involved as the key strategic step either a macrolactonisation or a RCM using the Grubbs 2 complex. Noteworthily, attempts to close the macrocycle employing the first-generation Grubbs' catalyst met with complete failure and no conversion of the RCM precursor was observed [39]. Biological investigations have shown that, depending on the nature of the acyl substituent attached to the backbone nitrogen atom, azathilones are highly potent growth inhibitors of drug-sensitive human cancer cells *in vitro* and, thus, are promising new lead structures for anticancer-drug discovery.

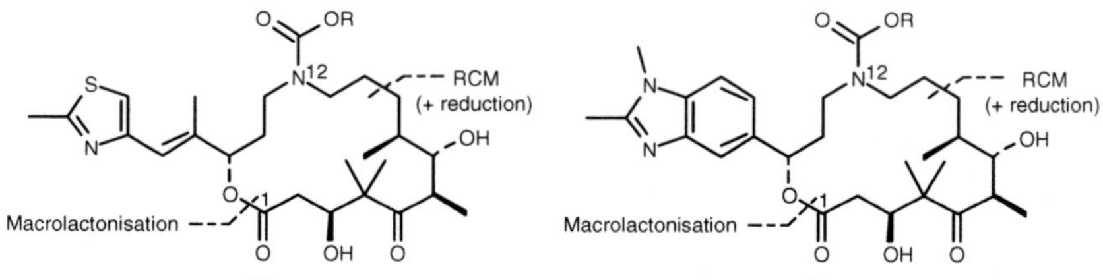

Scheme 15 Structure of representative azathilones

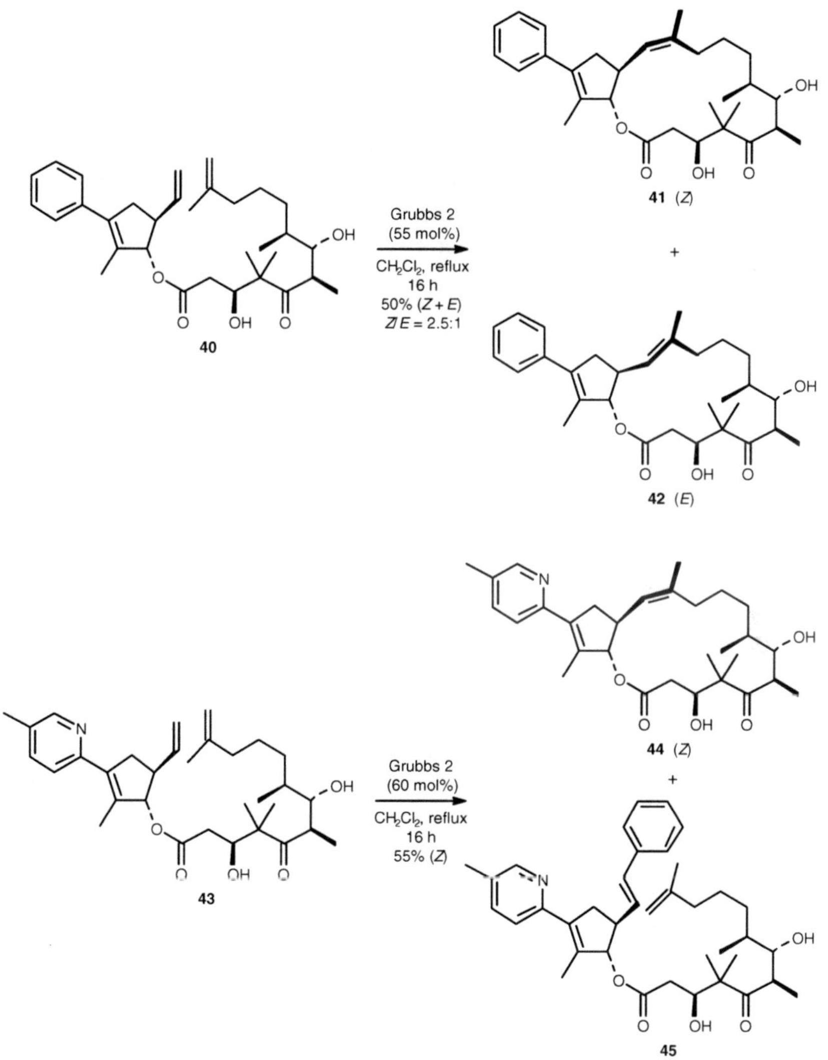

Scheme 16 Ring-closing metathesis in the synthesis of conformationally restrained epothilones

Recent investigations have also led to the synthesis of conformationally restrained epothilones. Different parts of the skeleton have been rigidified, including the western sector as illustrated by compounds **41** and **44** (Scheme 16) [40]. Again, RCM was the method of choice for the macrocyclisation. RCM of **40** by using the Grubbs 2 catalyst worked pretty well, affording the desired *Z*-alkene **41**, accompanied by the *E*-isomer (**42**) as a minor product. A similar approach to synthesise the pyridine analogue **44** was much more complicated. Indeed, in addition to the desired *Z*-product **44**, the phenyl analogue (**45**) and a dimer-like product from **43** were isolated as by-products. Formation of **45** was attributed to the high catalyst loading (60 mol%) required to overcome the sluggishness of the RCM reaction. Again, these results confirm the importance of neighbouring groups (phenyl (**40**) versus 5-methylpyridine-2-yl (**43**)) on the outcome of the RCM reaction [40]. Compound **44** exhibited strong and highly selective growth inhibitory activity against two leukaemia cell lines over solid human tumour cell lines *via* a mechanism of action that is currently under investigation.

In an attempt to constrain epothilone D, analogues **46** and **47** (Scheme 17) with a bridge between the C4-methyl and the C12-methyl carbons were prepared using RCM as the key step. In antiproliferative assays in the human ovarian cancer (A2780) and prostate cancer (PC3) cell lines, compounds **46** and **47** proved to be less active than epothilone D [41].

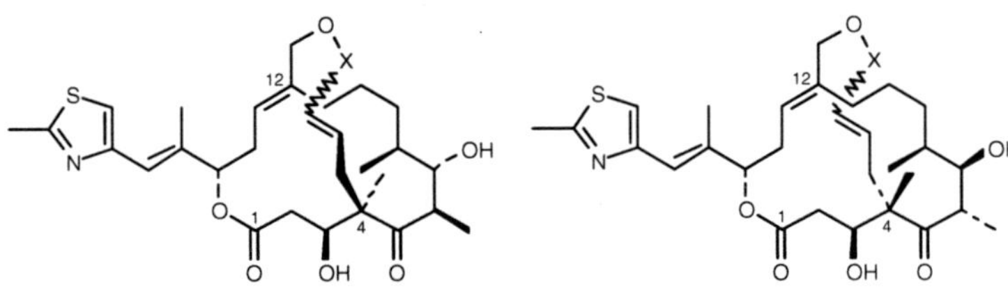

46 X = C(O), C(O)CH$_2$, C(O)(CH$_2$)$_2$, CH$_2$ **47**

Scheme 17 Structure of bridged epothilones

Three epothilones have been investigated in early clinical studies (phase I): Kos-1584 (Kosan/Roche), ABJ879 (Novartis), and BMS 310705 (BMS), but were finally rejected. Nowadays, four other epothilones are currently undergoing advanced clinical development (Scheme 18). The natural compounds epothilones B (**48**) and D (**49**) are produced by biosynthesis and the lactam BMS 247550 (**50**) by partial synthesis starting from epothilone B. In October 2007, the BMS–epothilone B-lactam **50**, now called ixabepilone, received FDA approval for the treatment of metastatic or advanced breast cancer [42]. Tumour types that have been investigated with **50** include breast, prostate, colorectal, non-small-cell lung, gastric, hepatobiliary, gynaecological, and pancreatic cancers. In addition, studies have been conducted for the treatment of sarcoma, melanoma, and non-Hodgkin's lymphoma [37].

On the other hand, ZK-EPO (**51**) is the first fully synthetic epothilone in clinical development [43]. Compared with its analogues, this compound combines high activity and efficacy, a fast and efficient cellular uptake, no recognition by efflux mechanisms, and an improved therapeutic window.

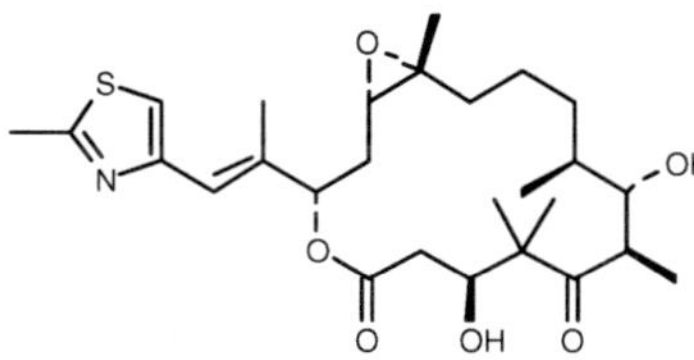

Epothilone B (**48**)
EPO-906, patupilone
phase II / III (Novartis)

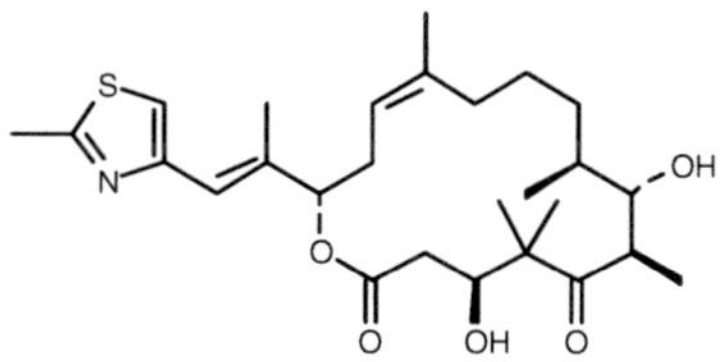

Epothilone D (**49**)
Kos-862
phase II (Kosan/Roche)

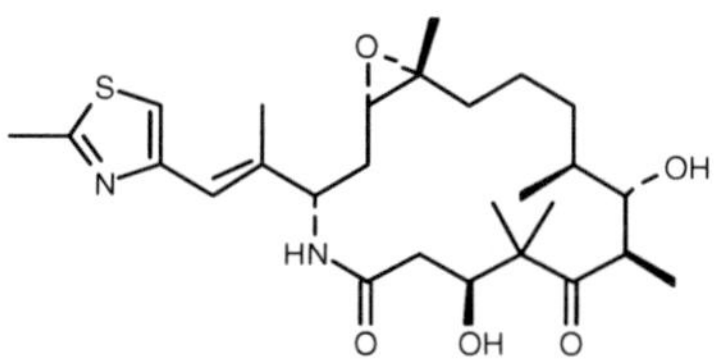

BMS 247550 (**50**)
ixabepilone
phase III (BMS)

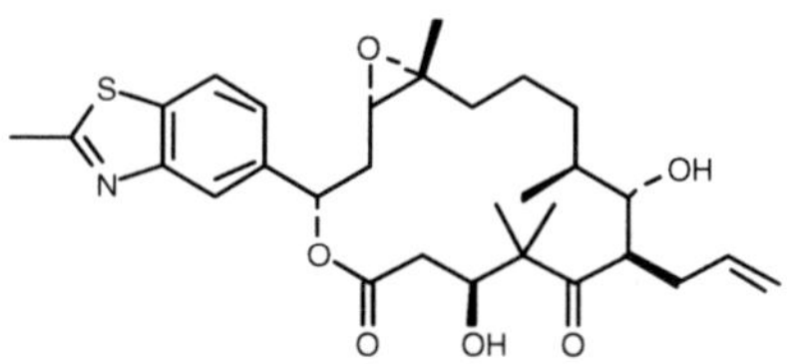

ZK-EPO (**51**)
phase II (Schering AG)

Scheme 18 Epothilones in advanced clinical trials

Interestingly, ZK-EPO emerged from about 350 active epothilone analogues synthesised in the laboratories of Schering AG (Berlin). Toward this aim, a highly convergent strategy was applied that offered good flexibility for introducing structural modifications at nearly every position of the 16-membered-ring skeleton. In particular, the macrocyclisation step of these syntheses was successfully accomplished using the Yamaguchi lactonisation (2,4,6-trichlorobenzoyl chloride, Et_3N, 4-DMAP, THF/toluene, 25°C), instead of a RCM. Unfortunately, indeed, most of the attempts at applying RCM to epothilone syntheses have been plagued by complete lack of stereocontrol in the generation of the desired olefin geometry in the products. Mixture of *Z*- and *E*-isomers were formed, requiring thereby a subsequent chromatographic purification. Surprisingly, as illustrated in Scheme 19 for the ring-formation of epothilone congeners through RCM, the *Z*:*E* ratio ranges from 1.7:1 to 1:9 simply by modifying the functionalities and the alcohol protecting groups, which indicates a remarkable sensitivity to permutations of functionality and stereochemistry at centres far remote from the site of olefin metathesis [44].

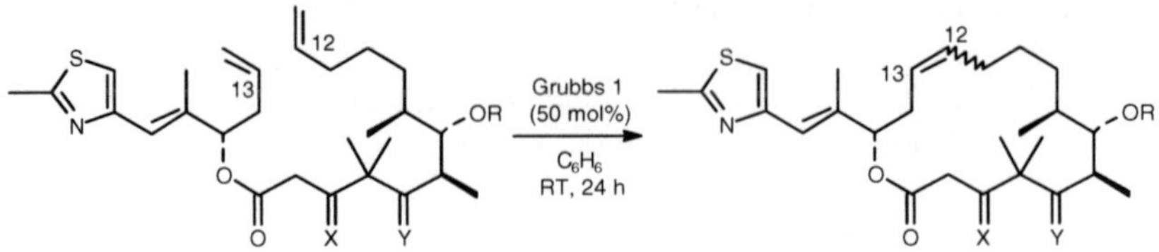

X	Y	R	Yield, %	*Z*:*E*
α-OH	α-OSiPh$_3$	Si*t*-BuMe$_2$	86	1 : 3
α-OSiEt$_3$	α-OSiPh$_3$	Si*t*-BuMe$_2$	80	1 : 5
α-OSi*t*-BuMe$_2$	O	Si*t*-BuMe$_2$	86	1.7 : 1
α-OH	O	H	65	1 : 2
β-OH	α-OSiPh$_3$	Si*t*-BuMe$_2$	81	1 : 9
β-OSi*t*-BuMe$_2$	O	Si*t*-BuMe$_2$	88	1 : 2

Scheme 19 Substrate effect on the ring-closing metathesis

A similar substrate effect was found for the diene-ene RCM leading to epothilone 490 (**55**, Scheme 20). Thus, exposure of the fully protected precursor **52** (R = CH_3, R′ = Troc ($C(O)OCH_2CCl_3$), R″ = TES ($SiEt_3$)) to the RCM reaction in methylene chloride gave a mixture of two compounds, **53** and **54**, in a 35:15 ratio, with a total yield of 50%. The major component of the product mixture was the desired *trans*-substituted diene **53**, along with the 14-membered macrolide **54** as a minor product, seemingly arising from a metathesis reaction involving the internal olefin. Quite surprisingly, RCM of partially protected **52** (R′ = Troc, R″ = H or R′ = H, R″ = TES) afforded the desired product **53** in 41% and 57% yield, respectively, with none of the 14-membered macrolide **54** being observed. More importantly, RCM of the fully deprotected substrate afforded a 64% yield of only the desired *E*-olefin. Interestingly, when the same series of reactions was carried out in refluxing toluene, this substrate effect was diminished, with 55–58% yields observed across the various substrates [45]. The origin of this substrate effect has not been determined. Nevertheless, the fact that both the C3 and the C7 alcohols are *β*- to carbonyl groups suggests a possible contribution of intramolecular hydrogen bondings in imparting some degree of rigidity to the cyclisation precursor [46].

It was also shown that a trifluoromethyl group in a vicinal relationship to the proposed RCM reaction centre has a detrimental impact on the desired reaction. Attempts to carry out the ring-closing metathesis reaction of **52** (R = CF_3, R′ = Troc, R″ = TES) led indeed primarily to apparent dimerisation of the starting material (Scheme 20) [47]. In contrast, the RCM reaction could be accomplished successfully when a carbon spacer was introduced between the RCM reaction centre and the CF_3 group (**57**, Scheme 21). As in the case of the conversion of the methyl analogue **57** (R = CH_3), the reaction provided exclusively the *trans* isomer **58**.

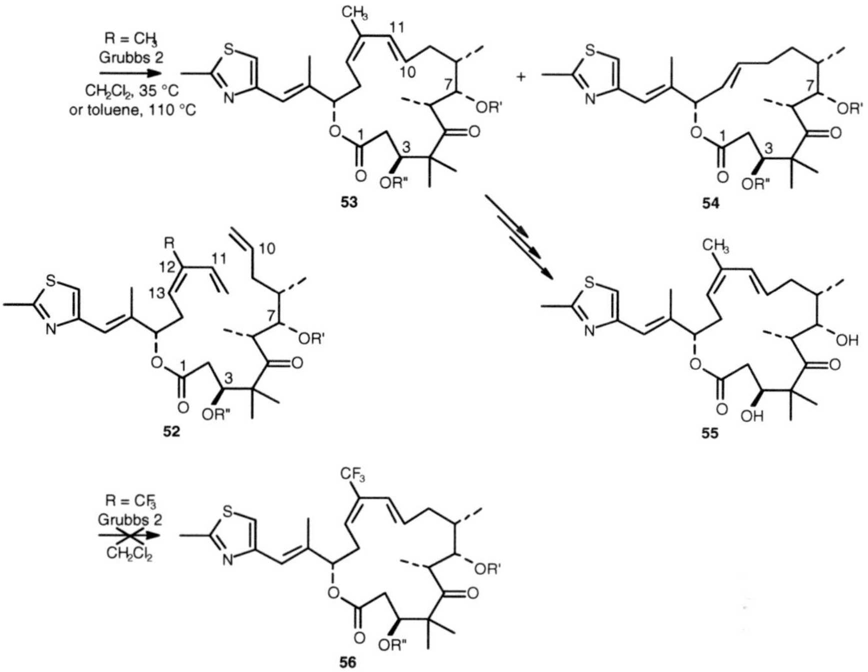

Effect of alcohol protection and solvent on RCM yield (%)

52 R'	R"	CH_2Cl_2 (35 °C, 5.5 h) **53** : **54**	toluene (110 °C, 25 min) **53** : **54**
Troc	TES	35 : 15	58 : 6
Troc	H	41 : 0	57 : 0
H	TES	57 : 0	
H	H	64 : 0	55 : 0

Troc = $C(O)OCH_2CCl_3$; TES = $SiEt_3$

Scheme 20 Substrate and solvent effects on ring-closing metathesis in the synthesis of epothilone 490

In alternative approaches to epothilones, the macrocycle was closed through a C9–C10 ring-closing metathesis. Unfortunately, the success of the RCM again strongly depended upon the protecting groups of the starting material. Thus, when diene **60** was exposed to RCM conditions using Grubbs 2 catalyst, compound **61** containing a seven-membered ring was formed in 91% yield, rather than the desired macrolactone (Scheme 23) [48]. It was suggested that the acetonide was the main cause of the complete failure of **60** to form any of the macrolactone *via* RCM. Possibly, the rigidity imposed by the acetonide did not allow the diene to readily adopt a conformation that allowed the C9 and C10 olefins to become sufficiently close to undergo RCM. With the compound unable to undertake the C9–C10 metathesis, the C10 terminal olefin is then left to undergo RCM with the trisubstituted C16–C17 olefin.

Other examples of such RCM reactions in the context of epothilone synthesis have been described (Scheme 23). Danishefsky first performed this reaction on a precursor that lacked the C3,C5 acetonide functionality [49]. When compounds **62**, which have a C3-TES ether and C5 ketone, were subjected to RCM, the desired macrolactones **63–65** were formed in 22–38% yield as a single *trans* isomer. However, the major product of the reaction was the seven-membered ring **66–68**, formed in 57–62% yield. To circumvent this unwanted side product, Danishefsky synthesised two new precursors, **69** and **70**, which lacked the C16–C17 trisubstituted olefin [49]. In these cases, RCM yielded the macrolactones **71** and **72** in 76–78% yield as a single (*E*)-isomer.

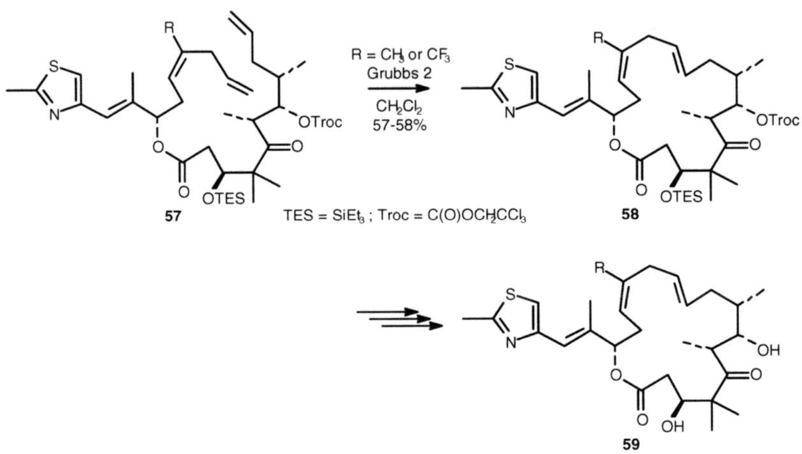

Scheme 21 Ring-closing metathesis in the synthesis of epothilone analogues

On the other hand, Sun and Sinha performed a RCM reaction on the advanced epothilone precursor **73** containing the C12–C13 epoxide functionality and obtained the macrolactone **74** in 89% yield as a 1:1 mixture of *E*/*Z* olefin isomers. From these results, it clearly appears that functionalities and/or protecting groups exert a strong influence on the outcome of the RCM reaction (Scheme 22) [50].

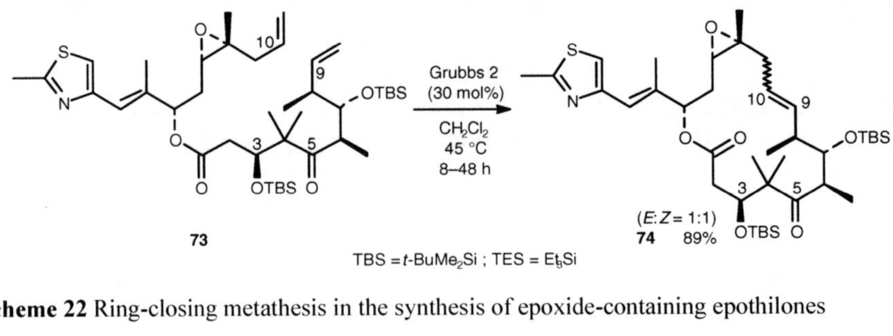

Scheme 22 Ring-closing metathesis in the synthesis of epoxide-containing epothilones

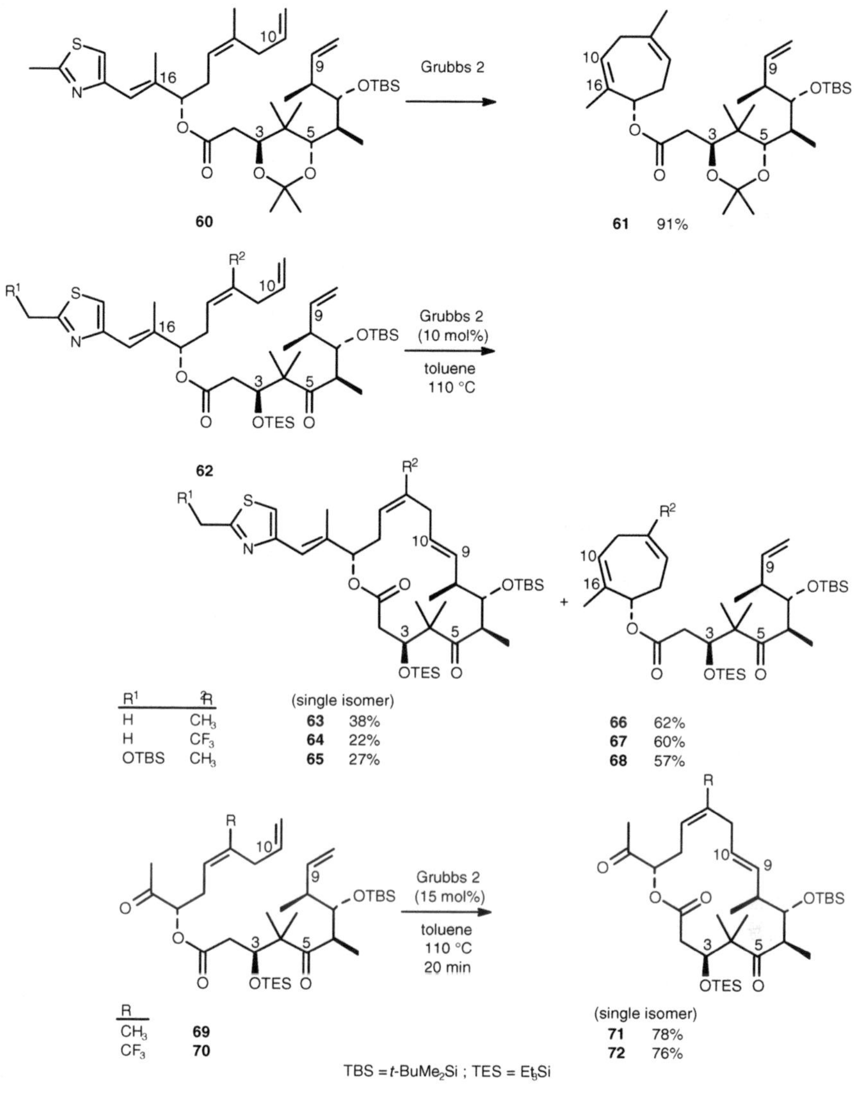

Scheme 23 Ring-closing metathesis in the synthesis of epothilones

3.2 Synthesis of Amphidinolides

The amphidinolides are a rapidly growing family of macrolides produced by marine dinoflagellates of the genus *Amphidinium*. Though structurally quite diverse, all members are distinguished by a pronounced cytotoxicity against various cancer cell lines, with some of them reaching potencies, which rank them amongst the most cytotoxic compounds known to date. For instance, the exceptional cytotoxicity of amphidinolide H (77) against KB human epidermoid carcinoma cells (IC_{50} value = 0.52 ng/ml) rivals that of many renowned anticancer agents.

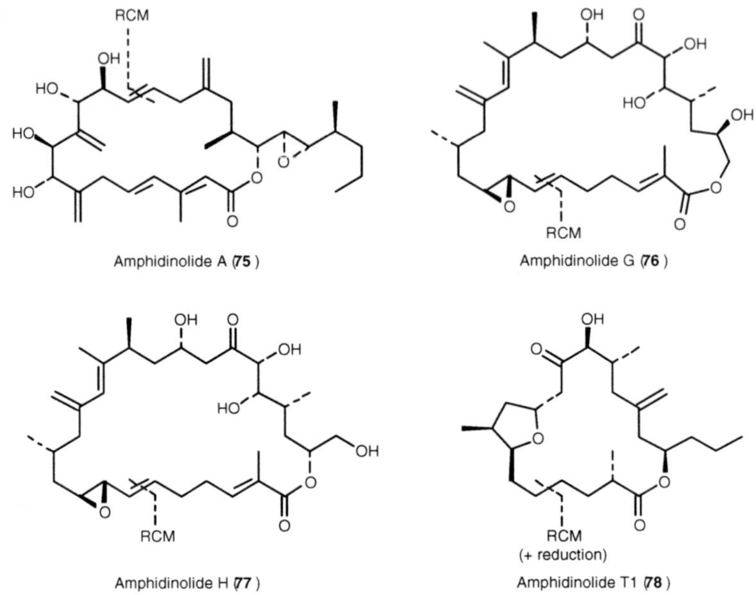

Scheme 24 Structure of representative amphidinolides

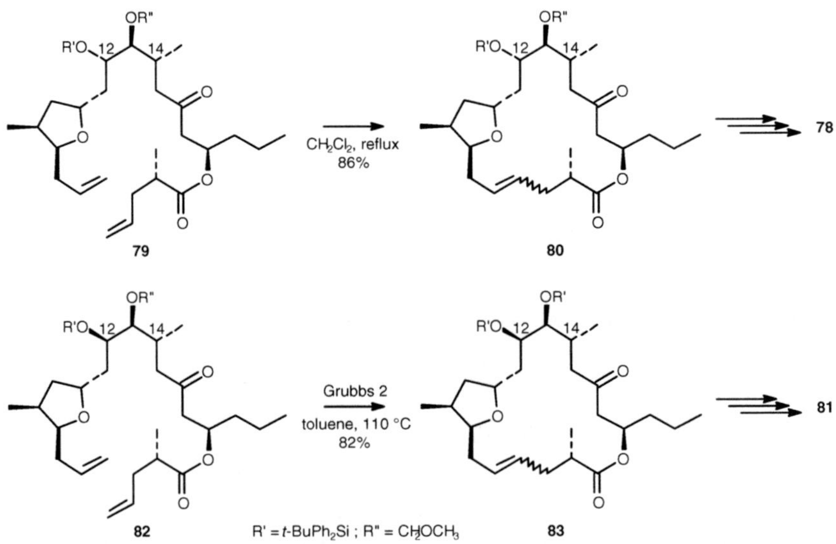

Scheme 25 Ring-closing metathesis in the synthesis of amphidinolides T1 (**78**) and T3 (**81**)

Several amphidonolides have been synthesised using different strategies [51]. As illustrated in Scheme 24 for the synthesis of amphidinolides A (**75**) [52], G (**76**), H (**77**) [53], T1 (**78**), and related T3, T4, and T5 [54], ring-closing metathesis is a highly versatile method for the formation of the macrocyclic scaffold. RCM of diene **79** worked exquisitely well when carried out under standard conditions in the presence of the Grubbs 2 complex as the precatalyst bearing an imidazol-2-ylidene ligand instead of the usual imidazolin-2-ylidene ligand (Scheme 25) [54a]. By contrast, in the total synthesis of amphidinolide T3 (**81**), the epimer **82** with the *anti–syn* stereotriad in the C12–C14 region was found to react much more reluctantly than its *syn–syn* configured counterpart **79**. Good conversion could only be attained by exchanging the solvent from CH_2Cl_2 to toluene and increasing the reaction temperature to 110°C (Scheme 25) [54b]. These results confirm the fact that heteroelements in proximity to the reacting alkenes as well as the configuration at these centres often strongly affect the efficiency of macrocyclisations by RCM. Furthermore, the relative configuration at C-12/C-13 of the cyclisation precursors also determines the stereochemical course of the RCM reaction. Thus, in the cyclisations of dienes **79** and **82**, which differ only in the configuration of the remote C-12 chiral centre, cycloalkene **80** derived from **79** was formed with an appreciable selectivity in favour of the *E*-isomer (*E*:*Z* = 6:1), while the corresponding *E*:*Z* ratio for its congener **83** was only 2:1.

The striking influence of remote substituents was also clearly evident in the construction by RCM of the 14-membered macrocycle found in **84**, a compound belonging to the class of cembrenes. The oxygenated cembrenes display amongst others anti-HIV activity, anti-inflammatory properties, and neuro- and cytotoxicity (Scheme 26).

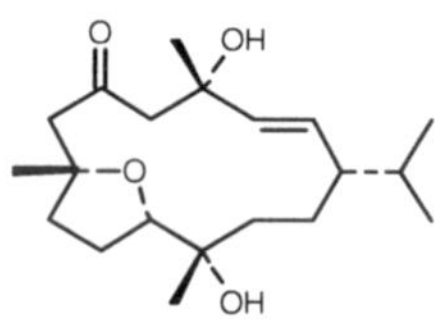

84

Scheme 26 Structure of polyoxygenated cembrene **84**

The key RCM of the precursor **85** with the Grubbs 2 catalyst provided the macrocycle **86** as the *Z* isomer in 89% yield (Scheme 27) [55]. Again, the protective groups of the hydroxy groups at C-8 and C-10 have a large influence upon the metathesis. Thus, when the acetyl protecting group of the alcohol at C-8 was replaced by a triisopropylsilyl group (**87**), no ring-closure could be achieved, presumably because of steric reasons. On the other hand, the use of other catalysts for the metathesis of **85**, such as the Fürstner's catalyst Neolyst M1 (**88**), afforded complex mixtures of dimers.

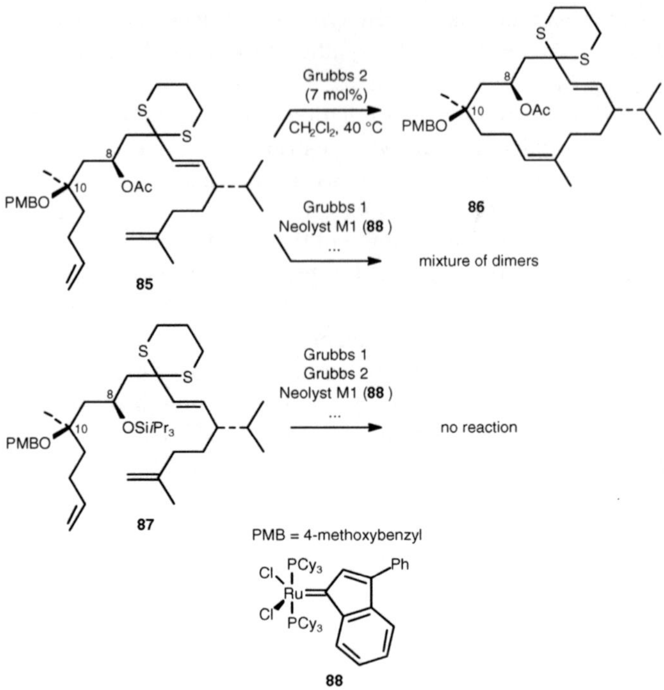

Scheme 27 Ring-closing metathesis in the synthesis of polyoxygenated cembrene **84**

The synthesis of amphidinolides H (**77**) and G (**76**) proved to be more challenging since the macrocyclisation was planned to occur *via* RCM of a labile vinyl epoxide (Scheme 28) [53, 56]. Inter- as well as intramolecular metathesis reactions of vinyl epoxides are indeed surprisingly rare in the literature [57]. Gratifyingly, exposure of **89** to catalytic amounts of the Grubbs 2 carbene complex in benzene resulted in a remarkably clean cyclisation, with formation of the 26-membered cycloalkene **90** in 72% yield as the *E*-isomer only. Noteworthily, the sensitivity of the material required to perform this macrocyclisation at ambient temperature.

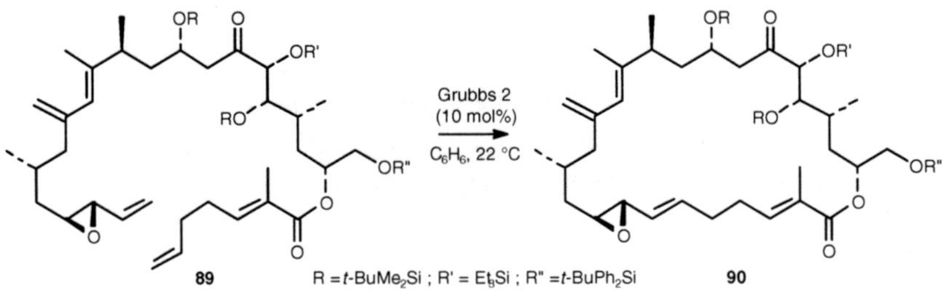

Scheme 28 Ring-closing metathesis in the synthesis of amphidinolides H and G

A similar approach involving the RCM of a vinyl epoxide and a conjugated diene as coupling partners was also successfully envisaged for the total synthesis of monocillin I (**91**, X = H) and radicicol (**91**, X = Cl) (Scheme 29) [58], two macrolides that exhibit a variety of antifungal and antibiotic properties. Radicicol additionally displays potent anticancer properties *in vitro* [59].

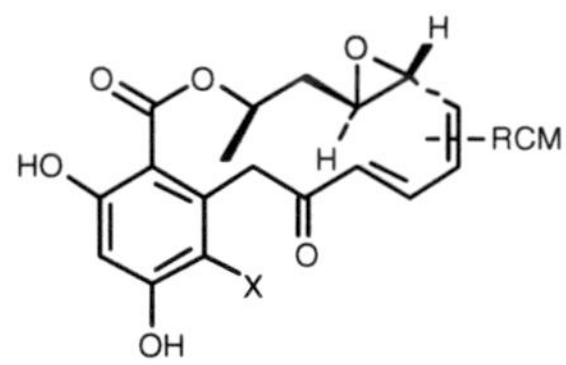

91

Scheme 29 Structure of monocillin I (X = H) and radicicol (X = Cl)

Likewise, RCM of compound **92** containing an activated epoxide was successful with Grubbs 2 catalyst, resulting in **93** with exclusive *Z*-configuration at the newly formed olefin (Scheme 30) [60]. Again, the success of RCM of compound **92** containing two free OH groups in addition to the oxirane ring highlights the high functional group tolerance of Grubbs 2 catalyst. On the other hand, in the key step of the total synthesis of multiplolide A (**96**, Scheme 31), a 10-membered lactone that exhibits antifungal activity against *Candida albicans* with a IC_{50} value of 7 μg/ml, RCM of the allylic substrate **94** resulted in the formation of compound **95** with the desired *E*-configuration. Interestingly, reversal of stereochemistry on one side of the diene (**97**), conversion of the alcohol into the tosylate (**98**), and protection as the PMB ethers (**99** and **100**) resulted in complex mixtures under any of the conditions attempted, confirming thereby the influence of conformation control on the outcome of the transformation (Scheme 32).

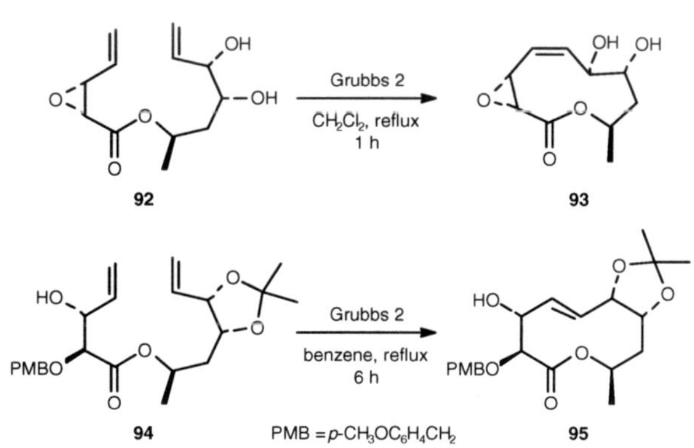

Scheme 30 Ring-closing metathesis in the synthesis of multiplolide A

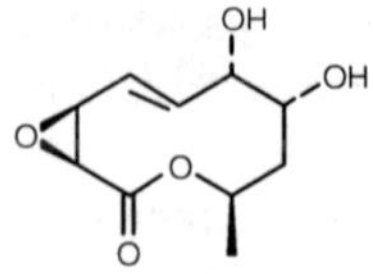

96

Scheme 31 Structure of multiplolide A

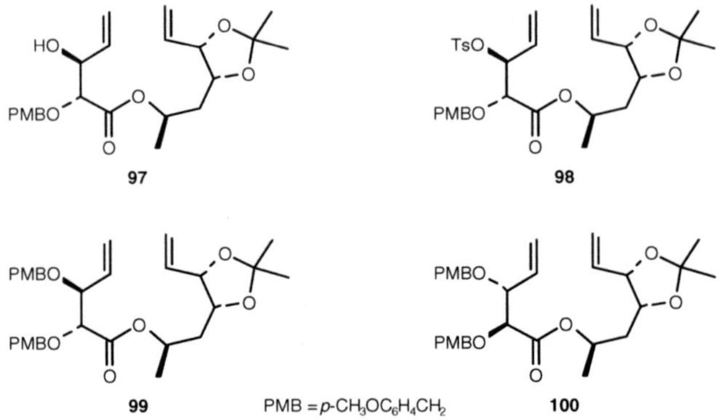

Scheme 32 Substrates that failed to undergo ring-closing metathesis

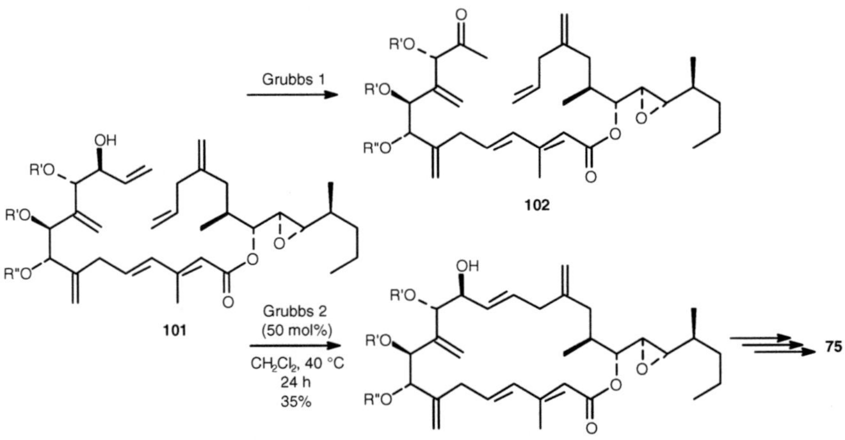

R' = p-$CH_3OC_6H_4CH_2$; R" = t-$BuMe_2Si$ **103**

Scheme 33 Ring-closing metathesis in the synthesis of amphidinolide A

The synthesis of amphidinolide A (**75**) posed another problem [52]. Given the array of olefinic functionality present in **101**, planning a late stage RCM was not without obvious risks. Of particular concern, indeed, was the ability to control which alkenes underwent metathesis and the geometry of the resultant olefin. Gratifyingly, the desired RCM occurred successfully using the Grubbs 2 catalyst (Scheme 33), although in only moderate yield and in the presence of a high catalyst loading. No other RCM products were detected and only the *E*-isomer (**103**) was observed.

Noteworthily, the first-generation Grubbs' catalyst truncated the allylic alcohol to give the methyl ketone (**102**). This side reaction was noted in other contexts [61] and investigated by using simpler substrates, such as allylic alcohol **104** (Scheme 34) [62]. Thus, in addition to the ring-closed cyclopent-2-en-1-ol (**105**), the methyl ketone **107** was formed in a 1:1.5 ratio, with concomitant consumption of the Grubbs 1 ruthenium carbene species. These results suggest that the initial carbene **108** undergoes tautomerisation to the enolyl ruthenium hydride species **109**, which can further undergo reductive elimination, either before or after tautomerisation to the oxoalkyl ruthenium hydride **110**, to produce the methyl ketone **107** (Scheme 34).

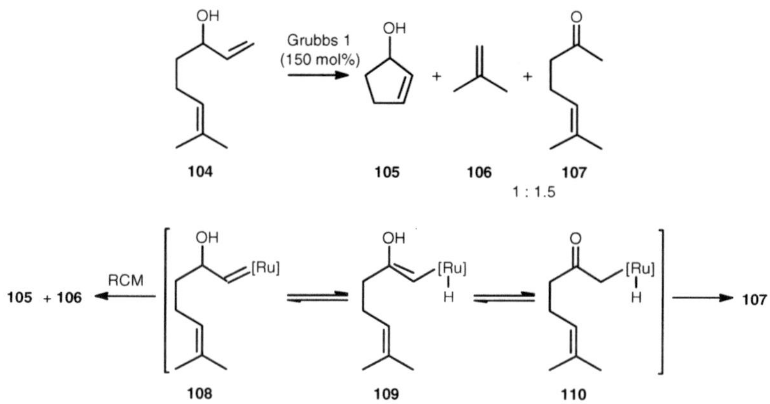

Scheme 34 Proposed mechanism for the activation of allylic alcohols

The synthesis of amphidinolide A (**75**) illustrated nicely the power of RCM for generating 20-membered macrocycles in a strictly regio- and chemoselective fashion from a precursor containing seven C=C bonds. A more exciting challenge was issued with the total synthesis of iejimalides, a family of novel polyene macrolides isolated from tunicates. Iejimalide B (**113**) – the most active member of this family – is known for its truly remarkable potency against a panel of 60 standard human cancer cell lines, with the concentrations required to inhibit growth by 50% and tumour gene index values in the low nanomolar range. Potent *in vivo* activity against 2,388 leukaemia was also reported. After it was demonstrated that all attempted lactonisation methods failed to afford the desired macrocycle [63], olefin metathesis appeared as a viable alternative for the macrocyclisation, though risky because application of RCM to the present case demands for no less than the selective activation of two out of ten double bonds in the cyclisation precursor **111** (Scheme 35). Gratifyingly, cyclisation of this polyene with the aid of the Grubbs 2 catalyst delivered the desired 24-membered macrocycle in good (**112b**) to excellent yield (**112a**) as a single *E*-isomer, with none of the ring-contraction products being observed [64]. Noteworthily, because of the exceptional thermal sensitivity of the precursors (**111**), it was instrumental that the RCM reaction proceeded at ambient temperature.

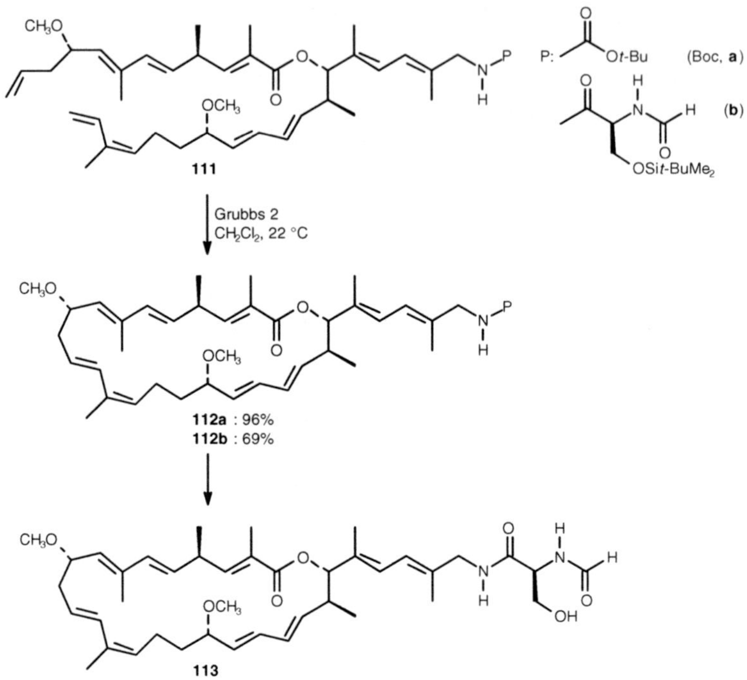

Scheme 35 Ring-closing metathesis in the synthesis of iejimalide B

By contrast, in the total synthesis of amphidinolide X (**114**), the challenging generation of a trisubstituted *E* alkene by RCM failed. None of the attempts to cyclise **115**, **116** (a diastereomer of **115**), and its keto derivative (**117**) using different amounts of the most promising Grubbs 2 and Hoveyda–Grubbs 2 reagents were successful (Scheme 36). >50% of unreacted starting material was recovered in most cases, whereas with diastereomer **116**, the *Z* isomer was isolated in 30–40% yield, exclusively [65]. Adding the catalyst in portions significantly improved the RCM process. Thus, the keto derivative **117** afforded exclusively (12*Z*)-amphidinolide X [(12*Z*)-**114**] in 85% yield [66]. Furthermore, in the presence of 20 mol% of Grubbs 1, substrate **118** bearing a protected alcohol did not form any RCM product with or without added $Ti(Oi\text{-}Pr)_4$. When the Grubbs 2 catalyst was used, **118** furnished the RCM products in 50% combined yield as a 71:29 (*Z*/*E*) mixture. In contrast to the protected alcohol **118**, the free alcohol **119** upon exposure to Grubbs 2 decomposed into unidentified by-products (Scheme 36).

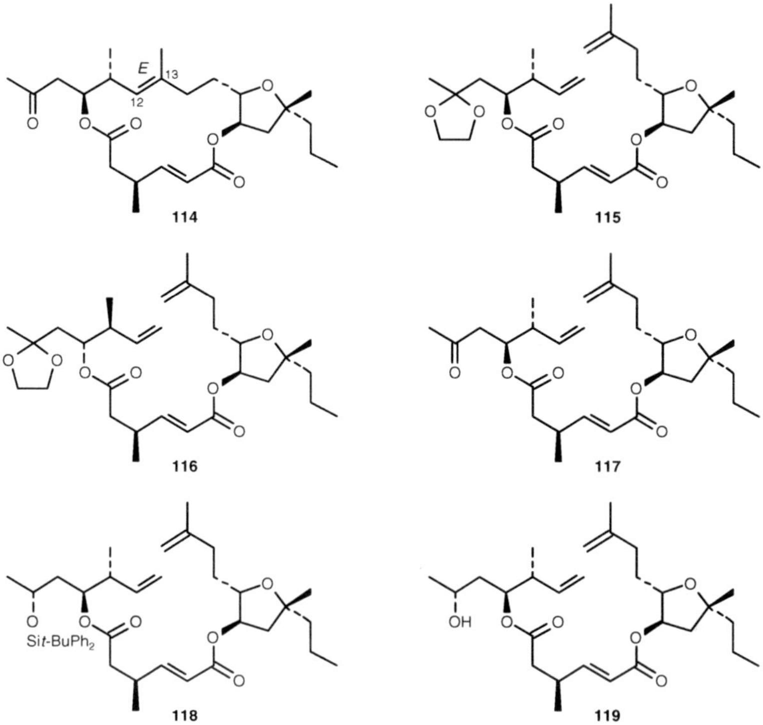

Scheme 36 Structure of amphidinolide X (**114**) and substrates that failed to undergo ring-closing metathesis

In an alternative strategy to amphidinolide X [65], RCM of the silicon-tethered diene **120** in the presence of the Schrock's catalyst (**4**) gave the desired siloxane **121** in 78% yield (Scheme 37). As already demonstrated in the literature for related compounds [67], diene **120** could not be cyclised using Ru initiators.

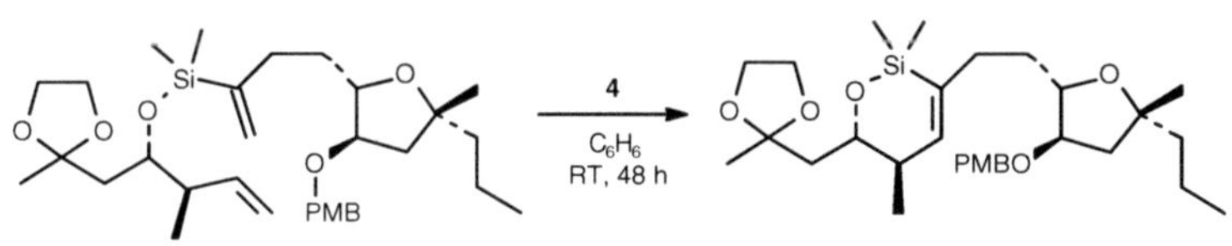

120 PMB = *p*-methoxybenzyl **121**

Scheme 37 Silicon-tethered ring-closing metathesis in the synthesis of amphidinolide X

It is generally agreed that the second-generation Grubbs' complex is superior to the first-generation Grubbs' complex. This is, however, not always the case. Thus, for reasons that are presently unclear, attempts to close the 19-membered macrocycle of amphidinolide E (**124**) failed when **122** was treated with the Grubbs 2 or Hoveyda–Grubbs 2 catalysts. Polyene **122**, indeed, decomposed under the different conditions investigated. These setbacks were overcome by using the Grubbs 1 catalyst (Scheme 38). The *E*,*E*-diene **123**, indeed, was isolated in 73% yield, together with an inseparable mixture of enyne metathesis products (10% yield) [68]. In an alternative approach, however, the triene side-chain (**127**) of amphidinolide E was prepared quite successfully using the Grubbs 2 catalyst *via* a sequential enyne metathesis between alkyne **125** and ethylene, and a cross-metathesis of the product thus obtained (**126**) with 2-methyl-1,4-pentadiene (Scheme 38) [51e].

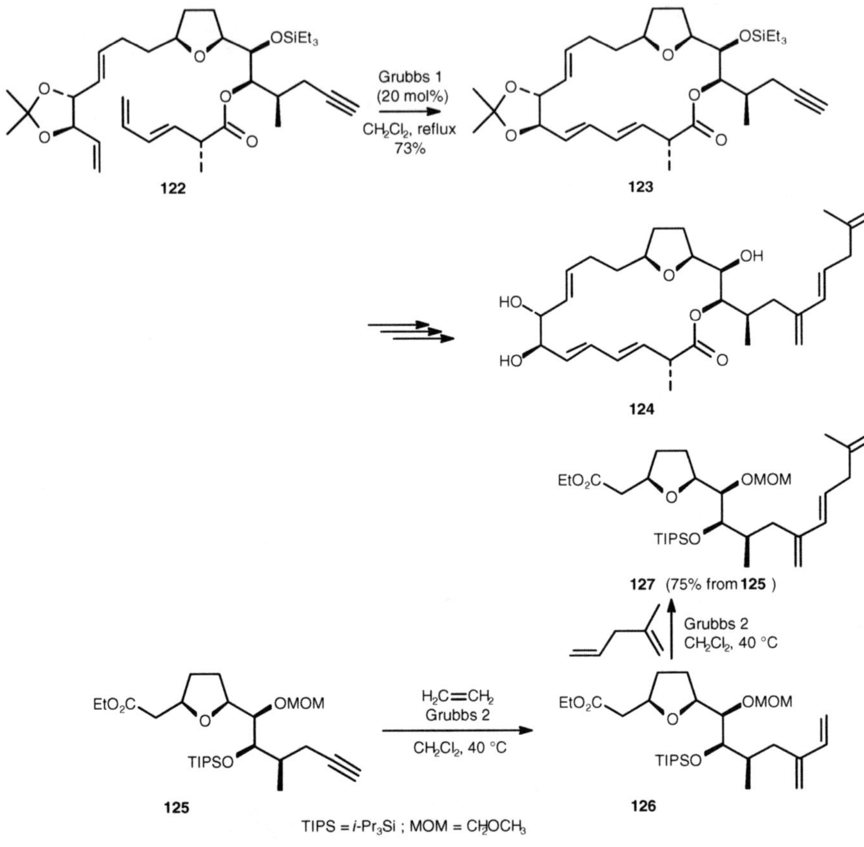

Scheme 38 Olefin metathesis approaches to amphidinolide E

In several cases, opposite trends were found when comparing the activity of Grubbs 1 and Grubbs 2 catalysts, as exemplified for the RCM of diene compounds **128** bearing different phenolic protecting groups (Scheme 39 and Table 1) [69]. In addition, the stereochemical outcome changed significantly depending on the catalyst that was used. Grubbs 1 produced exclusively the *E* isomer, whereas Grubbs 2 provided a mixture of the *E* (**129**) and *Z* (**130**) isomers under kinetic control (Table 1). It is worthy of note that the 12-membered salicylic macrolide scaffold thus obtained is found in several natural products, such as oximidines and salicylihalamides, that are potent cytotoxic agents against various human cancer cell lines.

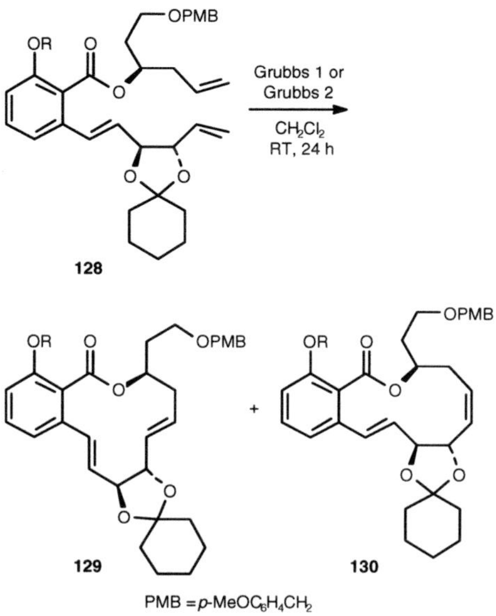

Scheme 39 Ring-closing metathesis in the synthesis of oximidines

Table 1 Ring-closing metathesis of diene **128** using Grubbs 1 or Grubbs 2 catalysts

128, R	Catalyst[a]	Yield (%)	129/130
H	Grubbs 1	64	64/0
	Grubbs 2	37	21/16
t-$BuMe_2Si$	Grubbs 1	47	47/0
	Grubbs 2	41	35/6
Me	Grubbs 1	38	38/0
	Grubbs 2	75	51/24
CH_3OCH_2	Grubbs 1	77	77/0
	Grubbs 2	80	51/29

[a]5 mol%.

Though powerful, RCM is far from being the panacea in the total syntheses of amphidinolides. Failures have been encountered, for instance for the projected total synthesis of amphidinolide N (**134**), a compound whose structure has so far not yet been fully determined. Despite numerous attempts and extensive variation of the reaction parameters, enyne **131** could not be cyclised directly to generate macrocyclic compound **133**, but was cleanly converted into diene **132** in 60% yield upon microwave irradiation in the presence of the second-generation Grubbs' ruthenium carbene under an ethylene atmosphere. Diene **132** was apparently all but inert to further productive metathesis events, failing either to cyclise to the corresponding macrocycle (**133**) or to undergo a significant degree of oligomerisation, despite prolonged exposure (under purely thermal conditions or microwave irradiation) to all commercially available ruthenium-based catalysts or the highly active Schrock's molybdenum-based catalyst (Scheme 40) [70].

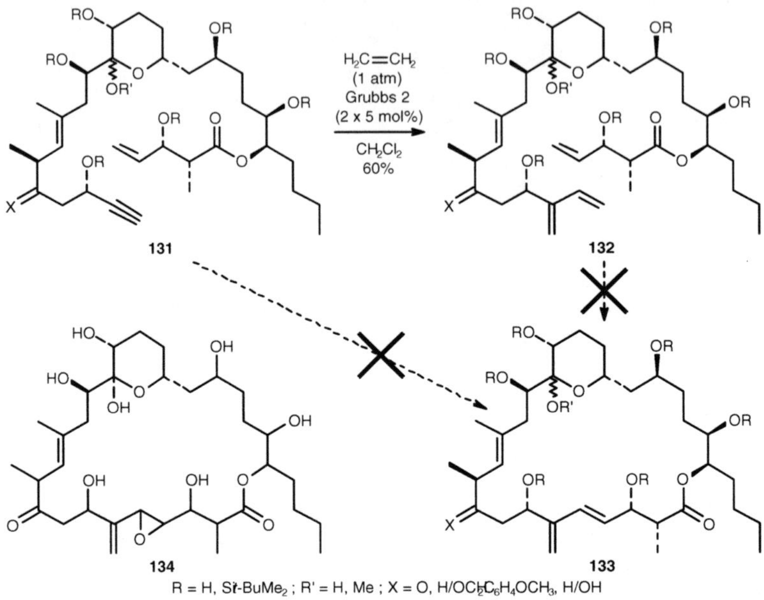

Scheme 40 Attempted ring-closing metathesis in the proposed synthesis of amphidinolide N

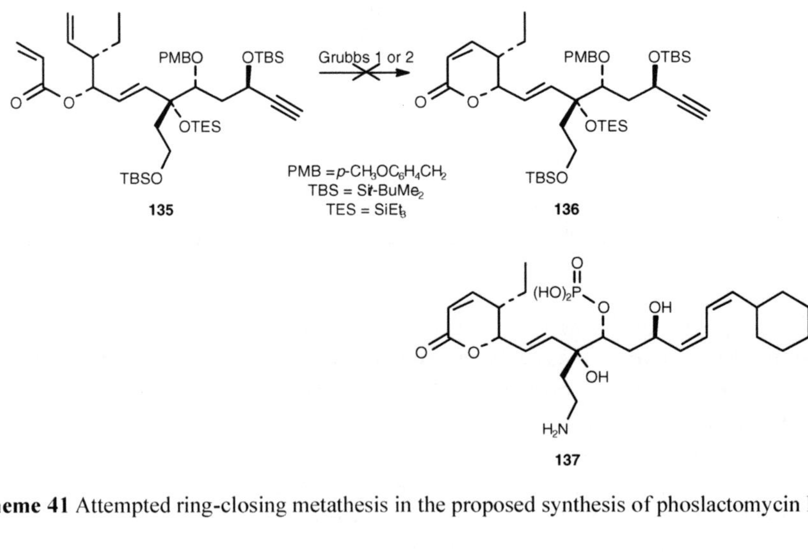

Scheme 41 Attempted ring-closing metathesis in the proposed synthesis of phoslactomycin B

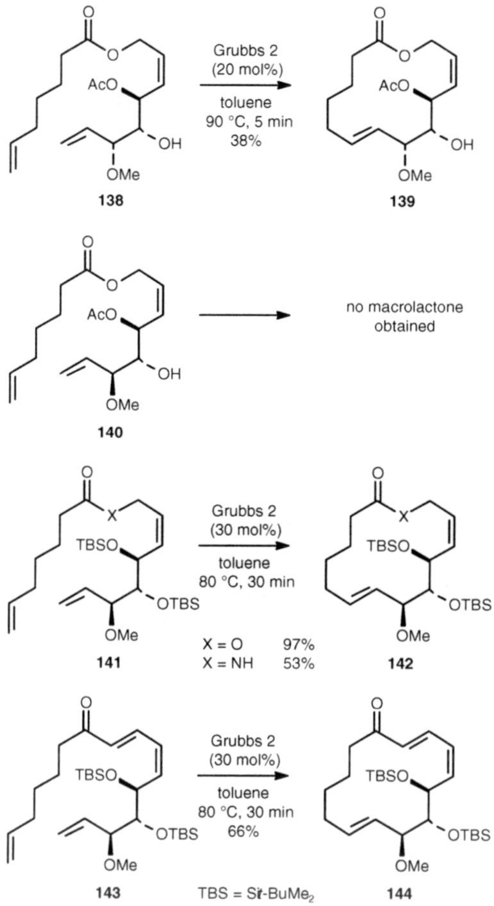

Scheme 42 Ring-closing metathesis in the synthesis of migrastatin analogues

These results were particularly galling in view of the fact that, had the ring-closure been successful, only very few further steps would have been required to complete the synthesis of the first stereoisomer of amphidinolide N (**134**). Alternative strategies to this target natural product involving cross-metathesis were also unsuccessful.

Likewise, attempts to synthesise phoslactomycin B (**137**) (an antitumour, antibacterial, and antifungal natural product) by applying the RCM reaction for the construction of δ-lactone **136** were vain using the first- and second-generation Grubbs' catalysts (Scheme 41) [71].

A recent report on the synthesis of a range of analogues of the migrastatin (an inhibitor of cell migration [72]) macrolide core confirmed that RCM is not so trivial. Thus, a number of attempts (varying the catalyst, concentration, solvent, temperature, and reaction time) to obtain macrocycle **139** in good yield were not successful (Scheme 42). The reaction of **138** with the Grubbs 2 catalyst in toluene at 90°C proved best and gave the desired product (**139**) after 5 min in 38% yield [73]. However, despite numerous attempts, RCM could not be achieved for **140**. The only product which could be identified was a homodimer of **140** and a number of other unidentified products were obtained. A reason for the low yields of macrolactone from **138** and no macrocyclisation from **140** was attributed to the presence of the *Z*-alkene group of **138** and **140**, which has the potential to undergo competing metathesis processes, consequently reducing the efficiency of the desired macrolactone formation. To block undesired metathesis processes at this site and facilitate RCM to give the desired macrolactone, steric bulk was increased in the environment of the reactive alkene in **140** by introducing two *O*-TBS protecting groups. Gratifyingly, the RCM of **141** (X = O) gave **142** in high yield (97%). On the other hand, when the related amide **141** (X = NH) was subjected to RCM under the same conditions, the yield (53%) was lower than observed during the synthesis of the macrolactone derivative **142** (Scheme 42), possibly due to the more polar amide group deactivating the catalyst somewhat [74]. Using the same protocol, RCM of **143** gave the macrocyclic dienone **144** in 66% yield.

Besides olefin metathesis, other organometallic methods, such as the sp^2-sp^3 Stille coupling [75] and the nickel-catalysed reductive coupling of alkynes with aldehydes [76] have also been employed successfully to form the macrocycle. Of significant interest, however, is the macrocycloisomerisation using the ruthenium-catalysed alkyne–alkene addition (Scheme 43), which culminated in the total synthesis of amphidinolide A [77]. Given the high degree of unsaturation present in the substrate (**145**), the chemoselectivity of this process is remarkable. In addition, this macrocyclisation proceeded in higher yields than the intramolecular Stille coupling [75] and the RCM [52].

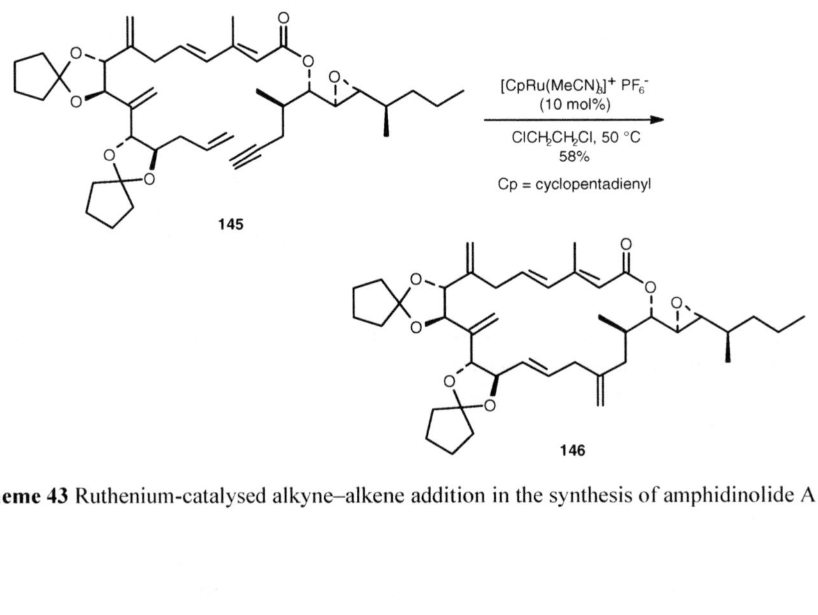

Scheme 43 Ruthenium-catalysed alkyne–alkene addition in the synthesis of amphidinolide A

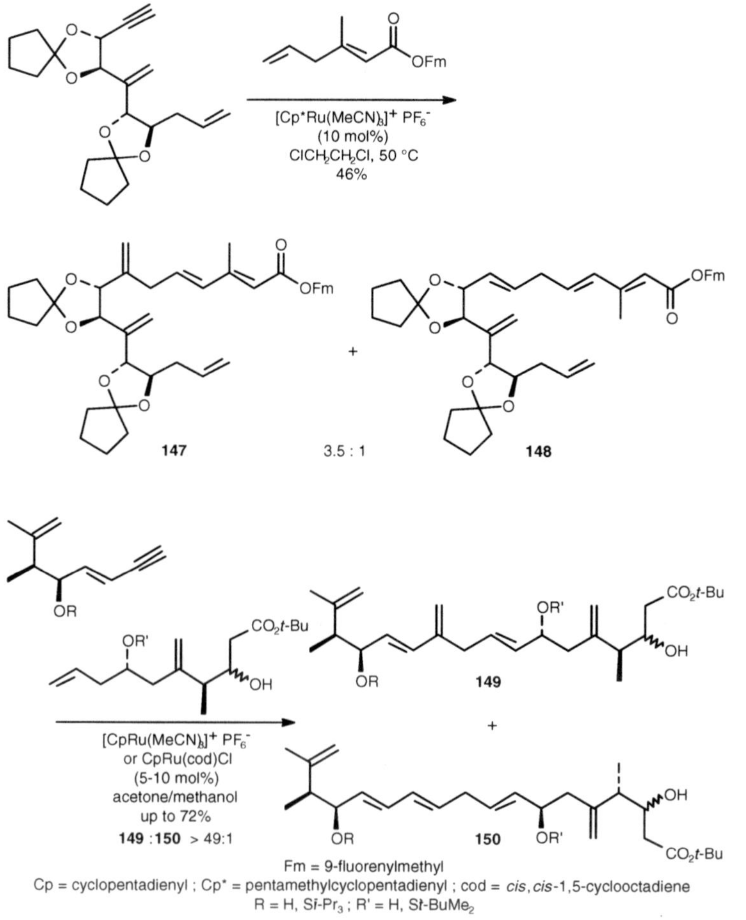

Scheme 44 Ruthenium-catalysed alkyne–alkene addition in the synthesis of amphidinolides A and P

In addition, the simpler intermolecular version of this process between two judiciously chosen alkyne and alkene partners provided the desired branched intermediate (**147**) for the synthesis of amphidinolide A [77], along with its linear isomer (**148**) as a minor product (Scheme 44). Similarly, enynes could also be engaged in the appropriate coupling with alkenes. Thus, the branched product **149** was formed regioselectively (Scheme 44) and finally transformed into amphidinolide P [78].

Equally successful was the synthesis of amphidinolide V (**155**), which took advantage of a sequence of ring-closing alkyne metathesis (RCAM) and subsequent intermolecular enyne metathesis (Scheme 45) [79]. RCAM proceeded nicely, even though the resulting product **153** is fairly strained, in the presence of the molybdenum-based catalyst formed *in situ* upon activation of complex **152** with CH_2Cl_2. The subsequent enyne metathesis between cycloalkyne **152** thus formed and ethylene gas installed the s-*trans* diene unit of **153**, spanned by the vicinal *exo*-methylene branches at C4 and C5, without any appreciable interference from the pre-existing double bonds.

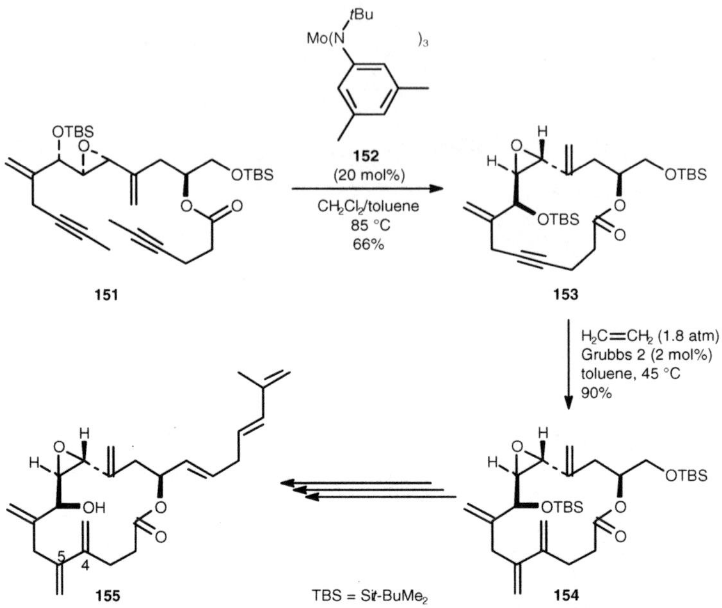

Scheme 45 Ring-closing alkyne metathesis/enyne metathesis sequence in the synthesis of amphidinolide V

4 Temporary Silicon-Tethered Ring-Closing Metathesis

The versatility of olefin metathesis was further nicely demonstrated by the total synthesis of spirofungin A (**156**) [80], a secondary metabolite from *Streptomyces violaceusniger* Tü 4113. Spirofungin A displays antifungal activity and notable antiproliferative activity in a panel of cancer cell lines, and selectively inhibits the activity of isoleucyl-tRNA synthetase in mammalian cells (Scheme 46).

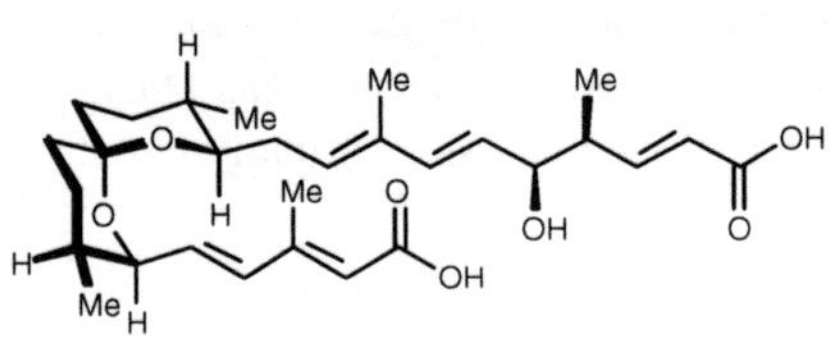

156

Scheme 46 Structure of spirofungin A

A recent, fully stereocontrolled and efficient synthesis of this natural product involves 20 steps in its longest linear sequence [80], six of them being catalysed by transition metal complexes: Three steps rely on olefin metathesis in the presence of the Grubbs 2 catalyst (Scheme 47), whereas three transformations are catalysed by palladium, viz. a concomitant hydrogenation of alkenes/hydrogenolysis of benzyl ethers, a Negishi and a Stille cross-coupling reaction.

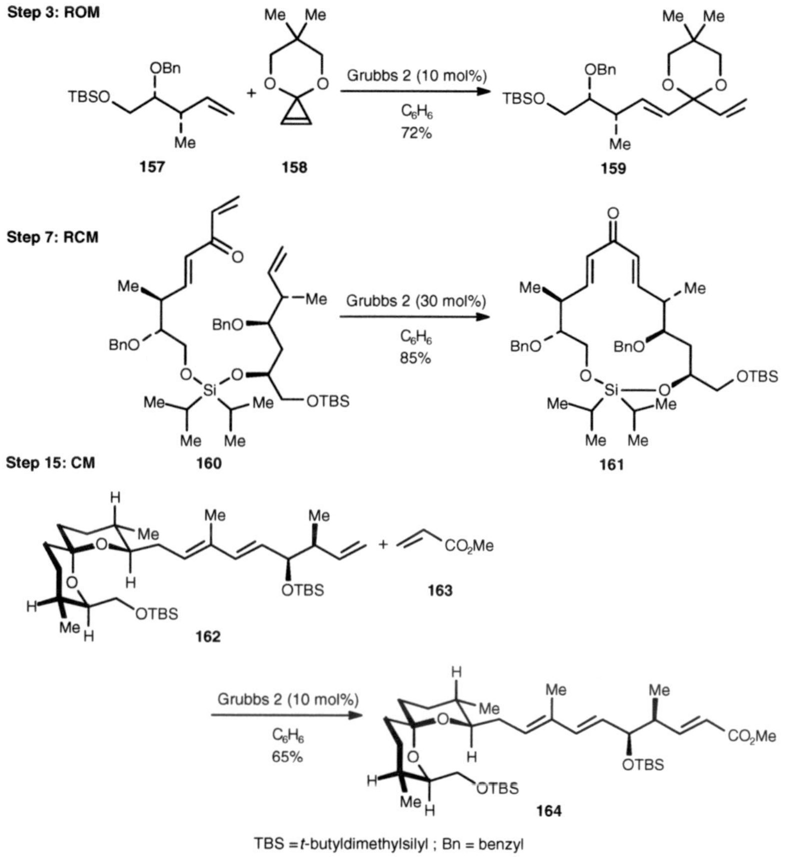

Scheme 47 Olefin metathesis steps in the total synthesis of spirofungin A

Thus, in the third step of the sequence (Scheme 47), alkene **157** reacted with cyclopropenone acetal **158** *via* a ring-opening metathesis to give diene **159**. In step 7, cyclisation of **160** afforded the 15-membered dienone **161**. Notably, the ring-closing metathesis proceeded with complete chemoselectivity only at the two terminal alkenes of trienone **160**. Finally, triene **162** was further elaborated *via* cross-metathesis with methyl acrylate (Step 15, Scheme 47).

Of utmost importance in this total synthesis was the formation of spiroketal **167** (Scheme 48). Thanks to the dialkoxysilane connector, spiroketalisation of the 15-membered silacyclic ketone **165** obtained after hydrogenation of **161** and concomitant hydrogenolysis of the two benzyl ethers, produced exclusively the desired tricyclic spiroketal **166**. Subsequent cleavage of the three O–Si bonds afforded the corresponding triol **167**, that, although favoured stereoelectronically, is sterically congested because of the axial disposition of the C-19 substituent (Scheme 48). Interestingly, in the absence of connection, the parent substrate **168** yielded a nearly equimolar mixture of two spiroketals, **169** and **170**, upon spontaneous spiroketalisation (Scheme 48) [81].

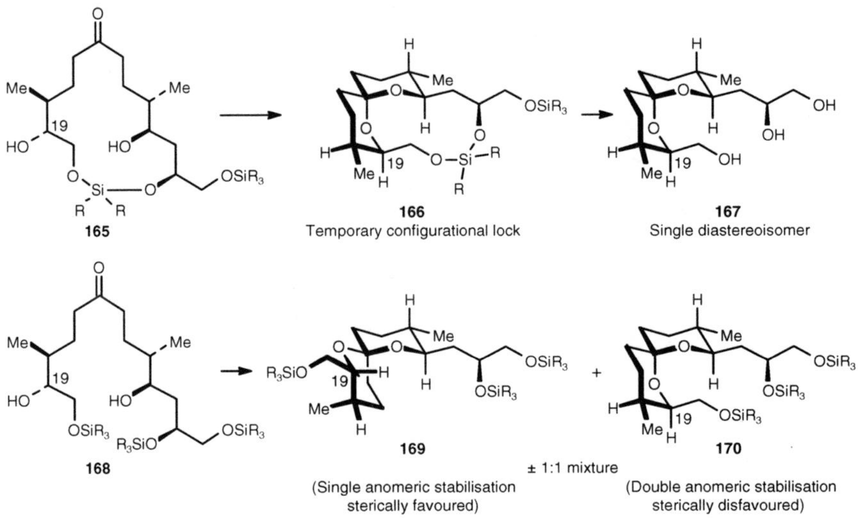

Scheme 48 Influence of the silicon tether on the spiroketalisation

A silicon connector is also used to facilitate the cross-metathesis reaction, *via* a strategy known as the temporary silicon-tethered ring-closing metathesis (TST RCM). As sketched in Scheme 49, this methodology ultimately leads to α,ω-diols (**173**) after cleavage of the O–Si bonds in the cyclic silaketal **172**.

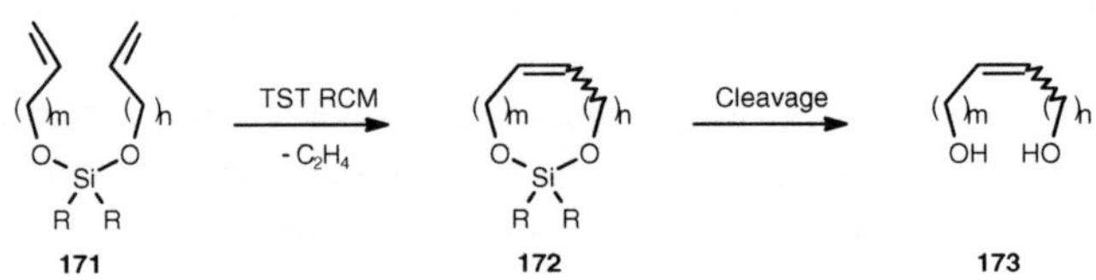

Scheme 49 Temporary silicon-tethered ring-closing metathesis

An added virtue of the silicon tether is the influence it imparts over the olefin geometry in the cyclic silaketal **172** and thereby in the product **173** obtained after deprotection. In most cases, the *Z*-olefin predominates (>90–95%). This represents a significant advantage over the cross-metathesis where the inability to control the *Z*/*E*-selectivity is often a serious limitation. The silicon-tethered ring-closing metathesis is also sensitive to the reaction conditions (catalyst loading, type of solvent, concentration, and temperature). Olefin isomerisation (**174** to **175** (both *E* and *Z*) to **176** (*EE*, *EZ*, and *ZZ*), and **177** to **178**, Scheme 50) has been reported as a side reaction, perhaps *via* metal (ruthenium [82] or molybdenum [83]) hydride species from background decomposition events. Subsequent RCM of **175** and **176** accounts for the formation of **179** and **180**, respectively (Scheme 50). As a result of these undesirable, competitive events, the TST RCM route is preferably applied on bis-alkoxysilanes derived from allylic and homoallylic alcohols.

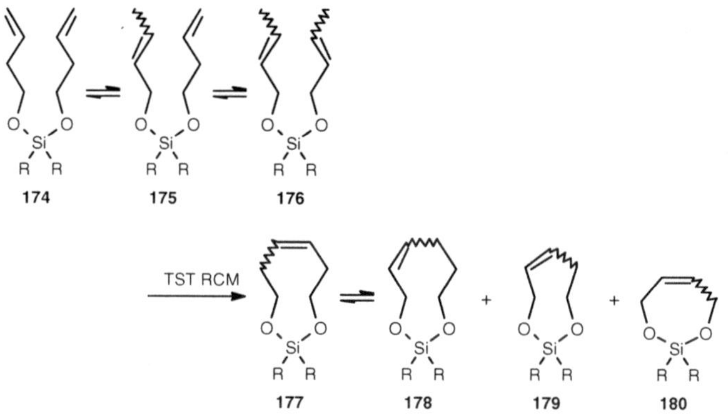

Scheme 50 Competitive olefin isomerisation–temporary silicon-tethered ring-closing metathesis

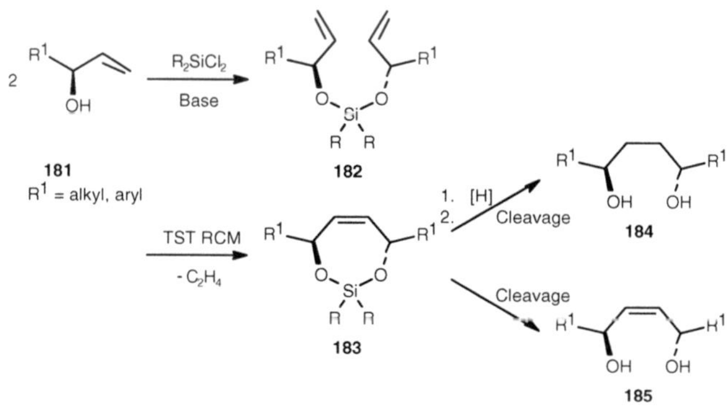

Scheme 51 Temporary silicon-tethered ring-closing metathesis applied to optically pure allylic alcohols

In particular, the TST RCM approach has been successfully adapted, utilising optically pure allylic alcohols **181**, to yield C_2-symmetrical 1,4-diols, **184** and **185** (Scheme 51) [84], which are useful as precursors to a variety of asymmetric catalysts, as chiral auxiliaries and as intermediates for two-directional synthesis.

The TST RCM methodology has also been applied to mixed bis-alkoxysilanes **186**, prepared from two different chiral allylic alcohols (Scheme 52). The advantage of this type of approach is the ability to couple a wide range of chiral allylic alcohols, which facilitates the preparation of useful intermediates for target directed synthesis.

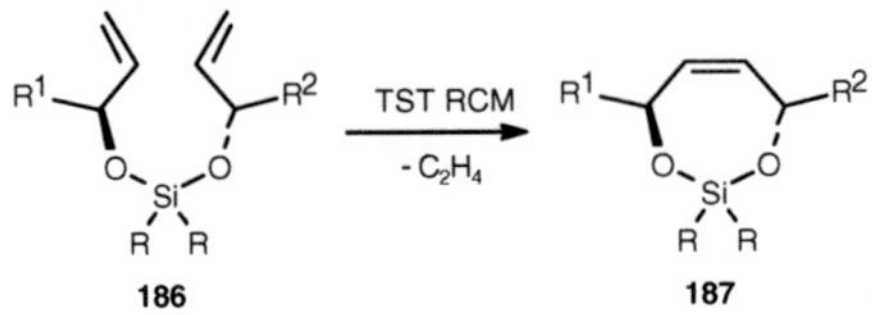

Scheme 52 Temporary silicon-tethered ring-closing metathesis applied to mixed bis-alkoxysilanes

188

189

190

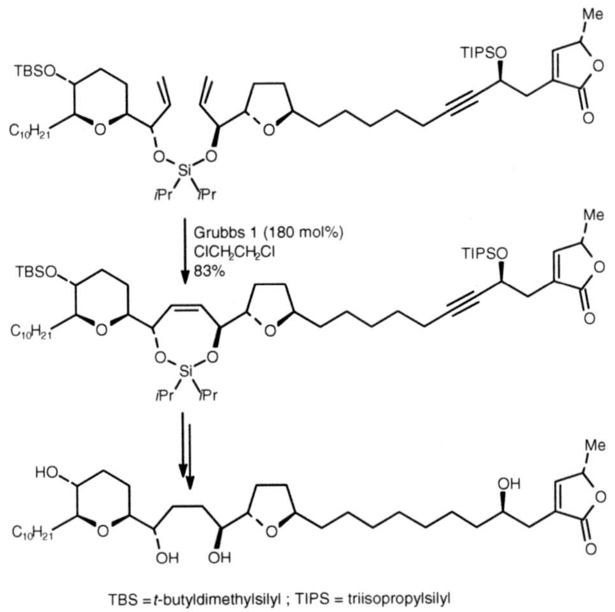

Scheme 53 Temporary silicon-tethered ring-closing metathesis in the synthesis of (–)-mucocin

Schemes 53, 54, and 55, for instance, illustrate the application of this protocol to the total synthesis of (–)-mucocin **(190)** [85] and (+)-*cis*-sylvaticin **(193)** [86], two potent antitumour annonaceous acetogenins, and (–)-attenol A **(197)** [87], a cytotoxic agent [88].

191

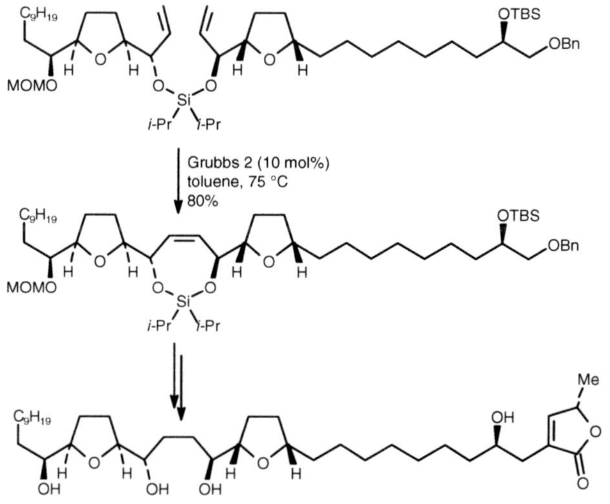

192

193

Bn = benzyl ; MOM = methoxymethyl ; TBS = t-butyldimethylsilyl

Scheme 54 Temporary silicon-tethered ring-closing metathesis in the synthesis of (+)-cis-sylvaticin

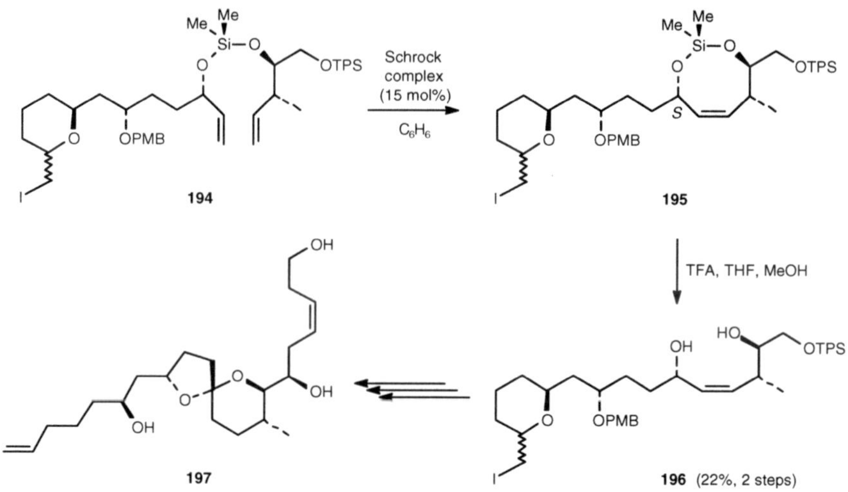

PMB = p-methoxybenzyl ; TPS = triphenylsilyl ; TFA = trifluoroacetic acid

Scheme 55 Temporary silicon-tethered ring-closing metathesis in the synthesis of attenol A

Noteworthily, other examples of tethered RCM reactions using substrates **198**–**200** (Scheme 56) have also been described.

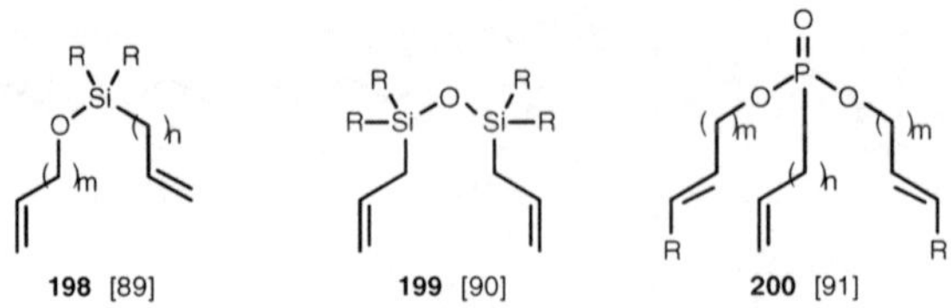

Scheme 56 Alternative tethered ring-closing metathesis precursors

5 Relay Ring-Closing Metathesis

The archazolids are a family of polyunsaturated 24-membered macrolides displaying powerful growth-inhibitory activity against a number of murine and human cancer cell lines at subnanomolar concentrations. Again, several methodologies can be employed to close the macrocycle. A Horner–Wadsworth–Emmons macrocyclisation, for instance, was preferred to the other methods for the synthesis of archazolid A (**201**, R′ = H, R″ = Me, Scheme 57) [92]. In principle, RCM could also be employed as an alternative for ring closure, but this approach was discarded to the profit of the so-called relay ring-closing metathesis (RRCM), as illustrated by the synthesis of archazolid B (**201**, R′ = R″ = H, Scheme 57).

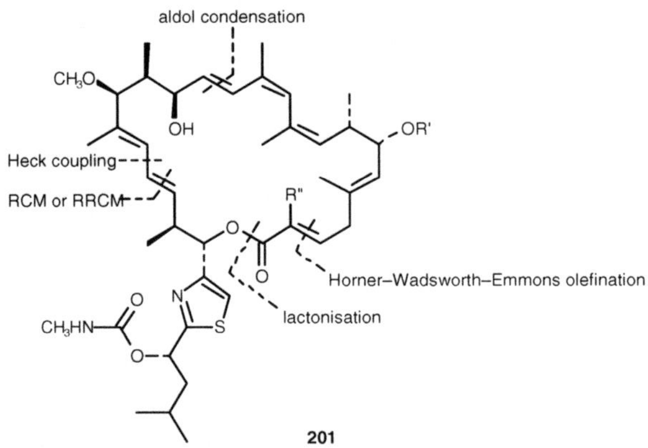

Scheme 57 Structure of archazolids

The rational behind the RRCM strategy rests on the fact that initiation of the RCM catalytic cycle between two sterically demanding alkenes as in **202** (Scheme 58) or between a sterically demanding alkene and an electronically deactivated alkene as in **207** (Scheme 59) is a challenge. Tetrasubstituted cycloalkenes such as **206** and **211** proved indeed extremely difficult to access *via* RCM using the Grubbs 1 catalyst. Although these shortcomings could be overcome using the Grubbs 2 catalyst, the formation of some tetrasubstituted olefins can still be less than satisfactory. On the other hand, contrary to the *intermolecular* reaction, the *intramolecular* version of a ruthenium alkylidene onto a sterically hindered and/or electronically deactivated alkene is attainable, as is the subsequent reaction of the previously inaccessible alkylidene.

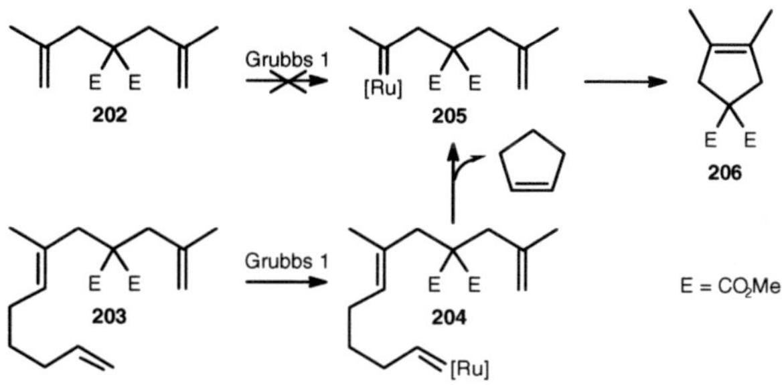

Scheme 58 Direct ring-closing metathesis between two sterically demanding alkenes versus relay ring-closing metathesis

As exemplified in Scheme 58, the relay RCM approach involves incorporation of a temporary tether containing a sterically unencumbered terminal olefin (**203**) for initiation of the catalytic cycle (**203–204**). The olefin is positioned such that a kinetically favourable formation of a five-membered ring is used to deliver the ruthenium onto the less accessible position (**205**). The desired RCM reaction then proceeds to give **206** [93]. Similarly, substrate **207** is unreactive to RCM conditions with the Grubbs 1 catalyst. The butenolide **211**, however, could be prepared from either of the starting materials **208** or **212** *via* initial reaction with the terminal olefins incorporated into the tethers (Scheme 59) [94]. In this instance, the relay procedure overcame both a steric and an electronic deactivation, as the intermolecular reaction of the Grubbs 1 catalyst with electron-deficient double bonds cannot usually be achieved even when they are unsubstituted.

Thus, archazolid B was successfully synthesised using the RRCM methodology instead of the more conventional RCM (Scheme 60) [95, 96].

It is also interesting to note that the synthesis of archazolid B proceeded in 19 steps for the longest linear sequence and involved several transition-metal-catalysed operations, including three very different reactions promoted by ruthenium catalysts. In addition to the RRCM reaction – the penultimate step of the synthesis –, the Trost alkene–alkyne coupling [77, 78, 97] and the Kita synthesis of enol ester **218** from carboxylic acid **217** and ethoxyacetylene [98] (Scheme 61) underscore the usefulness of ruthenium catalysis.

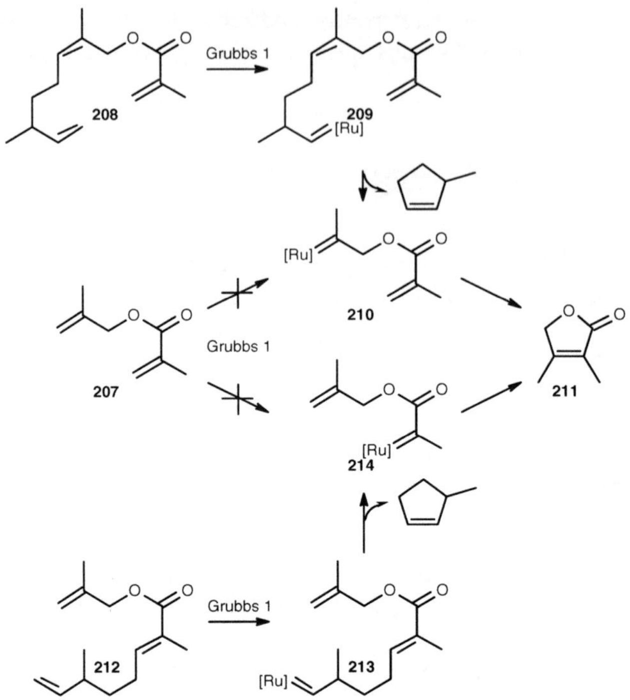

Scheme 59 Direct ring-closing metathesis between a sterically demanding alkene and an electronically deactivated alkene versus relay ring-closing metathesis

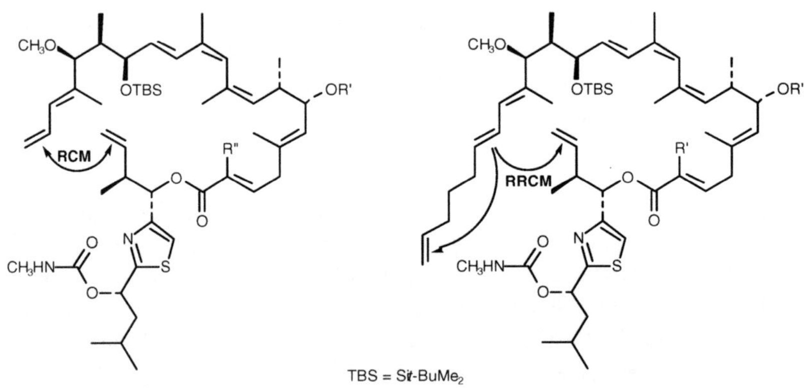

Scheme 60 Direct ring-closing metathesis versus relay ring-closing metathesis in the synthesis of archazolid B

In an alternative approach to phoslactomycin B (**137**), a relay RCM was applied on compound **219** using 18 mol% of the second-generation Grubbs' catalyst, during which dihydropyran **220** was expelled and the α,β-unsaturated lactone **221** was

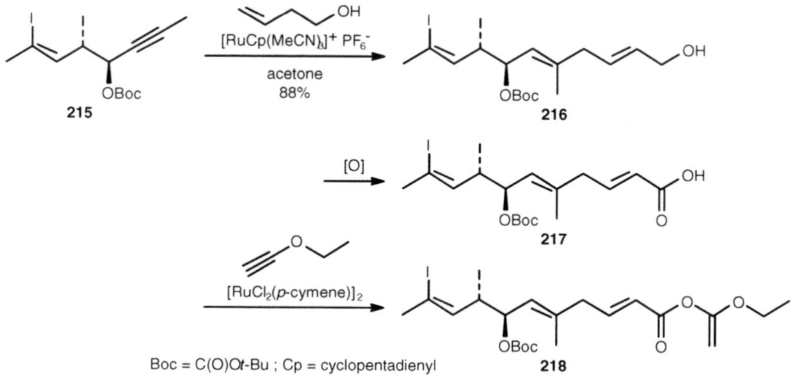

Scheme 61 Alkene–alkyne coupling and enol ester formation in the synthesis of archazolid B

produced in 88% yield (Scheme 62). When only 5 mol% of Grubbs 2 catalyst were used, the desired lactone **221** was formed in 75% yield, together with **222** (25%). Interestingly, the RCM of the latter compound subsequently proceeded at a much slower rate and required a higher catalyst loading to reach completion [99]. It is worth noting that in a previous approach to phoslactomycin B (Scheme 41), attempted RCM reactions for the formation of the unsaturated lactone were vain [71], thus highlighting the power of the relay RCM.

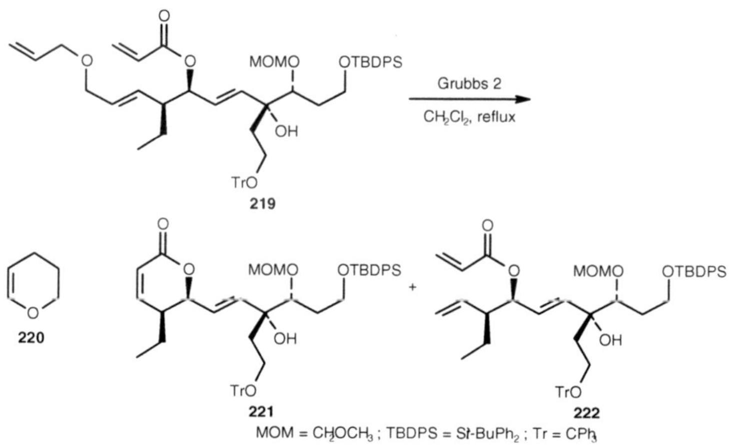

Scheme 62 Relay ring-closing metathesis in the synthesis of phoslactomycin B

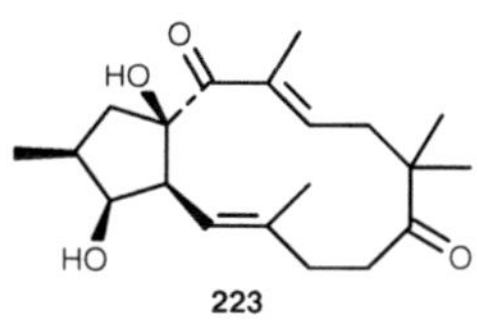

Scheme 63 Structure of characiol

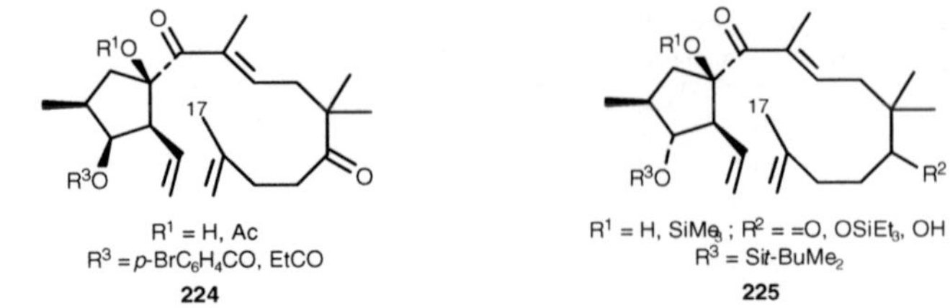

Scheme 64 Substrates that failed to undergo ring-closing metathesis

Relay RCM, however, is far from being the panacea in organic synthesis. For instance, in a research programme aimed at the total synthesis of characiol (**223**), RCM of trienes **224** (Schemes 63 and 64) was envisioned as a plausible key step to the macrocyclisation [100].

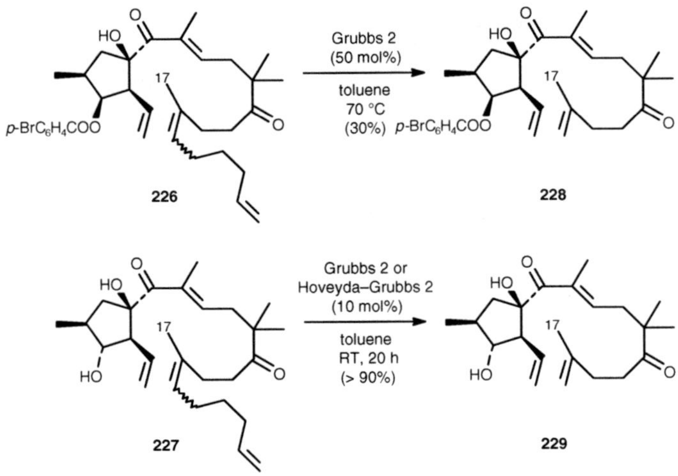

Scheme 65 Unsuccessful relay ring-closing metathesis reactions in the proposed synthesis of characiol

Despite many efforts using Grubbs' and Hoveyda–Grubbs' catalysts in different solvents at different temperatures, not even a trace of the desired cyclisation products could be identified. Also no RCM was observed with alternative substrates **225**. To overcome these failures, relay ring-closing metatheses of substrates **226** and **227** were attempted (Scheme 65). Unfortunately, the desired RRCM products did not formed. Instead, the relay tetherless trienes **228** and **229** were observed in almost quantitative yields, apparently formed by a RCM with release of cyclopentene followed by an intermolecular cross-metathesis process. The failure of the RCM and RRCM was attributed to the presence of the C17 methyl group, which might be responsible for the build-up of unacceptable steric strain during the metathesis process. This hypothesis was supported by the observation that trienes **230–232** lacking the C17 methyl group underwent RCM, albeit in low yields, upon exposure to

Grubbs 2 catalyst (Scheme 66). Furthermore, RCM of substrate **230** produced a mixture of double bond isomers **233** (*E*/*Z* = 2/1), whereas the partially protected analogues **231** and **232** gave a single *E*-configured double bond isomer, **234** and **235**, respectively, emphasising thus the beneficial effect of the benzoate group on the *E*/*Z* selectivity.

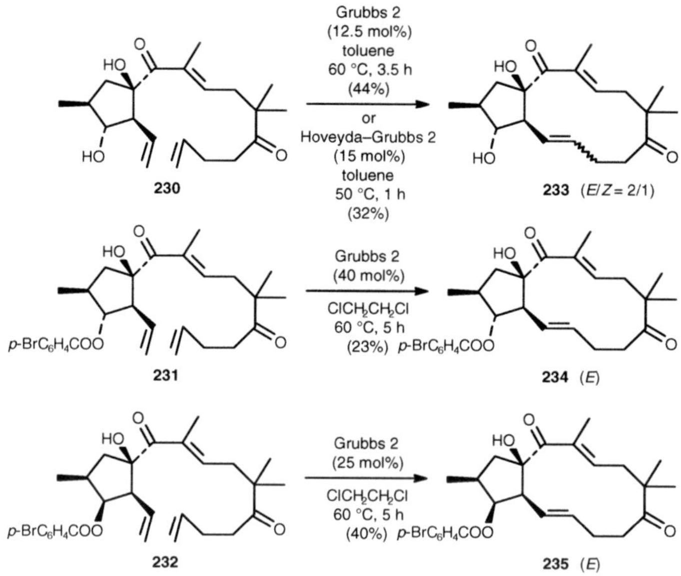

Scheme 66 Ring-closing metathesis in the synthesis of characiol derivatives

6 Sequential Reactions

6.1 One-Pot Sequential Reactions

6.1.1 Olefin metathesis/second transformation

Much interest has recently been attached to one-pot processes involving multiple (catalytic) transformations followed by a single workup stage [101]. A one-pot reaction is defined as the modification of an organic moiety *via* two elaborations, with addition of the second partner(s) (reagent and/or catalyst) only after the first transformation is complete. In other words, one-pot reactions involve isolated events.

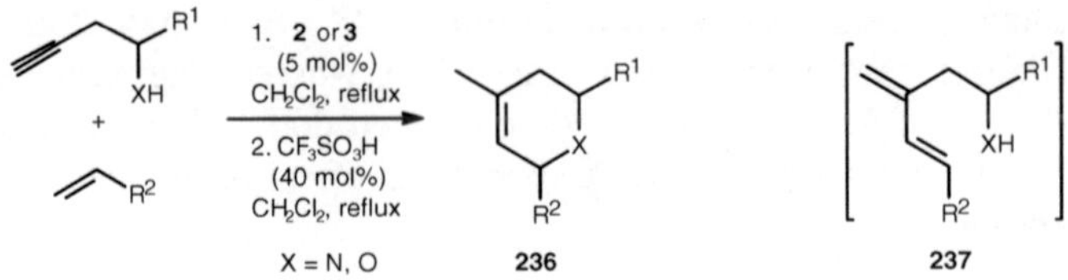

Scheme 67 One-pot enyne metathesis/acid-promoted cyclisation sequence to heterocycles

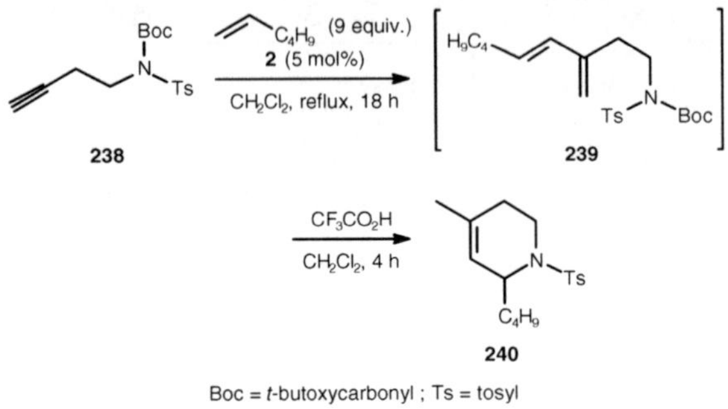

Scheme 68 One-pot enyne metathesis/acid-promoted Boc deprotection–cyclisation sequence to heterocycles

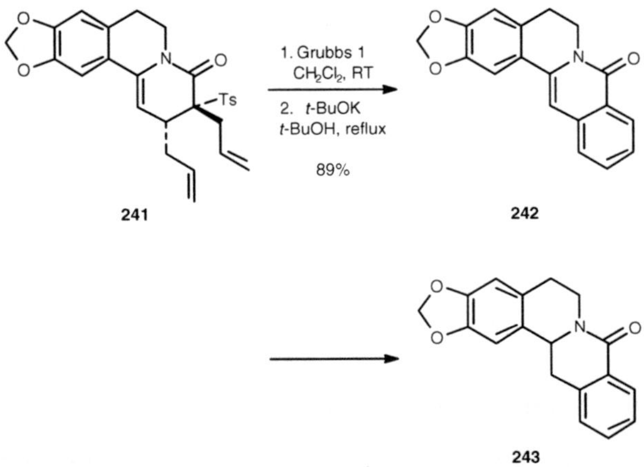

Scheme 69 One-pot ring-closing metathesis/dehydrosulfonation/oxidation sequence to (±)-gusanlung D

One-pot reactions involving a metathetic process in the first step may be considered as trivial when the second transformation is not affected by the ruthenium complex used in the previous step. This type of one-pot reactions is illustrated, for instance, by the synthesis of nitrogen and oxygen heterocycles (**236**) by enyne metathesis and *in situ* cyclisation (Scheme 67) [102]. Thus, once the enyne metathesis was complete, the mixture of the 1,3-diene **237**, excess olefin, and ruthenium catalyst was directly treated with an acid and heating was continued at 40°C. In the particular case of the enyne metathesis between the *N*-Boc protected alkyne **238** and 1-hexene (Scheme 68), exposure of diene *E*-**239** to TFA/CH_2Cl_2 resulted in cyclisation with concurrent Boc protecting group removal. Another example is illustrated in the total synthesis of (±)-gusanlung D (**243**), in which ring-closing metathesis of diene **241** was followed by dehydrosulfonation and a spontaneous oxidation (Scheme 69) [103].

6.1.2 Olefin metathesis/olefin metathesis

A stepwise metathesis/metathesis-based functionalisation of bicyclo[3.2.0]-heptenes **244** provided a stereodiverse library of eight known and anticipated lipid oxidation metabolites, 5-F_2-isoprostanes **248**. These compounds are interesting in that the levels of certain isoprostanes in bodily fluids are used as quantitative markers of disease-related oxidative stress. In addition, the isoprostanes are also linked to inflammatory events as well as smooth muscle regulation. In the key step of a recent total synthesis, ring-opening cross-metathesis (ROCM) of cyclobutenes **244** with ethylene, followed by a cross-metathesis of the resulting diene **245** with enone **246**, provided ketones **247** as single regio- and (*E*)-stereoisomers in 68–77% isolated yield in a single reaction vessel (Scheme 70) [11]. Interestingly, ring-opening of cyclobutene **244** with ethylene proceeded exquisitely well in the presence of the Grubbs 1 catalyst, whereas the second step involving the electron-deficient *α,β*-unsaturated ketone **246** was best performed using the Hoveyda–Grubbs 2 system (**3**). Alternative routes illustrated in Scheme 71, involving either a selective ring-opening cross-metathesis of the cyclobutene **249** with a suitably functionalised olefin (pathway A) or an intramolecular ring-opening/ring-closing metathesis (pathway B), were not satisfactory.

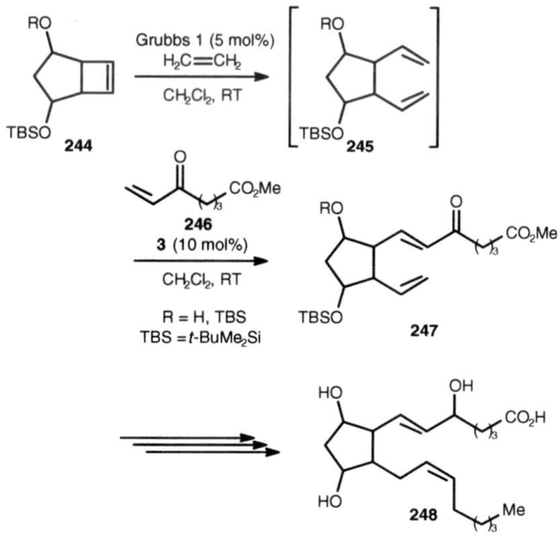

Scheme 70 One-pot ring-opening metathesis/cross-metathesis sequence in the synthesis of 5-F2-isoprostanes

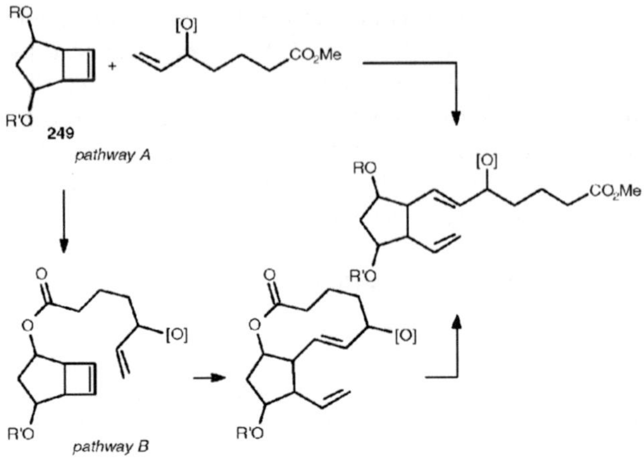

Scheme 71 Unsuccessful olefin metathesis routes to 5-F2-isoprostanes

6.1.3 First transformation/olefin metathesis

One-pot reactions in which olefin metathesis is not the first reaction of the sequence are more challenging, as the metathesis catalyst must tolerate the components (solvent, catalyst, excess reagent, …) used in the previous step(s), as well as the side-product(s) formed eventually.

Neopeltolide **250** (Scheme 72), a marine macrolide isolated from a deep-sea sponge, is an extremely potent inhibitor of the *in vitro* proliferation of the A-549 human lung adenocarcinoma, the NCI-ADR-RES human ovarian sarcoma, and the P388 murine leukaemia cell lines with nanomolar IC_{50} values.

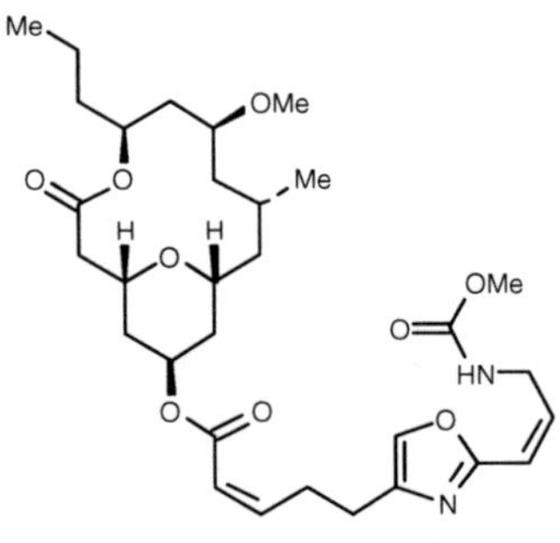

Scheme 72 Structure of neopeltolide

Additionally, this compound is a potent inhibitor of the growth of the fungal pathogen *Candida albicans*. An efficient total synthesis of (+)-neopeltolide was reported recently that exploits two one-pot reaction sequences involving olefin metathesis in the second step [104]. In the first one-pot reaction sequence (Scheme 73), protection of alcohol **251** as its PBM ether **253** using *p*-methoxybenzyl trichloroacetimidate **252** and lanthanum triflate, was followed by cross-metathesis with methyl acrylate, providing enoate **254** in 58% overall yield from **251**. The second reaction sequence exploited the Suzuki–Miyaura coupling and ring-closing metathesis (Scheme 74). Thus, lithiation of iodide **255** with *t*-BuLi in the presence of *B*-MeO-9-BBN generated the alkylborate **256**, which was reacted *in situ* with enol phosphate **257** using aqueous Cs_2CO_3 as a base and $[Pd(PPh_3)_4]$ as a catalyst in DMF at room temperature to give acyclic enol ether **258**. Subsequent RCM of **258** using the second-generation Grubbs' catalyst in toluene furnished the endocyclic enol ether **259** in 78% overall yield from **255**.

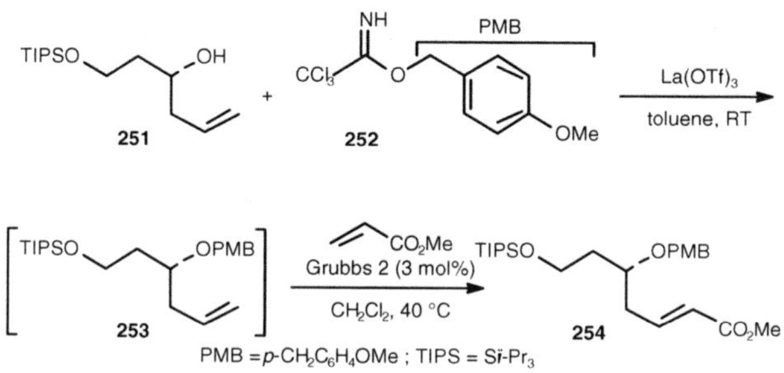

Scheme 73 One-pot alcohol protection/cross-metathesis sequence in the synthesis of neopeltolide

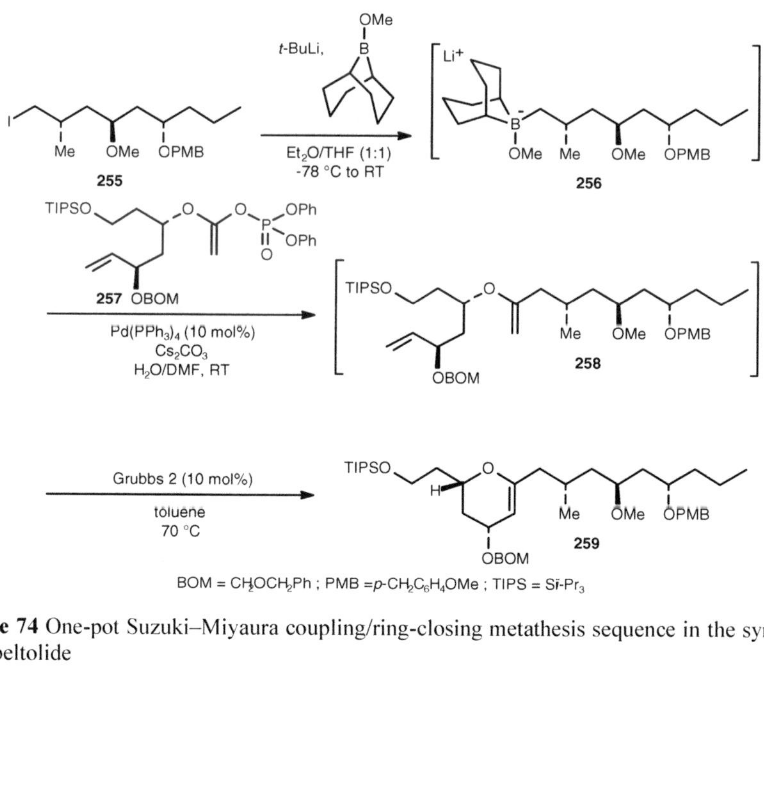

Scheme 74 One-pot Suzuki–Miyaura coupling/ring-closing metathesis sequence in the synthesis of neopeltolide

Cytotoxic marine metabolites (–)-attenol A **197** and (+)-attenol B **264** were synthesised using a similar strategy based on a Suzuki–Miyaura coupling/ring-closing metathesis sequence (Scheme 75) [88]. Thus, in the key step, hydroboration of **261** with 9-BBN-H generated an alkylborane, which was *in situ* coupled with enol phosphate **260**. Subsequent RCM of the intermediate **262** delivered endocyclic enol ether **263** in 76% yield.

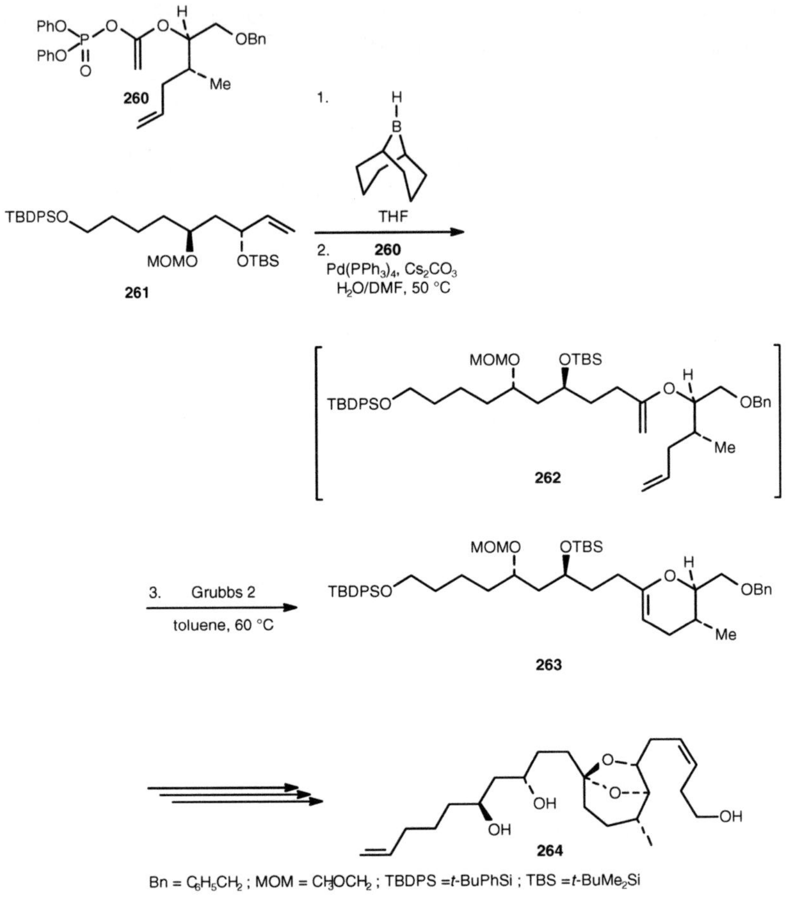

Scheme 75 One-pot Suzuki–Miyaura coupling/ring-closing metathesis sequence in the synthesis of (+)-attenol B

Bicyclic systems have been synthesised recently through a one-pot three-reactions sequence including RCM in the second step. Thus, enolisation and alkylation of compound **265**, followed by RCM with ruthenium alkylidene **266**, and decarboxylation generated the decalin system **267** in 43% isolated yield (Scheme 76) [105].

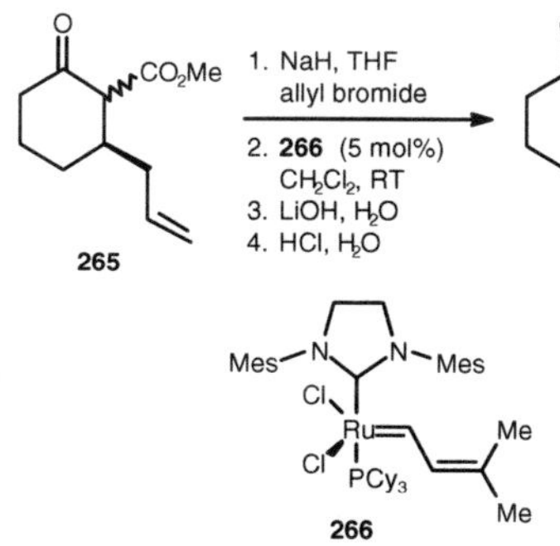

Scheme 76 One-pot alkylation/ring-closing metathesis/decarboxylation in the synthesis of decalin derivatives

6.2 Domino (Cascade) Reactions

The application of metathesis reactions in domino (cascade) sequences allow synthetic chemists to rapidly build complex molecular frameworks in a single operation. This type of synthesis has received a great deal of attention in recent years [106]. For instance, a domino ene–yne–ene RCM of an acyclic precursor has been used in the total synthesis of lepadin F (**268**) and G (**269**), natural compounds displaying significant and selective activity against malaria causing plasmodia and some trypanosomes, which are the main pathogens of sleeping sickness [107] (Scheme 77).

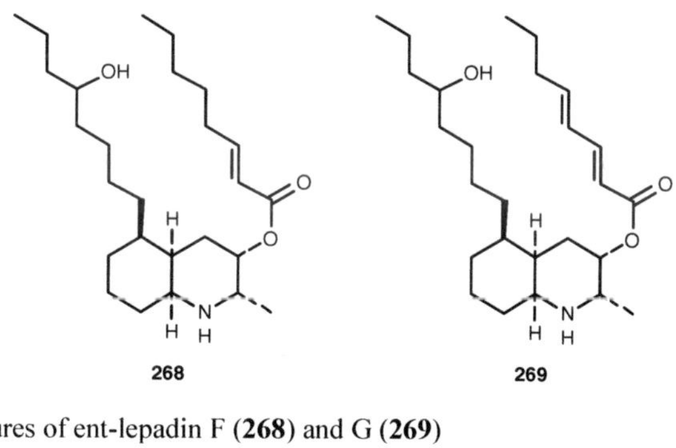

Scheme 77 Structures of ent-lepadin F (**268**) and G (**269**)

Starting from ene–yne–ene **270**, two different reaction pathways could be envisioned (Scheme 78): (a) metathesis may initiate at the most accessible, terminal double bond to produce Ru–carbene intermediate **271**, which on two consecutive RCMs gives rise to the desired bicycle **272** or (b) metathesis may initiate on the disubstituted alkenyl moiety to produce Ru–carbene species **273**. Domino RCM would then lead to the bicycle **274**. Gratifyingly, Grubbs 1 catalyst (10 mol%) at 60°C furnished **272** in 90% yield without significant formation of by-products as happened during reaction with Grubbs 2 and Hoveyda–Grubbs 2 catalysts (Scheme 78).

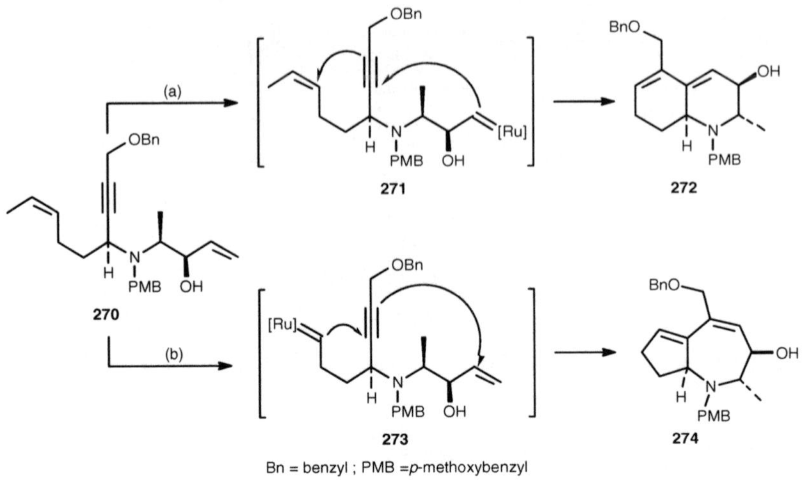

Scheme 78 Domino ring-closing metathesis in the total synthesis of lepadin F and G

In a series of elegant studies, the Blechert group has also demonstrated the synthetic utility of domino RCM–ROM–RCM reactions in the preparation of intermediates **276** and **279** en route to alkaloid natural products (+)-astrophylline (**277**) [108], (+)-dihydrocuscohygrine (**280**), and cuscohygrine (**281**) [109] (Scheme 79). In these processes, which convert carbocycles to heterocycles, the more strained unsaturated ring systems such as cyclopentene **275** and cycloheptene **278** derivatives undergo ring-opening relieving the embedded ring- strain followed by ring-closing metathesis into the more thermodynamically stable ring systems, **276** and **279**, respectively (Scheme 80).

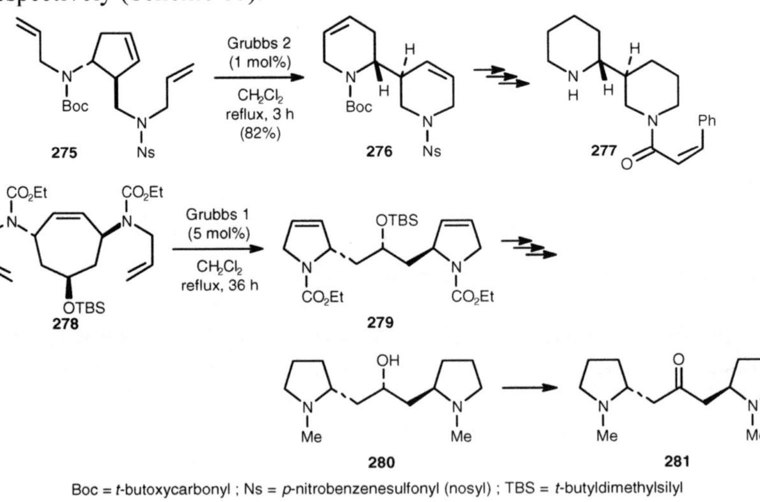

Scheme 79 Domino ring-opening metathesis/ring-opening metathesis/ring-closing metathesis in the synthesis of (+)-astrophylline (**277**), (+)-dihydrocuscohygrine (**280**), and cuscohygrine (**281**)

Some domino reactions also involve ethylene as a reaction partner, as in the total synthesis of norhalichondrin B (**281**), an impressive cytotoxic agent [110]. Thus, in a key step, the bridged bicyclic compound **282** was converted *via* a ROM–RCM sequence into pyranopyran **283** in 71% yield (Scheme 81).

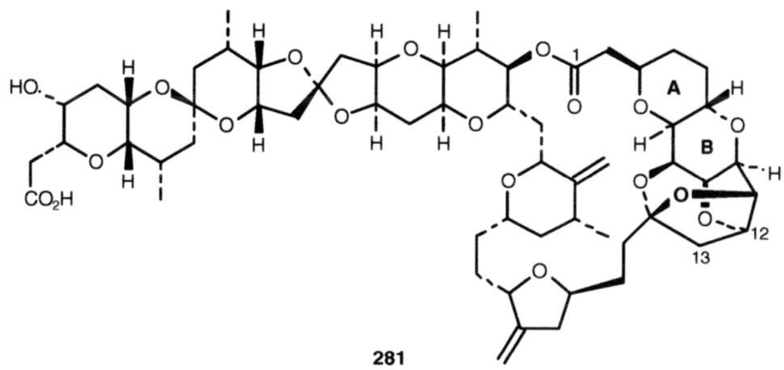

Scheme 80 Structure of norhalichondrin B

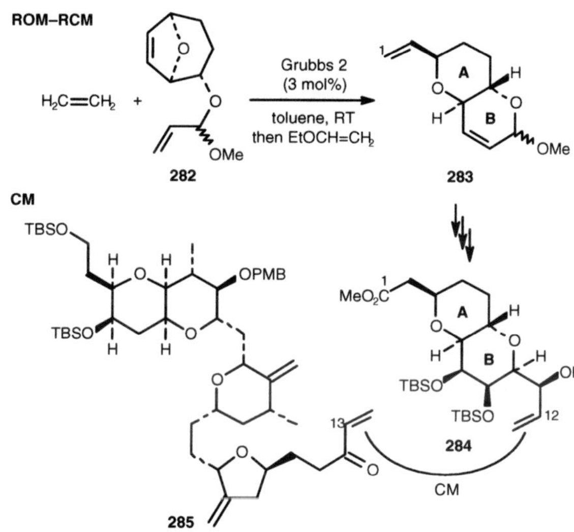

TBS = *t*-butyldimethylsilyl ; PMB = *p*-methoxybenzyl

Scheme 81 Domino ring-opening metathesis/ring-opening metathesis in the synthesis of norhalichondrin B

In a further step, allylic alcohol **284** and enone **285** were successfully cross-metathesised in the presence of 20 mol% of the recently reported catalyst **286** (Scheme 82) [111] to give the desired product in 62% yield. Again, this synthesis nicely highlights the power of cross-metathesis on highly functionalised intermediates. A similar strategy was applied for the synthesis of aburatubolactam A (**287**, Scheme 83), a compound displaying a diverse range of biological activities including cytotoxicity, antimicrobial activity, and the inhibition of superoxide generation [112]. The bicyclo[3.3.0]octene ring system **289** arose from a ring-opening/ring-closing metathesis of the functionalised bicyclo[2.2.1]heptene **288** using Grubbs 1 catalyst under an atmosphere of ethylene, whereas compound **292** was prepared through cross-metathesis between **290** and butene-1,4-diol derivative **291** catalysed by the Grubbs 2 catalyst (Scheme 84).

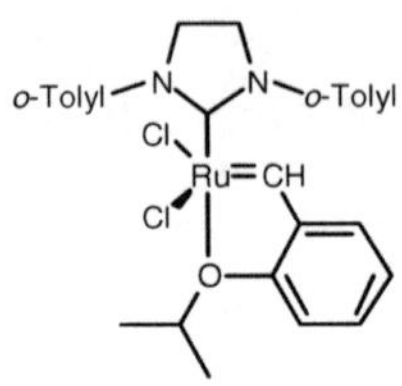

286

Scheme 82 Structure of Hoveyda–Grubbs' complex bearing *N*-tolyl groups

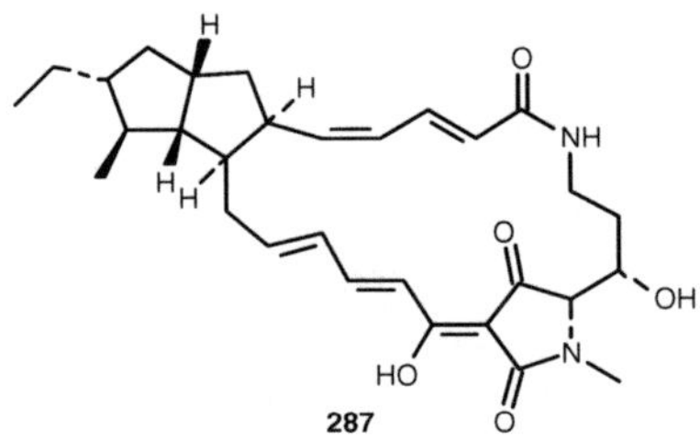

287

Scheme 83 Structure of aburatubolactam A

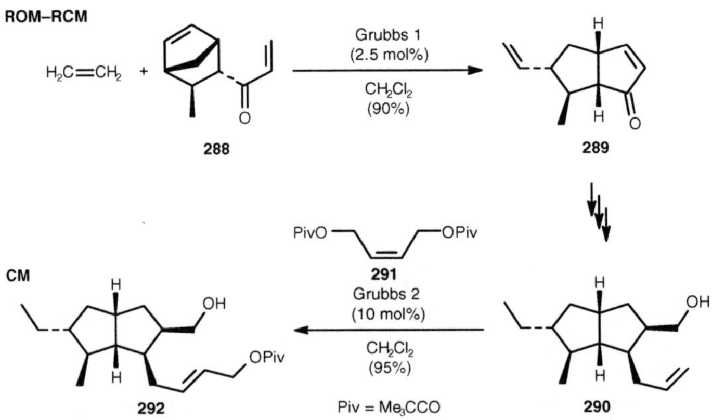

Scheme 84 Domino ring-opening metathesis/ring-closing metathesis in the synthesis of aburatubolactam A

As expected from the previous results, treatment of norbornene derivative **293** with Grubbs 1 catalyst under ethylene atmosphere led to the corresponding nine-membered carbocyclic derivative **294** *via* a domino ROM–RCM process (Scheme 85) [113]. By contrast, when the related compound **295** was treated under identical conditions, only the ring-opened product **296** was formed (Scheme 86), whereas attempted ROM–RCM with the more reactive Grubbs 2 catalyst caused extensive polymerisation of the norbornene derivative. However, when ring-opened product **296** was treated with Grubbs 2 catalyst, smooth ring closure took place in 6 h to produce tricycle **297** in 86% yield. On the basis of the above observation, a one-pot sequential protocol was used to accomplish ROM–RCM: The norbornene derivative was first treated with Grubbs 1 catalyst. Then, after disappearance of the starting material, Grubbs 2 catalyst was added to the reaction mixture allowing thereby the RCM process. In this way, a variety of bicyclo-annulated products were prepared in very good yields.

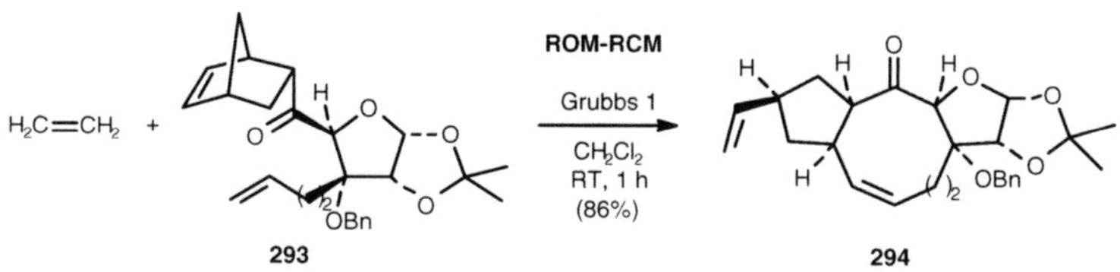

Scheme 85 Domino ring-opening metathesis/ring-closing metathesis in the synthesis of fused cyclic systems

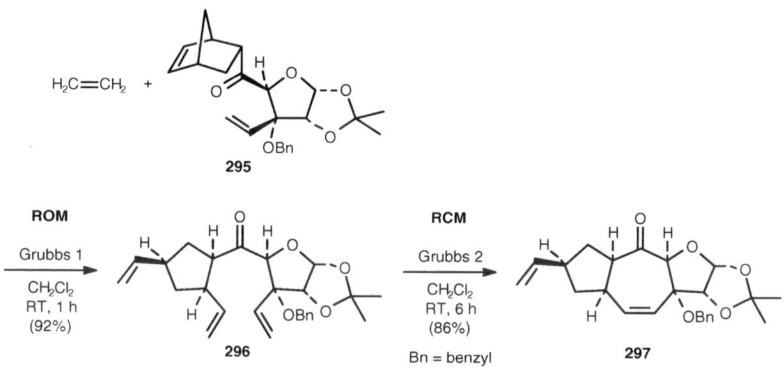

Scheme 86 One-pot ring-opening metathesis/ring-closing metathesis sequence in the synthesis of fused cyclic systems

Domino reactions also include relay ring-closing metathesis (RRCM, Section 5). The benefit of introducing a RRCM strategy was evident in the total synthesis of (+)-didemniserinolipid B (**298**, Scheme 87), in which the bicyclic ketal **302** (Scheme 88) was proposed as the key intermediate [114]. In a preliminary investigation, the direct RCM of substrate **299** was found to be problematic. Indeed, as reaction at the phenyl-substituted alkene is expected to be slow, initiation of metathesis is anticipated to occur at one of the vinyl groups, leading either to a productive species (**300**) leading to RCM, or to the unproductive species **301** leading to cross-metathesis or dimerisation (Scheme 88). In addition, the stoichiometric styrene released during RCM also underwent CM with **302**, providing **304** as its concentration increased during the progression of the reaction. To direct the course of the metathesis to the desired compound **302**, substrate **305** possessing the relay group was treated with Grubbs 1 catalyst, providing **302** in 82% yield (Scheme 89). Presumably, the catalyst initiated at the unhindered allyl group on the aryl ring to provide ruthenium alkylidene **306**. Subsequent ring-closure released 6-fluoroindene and produced ruthenium alkylidene **307** followed by RCM to produce the desired bicyclic ketal **302**, leaving thereby one vinyl group unreacted and available for later CM with **309** (Scheme 90).

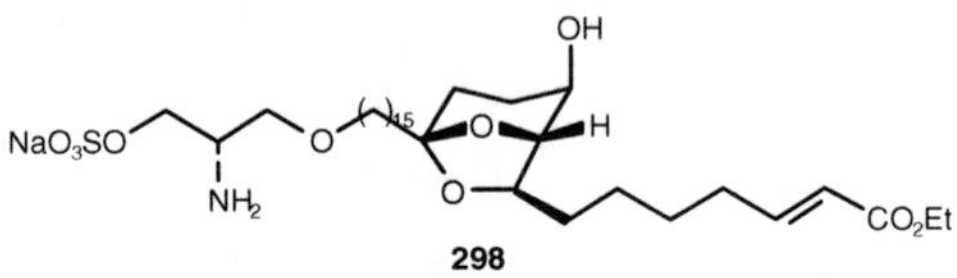

Scheme 87 Structure of (+)-didemniserinolipid B

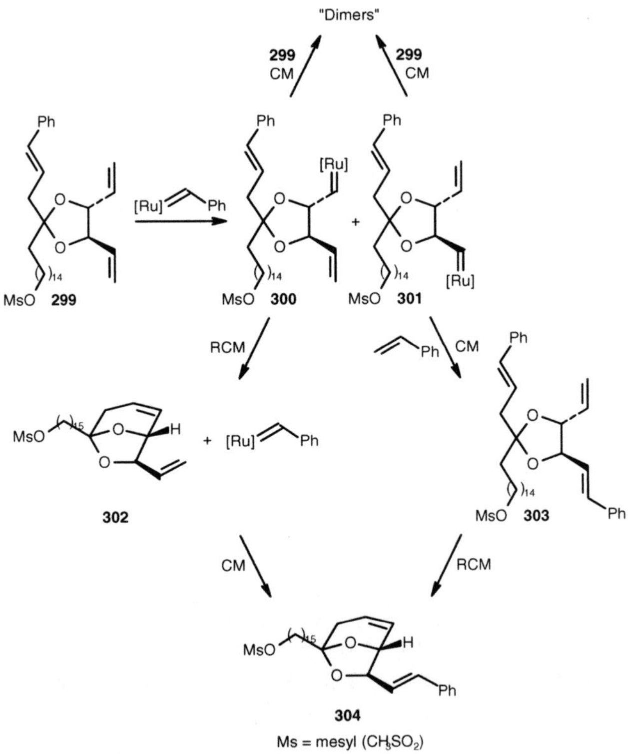

Scheme 88 Potential metathesis pathways for substrate **299**

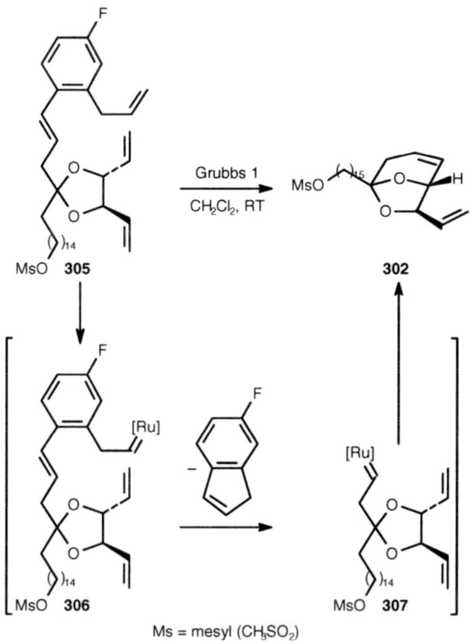

Scheme 89 Relay ring-closing metathesis in the synthesis of (+)-didemniserinolipid B

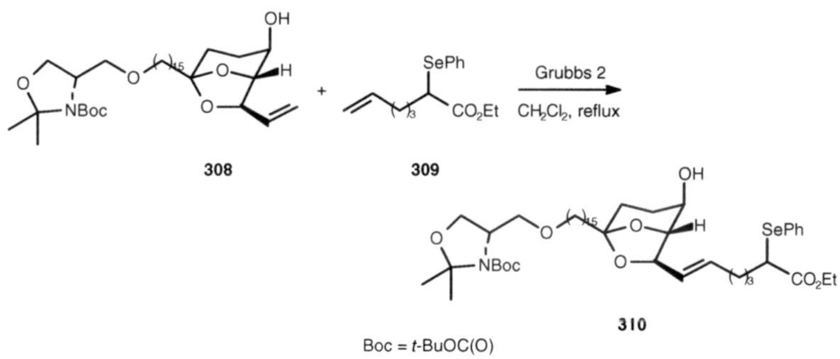

Scheme 90 Cross-metathesis in the synthesis of (+)-didemniserinolipid B

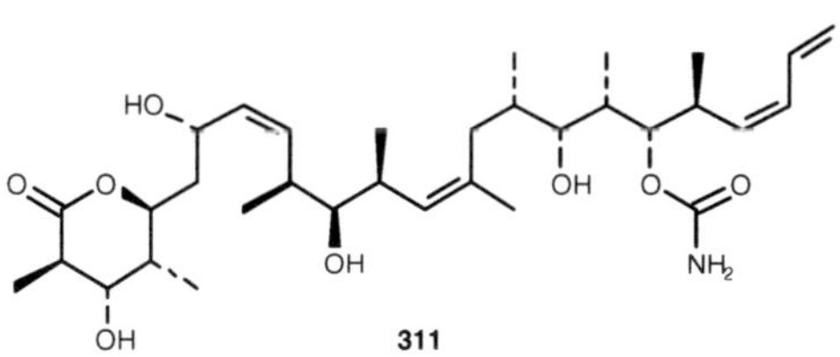

Scheme 91 Structure of (+)-discodermolide

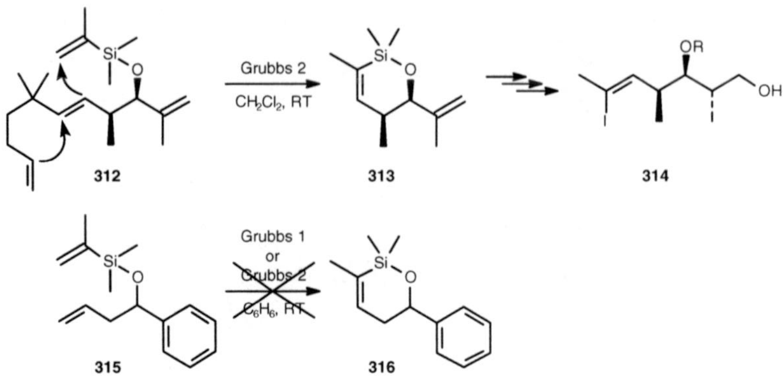

Scheme 92 Relay ring-closing metathesis in the synthesis of (+)-discodermolide

A relay RCM has also been recently applied to the synthesis of (+)-discodermolide (**311**), a natural product active against several multi-drug-resistant cancer cell lines [115]. The key intermediate **314** was prepared through RRCM of compound **312** followed by ring opening and iododesilylation (Schemes 91 and 92) [116]. Noteworthily, direct RCM of the model diene **315** failed using Grubbs' catalysts.

Although it is often accepted that one way for improving the efficiency of RCM of sterically hindered olefins is the introduction of relay ring-closing metathesis, exceptions do exist. Thus, in a preliminary investigation toward the total synthesis of jaspamide (**317**), a cyclodepsipeptide exhibiting a remarkable antifungal, anthelminthic, insecticidal and anti-cancer activity, only a moderate improvement in terms of rate and yield was witnessed using RRCM of **319** instead of RCM of **318** (Schemes 93 and 94). Consequently, the total synthesis of jaspamide was accomplished by conventional RCM [117].

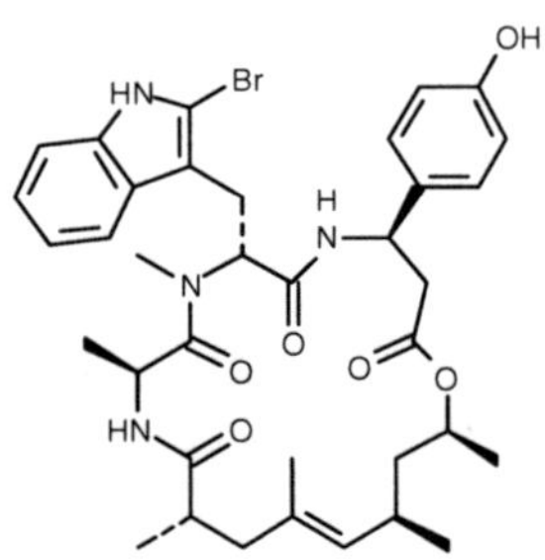

Scheme 93 Structure of jaspamide

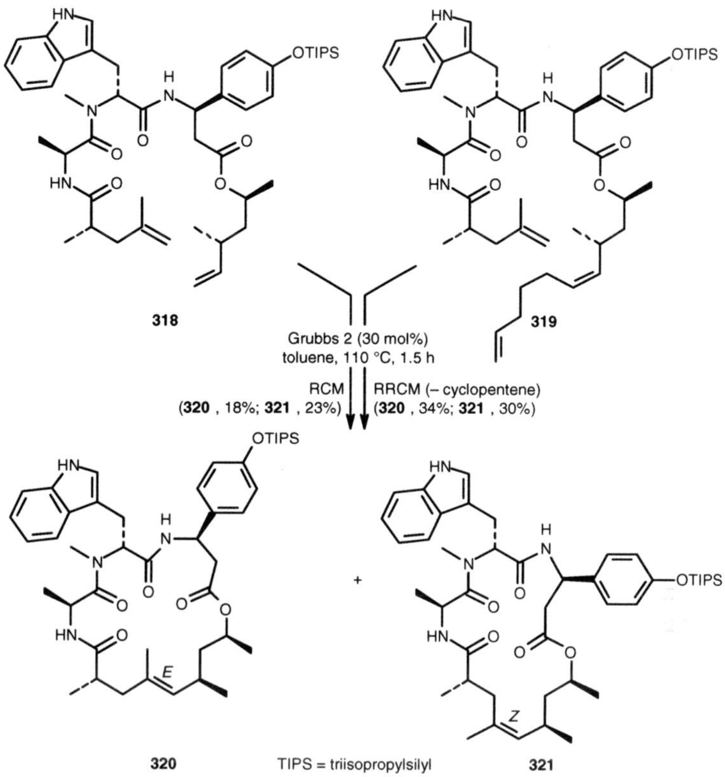

Scheme 94 Ring-closing metathesis versus relay ring-closing metathesis in the synthesis of jasplakinolide

6.3 Tandem Reactions

In tandem reactions, a given catalyst is transformed *in situ* into a new catalytically active species once the first transformation of the sequence has been completed. The junction of two catalytic cycles by using one common metal catalyst allows for a more efficient use of the expensive catalyst and the avoidance of unnecessary purification procedures. In recent years, tandem reactions combining olefin metathesis and cyclopropanation [118], double-bond isomerisation [119], allylic alcohol isomerisation [120], hydroarylation [121], and atom transfer radical reactions [122] have been developed [101, 123].

Of particular interest is the application of tandem reactions to the synthesis of muscone. Muscone (**326**) is a 15-membered macrocyclic ketone possessing a stereogenic centre at C3 that is *R* in the natural product isolated from the male musk deer. The *R* enantiomer is olfactively perceived as stronger with a very nice rich and powerful musky note, whereas the odor of the *S* enantiomer is less interesting and less strong. Taking advantage of the development in ruthenium-catalysed alkene metathesis, a very elegant synthesis of (*R*)-(–)-muscone was reported and relied on three tandem ruthenium-catalysed steps (Scheme 95) [124]. Starting from optically active (*R*)-citronellal (**322**), alcohol **323** was cyclised with the Grubbs 2 catalyst to the corresponding macrocyclic alcohol **324**. Finally, two one-pot and consecutive ruthenium-catalysed reactions, a transfer dehydrogenation followed by a C=C bond hydrogenation, gave the desired musk ingredient (**326**) in an overall yield of 56%.

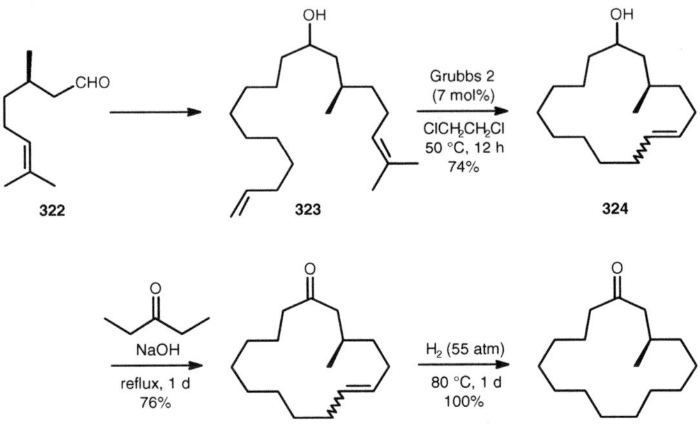

322 **323** **324** **325** **326**

Scheme 95 Tandem ring-closing metathesis/hydrogen transfer/hydrogenation reaction in the synthesis of (R)-(–)-muscone

A tandem sequence involving a cross-metathesis plus subsequent diastereoselective dihydroxylation [125] formed the key step of the total synthesis of anthopleurine (**330**), a sea anemone alarm pheromone (Scheme 96) [126]. The success of this tandem reaction rested first on the efficiency of the cross-metathesis. Thus, among the catalysts investigated, the Hoveyda–Grubbs' complex **3** and the Grela's catalyst **28** displayed the highest activity. Then, the feasibility of the sequence required the transformation of the metathesis catalyst into RuO_4, the dihydroxylation catalyst. This step was best performed using *n*-Bu_4NIO_4, and the Hoveyda–Grubbs' catalyst was found to be superior to the Grela's catalyst. Finally, the chiral auxiliary group also played a prominent role as it allowed for the diastereoselective RuO_4-catalysed dihydroxylation. Applying the cross-metathesis–diastereoselective dihydroxylation conditions to (*Z*)-1,4-dichlorobut-2-ene (**327**) and sulfamidate **328** followed by methanolysis afforded the enantiopure *vic*-diol (2*S*,3*S*)-**329** in 40% yield with excellent enantiopurity (>99% ee) (Scheme 96). Substitution of chloride by trimethylamine and saponification of the methyl ester led to the target compound, anthopleurine (**330**), in 87% yield.

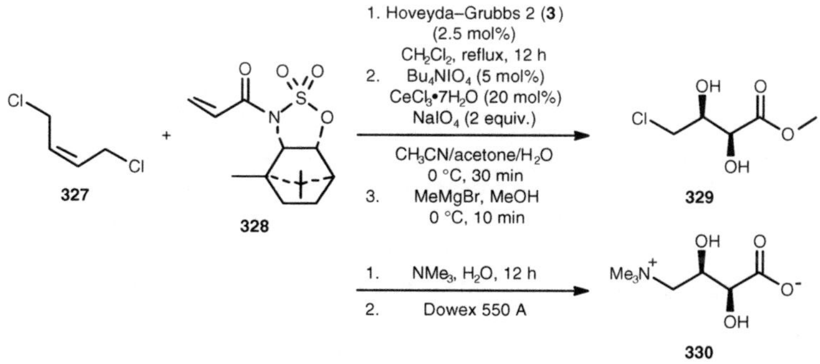

Scheme 96 Tandem cross-metathesis/diastereoselective dihydroxylation reaction in the synthesis of anthopleurine

6.3.1 Dynamic ring-closing metathesis

To the best of our knowledge, the first report on a dynamic ring-closing metathesis involving epimerisation appeared only very recently. Thus, for the total synthesis of oxazinidinyl platensimycin **331**, a simpler analogue of platensimycin **5**, Sintim and his coworkers envisioned the preparation of the bicyclic intermediate **334** through RCM of diene **332** (Schemes 97 and 98).

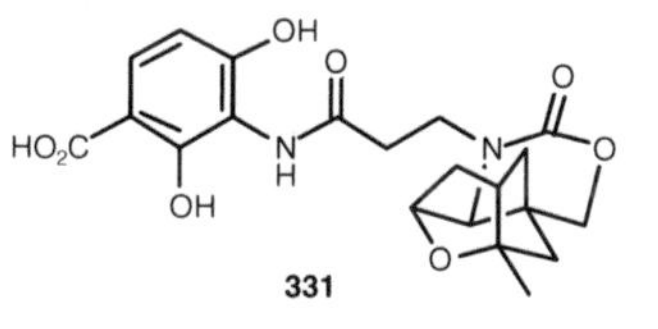

Scheme 97 Structure of oxazinidinyl platensimycin

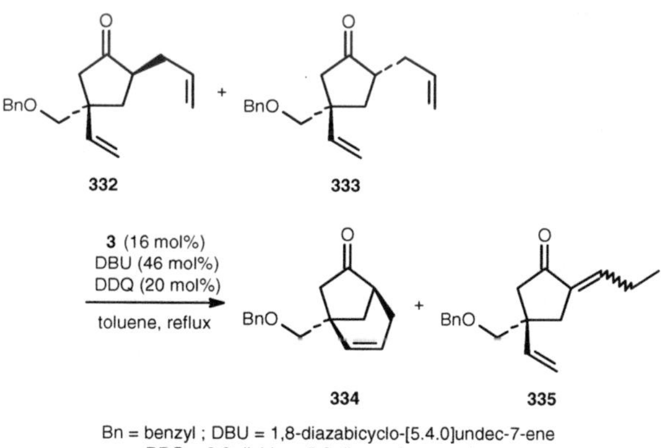

Bn = benzyl ; DBU = 1,8-diazabicyclo-[5.4.0]undec-7-ene
DDQ = 2,3-dichloro-5,6-dicyanobenzoquinone

Scheme 98 Dynamic ring-closing metathesis

Unfortunately, in the proposed synthetic route, the requisite *cis*-configured substrate **332** for the RCM reaction was formed as the minor product compared with the undesired *trans*-isomer **333** (**332**:**333** = 1:3.7). After having investigated the epimerisation of compound **333** to **332** in the presence of DBU (1,8-diazabicyclo-[5.4.0]undec-7-ene), as well as demonstrated the compatibility of the metathesis catalyst with the amine base, the following dynamic RCM protocol was applied: An epimeric mixture of compounds **332** and **333** (ratio of 1:3.7) was subjected to Hoveyda–Grubbs' catalyst **3** in the presence of DBU and 2,3-dichloro-5,6-dicyanobenzoquinone (DDQ), affording the ring-closed product **334** in 69% yield (83% based on the starting material). Noteworthily, in the absence of DBU, the desired product was obtained in a meager 20% yield. Furthermore, the benzoquinone additive was used for minimising the formation of the enone by-product **335** (10%) [127].

7 Conclusions

Ring-closing metathesis and, to a lesser extent, cross-metathesis have now become standard transformations in organic synthesis. As ruthenium catalysts tolerate most functional groups, protection–deprotection strategies, which often hinder the application of other reaction methodologies, are seldom required. In addition, ruthenium-catalysed olefin metathesis is ideal for use during the late stages of a total synthesis owing to the chemoselectivity exhibited by the catalyst and the mild reaction conditions required. Although numerous spectacular applications of olefin metathesis have been reported in recent years, the optimum catalyst and conditions tend to be highly substrate dependent and the *E*/*Z* isomeric ratio of the alkene thus obtained is often unpredictable. Ruthenium metathesis catalysts are also ideally suited for use in one-pot sequential reactions involving acetylene derivatives and ring systems, as they are stable toward a variety of reaction conditions and reagents and, often, their presence will not impede subsequent reaction transformations. This favourable combination of reactivity and functional-group compatibility has resulted in large number of citations to the use of ruthenium-based metathesis catalysts. Besides, in a modern industrial setting, the use of efficient chemical processes, preferably those with minimal impact on the environment, is essential. The term "green chemistry" has evolved as an umbrella concept to represent this general approach of conducting reactions. Metathesis attains the goals of green chemistry in three ways: (1) by providing a more efficient route over traditional methods to carbon–carbon bond formation and avoiding by-product formation; (2) through enabling the use of renewable resources; and (3) by providing a means to attain environmentally friendly products.

Acknowledgements The authors gratefully acknowledge support from the Commissariat général aux Relations internationales de la Communauté Wallonie-Bruxelles (CGRI), the Direction générale des Relations extérieures de la Région wallonne, the Fonds national de la Recherche scientifique (FNRS), the Romanian Ministry of Education and Research, and the Bulgarian Academy of Sciences. The authors also thank the Fonds pour la Formation à la Recherche dans l'Industrie et dans l'Agriculture (F.R.I.A.) for a fellowship to D. Bicchielli and F. Nicks.

References

[1] (a) Taylor RE, Zhao Z, Wünsch S (2008) Synthetic efforts towards the marine polyketide peloruside A. C R Chimie 11: 1369–1381; (b) Gerber-Lemaire S, Vogel P (2008) Spongistatins: Biological activity and synthetic studies. C R Chimie 11: 1382–1418; (c) Nicolaou KC, Chen JS, Edmonds DJ, Estrada AA (2009) Recent advances in the chemistry and biology of naturally occurring antibiotics. Angew Chem Int Ed 48: 660–719; (d) Nicolaou KC (2009) Inspirations, discoveries, and future perspectives in total synthesis. J Org Chem 74: 951–972

[2] For accounts of the development of the metathesis reaction, which include details of early applications, see: (a) Chauvin Y (2006) Olefin metathesis: The early days (Nobel lecture). Angew Chem Int Ed 45: 3741–3747; (b) Schrock RR (2006) Multiple metal–carbon bonds for catalytic metathesis reactions (Nobel lecture). Angew Chem Int Ed 45: 3748–3759; (c) Grubbs RH (2006) Olefin-metathesis catalysts for the preparation of molecules and materials (Nobel lecture). Angew Chem Int Ed 45: 3760–3765

[3] For a comprehensive review of metathesis reactions as well as applications, see: Grubbs RH (ed.) (2003) Handbook of metathesis, vols. 1, 2 & 3. Wiley-VCH, Weinheim

[4] For a comprehensive review of metathesis reactions in total synthesis, see: Nicolaou KC, Bulger PG, Sarlah D (2005) Metathesis reactions in total synthesis. Angew Chem Int Ed 44: 4490–4527

[5] For recent reviews on the utility of the ring-closing metathesis for macrocycle construction, see: (a) Deiters A, Martin SF (2004) Synthesis of oxygen- and nitrogen-containing heterocycles by ring-closing metathesis. Chem Rev 104: 2199–2238; (b) Gradillas A, Pérez-Castells J (2006) Macrocyclization by ring-closing metathesis in the total synthesis of natural products: Reaction conditions and limitations. Angew Chem Int Ed 45: 6086–6101

[6] Connon SJ, Blechert S (2003) Recent developments in olefin cross-metathesis. Angew Chem Int Ed 42: 1900–1923

[7] Naffziger MR, Ashburn BO, Perkins JR, Carter RG (2007) Diels–Alder approach for the construction of halogenated, *o*-nitro biaryl templates and application to the total synthesis of the anti-HIV agent siamenol. J Org Chem 72: 9857–9865

[8] Bisai A, West SP, Sarpong R (2008) Unified strategy for the synthesis of the "miscellaneous" *Licopodium* alkaloids: Total synthesis of (±)-lyconadin A. J Am Chem Soc 130: 7222–7223

[9] Gupta P, Kumar P (2008) An efficient total synthesis of decarestrictine D. Eur J Org Chem 7: 1195–1202

[10] Ghosh AK, Liu C (2003) Enantioselective total synthesis of (+)-amphidinolide T1. J Am Chem Soc 125: 2374–2375

[11] Pandya BA, Snapper ML (2008) A cross-metathesis route to the 5-F_2-isoprostanes. J Org Chem 73: 3754–3758

[12] Rahn N, Kalesse M (2008) The total synthesis of chlorotonil A. Angew Chem Int Ed 47: 597–599

[13] Crimmins MT, Christie HS, Long A, Chaudhary K (2009) Total synthesis of apoptolidin A. Org Lett 11: 831–834

[14] (a) Nicolaou KC, Li A, Edmonds DJ (2006) Total synthesis of platensimycin. Angew Chem Int Ed 45: 7086–7090; (b) Nicolaou KC, Edmonds DJ, Li A, Tria GS (2007) Asymmetric total syntheses of platensimycin. Angew Chem Int Ed 46: 3942–3945; (c) Nicolaou KC, Pappo D, Tsang KY, Gibe R, Chen DY-K (2008) A chiral pool based synthesis of platensimycin. Angew Chem Int Ed 47: 944–946

[15] Nicolaou KC, Tang Y, Wang J, Stepan AF, Li A, Montero A (2007) Total synthesis and antibacterial properties of carbaplatensimycin. J Am Chem Soc 129: 14850–14851

[16] Nicolaou KC, Lister T, Denton RM, Montero A, Edmonds DJ (2007) Adamantaplatensimycin: A bioactive analogue of platensimycin. Angew Chem Int Ed 46: 4712–4714

[17] Nicolaou KC, Tria GS, Edmonds DJ (2008) Total synthesis of platencin. Angew Chem Int Ed 47: 1780–1783

[18] For a recent review, see: Tiefenbacher K, Mulzer J (2008) Synthesis of platensimycin. Angew Chem Int Ed 47: 2548–2555

[19] Nicolaou KC, Stepan AF, Lister T, Li A, Montero A, Tria GS, Turner CI, Tang Y, Wang J, Denton RM, Edmonds DJ (2008) Design, synthesis, and biological evaluation of platensimycin analogues with varying degrees of molecular complexity. J Am Chem Soc 130: 13110–13119

[20] (a) Ferré-Filmon K, Delaude L, Demonceau A, Noels AF (2005) Stereoselective synthesis of (*E*)-hydroxystilbenoids by ruthenium-catalyzed cross-metathesis. Eur J Org Chem: 3319–3325; (b) Velder J, Ritter S, Lex J, Schmalz H-G (2006) A simple access to biologically important *trans*-stilbenes via Ru-catalyzed cross metathesis. Synthesis 2: 273–278

[21] (a) Tsantrizos YS, Ferland J-M, McClory A, Poirier M, Farina V, Yee NK, Wang X-J, Haddad N, Wei X, Xu J, Zhang L (2006) Olefin ring-closing metathesis as a powerful tool in drug discovery and development – potent macrocyclic inhibitors of the hepatitis C virus NS3 protease. J Organomet Chem 691: 5163–5171; (b) Yee NK, Farina V, Houpis IN, Haddad N, Frutos RP, Gallou F, Wang X-J, Wei X, Simpson RD, Feng X, Fuchs V, Xu Y, Tan J, Zhang L, Xu J, Smith-Keenan LL, Vitous J, Ridges MD, Spinelli EM, Johnson M, Donsbach K, Nicola T, Brenner M, Winter E, Kreye P, Samstag W (2006) Efficient large-scale synthesis of BILN 2061, a potent HCV protease inhibitor, by a convergent approach based on ring-closing metathesis. J Org Chem 71: 7133–7145

[22] Bieniek M, Bujok R, Stepowska H, Jacobi A, Hagenkötter R, Arlt D, Jarzembska K, Makal A, Wozniak K, Grela K (2006) New air-stable ruthenium olefin metathesis precatalysts derived from bisphenol S. J Organomet Chem 691: 5289–5297

[23] Nicola T, Brenner M, Donsbach K, Kreye P (2005) First scale-up to production scale of a ring closing metathesis reaction forming a 15-membered macrocycle as a precursor of an active pharmaceutical ingredient. Org Process Res Dev 9: 513–515

[24] Bajwa N, Jennings MP (2008) Syntheses of *epi*-aigialomycin D and *deoxy*-aigialomycin C via a diastereoselective ring closing metathesis macrocyclization protocol. Tetrahedron Lett 49: 390–393

[25] Sakaguchi H, Tokuyama H, Fukuyama T (2007) Stereocontrolled total synthesis of (–)-kainic acid. Org Lett 9: 1635–1638

[26] (a) Smith III AB, Basu K, Bosanac T (2007) Total synthesis of (–)-okilactomycin. J Am Chem Soc 129: 14872–14874; (b) Smith III AB, Bosanac T, Basu K (2009) Evolution of the total synthesis of (–)-okilactomycin exploiting a tandem oxy-Cope rearrangement/oxidation, a Petasis–Ferrier union/rearrangement, and ring-closing metathesis. J Am Chem Soc 131: 2348–2358

[27] Kwon MS, Woo SK, Na SW, Lee E (2008) Total synthesis of (+)-exiguolide. Angew Chem Int Ed 47: 1733–1735

[28] Barfoot CW, Burns AR, Edwards MG, Kenworthy MN, Ahmed M, Shanahan SE, Taylor RJK (2008) A convergent synthesis of the tricyclic core of the dictyosphaeric acids. Org Lett 10: 353–356

[29] Liang B, Negishi E (2008) Highly efficient asymmetric synthesis of fluvirucinine A_1 via Zr-catalyzed asymmetric carboalumination of alkenes (ZACA)–lipase-catalyzed acetylation tandem process. Org Lett 10: 193–195

[30] (a) Cong X, Yao Z-J (2006) Ring-closing metathesis-based synthesis of (3*R*,4*R*,5*S*)-4-acetylamino-5-amino-3-hydroxycyclohex-1-ene-carboxylic acid ethyl ester: A functionalized cycloalkene skeleton of GS4104. J Org Chem 71: 5365–5368; (b) Farina V, Brown JD (2006) Tamiflu: The supply problem. Angew Chem Int Ed 45: 7330–7334; (c) Shibasaki M, Kanai M (2008) Synthetic strategies for oseltamivir phosphate. Eur J Org Chem: 1839–1850

[31] (a) Zapf CW, Harrison BA, Drahl C, Sorensen EJ (2005) A Diels–Alder macrocyclization enables an efficient asymmetric synthesis of the antibacterial natural product abyssomicin C. Angew Chem Int Ed 44: 6533–6537; (b) Peters R, Fischer DF (2006) Total syntheses of the antibacterial natural product abyssomicin C. Angew Chem Int Ed 2006, 45: 5736–5739
[32] Nicolaou KC, Harrison ST (2006) Total synthesis of abyssomicin C and atrop-abyssomicin C. Angew Chem Int Ed 45: 3256–3260
[33] Höfle G, Bedorf N, Steinmetz H, Schomburg D, Gerth K, Reichenbach H (1996) Epothilone A and B – Novel 16-membered macrolides with cytotoxic activity: Isolation, crystal structure, and conformation in solution. Angew Chem Int Ed Engl 35: 1567–1569
[34] Balog A, Meng D, Kamenecka T, Bertinato P, Su D-S, Sorensen EJ, Danishefsky SJ (1996) Total synthesis of (–)-epothilone A. Angew Chem Int Ed Engl 35: 2801–2803
[35] (a) Yang Z, He Y, Vourloumis D, Vallberg H, Nicolaou KC (1997) Total synthesis of epothilone A: The olefin metathesis approach. Angew Chem Int Ed Engl 36: 166–168; (b) Nicolaou KC, Sarabia F, Ninkovic S, Yang Z (1997) Total synthesis of epothilone A: The macrolactonization approach. Angew Chem Int Ed Engl 36: 525–527
[36] Schinzer D, Limberg A, Bauer A, Böhm OM, Cordes M (1997) Total synthesis of (–)-epothilone A. Angew Chem Int Ed Engl 36: 523–524
[37] Mulzer J, Altmann K-H, Höfle G, Müller R, Prantz K (2008) Epothilones – A fascinating family of microtubule stabilizing antitumor agents. C R Chimie 11: 1336–1368
[38] Meng D, Bertinato P, Balog A, Su D-S, Kamenecka T, Sorensen EJ, Danishefsky SJ (1997) Total syntheses of epothilones A and B. J Am Chem Soc 119: 10073–10092
[39] Feyen F, Gertsch J, Wartmann M, Altmann K-H (2006) Design and synthesis of 12-aza-epothilones (azathilones) – "Non-natural" natural products with potent anticancer activity. Angew Chem Int Ed 45: 5880–5885
[40] Alhamadsheh MM, Gupta S, Hudson RA, Perera L, Tillekeratne LMV (2008) Total synthesis and selective activity of a new class of conformationally restrained epothilones. Chem Eur J 14: 570–581
[41] Chen Q-H, Ganesh T, Brodie P, Slebodnick C, Jiang Y, Banerjee A, Bane S, Snyder JP, Kingston DGI (2008) Design, synthesis and biological evaluation of bridged epothilone D analogues. Org Biomol Chem 6: 4542–4552
[42] http://www.cancer.gov/cancertopics/druginfo/fda-ixabepilone
[43] Klar U, Buchmann B, Schwede W, Skuballa W, Hoffmann J, Lichtner RB (2006) Total synthesis and antitumor activity of ZK-EPO: The first fully synthetic epothilone in clinical development. Angew Chem Int Ed 45: 7942–7948
[44] Meng D, Su D-S, Balog A, Bertinato P, Sorensen EJ, Danishefsky SJ, Zheng Y-H, Chou T-C, He L, Horwitz SB (1997) Remote effects in macrolide formation through ring-forming olefin metathesis: An application to the synthesis of fully active epothilone congeners. J Am Chem Soc 119: 2733–2734
[45] For examples and discussion of similar protecting-group effects on RCM reactions in the synthesis of salicylihalamides, see: (a) Fürstner A, Thiel OR, Blanda G (2000) Asymmetric synthesis of the fully functional macrolide core of salicylihalamide: Remote control of olefin geometry during RCM. Org Lett 2: 3731–3734; (b) Fürstner A, Dierkes T, Thiel OR, Blanda G (2001) Total synthesis of (–)-salicylihalamide. Chem Eur J 7: 5286–5298
[46] (a) Biswas K, Lin H, Njardarson JT, Chappell MD, Chou T-C, Guan Y, Tong WP, He L, Horwitz SB, Danishefsky SJ (2002) Highly concise routes to epothilones: The total synthesis and evaluation of epothilone 490. J Am Chem Soc 124: 9825–9832; (b) Rivkin A, Njardarson JT, Biswas K, Chou T-C, Danishefsky SJ (2002) Total syntheses of [17]- and [18]dehydrodesoxyepothilones B via a concise ring-closing metathesis-based strategy: Correlation of ring size with biological activity in the epothilone series. J Org Chem 67: 7737–7740
[47] Rivkin A, Biswas K, Chou T-C, Danishefsky SJ (2002) On the introduction of a trifluoromethyl substituent in the epothilone setting: Chemical issues related to ring forming olefin metathesis and earliest biological findings. Org Lett 4: 4081–4084

[48] Keck GE, Giles RL, Cee VJ, Wager CA, Yu T, Kraft MB (2008) Total synthesis of epothilones B and D: Stannane equivalents for β-keto ester dianions. J Org Chem 73: 9675–9691

[49] (a) Rivkin A, Yoshimura F, Gabarda AE, Chou T-C, Dong H, Tong WP, Danishefsky SJ (2003) Complex target-oriented total synthesis in the drug discovery process: The discovery of a highly promising family of second generation epothilones. J Am Chem Soc 125: 2899–2901; (b) Rivkin A, Yoshimura F, Gabarda AE, Cho YS, Chou T-C, Dong H, Danishefsky SJ (2004) Discovery of (*E*)-9,10-dehydroepothilones through chemical synthesis: On the emergence of 26-trifluoro-(*E*)-9,10-dehydro-12,13-desoxyepothilone B as a promising anticancer drug candidate. J Am Chem Soc 126: 10913–10922

[50] Sun J, Sinha SC (2002) Stereoselective total synthesis of epothilones by the metathesis approach involving C9–C10 bond formation. Angew Chem Int Ed 41: 1381–1383

[51] For approaches involving a macrolactonisation, see for instance: (a) Ghosh AK, Liu C (2003) Enantioselective total synthesis of (+)-amphidinolide T1. J Am Chem Soc 125: 2374–2375; (b) Lepage O, Kattnig E, Fürstner A (2004) Total synthesis of amphidinolide X. J Am Chem Soc 126: 15970–15971; (c) Fürstner A, Kattnig E, Lepage O (2006) Total syntheses of amphidinolide X and Y. J Am Chem Soc 128: 9194–9204; (d) Nicolaou KC, Bulger PG, Brenzovich WE (2006) Synthesis of *iso*-epoxy-amphidinolide N and *des*-epoxy-caribenolide I structures. Revised strategy and final stages. Org Biomol Chem 4: 2158–2183; (e) Kim CH, An HJ, Shin WK, Yu W, Woo SK, Jung SK, Lee E (2006) Total synthesis of (–)-amphidinolide E. Angew Chem Int Ed 45: 8019–8021; (f) Deng L-S, Huang X-P, Zhao G (2006) Stereocontrolled and convergent total synthesis of amphidinolide T3. J Org Chem 71: 4625–4635. For approaches involving a Wittig or Wittig-type olefination, see for instance: (g) Lu L, Zhang W, Carter RG (2008) Total synthesis of cytotoxic macrolide amphidinolide B_1 and the proposed structure of amphidinolide B_2. J Am Chem Soc 130: 7253–7255

[52] Maleczka Jr RE, Terrell LR, Geng F, Ward III JS (2002) Total synthesis of proposed amphidinolide A via a highly selective ring-closing metathesis. Org Lett 4: 2841–2844

[53] Fürstner A, Bouchez LC, Funel J-A, Liepins V, Porée F-H, Gilmour R, Beaufils F, Laurich D, Tamiya M (2007) Total syntheses of amphidinolides H and G. Angew Chem Int Ed 46: 9265–9270

[54] (a) Fürstner A, Aïssa C, Riveiros R, Ragot J (2002) Total synthesis of amphidinolide T4. Angew Chem Int Ed 41: 4763–4766; (b) Aïssa C, Riveiros R, Ragot J, Fürstner A (2003) Total syntheses of amphidinolide T1, T3, T4, and T5. J Am Chem Soc 125: 15512–15520; (c) Colby EA, Jamison TF (2005) A comparative analysis of the total syntheses of the amphidinolide T natural products. Org Biomol Chem 3: 2675–2684

[55] Tietze LF, Brazel CC, Hölsken S, Magull J, Ringe A (2008) Total synthesis of polyoxygenated cembrenes. Angew Chem Int Ed 47: 5246–529

[56] See also: Deng L, Ma Z, Zhang Y, Zhao G (2007) Synthetic studies toward amphidinolide H1: Segment C14–C26. Synlett: 87–90

[57] (a) Chatterjee AK, Morgan JP, Scholl M, Grubbs RH (2000) Synthesis of functionalized olefins by cross and ring-closing metatheses. J Am Chem Soc 122: 3783–3784; (b) Xiong Z, Corey EJ (2000) Simple total synthesis of the pentacyclic C_s-symmetric structure attributed to the squalenoid glabrescol and three C_s-symmetric diastereomers compel structural revision. J Am Chem Soc 122: 4831–4832; (c) McDonald FE, Wei X (2002) Concise, regioselective synthesis of the ABC tristetrahedropyran of thyrsiferol and venustatriol. Org Lett 4: 593–595

[58] Garbaccio RM, Stachel SJ, Baeschlin DK, Danishefsky SJ (2001) Concise asymmetric syntheses of radicicol and monocillin I. J Am Chem Soc 123: 10903–10908

[59] (a) Barluenga S, Dakas P-Y, Boulifa M, Moulin E, Winssinger N (2008) Resorcylic acid lactones: A pluripotent scaffold with therapeutic potential. C R Chimie 11: 1306–1317; (b) Hofmann T, Altmann K-H (2008) Resorcylic acid lactones as new lead structures for kinase inhibition. C R Chimie 11: 1318–1335

[60] Ramana CV, Khaladkar TP, Chatterjee S, Gurjar MK (2008) Total synthesis and determination of relative and absolute configuration of multiplolide A. J Org Chem 73: 3817–3822

[61] See for instance: Hoye TR, Zhao H (1999) Synthesis of a C(1)–C(14)-containing fragment of callipeltoside A. Org Lett 1: 169–171

[62] Hoye TR, Zhao H (1999) Some allylic substituent effects in ring-closing metathesis reactions: Allylic alcohol activation. Org Lett 1: 1123–1125

[63] Fürstner A, Aïssa C, Chevrier C, Teply F, Nevado C, Tremblay M (2006) Studies on iejimalide B: Preparation of the seco acid and identification of the molecule's "Achilles heel". Angew Chem Int Ed 45: 5832–5837

[64] (a) Fürstner A, Nevado C, Tremblay M, Chevrier C, Teply F, Aïssa C, Waser M (2006) Total synthesis of iejimalide B. Angew Chem Int Ed 45: 5837–5842; (b) Fürstner A, Nevado C, Waser M, Tremblay M, Chevrier C, Teply F, Aïssa C, Moulin E, Müller O (2007) Total synthesis of iejimalide A–D and assessment of the remarkable actin-depolymerizing capacity of these polyene macrolides. J Am Chem Soc 129: 9150–9161

[65] Rodríguez-Escrich C, Urpí F, Vilarrasa J (2008) Stereocontrolled total synthesis of amphidinolide X via a silicon-tethered metathesis reaction. Org Lett 10: 5191–5194

[66] Dai W-M, Chen Y, Jin J, Wu J, Lou J, He Q (2008) Total synthesis of amphidinolide X and its 12*Z*-isomer by formation of the C12–C13 trisubstituted double bond via ring-closing metathesis. Synlett: 1737–1741

[67] (a) Chang S, Grubbs RH (1997) A simple method to polyhydroxylated olefinic molecules using ring-closing olefin metathesis. Tetrahedron Lett 38: 4757–4760; (b) Barrett AGM, Beall JC, Braddock DC, Flack K, Gibson VC, Salter MM (2000) Asymmetric allylboration and ring closing alkene metathesis: A novel strategy for the synthesis of glycosphingolipids. J Org Chem 65: 6508–6514

[68] Va P, Roush WR (2006) Total synthesis of amphidinolide E. J Am Chem Soc 128: 15960–15961

[69] Matsuya Y, Takayanagi S-i, Nemoto H (2008) Kinetically controlled ring-closing metathesis: Synthesis of a potential scaffold for 12-membered salicylic macrolides. Chem Eur J 14: 5275–5281

[70] Nicolaou KC, Brenzovich WE, Bulger PG, Francis TM (2006) Synthesis of *iso*-epoxy-amphidinolide N and *des*-epoxy-caribenolide I structures. Initial forays. Org Biomol Chem 4: 2119–2157

[71] Wang Y-G, Takeyama R, Kobayashi Y (2006) Total synthesis of phoslactomycin B and its biosynthetic deamino precursor. Angew Chem Int Ed 45: 3320–3323

[72] Reymond S, Cossy J (2008) Migrastatin and analogues: New anti-metastatic agents. C R Chimie 11: 1447–1462

[73] Anquetin G, Rawe SL, McMahon K, Murphy EP, Murphy PV (2008) Synthesis of novel migrastatin and dorrigocin A analogues from D-glucal. Chem Eur J 14: 1592–1600

[74] Anquetin G, Horgan G, Rawe S, Murray D, Madden A, MacMathuna P, Doran P, Murphy PV (2008) Synthesis of novel macrolactam and macroketone analogues of migrastatin from D-glucal and comparison with macrolactone and acyclic analogues: A dorrigocin A congener is a potent inhibitor of gastric cancer cell migration. Eur J Org Chem: 1953–1958

[75] Lam HW, Pattenden G (2002) Total synthesis of the presumed amphidinolide A. Angew Chem Int Ed 41: 508–511

[76] Colby EA, O'Brien KC, Jamison TF (2005) Total syntheses of amphidinolides T1 and T4 via catalytic, stereoselective, reductive macrocyclizations. J Am Chem Soc 127: 4297–4307

[77] (a) Trost BM, Chisholm JD, Wrobleski ST, Jung M (2002) Ruthenium-catalyzed alkene–alkyne coupling: Synthesis of the proposed structure of amphidinolide A. J Am Chem Soc 124: 12420–12421; (b) Trost BM, Wrobleski ST, Chisholm JD, Harrington PE, Jung M (2005) Total synthesis of (+)-amphidinolide A. Assembly of the fragments. J Am Chem Soc 127: 13589–13597; (c) Trost BM, Harrington PE, Chisholm JD, Wrobleski ST (2005) Total synthesis of (+)-amphidinolide A. Structure elucidation and completion of the synthesis. J Am Chem Soc 127: 13598–13610

[78] Trost BM, Papillon JPN, Nussbaumer T (2005) Ru-catalyzed alkene–alkyne coupling. Total synthesis of amphidinolide P. J Am Chem Soc 127: 17921–17937

[79] Fürstner A, Larionov O, Flügge S (2007) What is amphidinolide V? Report on a likely conquest. Angew Chem Int Ed 46: 5545–5548

[80] Marjanovic J, Kozmin SA (2007) Spirofungin A: Stereoselective synthesis and inhibition of isoleucyl-tRNA synthetase. Angew Chem Int Ed 46: 8854–8857

[81] (a) La Cruz TE, Rychnovsky SD (2005) Synthesis of the spirofungin B core by a reductive cyclization strategy. Org Lett 7: 1873–1875; (b) Shimizu T, Satoh T, Murakoshi K, Sodeoka M (2005) Asymmetric total synthesis of (–)-spirofungin A and (+)-spirofungin B. Org Lett 7: 5573–5576

[82] Hoye TR, Promo MA (1999) Silicon tethered ring-closing metathesis reactions for self- and cross-coupling of alkenols. Tetrahedron Lett 40: 1429–1432

[83] (a) Clark JS, Kettle JG (1997) Synthesis of brevetoxin sub-units by sequential ring-closing metathesis and hydroboration. Tetrahedron Lett 38: 123–126; (b) Joe D, Overman LE (1997) An unexpected product arising from metal alkylidene mediated ring-closing diene metathesis. Tetrahedron Lett 38: 8635–8638

[84] Evans PA, Murthy VS (1998) Temporary silicon-tethered ring-closing metathesis approach to C_2-symmetrical 1,4-diols: Asymmetric synthesis of D-altritol. J Org Chem 63: 6768–6769

[85] Evans PA, Cui J, Gharpure SJ, Polosukhin A, Zhang H-R (2003) Enantioselective total synthesis of the potent antitumor agent (–)-mucocin using a temporary silicon-tethered ring-closing metathesis cross-coupling reaction. J Am Chem Soc 125: 14702–14703

[86] (a) Brown LJ, Spurr IB, Kemp SC, Camp NP, Gibson KR, Brown RCD (2008) Total synthesis of *cis*-sylvaticin. Org Lett 10: 2489–2492; (b) Bhunnoo RA, Hobbs H, Lainé DI, Light ME, Brown RCD (2009) Synthesis of the non-adjacent bis-THF core of *cis*-sylvaticin using a double oxidative cyclisation. Org Biomol Chem 7: 1017–1024

[87] Van de Weghe P, Aoun D, Boiteau J-G, Eustache J (2002) Silicon tether-aided coupling metathesis: Application to the synthesis of attenol A. Org Lett 4: 4105–4108

[88] For an alternative synthesis of (–)-attenol A and (+)-attenol B exploiting a cross-metathesis and a ring-closing metathesis, see: Fuwa H, Sasaki M (2008) An efficient strategy for the synthesis of endocyclic enol ethers and its application to the synthesis of spiroacetals. Org Lett 10: 2549–2552

[89] (a) Chang S, Grubbs RH (1997) A simple method to polyhydroxylated olefinic molecules using ring-closing olefin metathesis. Tetrahedron Lett 38: 4757–4760; (b) Meyer C, Cossy J (1997) Synthesis of oxygenated heterocycles from cyclic allylsiloxanes using ring-closing olefin metathesis. Tetrahedron Lett 38: 7861–7864; (c) Cassidy JH, Marsden SP, Stemp G (1997) Stereoselective synthesis of 2,3,5-trisubstituted tetrahydrofurans by an allyl silane metathesis – Nucleophilic addition sequence. Synlett: 1411–1413

[90] Forbes MDE, Patton JT, Myers TL, Maynard HD, Smith Jr DW, Schultz GR, Wagener KB (1992) Solvent-free cyclization of linear dienes using olefin metathesis and the Thorpe–Ingold effect. J Am Chem Soc 114: 10978–10980

[91] Hanson PR, Stoianova DS (1998) Ring closing metathesis on a phosphonate template. Tetrahedron Lett 39: 3939–3942

[92] Menche D, Hassfeld J, Li J, Rudolph S (2007) Total synthesis of archazolid A. J Am Chem Soc 129: 6100–6101

[93] Hoye TR, Jeffrey CS, Tennakoon MA, Wang J, Zhao H (2004) Relay ring-closing metathesis (RRCM): A strategy for directing metal movement throughout olefin metathesis sequences. J Am Chem Soc 126: 10210–10211

[94] Wallace DJ (2005) Relay ring-closing metathesis – A strategy for achieving reactivity and selectivity in metathesis chemistry. Angew Chem Int Ed 44: 1912–1915

[95] Roethle PA, Chen IT, Trauner D (2007) Total synthesis of (–)-archazolid B. J Am Chem Soc 129: 8960–8961

[96] For another application of RRCM in synthesis, see: Wang X, Bowman EJ, Bowman BJ, Porco Jr JA (2004) Total synthesis of the salicylate enamide macrolide oximidine III: Application of relay ring-closing metathesis. Angew Chem Int Ed 43: 3601–3605

[97] Trost BM, Toste FD (1999) A new Ru catalyst for alkene–alkyne coupling. Tetrahedron Lett 40: 7739–7743

[98] Kita Y, Maeda H, Omori K, Okuno T, Tamura Y (1993) A novel efficient synthesis of 1-ethoxyvinyl esters and their use in acylation of amines and alcohols: Synthesis of water-soluble oxaunomycin derivatives. Synlett: 273–274

[99] Druais V, Hall MJ, Corsi C, Wendeborn SV, Meyer C, Cossy J (2009) A convergent approach toward the C1–C11 subunit of phoslactomycins and formal synthesis of phoslactomycin B. Org Lett 11: 935–938

[100] Helmboldt H, Hiersemann M (2009) Synthetic studies toward jatrophane diterpenes from *Euphorbia characias*. Enantioselective synthesis of (–)-15-*O*-acetyl-3-*O*-propionyl-17-norcharaciol. J Org Chem 74: 1698–1708

[101] Fogg DE, dos Santos EN (2004) Tandem catalysis: A taxonomy and illustrative review. Coord Chem Rev 248: 2365–2379

[102] Kalbarczyk KP, Diver ST (2009) Enyne metathesis/Brønsted acid-promoted heterocyclization. J Org Chem 74: 2193–2196

[103] Chang J-K, Chang N-C (2008) Total synthesis of (±)-gusanlung D. Tetrahedron 64: 3483–3487

[104] Fuwa H, Naito S, Goto T, Sasaki M (2008) Total synthesis of (+)-neopeltolide. Angew Chem Int Ed 47: 4737–4739

[105] Shizuka M, Snapper ML (2008) Catalytic enantioselective Hosomi–Sakurai conjugate allylation of cyclic unsaturated ketoesters. Angew Chem Int Ed 47: 5049–5051

[106] (a) Nicolaou KC, Montagnon T, Snyder SA (2003) Tandem reactions, cascade sequences, and biomimetic strategies in total synthesis. Chem Commun: 551–564; (b) Nicolaou KC, Edmonds DJ, Bulger PG (2006) Cascade reactions in total synthesis. Angew Chem Int Ed 45: 7134–7186

[107] Niethe A, Fischer D, Blechert S (2009) Total synthesis of ent-lepadin F and G by a tandem ene–yne–ene ring closing metathesis. J Org Chem 73: 3088–3093

[108] Schaudt M, Blechert S (2003) Total synthesis of (+)-astrophylline. J Org Chem 68: 2913–2920

[109] Stapper C, Blechert S (2002) Total synthesis of (+)-dihydrocuscohygrine and cuscohygrine. J Org Chem 67: 6456–6460

[110] Jackson KL, Henderson JA, Motoyoshi H, Phillips AJ (2009) A total synthesis of norhalichondrin B. Angew Chem Int Ed 48: 2346–2350

[111] Stewart IC, Douglas CJ, Grubbs RH (2008) Increased efficiency in cross-metathesis reactions of sterically hindered olefins. Org Lett 10: 441–444

[112] Henderson JA, Phillips AJ (2008) Total synthesis of aburatubolactam A. Angew Chem Int Ed 47: 8499–8501

[113] Malik CK, Yadav RN, Drew MGB, Ghosh S (2009) Synthesis of fused cyclic systems containing medium-sized rings through tandem ROM–RCM of norbornene derivatives embedded in a carbohydrate template. J Org Chem 74: 1957–1963

[114] Marvin CC, Voight EA, Suh JM, Paradise CL, Burke SD (2008) Synthesis of (+)-didemniserinolipid B: Application of a 2-allyl-4-fluorophenyl auxiliary for relay ring-closing metathesis. J Org Chem 73: 8452–8457

[115] Smith III AB, Freeze BS (2008) (+)-Discodermolide: Total synthesis, construction of novel analogues, and biological evaluation. Tetrahedron 64: 261–298

[116] Xie Q, Denton RW, Parker KA (2008) A relay ring-closing metathesis synthesis of dihydrooxasilines, precursors of (*Z*)-iodo olefins. Org Lett 10: 5345–5348

[117] Tannert R, Hu T-S, Arndt H-D, Waldmann H (2009) Solid-phase based total synthesis of jasplakinolide by ring-closing metathesis. Chem Commun: 1493–1495

[118] (a) Kim BG, Snapper ML (2006) Preparation of alkenyl cyclopropanes through a ruthenium-catalyzed tandem enyne metathesis–cyclopropanation sequence. J Am Chem Soc 128: 52–53; (b) Murelli RP, Catalán S, Gannon MP, Snapper ML (2008) Ruthenium-catalyzed tandem enyne-cross metathesis–cyclopropanation: Three-component access to vinyl cyclopropanes. Tetrahedron Lett 49: 5714–5717

[119] (a) Schmidt B (2004) *In situ* conversion of a Ru metathesis catalyst to an isomerization catalyst. Chem Commun: 742–743; (b) Schmidt B (2004) Ruthenium-catalyzed olefin metathesis double-bond isomerization sequence. J Org Chem 69: 7672–7687

[120] Finnegan D, Seigal BA, Snapper ML (2006) Preparation of aliphatic ketones through a ruthenium-catalyzed tandem cross-metathesis/allylic alcohol isomerization. Org Lett 8: 2603–2606

[121] Chen J-R, Li C-F, An X-L, Zhang J-J, Zhu X-Y, Xiao W-J (2008) Ru-catalyzed tandem cross-metathesis/intramolecular-hydroarylation sequence. Angew Chem Int Ed 47: 2489–2492

[122] (a) Schmidt B, Pohler M, Costisella B (2004) Ring-closing olefin metathesis and radical cyclization as competing pathways. J Org Chem 69: 1421–1424; (b) Schmidt B, Pohler M (2005) Ruthenium-catalyzed tandem ring closing metathesis (RCM) – atom transfer radical cyclization (ATRC) sequences. J Organomet Chem 690: 5552–5555

[123] Dragutan V, Dragutan I (2006) A resourceful new strategy in organic synthesis: Tandem and stepwise metathesis/non-metathesis catalytic processes. J Organomet Chem 691: 5129–5147

[124] Louie J, Bielawski CW, Grubbs RH (2001) Tandem catalysis: The sequential mediation of olefin metathesis, hydrogenation, and hydrogen transfer with single-component Ru complexes. J Am Chem Soc 123: 11312–11313

[125] Beligny S, Eibauer S, Maechling S, Blechert S (2006) Sequential catalysis: A metathesis/dihydroxylation sequence. Angew Chem Int Ed 45: 1900–1903

[126] Neisius NM, Plietker B (2008) Diastereoselective Ru-catalyzed cross-metathesis–dihydroxylation sequence. An efficient approach toward enantiomerically enriched *syn*-diols. J Org Chem 73: 3218–3227

[127] Wang J, Lee V, Sintim HO (2009) Efforts towards the identification of simpler platensimycin analogues – The total synthesis of oxazinidinyl platensimycin. Chem Eur J 15: 2747–2750

Probing the Mechanism of the Double C–H (De)Activation Route of a Ru-Based Olefin Metathesis Catalyst

Albert Poater, Luigi Cavallo*

Department of Chemistry, University of Salerno, via Ponte don Melillo, Fisciano, I-84084, Italy

*E-mail: lcavallo@unisa.it

Abstract A theoretical study of a double C–H activation mechanism that deactivates a family of second generation Ru-based catalysts is presented. DFT calculations are used to rationalize the complex mechanistic pathway from the starting precatalyst to the experimentally characterized decomposition products. In particular, we show that all the intermediates proposed by Grubbs and coworkers are indeed possible intermediates in the deactivation pathway, although the sequence of steps is somewhat different.

Keywords Computational chemistry · Olefin metathesis · Decomposition reactions · Ruthenium catalysts · N-heterocyclic carbenes (NHC) · Density functional theory

1 Introduction

Ru-based catalysts for olefin metathesis have acquired a prominent role in modern organic synthesis [1], and are expanding their area from lab scale to industrial production. This step forward requires that very active and stable catalysts have to be designed. Along this direction, replacement of a phosphane ligand with a N-heterocyclic carbene (NHC) ligand in Grubbs first generation catalysts [2] lead to NHC-based second generation catalysts that proved more active and in many cases even more stable [3]. The origin of the greater activity of NHC-based catalysts is now rather well understood and, although still incomplete [4, 5], this knowledge is currently used to approach the ambitious "rational design" of new catalysts. Unfortunately, stability remains an issue, and very little is known about deactivation pathways [6]. In one case, however, detailed experiments have characterized the deactivation products [6f], which opened the route to the rationalization of catalysts deactivation.

V. Dragutan et al. (eds.), *Green Metathesis Chemistry: Great Challenges in Synthesis, Catalysis and Nanotechnology*,

Precatalyst **3** heated in benzene at 60°C under inert condition decomposes to complex **5** after 3 days (58% yield) with traces of complex **6** (see Chart 1). Replacing benzene with CD_2Cl_2 leads to much faster decomposition. However, after 12 h at 40°C the main decomposition product is **6** (38% yield), while **5** is the minor decomposition product (24% yield). Complexes **5** and **6** were fully characterized by X-ray crystallography [6f].

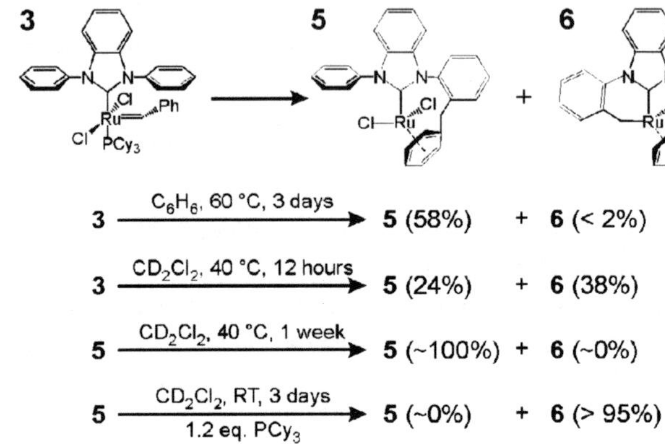

Chart 1 Decomposition of precatalyst **3**

Further experiments revealed that complex **5** is stable at 40°C for a week, while it evolves to complex **6** at RT in only 3 days if PCy_3 is added to reaction media. This suggested that **5** is a precursor of **6**, and that activation of the *ortho* C–H bond of a N-phenyl ring of the NHC ligand likely is the first step. On this basis, the deactivation mechanism shown in Chart 2 was proposed [6f]. Intermediates **8** and **9** are only hypothetical species along the deactivation pathway, since no intermediates could be characterized.

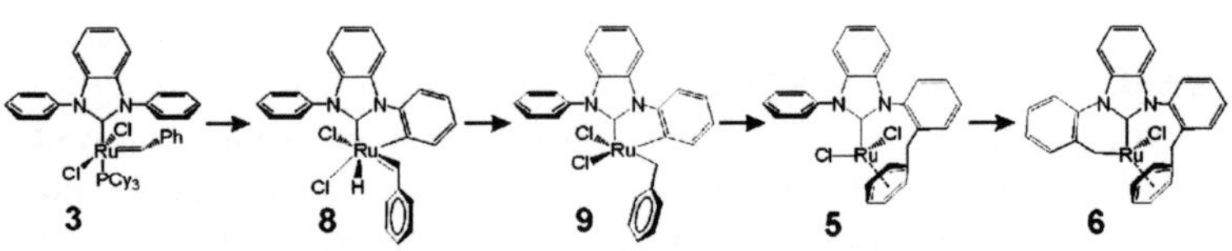

Chart 2 Deactivation mechanism of precatalyst **3**

Assisted by DFT calculations we propose here a complete theoretical characterization of this deactivation mechanism. To allow for an easy comparison with the experimental paper, we denote species that were described in the experimental paper with the same labels, see (Scheme 1). We show here that **8** and **9** are indeed intermediates along the deactivation pathway, and that **5** is a precursor of **6**. However, we also think that **9** could precede **8**, rather than follow it.

2 Computational Details

All the Density Functional Theory (DFT) calculations were performed using the Gaussian03 package [7]. The BP86 GGA functional of Becke and Perdew was used [8]. The SVP triple-ζ basis set with one polarization function was used for main group atoms [9], while the relativistic SDD effective core potential in combination with a triple-ζ basis set was used for the Ru atom [10]. All geometries were verified by frequency calculations that resulted in 0 and 1 imaginary frequency for intermediates and transition states, respectively. The reported energies include the vibrational gas-phase zero-point energy term, and a solvation term that was obtained through single-point calculations on the gas-phase optimized geometries. The polarizable continuous solvation model IEF-PCM as implemented in the Gaussian03 package has been used [11]. CH_2Cl_2 was chosen as model solvent, with a dielectric constant $\varepsilon = 8.93$. Standard non-electrostatic terms were also included.

3 Results and Discussion

The first part of the deactivation intermediate, from the naked 14e species **7** to the intermediate **10** is shown in Figure 1. The first step of the deactivation path starts from complex 7 (formed by phosphane dissociation from **3**, which costs 16.6 kcal/mol) and leads to complex **6**. The electron-deficient 14-electron species **7** can engage one of the *ortho* C-H bonds of a N-phenyl ring of the NHC-ligand into a rather stable agostic interaction through transition state **7-7′** and a barrier of 11.6 kcal/mol. The resulting intermediate **7′**, which is 4.1 kcal/mol higher in energy than 7, is a text-book example of agostic complexes, with a clearly elongated C–H bond, 1.17 Å with respect to 1.10 Å in **7**, and a Ru-C-H angle of only 53.3°.

The next step is an oxidative addition consisting in the transfer of the agostic hydrogen of **7′** to the nearby α C-atom of the benzylidene group, through transition state **7′-9** and a barrier of 19.9 kcal/mol, to reach intermediate **9**, in which the benzylidene group is transformed into a benzyl group, and the Ru atom forms a new σ-bond with the *ortho* C atom of the N-phenyl ring of NHC-ligand involved in the agostic interaction. Intermediate **9**, originally suggested by Grubbs and coworkers, is 6.1 kcal/mol higher in energy than **7**, and it is stabilized by an interaction between the Ru atom and the *ipso* C atom of the benzyl group.

The other key intermediate proposed by Grubbs and coworkers, the Ru-hydride species **8**, is reached from **9** via the transfer of a α hydrogen from the benzyl group to the Ru atom through transition state **9-8** and a barrier of 11.7 kcal/mol.

This H-transfer step restores the benzylidene group in **8**. Intermediate **8**, 17.7 kcal/mol above 7, is rather unstable since the newly formed benzylidene group can inserts into the σ-bond between the Ru atom and the *ortho* C atom of the N-phenyl ring of NHC-ligand through transition state **8-10** and the almost negligible barrier of 0.2 kcal/mol only. The Ru-hydride intermediate **10**, which is 9.4 kcal/mol above 7, was not postulated by Grubbs and coworkers, and it is the only additional "chemically different" piece that is needed to connect **7** to **5** and **6**.

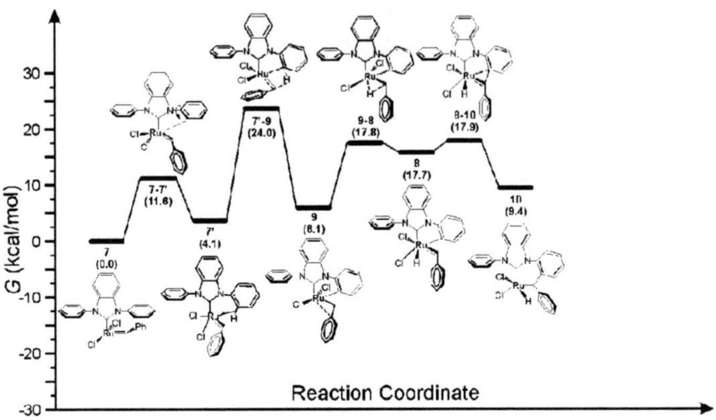

Figure 1 First part of the deactivation pathway

The second part of the deactivation intermediate, from intermediate **10** to the experimentally characterized products is shown in Figure 2. Intermediate **10** can evolve through transition **10-5'''** and a barrier of 5.9 kcal/mol to intermediate **5'''**. This reductive elimination step breaks the Ru-hydride bond. Intermediate **5'''**, which is 1.6 kcal/mol above **7**, is stabilized by a strong interaction with the *ipso* C atom of the Ph group of the formerly benzylidene moiety [12]. Intermediate **5'''** is only a kinetic intermediate to the decomposition product **5**, which can be reached from **5'''** through a series of haptotropic shifts that involve the Ru atom and the Ph group of the formerly benzylidene moiety. The decomposition product **5** is remarkably stable, since it is 21.5 kcal/mol below the starting species **7**.

Nevertheless, one of the Cl atoms of **5** can interact with an *ortho* H atom of the other N-phenyl ring of the NHC ligand to eliminate one HCl molecule, through transition state **5–6** and a barrier of 28.4 kcal/mol, to yield the other observed decomposition product, **6**. Complex **6** + HCl is 4.9 kcal/mol higher in energy than **5**. On the other hand, if the extracted HCl molecule is trapped by a PCy_3 molecule, the resulting product **6** + $[HPCy_3^+][Cl^-]$ is 3.1 kcal/mol below **5**, which is in quailtative agreement with the experimental finding that **5** is thermodynamically stable unless a Lewis base such PCy_3 is added, which shifts the equilibrium towards **6** [13].

The decomposition pathway discussed so far is the one that seems to be preferred according to our calculations. Of course, we also tested other possibilities. For example, we started with the **7** → **8** → **9** pathway postulated by Grubbs and coworkers. However, transition state **7′-8**, which corresponds to a direct transfer of the *ortho* H atom from the N-phenyl ring of the NHC ligand to the Ru atom, and that connects **7′** to **8** in a single step, was found to be 7.9 kcal/mol above transition state **7′-9**, which rules out the **7** → **8** → **9** pathway in favor of the **7** → **9** → **8** pathway. Similarly, we tested if **9** can be directly transformed into one of the hapto-isomers of **5**. If feasible, this step would simplify the decomposition pathway by removing intermediate **8**. However, the best transition state we found, **9-5'''**, is 14.3 kcal/mol higher in energy than transition state **9-8**, which suggests that **8** is a real intermediate along the decomposition pathway.

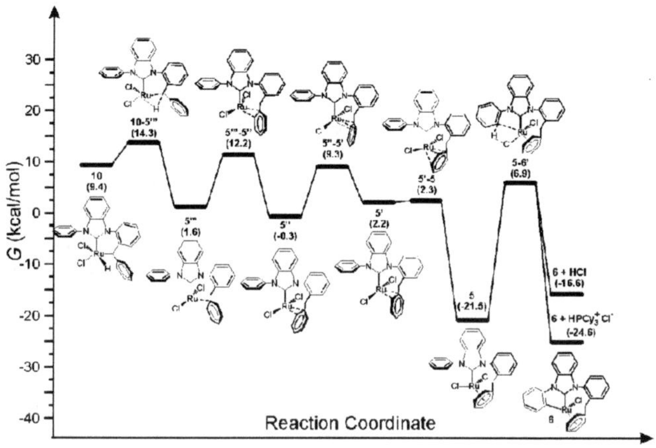

Figure 2 Second part of the deactivation pathway

4 Conclusions

In conclusion, aided by DFT calculations we have characterized the full deactivation mechanism of a second generation Ru-based catalyst from the starting precatalyst to the experimentally determined deactivation products. In particular, we have shown that all the intermediates proposed by Grubbs and coworkers are indeed possible intermediates in the deactivation pathway, although the sequence of steps is somewhat different.

Acknowledgments L.C. thanks the INSTM (CINECA Grant) for financial support. A.P. thanks the Spanish Ministerio de Educación y Ciencia (MEC) for a postdoctoral contract.

References

[1] Grubbs RH (ed.) (2003). Handbook of metathesis. Wiley-VCH, Weinheim, Germany

[2] (a) Nguyen ST, Grubbs RH, Ziller JW (1993) J Am Chem Soc 115:9858–9859

[3] (a) Scholl M, Ding S, Lee CW, Grubbs RH (1999) Org Lett 1:953–956; (b) Huang J, Stevens ED, Nolan SP, Peterson JL (1999) J Am Chem Soc 121:2674–2678; (c) Weskamp T, Kohl FJ, Hieringer W, Gleich D, Herrmann WA (1999) Angew Chem Int Ed 38:2416–2419; (d) Bielawski CW, Grubbs RH (2000) Angew Chem Int Ed 39:2903–2906

[4] (a) Hoveyda AH, Schrock RR (2001) Chem-Eur J 7:945–950; (b) Schrock RR, Hoveyda AH (2003) Angew Chem Int Ed 42:4592–4633; (c) Fürstner A (2000) Angew Chem Int Ed 39:3012–3043; (d) Trnka TM, Grubbs RH (2001) Acc Chem Res 34:18–29

[5] (a) Dias EL, Nguyen ST, Grubbs RH (1997) J Am Chem Soc 119:3887–3897; (b) Ulman M, Grubbs RH (1998) Organometallics 17:2484–2489; (c) Adlhart C, Hinderling C, Baumann H, Chen P (2000) J Am Chem Soc 122:8204–8214; (d) Michrowska A, Bujok R, Harutyunyan S, Sashuk V, Dolgonos G, Grela K (2004) J Am Chem Soc 126:9318–9325; (e) Getty K, Delgado-Jaime MU, Kennepohl P (2007) J Am Chem Soc 129:15774–15776; (f) van der Eide EF Romero PE, Piers WE (2008) J Am Chem Soc ASAP; (g) Cavallo L (2002) J Am Chem Soc 124:8965–8973; (h) Correa A, Cavallo L (2006) J Am Chem Soc 128:13352–13353

[6] (a) Jazzar RFR, Macgregor SA, Mahon MF, Richards SP, Whittlesey MK (2002) J Am Chem Soc 124:4944–4945; (b) Giunta D, Hölscher M, Lehmann CW, Mynott R., Wirtz C, Leitner W (2003) Adv Synth Catal 345:1139–1145; (c) Abdur-Rashid K, Fedorkiw T, Lough AJ Morris RH (2004) Organometallics 23:86–94; (d) Dorta R, Stevens ED, Nolan SP (2004) J Am Chem Soc 126:5054–5055; (e) Scott NM, Dorta R, Stevens ED, Correa A, Cavallo L, Nolan SP (2005) J Am Chem Soc 127:3516–3526; (f) Hong SH, Chlenov A, Day MW, Grubbs RH (2007) Angew Chem Int Ed 46:5148–5151; (g) Berlin JM, Campbell K, Ritter T, Funk TW, Chlenov A, Grubbs RH (2007) Org Lett 9:1339–1342; (h) Hong SH, Day MW Grubbs RH (2004) J Am Chem Soc 126:7414–7415; (i) Ulman M, Grubbs RH (1999) J Org Chem 64:7202–7207; (j) Hong SH, Wenzel AG, Salguero TT, Day MW, Grubbs RH (2007) J Am Chem Soc 129:7961–7968

[7] Gaussian 03 (2003) Gaussian Inc. Pittsburgh, PA

[8] (a) Becke AD (1988) Phys Rev A 38:3098–3100; (b) Perdew JP (1986) Phys Rev B 33:8822–8824; (c) Perdew JP (1986) Phys Rev B 34:7406

[9] (a) Schaefer A, Horn H, Ahlrichs R (1992) J Chem Phys 97:2571–2577; (b) Schaefer A, Huber C, Ahlrichs R (1994) J Chem Phys 100:5829–5835

[10] (a) Haeusermann U, Dolg M, Stoll H, Preuss H (1993) Mol Phys 78:1211–1224; (b) Kuechle W, Dolg M, Stoll H, Preuss H (1994) J Chem Phys 100:7535–7542; (c) Leininger T, Nicklass A, Stoll H, Dolg M, Schwerdtfeger P (1996) J Chem Phys 105:1052–1059

[11] (a) Cossi M, Barone V, Cammi R, Tomasi J (1996) Chem Phys Lett 255:327–335; (b) Cancès MT, Mennucci B, Tomasi J (1997) J Chem Phys 107:3032–3041; (c) Cossi M, Barone V, Mennucci B, Tomasi J (1998) Chem Phys Lett 286:253–260

[12] For examples of η^2-coordinated benzyl groups see: (a) Cotton FA, Murillo CA, Petrukhinaa MA (1999) J Organomet Chem 573:78–86; (b) Clegg W, Elsegood MRJ, Dyer PW, Gibson VC, Marshall EL (1999) Acta Cryst C 55:916–918

[13] Attempts to find a transition state in which PCy_3 assists H-extraction from **5** were unsuccessful

A Comparison of the Performance of the Semiempirical PM6 Method Versus DFT Methods in Ru-Catalyzed Olefin Metathesis

Andrea Correa,* Albert Poater, Francesco Ragone, Luigi Cavallo

Department of Chemistry, University of Salerno, via Ponte don Melillo, Fisciano, I-84084, Italy, E-mail: acorrea@unisa.it

*E-mail: lcavallo@unisa.it

Abstract In this work we compare the performance of the semiempirical PM6 method with a more accurate DFT method when applied to Ru-catalyzed olefin metathesis. We demonstrate that the PM6 method reproduces with interesting accuracy the geometries located with a DFT approach. As for the energetics, the relative DFT stability of the metallacycle with respect to the coordination intermediate is reproduced with reasonable accuracy by the PM6 method, whereas the olefin coordination energy and the energy barrier of the metathesis step are overestimated. Further, for the same system we performed a PM6-based meta-dynamics study of the olefin metathesis reaction, which indicated a reasonable good behavior of the system also under dynamic conditions. In conclusion, the obtained results validate the use of the semiempirical PM6 method for preliminary and computationally fast screening on new ligands/substrates in Ru catalyzed olefin metathesis.

Keywords Homogenous catalysis · Density functional · Theory calculations · Metathesis · Semiemipirical methods · ab-initio Molecular dynamics

1 Introduction

Recent decades have seen ruthenium catalyzed olefin metathesis emerging as one of the most useful and versatile tools for the synthesis of C–C bonds [1–4]. Indeed, this reaction allows the synthesis of complicated molecules in fewer and simpler steps relative to classical synthetic approaches, and it is particularly suited for the synthesis of medium-size rings, which are otherwise difficult to close.

The discovery by Grubbs and coworkers of well-defined Ru-based (pre)catalysts, such as $(PCy_3)_2Cl_2Ru{=}CHR$, broadened its scope significantly: they benefit from air and moisture resistence and they operate in mild condition with very high tolerance toward heteroatoms-containing functional groups. Substitution of a

V. Dragutan et al. (eds.), *Green Metathesis Chemistry: Great Challenges in Synthesis, Catalysis and Nanotechnology*,

single phosphine by an N-heterocyclic carbine, NHC, ligand led to heteroleptic (pre)catalysts whose activity is not only higher than that of early transition-metal catalysts [5–8].

The discovery by Hofmann and coworkers of Ru complexes with chelating bisphosphine ligands [9, 10] and by Fuerstner and coworkers of cationic allenylidene Ru complexes [11] opened new routes. Tuning and new use of this reaction contribute to its relevance as an effective tool in organometallic chemistry [12–19]. The development of such effective catalysts was considerably accelerated when fundamental mechanistic studies elucidated the real nature of the active species, and offered a more or less detailed picture of the overall mechanism.

Experimental [20–24] and theoretical [22, 25–28] studies converged to the mechanism briefly reported in Figure 1 as the most probable. Substitution of a phosphine from the starting complex by the olefinic substrate, through a naked 14-electrons intermediate, generates the 16-electrons intermediate in which the olefin is cis coordinated to the alkylidene.

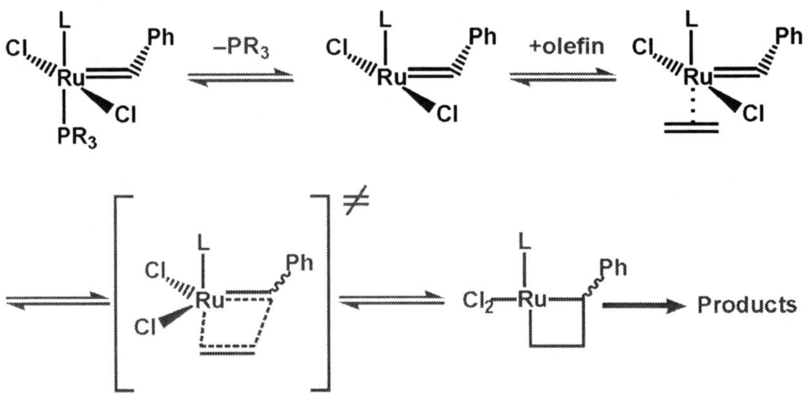

Figure 1 Olefin metathesis mechanism

Reaction of the olefin with the alkylidene moiety, then, leads to the metallacycle intermediate that rapidly evolves toward products. The formation of the phosphine-dissociated but olefin-bound intermediate is supported by kinetic data [20], isolation and characterization of a monophosphine Ru catalyst "caught in the act" [29], detection of monophosphine intermediates by electrospray ionization tandem mass spectrometry [22, 30], and quantum mechanics studies [7, 22, 24–28].

Although the basics of the mechanism have been clarified, a full understanding of these catalysts has not been achieved yet, and a clear structure–function relationship is not established as well. On the other hand, an understanding of the details of the mechanisms operative with both classes of catalysts is the key to designing both better performing catalysts and catalysts with new scope. In this respect, computational chemistry techniques are a very powerful tool to shed light on this class of reactions, which can contribute to a detailed understanding of the laws that rule these systems.

Over the years, quantum chemistry has made significant contribution to the study of organometallic reaction mechanisms [31].

One limitation of conventional computational studies (static methods) is that only the potential energy minima and transition state (TS) are characterized. Moreover, in many cases finding the right pathway connecting two intermediates is not a trivial question. Sometimes, to find the lowest energy surface on which a chemical process occurs can request many attempts. Apart from this, finite temperature effects arising from the dynamical behaviour of these complexes, such as vibrational and fluxional effects [32], are difficult to investigate exhaustively using these conventional "static" methods. On the other hand, Molecular dynamics (MD) can provide an elegant and direct means to examine the real free energy surface including finite temperature effects through the simulation of atoms motions during the time.

In the last few years, MD in combination with Density Functional Theory (DFT) has made significant contributions to our understanding of fundamental chemical systems [33], including some notable examples in organometallic chemistry [34]. Unfortunately, the huge computational cost of these DFT-MD methods greatly limits their use as an effective sampling technique to study complex reaction mechanisms.

The recent extension of semiempirical methods, such as the PM6 method [35], to transition metals has opened a new route in computational chemistry. Semiempirical methods are very less demanding, in terms of computing power, with respect *ab initio* and DFT methods. Of course, this increase in computing performances is traded with a decrease in terms of accuracy, which is more severe in the case of transition metals compared to main group atoms. Anyway, the low computational cost of semiempirical methods in conjunction with their implementation in MD packages like CP2K [36] gives the possibility to use them for fast and preliminary dynamical study of an organometallic reaction mechanism.

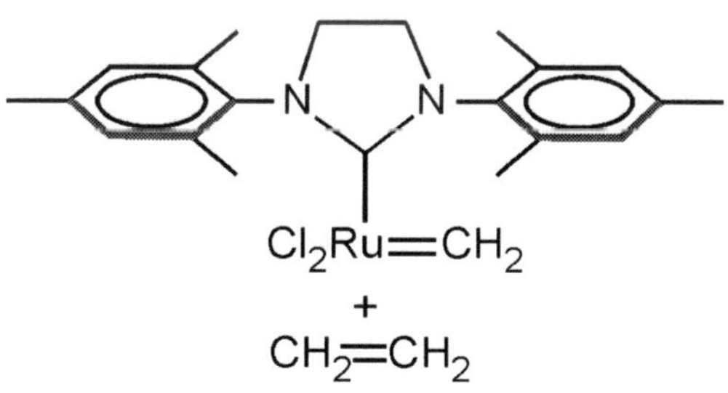

Figure 2 The catalytic system investigated in this study

In this paper, focusing on the system reported in Figure 2, we tested the performances of the PM6 method to investigate the mechanism of Ru-catalyzed metathesis reactions. In particular, we will compare results obtained with the PM6 method with results obtained with a DFT approach based on the BP86 functional, which has been used successfully in the past to investigate this class of reactions [29–31]. Finally, we will also explore if the PM6 approach can be used in a molecular dynamics study of the metathesis reaction with the same system.

2 Computational Details

Static Calculations: The DFT calculations were performed at the BP86 [37] level of theory, with the Gaussian03 program package [38]. The SDD/ECP triple-ζ basis set was used for Ruthenium, while the SVP double-ζ basis set with a polarization function was used for main group atoms. Stationary points on the potential energy surface were tested with frequency calculations that resulted in 0 and 1 imaginary frequency for minima and transition states, respectively.

The DFT optimized geometries have been used as input structures for PM6 [35] calculations, performed with CP2K package [36].

Molecular Dynamics. We performed PM6-based molecular dynamics simulations using Quickstep [36], which is part of the CP2K program package [36]. A Born-Oppenheimer molecular dynamics (BOMD) approach was used, which means that at every step the wavefunction was optimized. The system has been simulated in the NVT ensemble with a time step of 0.5 fs. A Nosé-Hoover thermostat with a chain length of 4 was used to fix the temperature at 300 K. The optimized bottom-bound coordination intermediate has been used as input structure for MD. Before metadynamics simulation, the input structure has been equilibrated at 300 K for 500 fs.

Metadynamics. The available computer power does not typically allows for adequate sampling in molecular dynamics simulations to observe rare events such as chemical reactions (barrier energy >> k_bT, where k_b is the Bolzmann constant). Metadynamics is a nonequilibrium method that allows the system to escape minima in order to sample the rest of the free energy surface on a time scale that is accessible by present day computers [39, 40]. The metadynamics method has been successful used in a number of different applications, including the investigation of bacterial chloride channels [41], deprotonation of formic acid [42], flexible ligand docking [43] and organometallic systems [44]. The method is based on the assumption that it is possible to define one or more collective coordinates that can distinguish between reactants and products and can sample the low-energy reaction paths. Collective coordinates must be functions of the ionic coordinates (i.e., bond lengths, dihedral angles, coordination number, etc.). A series of small repulsive Gaussian potentials (hills), centered on the values of the collective coordinates, are added during the dynamics, preventing the system from revisiting the same points in phase space and creating a history dependent biasing potential [39].

In the present study, metadynamics has been performed with the same settings used to equilibrate the system, adding a hill every 100 molecular dynamic steps. The height and width of the hills was set as 0.0005 and 0.15 au respectively. The distance between the carbon atom of the methyldene group and one carbon atom of the coordinated ethene molecule (CA, see Figures 3–5 for labels) has been chosen as collective variable. The mass of the collective variable was set to 10 amu and the force constant of the spring was set to 0.05 au. Phase space boundaries in the form of repulsive potentials were located as a quadratic wall at a distance of 6.5 Bohr.

3 Results and Discussion

We start this section with a comparison between the PM6 and DFT results. We first discuss and compared the optimized structures, to check the performance of the semiempirical PM6 method versus the more reliable DFT approach. Considering the ethene coordination intermediate, the PM6 method reproduces very well the DFT optimized structure. The rmsd between the PM6 optimized structure and the DFT one is only 0.01 Å. With both method the π-complex presents a geometry with the C=C double bond of the ethylene almost perpendicular to the Ru=CH_2 (alkylidene) bond, see Figure 3.

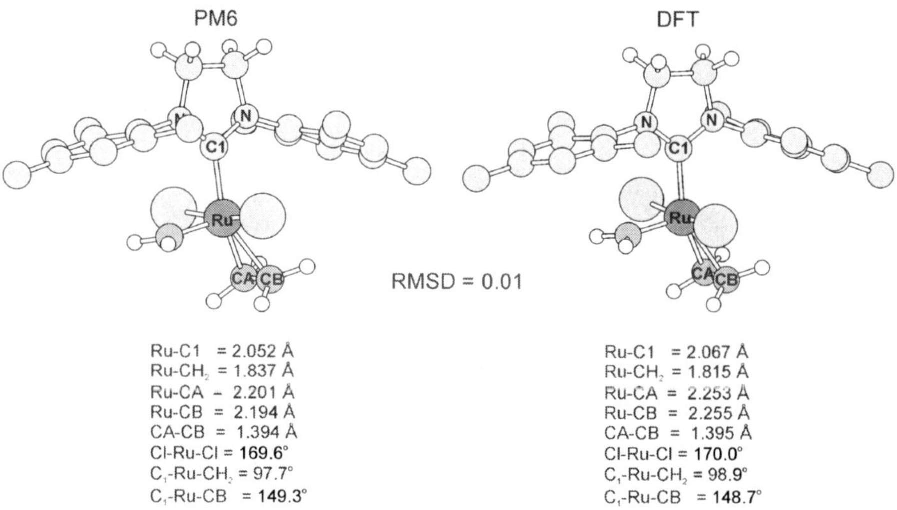

Figure 3 Optimized geometries for coordination intermediates calculated by PM6 (*left* side) and DFT (*right* side)

According with previous theoretical study [26], the olefin is slightly asymmetrically coordinated, while the C=C bond is slightly elongated from the value assumed in the free C_2H_4 molecule (1.34 Å both at the PM6 and DFT levels of theory). The DFT NHC and olefin coordination distances, as well as the Ru=CH_2 and the Ru bond distance also are very nicely reproduced, and the small differences between the two methods could also be the differences between two different functionals. Focusing on bond angles, the Cl–Ru–Cl angle (169.6° and 170.3° at the PM6 and DFT levels, respectively) are as well.

The computed transition state structures for the ethylene metathesis step are displayed in Figure 4, and some important geometrical parameters are also reported. The agreement between DFT and PM6 is quite good also in this case, with the olefin that has rotated by almost 90° assuming an orientation roughly parallel to the Ru=CH_2 bond. The rmsd between the two structures is 0.03 Å, which is slightly higher with respect to the rmsd we have found for the π-complex, but still is very small. In both structures the reacting atoms assume almost planar four-center geometry, which is typical for olefin metathesis [45–49]. The deviation of the C atom of the ethylene which is going to form the new C–C bond from the plane defined by the forming C–Ru–C bonds is 0.6 Å in the PM6 geometry, which compares very well with the value of 0.4 Å in the DFT transition state.

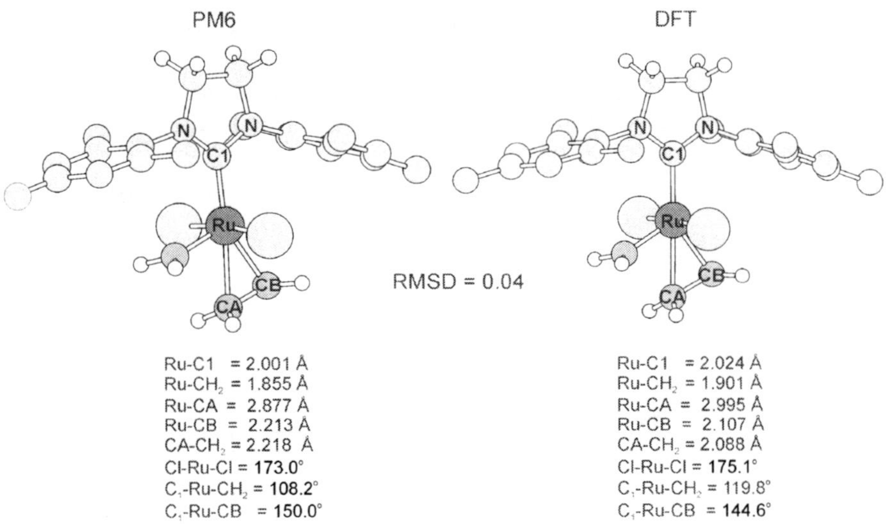

Figure 4 Transition state structures calculated by PM6 (*left* side) and DFT (*right* side)

The main difference is in the length of the forming C–C bond, which is roughly 0.13 Å longer in the PM6 geometry, indicating a slightly earlier transition state. As a consequence, the Ru–CB distance of the incipient Ru–C bond is calculated to be slightly longer in the PM6 geometry. Focusing on the bond angles, both the C1–Ru–CH2 and C1–Ru–CB angles calculated at the PM6 level (108.2° and 150.0°), respectively, nicely reproduce the DFT values (119.8° and 144.6°). In conclusion, despite these small differences in the position of the emerging C–C bond, the PM6 method reproduces very well the DFT transition state geometry.

Very shortly, also for the metallacycle structures we found a very nice agreement between the PM6 and DFT geometries (see Figure 4), with a rmsd of only 0.02 Å. The former Ru=C(alkylidene) bond, which evolved into a Ru–C single bond, is about 0.16 Å longer in the metallacycles relative to the coordination intermediates with both methods, and a small shortening in the Ru–C1(NHC) distance was also detected with both methods, compare Figures 3 and 5.

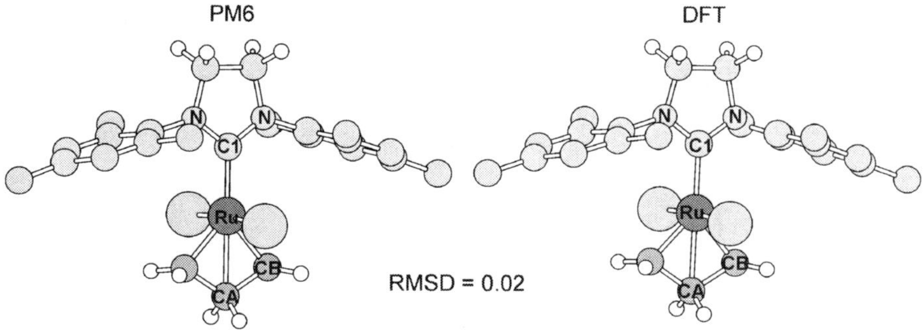

PM6	DFT
Ru-C1 = 1.981 Å	Ru-C1 = 2.012 Å
Ru-CH_2 = 1.995 Å	Ru-CH_2 = 1.975 Å
Ru-CA = 2.304 Å	Ru-CA = 2.278 Å
Ru-CB = 1.995 Å	Ru-CB = 1.975 Å
CA-CH_2 = 1.560 Å	CA-CH_2 = 1. 588 Å
Cl-Ru-Cl = 174.2°	Cl-Ru-Cl = 174.7°
C_1-Ru-CH_2 = 138.2°	C_1-Ru-CH_2 = 137.2°
C_1-Ru-CB = 180.0°	C_1-Ru-CB = 179.9°

Figure 5 Optimized geometries for metallacycle intermediates calculated by PM6 (*left* side) and DFT (*right* side)

With regards to the energetic of metathesis reaction, Figure 6 displays the energy profiles obtained with the DFT and PM6 methods. In both cases the coordination intermediate has been taken as zero of energy. In agreement with previous theoretical investigations, the DFT transition state is roughly 2 kcal/mol higher in energy with respect to the coordination intermediate, while the DFT-metallacycle intermediate is roughly 15 kcal/mol lower in energy with respect to the coordination intermediate. Remarkably different, instead, is the ethylene co-ordination energy to the naked 14-electrons species to form the coordination intermediate. In fact, the PM6 coordination energy, roughly −22.1 kcal/mol, is remarkably higher than the DFT coordination energy, roughly −11.3 kcal/mol.

With regards to shape of the energy profiles, it is easy to see that the agreement between the stability of the metallacycle intermediate calculated with the two methods is almost perfect (−14.5 kcal/mol by DFT and −12.8 kcal/mol by PM6), whereas the energy barrier leading to the transition state obtained by PM6 methods (about 8 kcal/mol) is quite greater than that calculated by the DFT approach (about 2 kcal/mol). Anyway, the overall shape of the metathesis reaction is substantially similar, which indicated that the semiempirical PM6 method can be considered as an effective tool to investigate the main features of olefin metathesis, such as in a preliminary screening of a family of new ligands.

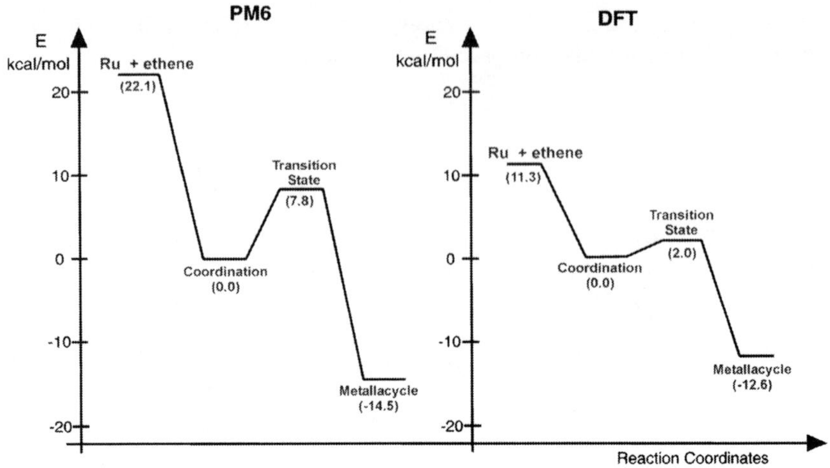

Figure 6 Energy profiles for olefin metathesis calculated by DFT (*right* side) and PM6 (*left* side)

To analyze for dynamical effects in the metathesis reaction, we have performed metadynamics simulations at 300 K using as collective variable the distance between C(alkylidene) and CA of ethylene. The simulation has been stopped after 32,000 fs, which corresponds to 18,000 fs after the metallacycle was observed for the first time. The resulting Free Energy Surface (FES) is reported in Figure 7.

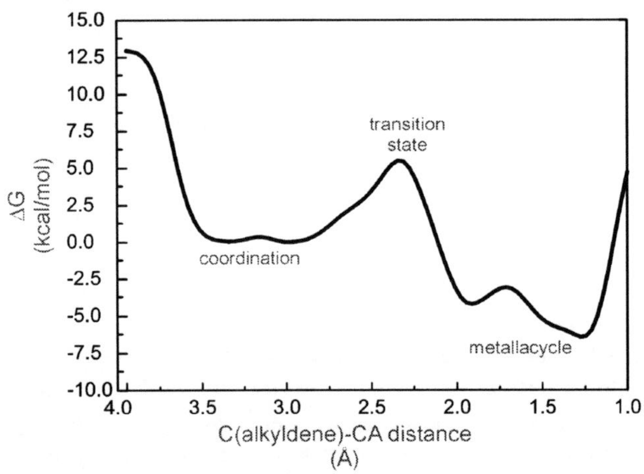

Figure 7 Free energy surface reconstructing by metadynamics at 300 K

The FES shows a small pick close to metallacycle well around 1.7 Å. Detailed structural analysis evidenced that this peak is just an artifact due to the choice of using just one collective variable. The coordination intermediate minimum is quite broad, which is consistent with its relatively high conformational freedom. The activation free energy is about 6.8 kcal/mol, very close to the static calculation result (7.8 kcal/mol, see Figure 6).

Differently, the stability of the metallacycle intermediate with respect to the coordination intermediate is reduced by roughly 5 kcal/mol according to the free energy simulations. It is tempting to suggest that the decreased stability of metallacycle is due to entropic contributions. In fact, the reduced conformational freedom in the metallacycle with respect to coordination intermediate gives an unfavorable entropic contribution to free energy, which would have been impossible to detect by static calculations. However, it must be noted that it could also be an artifact related to the limited number of distances used in the definition of the collective variable. Longer and more detailed simulations would solve this issue, but this is out of the scope of the present work.

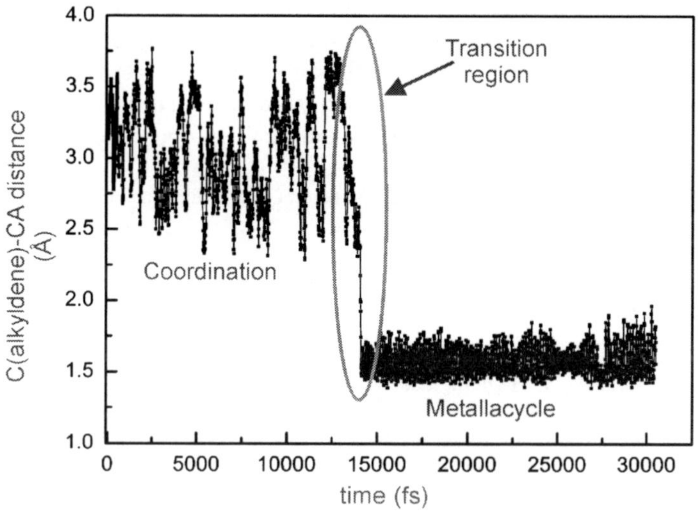

Figure 8 Time evolution of C(alkylidene)-CA distance during metadynamics at 300 K

Figure 8 displays the time evolution of the C(alkylidene)-CA distance during the metathesis reaction. The coordination intermediate is characterized by an average C(alkylidene)-CA distance of 3.1 Å with broad oscillations due to rotations of the olefins around the axis connecting the Ru atom to the center of the C=C double bond. The transition state is reached around 14,100 fs, and after this the system collapses into the metallacycle, which is characterized by an average C(alkylidene)-CA distance of 1.6 Å. Of course, as soon the metallacycle is reached the oscillations in the C(alkylidene)-CA distance are reduced in amplitude.

4 Conclusions

In summary, we have presented a comparative study in which we tested the performances of the semiempirical PM6 method versus a more accurate DFT method based on the BP86 functional. Our calculations suggest that the PM6 geometries reproduce with interesting accuracy the DFT geometries of almost all the steps of the olefin metathesis reaction, including the transition state for metallacycle formation. As regards the energetic of the reaction, the PM6 approach severely overestimate the ethylene coordination energy, while the relative energy difference between the metallacycle and the coordination intermediate is reproduced with good accuracy. Focusing on the energy barrier for the metathesis reaction, the PM6 value overestimate the DFT value, although not dramatically. In conclusion, our results indicate that the PM6 approach could be used as a very fast tool to pre-optimize geometries that can be subsequently refined by more expensive DFT calculations or, more interestingly, could be used in the fast screening of a large library of different ligand/substrates to select those systems that could be subsequently tested with more accurate DFT methods and finally proposed to experimentalists.

Acknowledgments We thank the Regione Campania Legge 5 for financial support.

References

[1] Ivin KJ, Mol JC (1997) Olefin metathesis and metathesis polymerization. Academic Press, San Diego, CA
[2] Fürstner A (2000) Angew Chem Int Ed 39:3012
[3] Buchmeiser MR (2000) Chem Rev 100:1565
[4] Trnka TM, Grubbs RH (2001) Acc Chem Res 34:18
[5] Scholl M, Ding S, Lee CW, Grubbs RH (1999) Org Lett 1:953
[6] Huang J, Stevens ED, Nolan SP, Peterson JL (1999) J Am Chem Soc 121:2674
[7] Weskamp T, Kohl FJ, Hieringer W, Gleich D, Herrmann WA (1999) Angew Chem Int Ed 38:2416
[8] Bielawski CW, Grubbs RH (2000) Angew Chem Int Ed 39:2903
[9] Hansen SM, Volland MAO, Rominger F, Eisentrager F, Hofmann P (1999) Angew Chem Int Ed 38:1273
[10] Hansen SM, Rominger F, Metz M, Hofmann P (1999) Chem Eur J 5:557
[11] Fürstner A, Liebl M, Lehmann CW, Picquet M, Kunz R, Bruneau C, Touchard D, Dixneuf PH (2000) Chem Eur J 6:1847
[12] Fürstner A, Ackermann L, Gabor B, Goddard R, Lehmann CW, Mynott R, Stelzer F, Thiel OR (2001) Chem Eur J 7:3236
[13] La DS, Sattely ES, Ford JG, Schrock RR, Hoveyda AH (2001) J Am Chem Soc 123:7767
[14] Louie J, Bielawski CW, Grubbs RH (2001) J Am Chem Soc 123:11312
[15] Louie J, Grubbs RH (2001) Angew Chem Int Ed 40:247
[16] Choi T-L, Woo Lee C, Chatterjee AK, Grubbs RH (2001) J Am Chem Soc 123:10417
[17] Furstner A, Ackermann L, Beck K, Hori H, Koch D, Langemann K, Liebl M, Six C, Leitner W (2001) J Am Chem Soc 123:9000
[18] Liu L, Postema MHD (2001) J Am Chem Soc 123:8602

[19] Choi T-L, Chatterjee AK, Grubbs RH (2001) Angew Chem Int Ed 40:1277
[20] Dias EL, Nguyen ST, Grubbs RH (1997) J Am Chem Soc 119:3887
[21] Ulman M, Grubbs RH (1998) Organometallics 17:2484
[22] Adlhart C, Hinderling C, Baumann H, Chen P (2000) J Am Chem Soc 122:8204
[23] Adlhart C, Volland MAO, Hofmann P, Chen P (2000) Helv Chim Acta 83:3306
[24] Adlhart C, Chen P (2000) Helv Chim Acta 83:2192
[25] Aagaard OM, Meier RJ, Buda F (1998) J Am Chem Soc 120:7174
[26] Cavallo L (2001) J Am Chem Soc 123:8965
[27] Costabile C, Cavallo L (2004) J Am Chem Soc 126:9592
[28] Correa A, Cavallo L (2006) J Am Chem Soc 128:13352
[29] Tallarico JA, Bonitatebus PJ Jr, Snapper ML (1997) J Am Chem Soc 119:7157
[30] Hinderling C, Adlhart C, Baumann H, Chen P (1998) Angew Chem Int Ed 37:7
[31] (a) Niu SQ, Hall MB (2000) Chem Rev 100:453 (b) Ziegler T, Autschbach J (2005) Chem Rev 105:2695
[32] (a) Michalak A, Ziegler T (2001) J Phys Chem A 105:4333–4343; (b) Seth M, Senn HM, Ziegler T (2005) J Phys Chem A 109:5136
[33] (a) Marx D, Tuckerman ME, Hutter J, Parrinello M (1999) Nature 397:601; (b) Raugei S, Klein ML (2001) J Am Chem Soc 123:9484; (c) Kuo IFW, Mundy CJ (2004) Science 303:658
[34] (a) Margl P, Lohrenz JCW, Woo TK, Ziegler T, Blochl PE (1996) Polym Mater Sci Eng 74:397; (b) Woo TK, Margl PM, Ziegler T, Bloechl PE (1997) Organometallics 16:3454; (c) Margl PM, Woo TK, Bloechl PE, Ziegler T (1998) J Am Chem Soc 120:2174; (d) Woo TK, Margl PM, Deng L, Cavallo L, Ziegler T (1999) Catal Today 50:479; (e) De Angelis F, Fantacci S, Sgamellotti A (2006) Coord Chem Rev 250:1497; (f) Magistrato A, Togni A, Rothlisberger U (2006) Organometallics 25:1151; (g) Buhl M, Golubnychiy V (2007) Organometallics 26:6213; (h) Gossens C, Tavernelli I, Rothlisberger U (2008) J Am Chem Soc 130:10921
[35] Stewart JJP (2007) J Mol Modeling 13:1173
[36] VandeVondele J, Krack M, Mohamed F, Parrinello M, Chassaing T, Hutter J (2005) Comput Phys Commun 167:103
[37] (a) Becke A (1988) Phys Rev A 38:3098; (b) Perdew JP (1986) Phys Rev B 33:8822; (c) Perdew JP (1986) Phys Rev B 34:7406
[38] Frisch MJ, Trucks CW, Schlegel HB, Scuseria GE, Robb MA, Cheeseman JR, Montgomery JA Jr, Vreven T, Kudin KN, Burant JC, Millam JM, Iyengar SS, Tomasi J, Barone V, Mennucci B, Cossi M, Scalmani G, Rega N, Petersson GA, Nakatsuji H, Hada M, Ehara M, Toyota K, Fukuda R, Hasegawa J, Ishida M, Nakajima T, Honda Y, Kitao O, Nakai H, Klene M, Li X, Knox JE, Hratchian HP, Cross JB, Adamo C, Jaramillo J, Gomperts R, Stratmann RE, Yazyev O, Austin AJ, Cammi R, Pomelli C, Ochterski JW, Ayala PY, Morokuma K, Voth GA, Salvador P, Dannenberg JJ, Zakrzewski VG, Dapprich S, Daniels AD, Strain MC, Farkas O, Malick DK, Rabuck AD, Raghavachari K, Foresman JB, Ortiz JV, Cui Q, Baboul AG, Clifford S, Cioslowski J, Stefanov BB, Liu G, Liashenko A, Piskorz P, Komaromi I, Martin RL, Fox DJ, Keith T, Al-Laham MA, Peng CY, Nanayakkara A, Challacombe M, Gill PMW, Johnson B, Chen W, Wong MW, Gonzalez C, Pople JA (2003) Gaussian 03, B1. Gaussian Inc, Pittsburgh PA
[39] Laio A, Parrinello M (2002) Proc Natl Acad Sci USA 99:12562
[40] Micheletti C, Laio A, Parrinello M (2004) Phys Rev Lett 92:170601
[41] Gervasio FL, Parrinello M, Ceccarelli M, Klein ML (2006) J Mol Biol 390:1212
[42] Lee JG, Asciutto E, Babin V, Sagui C, Darden T, Roland C (2006) J Phys Chem B 110:2325
[43] Gervasio FL, Laio A, Parrinello M (2005) J Am Chem Soc 127:2600
[44] Michel C, Laio A, Mohamed F, Krack M, Parrinello M, Milet A (2007) Organometallics 26:1241

[45] Lohrenz JCW, Woo TK, Ziegler T (1995) J Am Chem Soc 117:12793
[46] Deng L, Woo TK, Cavallo L, Margl PM, Ziegler T (1997) J Am Chem Soc 119:6177
[47] Lauher JW, Hoffmann R (1976) J Am Chem Soc 98:1729
[48] Kawamura-Kuribayashi H, Koga N, Morokuma K (1992) J Am Chem Soc 114:8687
[49] Yoshida T, Koga N, Morokuma K (1995) Organometallics 14:746

Mechanism of Gold-Catalyzed Cycloisomerization of Enynyl Esters

Andrea Correa, Luigi Cavallo*

Department of Chemistry, University of Salerno, via Ponte don Melillo, Fisciano, I-84084, Italy

*E-mail: lcavallo@unisa.it

Abstract A theoretical study on the Au-catalyzed cycloisomerization of a branched dienyne possessing an acetate at the propargylic position is presented. The peculiar architecture of the dienyne precursor, exhibiting both a 1,6- and a 1,5-enyne skeleton, leads, in the presence of alkynophilic gold catalysts, to a mixture of bicyclic compounds. DFT calculations are presented that rationalize in full THE manifold of reactions that lead to the different products.

Keywords Homogeneous catalysis · Density functional theory calculations · Gold catalysis · Cycloisomerization · Enynes · Propargylic esters · Allenes

1 Introduction

According to Trost's definition [1], atom economy reactions are of primary importance in modern organic chemistry, and polyunsaturated substrates are ideal substrates because they allow for a number of unprecedented transformations in the presence of late transition metal catalysts [2]. In this peculiar field [3], the cycloisomerization of enynes remains unique because of the increase of molecular complexity achieved in one chemical step [4].

After extensive studies with Pd [5] and Pt [6], Au salts have emerged lately as powerful catalysts for a myriad of transformations involving enynes [7, 8] and enyne compounds possessing an ester group at the propargylic position are a particular class of substrates due to their peculiar reactivity pattern. In fact, an ester group adequately placed will perform an internal 1,2- or 1,3-shift upon electrophilic activation of the C≡C bond (Figure 1, **I**), leading to rearranged products **II** and **III**, which can then further evolve as a function of the remaining pendant groups (Figure 1) [9].

The wide diversity- and complexity- of accessible structures from readily assembled precursors is probably the most appealing feature of the cycloisomerization of enynes. But it can as well be seen as a disadvantage since minute

V. Dragutan et al. (eds.), *Green Metathesis Chemistry: Great Challenges in Synthesis, Catalysis and Nanotechnology*,

changes in the starting compound can lead to unprecedented outcomes, rendering the reactivity of a specific substrate often unpredictable. For example, the reactivity of dienyne **1**, which formally exhibits a 1,5- and a 1,6-enyne scaffold, was recently investigated experimentally, and different selectivity as a function of the nature of the catalyst employed was observed (Figure 2) [10].

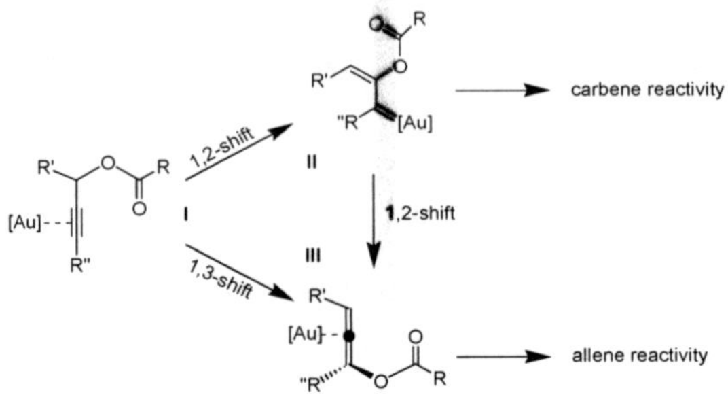

Figure 1 Reactivity of propargylic acetates in the presence of gold

Differently from Pt-based catalysts, which mainly yield the bicyclo[4.1.0]heptene **2** over the bicyclco[3.1.0]hexene **3**, Au-based catalysts of the type $[(NHC)\text{-}Au]^+$ (NHC = N-heterocyclic carbene) [11, 12] afforded more contrasted ratios. More interestingly, the use of Au^I salts instead of Pt^{II} ones led to the formation of an unexpected bicyclo[3.1.0]hexene compound **4**.

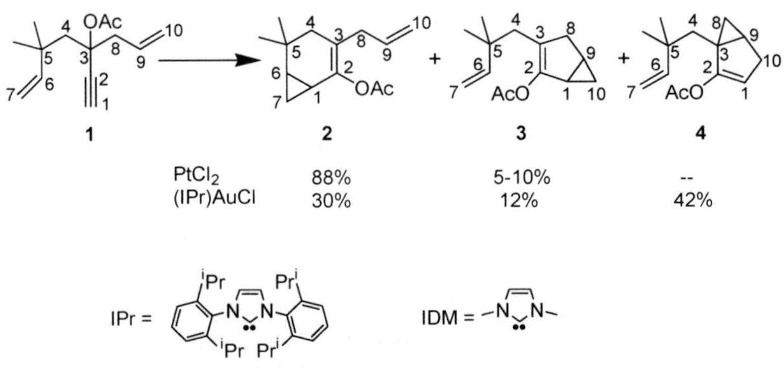

Figure 2 Pt- and Au-catalyzed cycloisomerization of **1**

Intrigued by this unprecedented mode of cyclization, we decided to engage in a detailed theoretical examination of the reaction parameters that could influence the formation of this type of carbocycles. Herein, we disclose the results of these studies, which have allowed for the proposal of a new type of reactivity in the enynyl ester series. Hence, we have shown that the most likely pathway to **4** involves, after formation of an allenyl ester, carbocyclization of an allenene core followed by "retro-migration" of the ester moiety. The process of "retro-migration" was further studied on an extended set of substrates. We also focused on the reversibility of the OAc-migration between the enyne esters and their corresponding [3,3] rearranged counterparts, namely the allenyl esters.

2 Computational Details

All the Density Functional Theory (DFT) calculations were performed using the Gaussian03 package [13]. The BP86 GGA functional of Becke and Perdew was used [14]. The TZVP triple-ζ basis set with one polarization function was used for main group atoms [15], while the relativistic SDD effective core potential in combination with a triple-ζ basis set was used for the Au atom [16]. All geometries were verified by frequency calculations that resulted in 0 and 1 imaginary frequency for intermediates and transition states, respectively. The reported energies include the vibrational gas-phase zero-point energy term, and a solvation term that was obtained through single-point calculations on the gas-phase optimized geometries. The polarizable continuous solvation model IEF-PCM as implemented in the Gaussian03 package has been used [17]. CH_2Cl_2 was chosen as model solvent, with a dielectric constant $\varepsilon = 8.93$. Standard non-electrostatic terms were also included.

3 Results and Discussion

The numbering of atoms in the starting dienyne **1** and in the products **2**, **3** and **4** is shown in Figure 2. For these species, we use labels such as **1Au**, **3Au** and so on, to indicate their coordination to the Au catalyst. For all the structures connecting the reactant and the various products, we do not add the Au label, since coordination to the metal is assumed. Considering the great number of structures we had to calculate, we had to reduce the NHC ligand to the simple 1,3-dimethylimidazol-2-ylidene, see Figure 2.

The catalytic cycle starts with displacement of a BF_4^- counterion from the $[(IDM)Au^+][BF_4^-]$ species by the alkyne group of **1**, to furnish intermediate **1Au** [18] In CH_2Cl_2 displacement of the BF_4^- counterion by the substrate is exergonic by 26 kJ/mol. Structure **1Au** is then the branching point for three different reaction paths (Figure 2).

The first consists of a nucleophilic attack of the C6–C7 double bond to the C1 atom [19], see Figure 1. This cyclopropanation step leads to intermediate **5**, which already presents the bicyclic skeleton of product **2**. The Au-carbene intermediate **5** is 60 kJ/mol lower in energy than the starting alkyne coordinated intermediate **1Au**, and is reached through transition state **1Au-5**, with a rather low energy barrier of 32 kJ/mol (Figure 3).

Formation of **2** from **5** involves a 1,2-shift of the ester group from C3 to C2. After a first step leading to the formation of intermediate **6**, through transition state **5–6**, exhibiting a barrier of 25 kJ/mol, the 1,2-shift is then completed via transition state **6-2Au**, with a barrier of 13 kJ/mol, and finally affords the coordinated product **2Au**. Both the intermediate **6** and **2Au** are lower in energy than intermediate **1Au** (by 82 and 147 kJ/mol, respectively) and thus formation of **2Au** is a substantially downhill path from the starting alkyne coordinated species **1Au** to the product coordinated species **2Au**. Product release from **2Au** assisted by a coordinating BF_4^- counterion is endergonic by 6 kJ/mol, leads to product **2** and closes the catalytic cycle by forming the starting $[(IDM)Au^+][BF_4^-]$ species **0**.

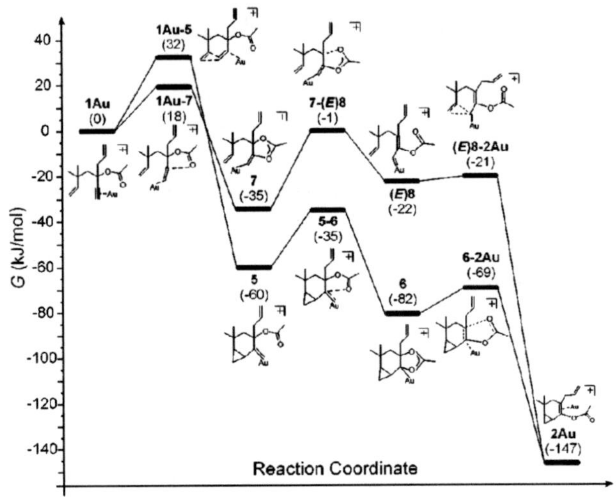

Figure 3 Energy profiles leading to **2**

However, product **2** can be reached from **1Au** through an alternative reaction pathway. Indeed, as mentioned before, in addition to the cyclopropanation *then* migration sequence that was just examined, the migration *then* cyclopropanation sequence has to be considered. Hence, starting from the alkyne coordinated species **1Au**, a 1,2-shift of the ester group corresponding to attack of the carboxylic O atom to C2 leads to intermediate **7** through transition state **1Au-7**, with an energy barrier of 18 kJ/mol (Figure 3). Intermediate 7, which is 35 kJ/mol lower in energy than **1Au**, is another branching point in this complex reaction manifold. The branch that leads to **2** involves the breaking of the C3–O bond to furnish intermediate **(*E*)8**, with a barrier of 34 kJ/mol, and a subsequent cyclopropanation step, corresponding to attack of the C6–C7 double bond to C1, through transition state **(*E*)8-1Au**. A last step with an almost negligible energy barrier of only 1 kJ/mol leads to the product coordinated species **2Au**, from which **2** is released, closing the catalytic cycle. Intermediate **(*E*)8** is 22 kJ/mol lower in energy than **1Au**. In the framework of the Curtin–Hammett principle, the actual pathway that is followed to reach **2** is determined by the energy difference between the transition states of highest energy along the two alternative pathways, that is transition states **1Au-5** and **1Au-7**. According to our calculations, transition state **1Au-7** is lower in energy that **1Au-5** by 14 kJ/mol, which suggests that the main reaction channel leading to **2** involves the 1,2-shift of the ester group first and then a cyclopropanation step, although the alternative path corresponding to the cyclopropanation *then* 1,2-shift sequence is competitive.

Next, we focused on the formation of **3**, which can also be reached through two alternative pathways (cyclopropanation *then* 1,2-shift or the reverse sequence) that are very similar to those just described to rationalize the formation of **2**. The first path involves again the nucleophilic attack of a C–C double bond to C1 but, in this case, it is the C9–C10 double bond that we have to consider (Figure 4). This cyclopropanation step leads to intermediate **9**, which already presents the bicyclic skeleton of product **3**. The Au-carbene intermediate **9** is 26 kJ/mol lower in energy than the starting alkyne coordinated intermediate **1Au**, and is obtained via transition state **1Au-9**, with the rather low energy barrier of 36 kJ/mol.

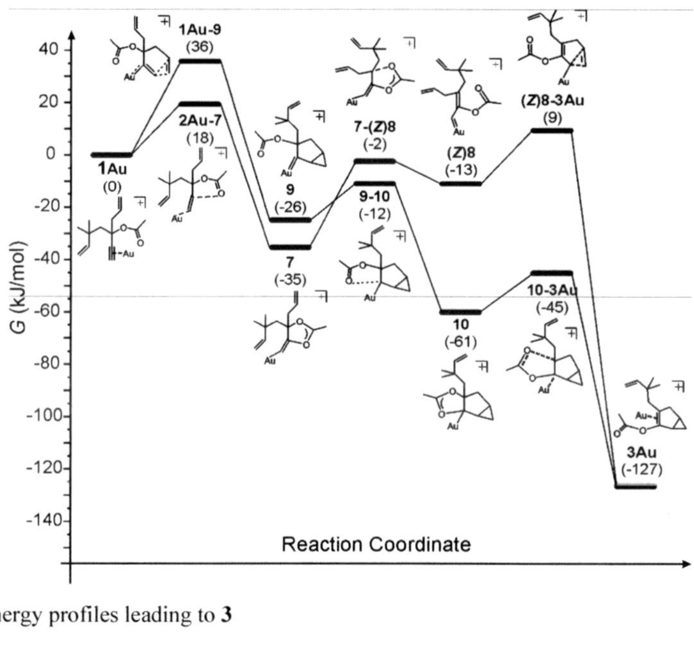

Figure 4 Energy profiles leading to **3**

Transition state **1Au-9** is slightly higher in energy than the similar transition state **1Au-5** because of the higher steric strain associated with the formation of a more strained 5-membered ring in **1Au-9**, compared to formation of a 6-membered ring in **1Au-5**. Formation of **3** from **9** involves then the 1,2-shift of the ester group from C3 to C2. The first step of this 1,2-shift is the formation of intermediate **10**, through transition state **9-10**, with a barrier of 14 kJ/mol. Final formation of the coordinated product **3Au** occurs through transition state **10-3Au** with an energy barrier of 16 kJ/mol. Both intermediates **10** and **3Au** are lower in energy than intermediate **1Au** (by 61 and 127 kJ/mol, respectively) and thus formation of **3Au**, similarly to that of **2Au**, is a substantially downhill path from the starting alkyne coordinated reactant **1Au** to the product coordinated structure **3Au** (Figure 4). Product release from **3Au** assisted by a coordinating BF_4^- counterion is endergonic by 13 kJ/mol, leads to product **3** and closes the catalytic cycle by forming the starting $[(IDM)Au^+][BF_4^-]$ species **0**.

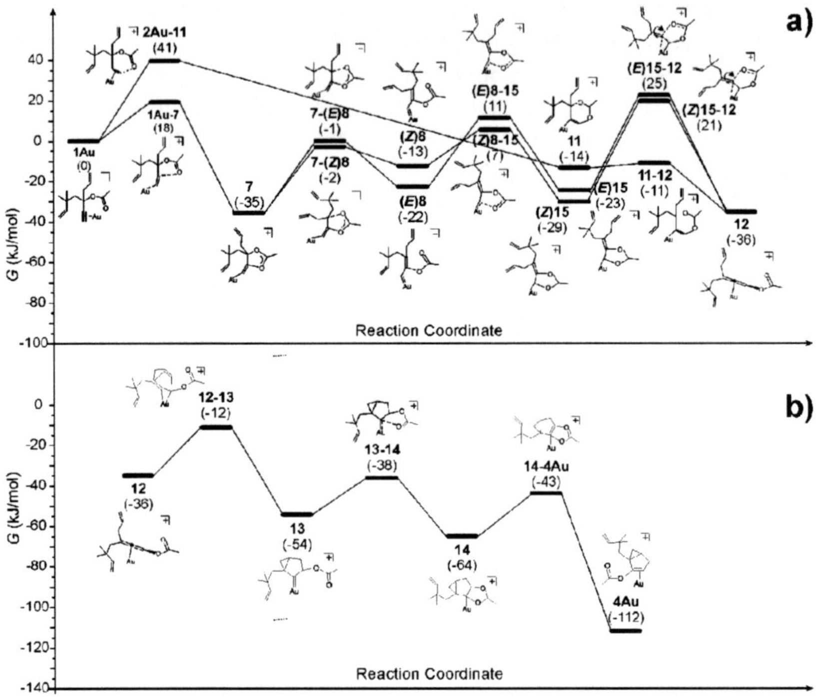

Figure 5 Energy profiles leading to **4**. Part a, from **1Au** to **12**. Part b, from **12** to **4Au**

As in the case of **2**, **3** can be reached from **1Au** through the alternative path that involves initially the 1,2-shift of the ester group. As discussed above, the branching point is intermediate **7** (see Figure 4). In fact, breaking of the C3–O13 bond can alternatively lead to intermediate ***(Z)*8**, with a barrier of 33 kJ/mol, which, upon attack of the C11–C12 double bond to C1, leads to the product coordinated species **3Au** via transition state ***(Z)*8-3Au**, with an energy barrier of 22 kJ/mol. Bicyclo [3.1.0]hexene **3** is then released from **3Au** by coordination of a BF_4^- anion and the catalytic cycle closed. Intermediate ***(Z)*8** is 13 kJ/mol lower in energy than **1Au**, and it is less stable than the ***(E)*8** isomer by 9 kJ/mol. Considering that transition state **1Au-7** is 18 kJ/mol lower in energy than transition state **1Au-9**, we believe that formation of **3** proceeds to a great extent through a 1,2-shift followed by a cyclopropanation step.

While the possible mechanistic pathways explaining the formation of **2** and **3** from **1** were already proposed in the literature [10, 21–23], the formation of **4** had only little mechanistic explanation prior to this study [10, 20]. We anticipate that **4** could be formed following, at least, three different reaction paths. In all cases the allene coordinated intermediate **12** is the key intermediate in order to rationalize the formation of **4** (see Figure 5). The simplest pathway involves a 1,3-shift of the ester group from **1Au**, leading to intermediate **11** through transition state **1Au-11**, with an energy barrier of 41 kJ/mol (see Figure 5). Intermediate **11**, which is 14 kJ/mol lower in energy than **1Au**, then evolves, via transition state **11-12** and the almost negligible energy barrier of 3 kJ/mol, to the allene coordinated species **11**, which is 36 kJ/mol lower in energy than **1Au**.

We remark here that several geometries, each rather similar in energy, can be adopted by **12**. For the sake of simplicity, in all reaction pathways presented, the most stable isomer of **12** is discussed. The allene coordinated species **12** can evolve towards the product coordinated species **4Au** through simultaneous attack of the C9–C10 double bond to C1 and C3 to form the bicyclo[3.1.0]hexene skeleton of **4** in a single step, leading to intermediate **13** through transition state **12-13** and an energy barrier of 24 kJ/mol.

Finally, a 1,2-shift of the ester group of **13**, through intermediate **14** and transition states **13-14** and **14-4Au**, with energy barriers of 16 and 21 kJ/mol respectively, leads to the Au coordinated product **4Au**. Product release from **4Au** assisted by a coordinating BF_4^- counterion is endergonic by 14 kJ/mol, leads to product **4** and closes the catalytic cycle by forming the starting $[(IDM)Au^+][BF_4^-]$ species. Incidentally, it should be noted that the BF_4^- counterion can also displace the allene from **12**, yielding the starting $[(IDM)Au^+][BF_4^-]$ species, and releasing the allene in the reaction media.

Formation of **4** can be also explained by two other pathways that branch from intermediates ***(E)*8** and ***(Z)*8**; intermediates that we already introduced to rationalize the formation of **2** and **3** respectively. We first discuss branching from intermediate ***(E)*8**. Instead of the cyclopropanation step corresponding to nucleophilic attack of the C6–C7 double bond to C1, ***(E)*8** can undergo a 1,2-shift of the ester group, leading initially to intermediate ***(E)*15**, via transition state ***(E)*8-*(E)*15** with an energy barrier of 33 kJ/mol, and then to the Au-allene species **12** through transition state ***(E)*15-12** and an energy barrier of 48 kJ/mol. Once intermediate **12** has been reached, the reaction can evolve to **4** as described before (Figure 5). However, this branching is quite unlikely considering that the 1,2-shift transition state ***(E)*8-*(E)*15** has to compete with the cyclopropanation transition state ***(E)*8-2Au**, which is 32 kJ/mol lower in energy. Considering the level of accuracy of this kind of calculations, we can state that once intermediate ***(E)*8** is reached it cannot evolve to intermediate **12**.

We now discuss branching from intermediate **(*Z*)8**. Instead of the cyclopropanation step corresponding to nucleophilic attack of the C11–C12 double bond to C1, **(*Z*)8**, similarly to **(*E*)8**, can undergo a 1,2-shift of the ester group leading initially to intermediate **(*Z*)15**, via transition state **(*Z*)8-(*Z*)15** with an energy barrier of 20 kJ/mol, and then to the Au-allene species **12** through transition state **(*Z*)15-12** and an energy barrier of 50 kJ/mol. Once intermediate **12** has been reached, the reaction can evolve to **4** as described before. In this case, we found that the 1,2-shift transition state **(*Z*)8-(*Z*)15** has to compete with the cyclopropanation transition state **(*Z*)8-3Au**, which is only 2 kJ/mol higher in energy. This implies that branching from **(*Z*)8** along both reaction pathways is a very likely event.

Our calculations clearly indicate that this catalysis is characterized by numerous highly competitive reaction pathways. As a warning, we cannot exclude that there are other reaction pathways that we were not able to envisage. Nevertheless, calculations clearly demonstrate that the high reactivity of the starting structure **1Au** is at the origin of this diversity. As we already indicated, the alkyne coordinated species can undergo both 1,2 and 1,3-shift of the ester group, leading respectively to the Au-carbene species **(*E*)8** and **(*Z*)8**, and to the Au-allene species **12**. All these species are connected in a catalytic cycle we labeled as "Golden Carousel". To better understand this point, the most relevant sections of the numerous reaction pathways are shown in Figure 6.

It appears clearly that the alkyne coordinated species **1Au** is the intermediate of highest energy in the cycle, while the reservoir of active species are intermediate **5** (which is not a way off the cycle) and the allene species **12**. Intermediate **5** is easily formed from **1Au**. The Au-carbene species **(*E*)8** and **(*Z*)8** represent the way off the carousel to products **2** and **3**, respectively, while the Au-allene species **12** is the way off to product **4**. Evolution of **(*E*)8**, **(*Z*)8** and **12** depends on the relative energy of the three transition states (two of them correspond to clockwise and counter-clockwise movements in the golden-carousel, the third to a way off from the carousel) that can be reached from each of these intermediates.

Intermediate **(*E*)8** is an excellent way off the carousel, since the transition state that connects intermediate **(*E*)8** and product **2**, **(*E*)8-2Au**, is quite lower in energy (by 20 and 32 kJ/mol, respectively) than transition states **5-(*E*)8** and **(*E*)8-(*E*)15**, which are the transition states to be reached by **(*E*)8** in order to move on the carousel. This explains the easy formation of **2**. Differently, **(*Z*)8** is not a good way off the carousel, since the most likely event is a counter-clockwise move in the carousel, to yield the highly stable intermediate **5**. The two other transition states of highest energy accessible from intermediate **(*Z*)8**, **(*Z*)8-3Au** and **(*Z*)15-12**, correspond respectively to the way off the carousel leading to **3Au** and to a clockwise move towards the allene intermediate **12** (Figure 6). They are 11 and 23 kJ/mol higher in energy than transition state **5-(*Z*)8**, which explains the low amount of products **3** and **4** formed with NHC ligands of low steric bulkiness.

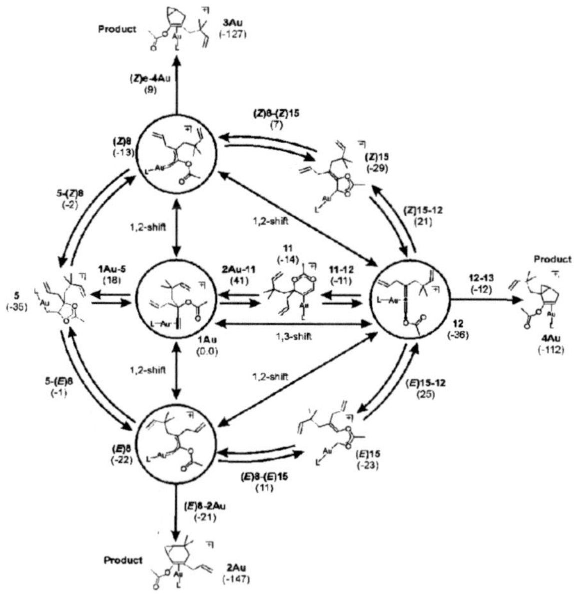

Figure 6 Golden carousel

Finally, **12** is another excellent way off the carousel, since the transition state that connects intermediate **12** and product **4**, **12-13**, is quite lower in energy (by 33 and 37 kJ/mol, respectively) than transition states **12-(*Z*)15** and **12-(*E*)15**, which are the transition states to be reached by **12** in order to move in the carousel. Thus, the relatively small amount of **4** produced with **1Au** as an entry point into the Golden Carousel is explained by the relatively high-energy transition state **(*Z*)15-12**, which does not allow an easy formation of **12** from **(*Z*)8**. On the other hand, the direct 1,3-shift pathway from **1Au** to **12** is blocked by the relatively high-energy transition state **1Au-11**. Finally, the Golden Carousel we proposed clearly explains the high amount of **4** that is formed when the allene **12** species is used as entry point.

Overall, the theoretical results presented here have notably allowed us to rationalize the formation of the unprecedented bicyclo[3.1.0]hexene **4**. Hence, according to our calculations, the cyclization would occur after formation of the allenyl ester, between the allene and the ene part of the 1,4-allenene core [24].

Acknowledgments We thank the Regione Campania Legge 5 for financial support.

References

[1] Trost BM (1991) Science 254:1471–1477

[2] For recent reviews on the chemistry of polyunsaturated substrates, see: (a) Malacria M, Goddard J-P, Fensterbank L (2006) In: Crabtree R, Mingos M, Ojima I (eds.) Comprehensive organometallic chemistry III, Vol. 10, Chap. 7, Pergamon, Oxford, England, pp. 299–368; (b) Aubert C, Fensterbank L, Gandon V, Malacria M (2006) Top Organomet Chem 19:259–294

[3] (a) Trost BM (1995) Angew Chem Int Ed 34:259–281; (b) Trost BM (2002) Acc Chem Res 35:695–705

[4] For reviews on enyne cycloisomerization, see: (a) Nieto-Oberhuber C, López S, Jiménez-Núñez E, Echavarren AM (2006) Chem Eur J 12:5916–5923; (b) Zhang Z, Zhu G, Tong X, Wang F, Xie X, Wang J, Jiang L (2006) Curr Org Chem 10:1457–1478; (c) Bruneau C (2005) Angew Chem Int Ed 44:2328–2334; (d) Diver ST, Giessert AJ (2004) Chem Rev 104:1317–1382; (e) Echavarren AM, Nevado C (2004) Chem Soc Rev 33:431–436; (f) Echavarren AM, Méndez M, Muñoz MP, Nevado C, Martín-Matute B, Nieto-Oberhuber C, Cárdenas DJ (2004) Pure Appl Chem 76:453–463; (g) Lloyd-Jones G (2003) Org Biomol Chem 1:215–236; (h) Aubert C, Buisine O, Malacria M (2002) Chem Rev 102:813–834

[5] For a review, see: Trost BM (1990) Acc Chem Res 13:385–393

[6] For reviews, see: (a) Añorbe L, Domínguez G, Pérez-Castells J (2004) Chem Eur J 10:4938–4943; (b) Méndez M, Mamane V, Fürstner A (2003) Chemtracts 16:397–425

[7] For recent general reviews on homogeneous Au-catalysis, see: (a) Hashmi ASK (2007) Chem Rev 107:3180–3211; (b) Fürstner A, Davies PW (2007) Angew Chem Int Ed 46:3410–3449; (c) Gorin DJ, Toste FD (2007) Nature 446:395–403; (d) Patil NT, Yamamoto Y (2007) ARKIVOC:6–19; (e) Jiménez-Núñez E, Echavarren AM (2007) Chem Commun :333–346; (f) Hashmi ASK, Hutchings GJ (2006) Angew Chem Int Ed 45:7896–7936; (g) Hoffmann-Röder A, Krause N (2005) Org Biomol Chem 3:387–391; (h) Arcadi A, Di Giuseppe S (2004) Curr Org Chem 8:795–812

[8] For reviews focused on enynes in Au-catalysis, see: (a) Zhang L, Sun J, Kozmin SA (2006) Adv Synth Catal 348:2271–2296.; (b) Ma S, Yu S, Gu Z (2006) Angew Chem Int Ed 45:200–203

[9] For reviews on propargylic esters in Pt- and Au-catalysis, see: (a) Marion N, Nolan SP (2007) Angew Chem Int Ed 46:2750–2752; (b) Marco-Contelles J, Soriano E (2007) Chem Eur J 13:1350–1357

[10] Marion N, de Frémont P, Lemière G, Stevens ED, Fensterbank L, Malacria M, Nolan SP (2006) Chem Commun :2048–2050

[11] For a review focused on N-heterocyclic carbenes in gold catalysis, see: Marion N, Nolan SP (2008) Chem Soc Rev 37:1776–1782

[12] For general reviews on N-heterocyclic carbenes, see: (a) Glorius F (ed.) (2007) N-heterocyclic carbenes in transition metal catalysis (Top Organomet Chem Vol. 28). Springer, Berlin/Heidelberg, Germany; (b) Nolan SP (ed.) (2006) N-heterocyclic carbenes in synthesis. Wiley-VCH, New York; (c) Bourissou D, Guerret O, Gabbai FP, Bertrand G (2000) Chem Rev 100:39–92

[13] Gaussian 03 (2003). Gaussian Inc, Pittsburgh, PA

[14] (a) Becke AD (1988) Phys Rev A 38:3098–3100; (b) Perdew JP (1986) Phys Rev B 33:8822–8824; (c) Perdew JP (1986) Phys Rev B 34:7406

[15] (a) Schaefer A, Horn H, Ahlrichs R (1992) J Chem Phys 97:2571–2577; (b) Schaefer A, Huber C, Ahlrichs R (1994) J Chem Phys 100:5829–5835

[16] (a) Haeusermann U, Dolg M, Stoll H, Preuss H (1993) Mol Phys 78:1211–1224; (b) Kuechle W, Dolg M, Stoll H, Preuss H (1994) J Chem Phys 100:7535; (c) Leininger T, Nicklass A, Stoll H, Dolg M, Schwerdtfeger P (1996) J Chem Phys 105:1052–1059

[17] (a) Cossi M, Barone V, Cammi R, Tomasi J (1996) Chem Phys Lett 255:327–335; (b) Cancès MT, Mennucci B, Tomasi J (1997) J Chem Phys 107:3032–3041; (c) Cossi M, Barone V, Mennucci B, Tomasi J (1998) Chem Phys Lett 286:253–260

[18] For references on isolated and structurally characterized [(η2-RC≡CR)AuI] complexes, see: (a) Schulte P, Behrens U (1998) Chem Commun :1633–1634; (b) Shapiro ND, Toste FD (2008) Proc Natl Acad Sci USA 105:2779–2782

[19] Even though 5-*exo*-dig cyclizations are more common for 1,6-enynes, there are numerous precedents of 6-*endo*-dig cyclizations in the literature, see: (a) Nevado C, Cárdenas DJ, Echavarren AM (2003) Chem–Eur J 9:2627–2635; (b) Nieto-Oberhuber C, Muñoz M P, Buñuel E, Nevado C, Cárdenas DJ, Echavarren AM (2004) Angew Chem Int Ed 43:2402–2406; (c) Lee SI, Kim SM, Choi MR, Chung YK (2006) J Org Chem 71:9366–9372; (d) Ferrer C, Raducan M, Nevado C, Claverie CK, Echavarren AM (2007) Tetrahedron 63:6306–6316; (e) Kim SM, Park JH, Choi SY, Chung YK (2007) Angew Chem Int Ed 46:6172–6175

[20] In a review on Au-catalyzed transformations, Echavarren proposed, for the transformation **2** → **5**, an alternative and interesting cyclopropanation/1,2-OAc shift/cationic rearrangement sequence, see Ref. [7e]

[21] Mainetti E, Mouriès V, Fensterbank L, Malacria M, Marco-Contelles J (2002) Angew Chem Int Ed 41:2132–2135

[22] Johansson M, Gorin DJ, Staben ST, Toste FD (2005) J Am Chem Soc 127:18002–18003

[23] (a) Fürstner A, Hannen P (2006) Chem Eur J 12:3006–3019; (b) Fehr C, Galindo J (2006) Angew Chem Int Ed 45:2901–2904. For a theoretical approach, see: (c) Soriano E, Marco-Contelles J (2007) J Org Chem 72:2651–2654

[24] Au-Catalyzed cyclization of 1,4-allenenyl esters has been described previously on related systems but no migration of the ester was observed: (a) Buzas A, Gagosz F (2006) J Am Chem Soc 128:12614–12615. For other reports on 1,4-allenenes cyclization in the presence of gold catalysts, see: (b) Huang X, Zhang L (2007) J Am Chem Soc 129:6398–6399; (c) Huang X, Zhang L (2007) Org Lett 9:4627–4630

Carbonyl-Olefin Exchange Reaction and Related Chemistry

Christo Jossifov,* **Radostina Kalinova**

Institute of Polymers, Bulgarian Academy of Sciences Acad.G Bonchev 103A, 1113 Sofia, Bulgaria

*Fax: (+359)2870-03-09; e-mail: jossifov@polymer.bas.bg

Abstract A new carbon–carbon double bond forming reaction (carbonyl olefin exchange reaction) mediated by transition metal catalytic systems has been discovered. The catalytic systems used (transition metal halides or oxohalides alone or in combination with Lewis acids) are active only in the case when the two reacting groups are in one molecules and are conjugated. In addition these systems accelerate other reactions which run simultaneously with the carbonyl olefin metathesis rendering a detailed investigation of the process very complicated.

Keywords Carbonyl–olefin exchange reaction · Olefin metathesis · Polymerization · Reductive coupling

1 Introduction

Synthesis is a discipline that is a central to all areas of chemistry. It encompasses the unique ability of chemists to develop new reactions and to design molecules or molecular systems with a desired (or anticipated) set of properties – be they enzyme inhibitors, receptor agonists, fluorescent dyes, transition metal catalysts, molecular devices, nanotubes, modified surfaces, solid-state compositions, or novel polymers. There is an incredible advance in the development of new synthetic methods and new strategies for synthesis of structurally and stereochemically complex molecules during the last century. Even a cursory examination of the workhorse bond forming reactions in use today – ranging from the multitude palladium(0)- and nickel(0)-catalyzed coupling reactions, to ruthenium-mediated olefin metathesis and other transition metal mediated C–C bond-forming reactions, to C–H activation, to the many important end newly emerging methods for asymmetric synthesis and asymmetric catalysis, among many – reveals that most were not available to the practicing organic chemists even 2 decades ago. In essence, efforts to apply new methods to complex molecules provide a Darwinian selection pressure that insures that the methodology platforms available to the bench chemist continue to evolve in a highly productive direction. Every new

V. Dragutan et al. (eds.), *Green Metathesis Chemistry: Great Challenges in Synthesis, Catalysis and Nanotechnology*,

reaction has a strong impact in the field of organic chemistry and catalysis, both in academia as well in industry.

The present paper deals with a new carbon–carbon double bond forming reaction. Information about the new reaction, the related chemistry, and some arguments pro and con the validity of the carbene mechanism of the new reaction are presented.

2 Carbonyl–Olefin Exchange Reaction and Related Chemistry

The carbonyl olefin metathesis (COM) is a new reaction discovered in the Institute of Polymers, Bulgarian Academy of Sciences. It was proposed for the first time to account for the polymerization of α,β-unsaturated carbonyl compounds in the presence of WCl_6 [1]. It was shown indeed that treatment of the α,ß-unsaturated carbonyl compound 1,3-diphenylprop-2-en-1-one (chalcone), with WCl_6 resulted in the formation of polyphenylacetylene and benzaldehyde. Using the same procedure several differently substituted α,ß-unsaturated carbonyl compounds were polymerized affording substituted polyacetylenes (polyphenylacetylene, polydiphenylacetylene, polymethylacetylene, polycamphor, etc.) [2, 3]. The general scheme of these transformations can be represented as follows (Scheme 1). The single C–C bond can be part of a cycle.

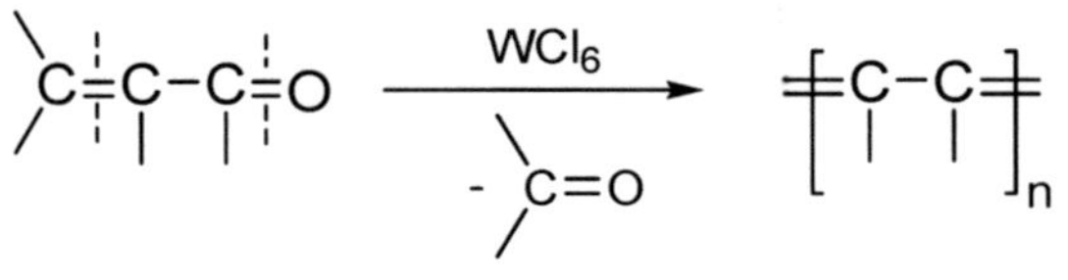

Scheme 1 Polymerization of α,ß-unsaturated carbonyl compounds – general scheme

The chain propagation step of the carbonyl olefin metathesis polymerization is described in Scheme 2.

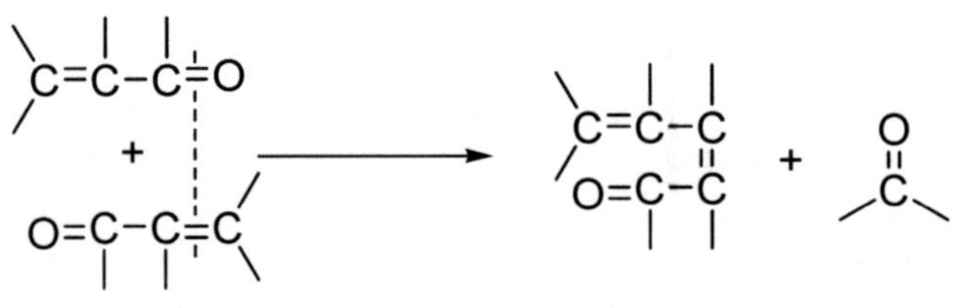

Scheme 2 Polymerization of α,ß-unsaturated carbonyl compounds – chain propagation

This process has been denoted as carbonyl olefin exchange reaction. This name has been used until the hypothesis about the carbene mechanism has been launched [4].

The dimer thus obtained possesses carbonyl and olefin end-groups and is able to undergo further similar transformations resulting into chain propagation. Every step is accompanied by the evolution of a low-molecular weight product. The dimer formation was proved by comparing the gas phase chromatogram of the reaction mixture to that of an authentic sample prepared separately [3, 5]. The dimer polymerizes under the influence of WCl_6 giving again substituted polyacetylene and carbonyl compound. All these facts show that the polymerization of α,ß-unsaturated carbonyl compounds is a step process and a true polycondensation reaction with the formation of the conjugated system as the likely driving force.

The results of further investigations of the polymerization of α,ß-unsaturated carbonyl compounds can be described shortly as follows [6]:

(a) The reaction can be carry out either without solvents or in aromatic solvents. Better results were obtained in chlorobenzene.
(b) The products are oligomers. Their yield and molecular weight rise with the quantity of WCl_6, the temperature and the reaction time. The molar ratio monomer/WCl_6 varies from 2 to 1. Good results (high yield [90% of the theoretical value] and molecular weight of about several thousands) are obtained when this ratio is 1. It should be mentioned that with an equimolar amount of WCl_6 the reaction took place even at ambient temperature.
(c) During the reaction the transition metal changes its degree of oxidation.
(d) The more substituted the propenone is, the higher is the polymer yield.
(e) The polyacetylenes are obtained as complexes containing up to 40 mol% WCl_6, which is in agreement with the fact that several conjugated polymers can be partially oxidized ("doped") with WCl_6. In order to obtain pure polyacetylene the polymer thus obtained has to be treated with concentrated sodium hydroxide.
(f) Some of the low molecular weight carbonyl compounds (benzaldehyde, acetophenone), formed in the main reactions, interact with WCl_6 evolving HCl. Others (benzophenone) do not react with WCl_6. In the latter cases higher yields of polymers (up to 100%) can be achieved using lower amounts of WCl_6 (WCl_6/monomer $<1:1$).

The polymerization of α,ß-unsaturated carbonyl compounds is very similar to the Friedel-Crafts acylation of benzene, where the catalyst forms a complex with the reaction products, necessitating thereby to use WCl_6 in an equimolecular amount. In order to avoid the consumption of WCl_6 due to polyacetylene doping carbonyl olefin metathesis starting from highly substituted olefins (tetraphenylethylene, stilbene, 1,1-diphenylethylene) and simple carbonyl compounds (benzophenone, terephthalaldehyde) has been attempted. In these cases the two functional groups are in different molecules and formation of new olefins and carbonyl compounds is expected instead of a conjugated polymer. These attempts were unsatisfactory [7], most probably because WCl_6 is not active enough for carbonyl olefin metathesis in general. It is active only when the two functional groups are in one molecule and are conjugated.

WCl_6 is a representative of the classical catalytic systems for olefin metathesis (OM) [8]. Because of this and because of the formal similarity between COM and OM we looked among the above mentioned systems for a more effective catalyst for COM. We were not able to promote COM when the two functional groups are not conjugated, but we have found that some of them are active in the polymerization of α,ß-unsaturated carbonyl compounds [9]. The best results are obtained when so called Friedel-Crafts type metathesis type catalytic systems (transition metal halides or oxohalides + Lewis acids) are used. Performing these investigations we ran into very striking unexpected results: Using $WOCl_4$ (alone or in combination with $AlCl_3$ as co-catalyst) as a catalyst for polymerization of 1,2,3,3-tetraphenylprop-2-en-1-one, the yield of polydiphenylacetylene was higher than the expected one with regard to the stoichiometry presented in Scheme 2 (Table 1) [10].

Table 1 Experimental results from the polymerization of 1,2,3,3-tetraphenylprop-2-en-1-one mediated by $WOCl_4$+$AlCl_3$

Experiments[a]	1	2	3
$WOCl_4$ in g	0.66	0.60	0.016
$AlCl_3$ in g	–	0.28	0.10
Polymer yield in g	0.94	0.86	0.95

[a]Reaction conditions: monomer 1,2,3,3-tetraphenylprop-2-en-1-one (1 g); solvent: chlorobenzene; time: 8 h; temperature: 90°C; theoretical yield of oligomeric and polymeric products according to Scheme 3 – 0.49 g

When $AlCl_3$ is used as co-catalyst, the polymer has more pronounced carbonyl IR absorption than that of the polymer obtained with $WOCl_4$ alone, both polymers having almost the same molecular weight. This phenomenon could be explained assuming that the polymer molecules possess different end groups. Traces of molecular oxygen, benzophenone, tetraphenylethylene, and benzyl were identified among the reaction products [10]. Quantitative transformation of the monomer can be achieved even with catalytic amounts of $WOCl_4$ (Table 1, Scheme 3). All products of the reaction are shown in Scheme 3 [4].

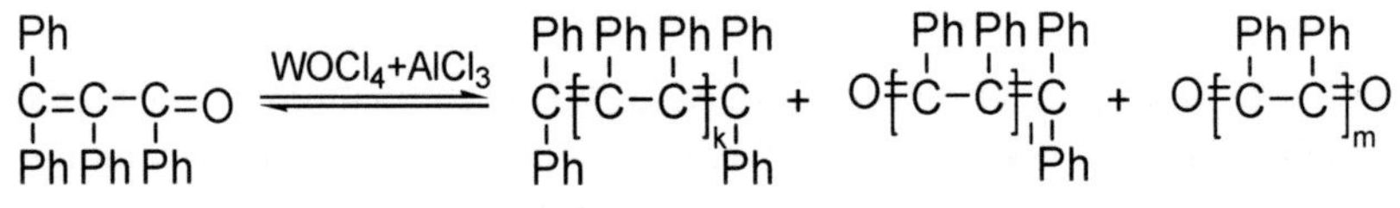

A B C

k,l,m=0,1,2...

A(k=0) - tetraphenylethylene
A(l=0) - benzophenone
A(m=1) - benzil
A(m=0) - molecular oxygen

Scheme 3 Polymerization of 1,2,3,3-tetraphenylprop-2-en-1-one

Obviously several double bond formation reactions have taken place simultaneously and these reactions should be catalytic ones. We presume these are: carbonyl olefin metathesis (Scheme 4), olefin metathesis (Scheme 5), and reductive coupling (RC) of carbonyl compounds, accompanied with molecular oxygen evolution (Scheme 6).

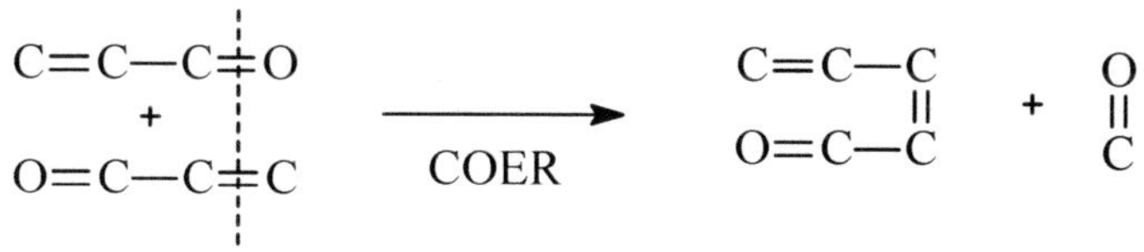

Scheme 4 Carbonyl olefin exchange reaction (carbonyl olefin metathesis)

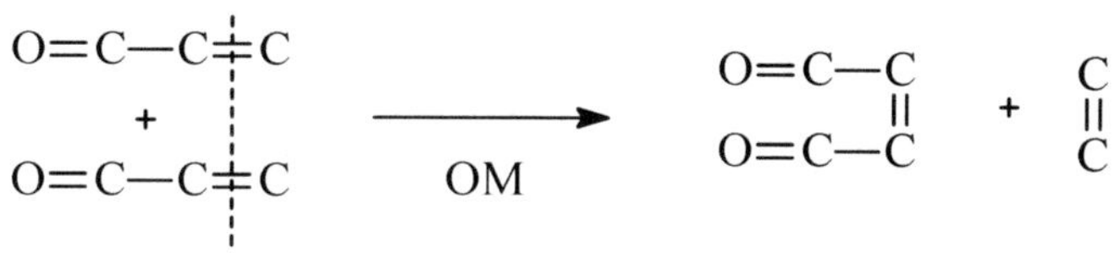

Scheme 5 Olefin metathesis

Scheme 6 Reductive coupling

In the case of 1,2,3,3-tetraphenylprop-2-en-1-one polymerization the new double bonds are always the same (polydiphenylacetylene repeating unit) regardless of the type of the reaction. The low molecular weight products however are different– carbonyl compound, olefin and molecular oxygen, respectively. The only polymer formation reaction is the COM (the dimer and the oligomer products possess carbonyl and olefin end-groups). The other two reactions either produce dimers or double the molecular weights of the oligomers obtained via COM. The higher yield than stoichiometry of COM could be explained by the participation of all double bonds (not only the end-ones) and the low molecular-weight products in the discussed reactions, i.e. by accepting these reactions are reversible. An example explaining the benzil formation is given in Scheme 7.

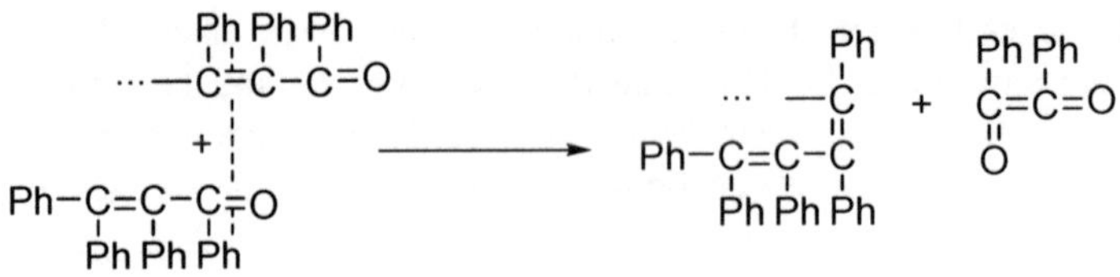

Scheme 7 Benzil formation

If this explanation is correct, one can suggest that the above mentioned reactions could run separately. Proving this assumption, we have succeeded to carry out reductive coupling polymerization of the conjugated dicarbonyl compound benzil under the influence of the Friedel-Crafts type metathesis catalytic system WCl_6 + $AlCl_3$ [11]. In this case the quantity of the transition metal compound is less than the quantity of the Ti reagent used for Mc Murry reaction. The very new moment here is the absence of reducing agent. The result is polydiphenylacetylene with carbonyl end-groups.

If this reaction proceeds in the presence of tetraphenylethylene the polymer obtained has olefin end groups, which is a result of COM between the carbonyl end groups and the olefin tetraphenylethylene (Scheme 8).

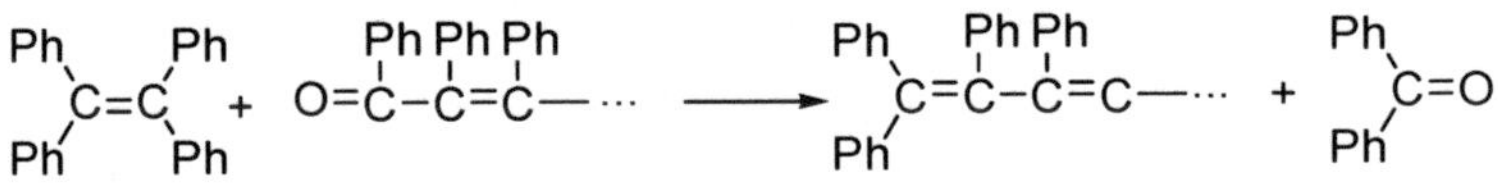

Scheme 8 COM between the carbonyl end groups and the olefin tetraphenylethylene

Thus we succeeded for the first time to perform COM when the two functional groups (olefin and carbonyl) are in two different molecules [11, 12]. The carbonyl group in this case however is again conjugated. If the double bond of the olefin is included into a cycle the cycloolefins could copolymerize with conjugated dicarbonyl compounds. We succeed indeed for the first time in copolymerizing benzyl, terephthalaldehyde, and benzoquinon with cycloolefins (cyclopentene, cyclooctene, norbornene) [13–15]. The result of these copolymerizations are oligomer products possessing conjugated blocks, originating from the dicarbonyl compounds, linked by soft segments, originating from the cycloolefins. The latter however do not contain unconjugated olefin double bonds. These facts show that the situation is more complicated than we have expected.

3 Possible Mechanisms of Carbonyl-Olefin Exchange Reaction

Proposing a possible mechanism for the new reaction is indispensable for determination of the possible structure of the catalytically active species. The idea about the carbene mechanism of the carbonyl olefin metathesis, we have launched, is based on the following arguments.

The transition metal halides and oxohalides as well as their combinations with Lewis acids are representatives of the classical catalytic systems for the olefin metathesis, which can be described as follows: the alkylidene fragments of two olefins can be exchanged to give another combination [16] (Scheme 9).

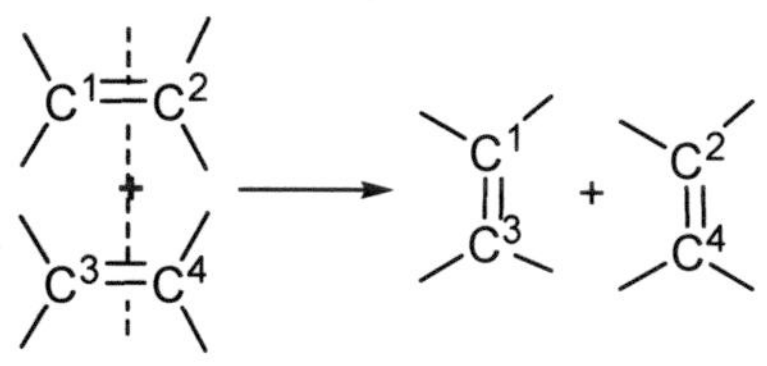

Scheme 9 General scheme of the olefin metathesis

The mechanism of the olefin metathesis involves a metal-carbene species (or more precisely a metal-alkylidene species), the coordination of the olefin onto the metal, followed by the shift of the coordinated olefin to form the metallacyclobutane intermediate. Finally, the topologically identical shift of the new coordinated olefin in the metallacyclobutane in a direction perpendicular to the initial olefin shift generates a new metal-alkylidene to which the new olefin is coordinated and then liberated. This new olefin contains an alkylidene fragment from the catalyst and the another one from the starting olefin. The new metal-alkylidene complex contains the remaining fragment of the starting olefin and can re-enter into a catalytic cycle in the same way as the first one (Scheme 10).

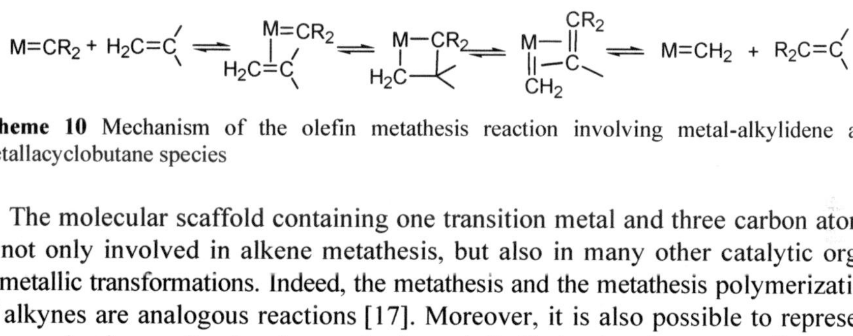

Scheme 10 Mechanism of the olefin metathesis reaction involving metal-alkylidene and metallacyclobutane species

The molecular scaffold containing one transition metal and three carbon atoms is not only involved in alkene metathesis, but also in many other catalytic organometallic transformations. Indeed, the metathesis and the metathesis polymerization of alkynes are analogous reactions [17]. Moreover, it is also possible to represent the mechanisms of σ-bond metathesis and β-elimination by a metallo-square scheme [18].

The Wittig-like reaction of transition metal-carbene complexes [19] can be considered as an extension of afore mentioned chemistry (Scheme 11).

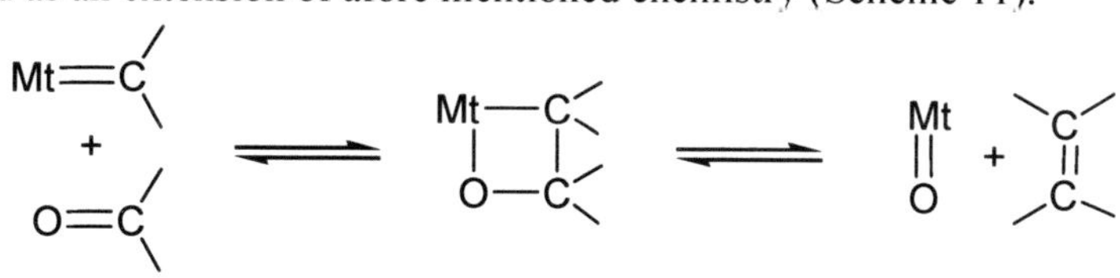

Scheme 11 Wittig-like reaction of transition metal carbene complexes

In this case the intermediate is an oxometallacyclobutane. This reaction is usually irreversible because of the strong Mt=O bonds. The backward reaction, however, is considered as a part of the catalytic cycle of the pinacolisation [20] and epoxidation [21] of olefins. There are some experimental evidences that this reaction is an initiation step in olefin metathesis when a rhenium heterogeneous catalytic system is used [22]. Accepting that the Wittig-like reaction of the transition metal-carbene complexes can be reversible and substituting one carbon atom in the description of the mechanism of the olefin metathesis for oxygen, a new catalytic cycle can be drawn, which is described shortly as follows: A transition metal-carbene complex reacts with a carbonyl compound generating an olefin and a transition metal oxocomplex. Then the oxocomplex reacts with an olefin generating a new carbonyl compound and a new transition metal-carbene complex. The intermediates in both steps are oxometallacyclobutanes. The overall result is represented in Scheme 12.

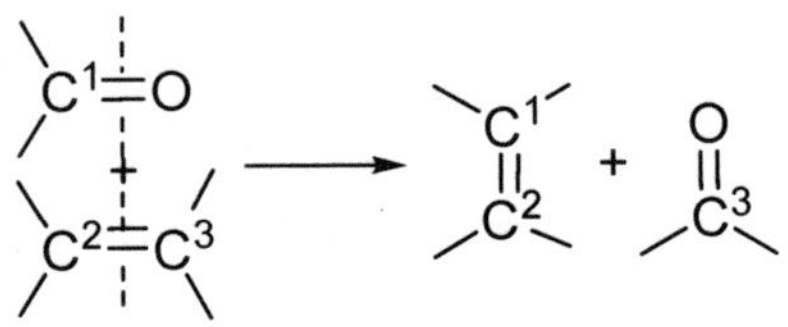

Scheme 12 The general scheme of carbonyl olefin metathesis

The formal similarity between this scheme and that of olefin metathesis is obvious in that one carbon atom in the latter is replaced by an oxygen atom. That is why the new reaction represented in Scheme 4 has to be denoted carbonyl olefin metathesis (the etymology of the word "metathesis" comes from the Greek μεταθεσίζ [metathesis] that means "transposition").

The transition metal halides and oxohalides are classical catalysts for olefin metathesis because they are precursors of transition metal alkilydene and oxo-alkilydene species. There are, however, a few papers describing OM catalyzed by non-transition elements [23], and platinum and acid catalyzed enyne metathesis [24]. The reported mechanisms of these transformations do not include participation of carbene species. Bickelhaupt at all reported for the acid catalyzed olefination of benzaldehyde with an olefin [25]. The overall result of this transformation can be represented with the scheme of the carbonyl olefin metathesis, but the reported mechanism again do not include participation of carbene species.

Because WCl_6 is a Lewis acid itself a noncarbene mechanism of COM cannot be ruled out.

4 Summary and Outlook

In summary a new carbon–carbon double bond forming reaction (carbonyl olefin exchange reaction) mediated by transition metal catalytic systems has been discovered. However the catalytic systems used till now (transition metal halides or oxohalides alone or in combination with Lewis acids) are active only in the case when the two reacting groups are in one molecules and are conjugated. In addition these systems accelerate other reactions which run simultaneously with the carbonyl olefin metathesis rendering a detailed investigation of the process very complicated.

Having in mind the aforementioned hypothesis about the possible mechanisms of the hypothesis about the possible mechanisms of the reaction a conclusion can be drown: The species catalytically active in respect to the carbonyl olefin metathesis are transition metal alkylidene- or oxo-complexes.

Everyone dealing with transition metal catalytic systems faces a good number of hardships. The systems under question can mediate different reactions. That is why the adjustment of reaction conditions (temperature, reaction time, concentration, molar ratio of the reactants and etc.) for a given reaction requires laborious and precise work. In some cases the final result depends on subtle factors like "the shape of glasswear" [26] or "coworker dependence" [27]. The McMurry reaction is a classical representative of such delicate reactions. No doubt these difficulties do not discourage chemists, but rather incite them to consider the new reactions with a more critical view.

The use of well defined transition metal complexes is a well established way to overcome these difficulties. There are numberless ligands that facilitate tuning the reactivity of a given metal. Almost all the reactions discovered during the last quarter of the 20th century are illustrations of this strategy. In addition some of the well defined transition metal complexes are able to mediate different reactions. For example the carbene complexes accelerate the Kharash reaction and atom transfer radical polymerization [28] and the Mn and Cr salen complexes accelerate the cyclopropanation and epoxidation of olefins [29]. We believe that it would be possible to synthesize transition metal complexes able to catalyze the COER when the two functional groups are not in one molecule and are not conjugated. It is worth working in this direction. If such complexes will be synthesized, then the COER will be an alternative to some of the existing carbon–carbon bond formation reactions, especially to the carbonyl olefination reactions (Wittig, Peterson, Julia, etc. reactions).

Acknowledgements The financial support of this work by the National Science Fund of the Bulgarian Ministry of Education and Science, project X-1413, is gratefully acknowledged.

References

[1] Schopov I, Jossifov C (1983) Macromol Chem Rapid Commun 4:659–662
[2] Schopov I, Mladenova L (1985) Macromol Chem Rapid Commun 6:659–663
[3] Schopov I, Mladenova L (1982) Synthetic Metals 48:249–258
[4] Jossifov C (1993) Eur Polym J 29:9–13
[5] Schopov I (1988) Acta Polymerica 39:91–94
[6] Iossifov C (2002) In: Khosravi E, Szymanska-Buzar T (eds.) Ring opening metathesis polymerization and related chemistry. NATO Science Series II.56:425–436. Kluwer, Dordrecht
[7] Iossifov C, Unpublished results
[8] Dragutan V, Balaban AT, Dimonie M (1985) Olefin metathesis and ring opening polymerization of cycloolefins. Wiley, Chichester
[9] Jossifov C, Schopov I (1991) Macromol Chem 192:857–861
[10] Jossifov C, Schopov I (1991) Macromol Chem 192:863–866
[11] Jossifov C (1998) Eur Polym J 34:883–885
[12] Pavlic M, Pfleger J, Iossifov C, Vohlidal J (2004) Macromol Symp 211:555–562
[13] Jossifov C (1998) J Mol Catal A: Chem 135:263–267
[14] Jossifov C, Ilieva O (2002) J Mol Catal A: Chem 190:235–239
[15] Dobricov G, Kolencov K, Zhechev D, Yourukova L, Rassovska M, Jossifov C, Parvanova N (2004) Vacuum 78:227–230
[16] Grubbs RH (ed.) (2003) Hanbook of metathesis. Wiley-VCH, Weinheim
[17] Katz TJ, Lee S (1980) J Am Chem Soc 102:422
[18] Astruc D (2005) New J Chem 29:42–56
[19] Takeda T (ed.) (2003) Modern carbonyl olefination. Wiley-VCH, Weinheim, pp. 151–159
[20] Hentges G, Sharpless B (1980) J Am Chem Soc 102:4263–4265
[21] McGarrigle M, Gilheany G (2005) Chem Rev 105:1563–1602
[22] Salameh A, Coperet C, Basset J-M, Böhm W, Röper M (2007) Adv Synth Catal 349: 238–242
[23] Ivin KJ, Mol JC (1997) Olefin metathesis and metathesis polymerization. Academic Press, London, p. 12
[24] Furstner A, Szillat H, Gabor B, Mynott R (1998) J Am Chem Soc 120:8305–8314
[25] Van Schaik H-P, Vijen R-J, Bickelhaupt F (1994) Angew Chem 106:1703–1704
[26] Bedford RB, Cazin C, Holder D (2004) Coord Chem Rev 248:2283–2321
[27] Ephritikhine M (1998) Chem Commun :2549–2554
[28] Simal F, Demonceau A, Noels AF (1999) Tetrahedron Lett 40:5689–5693
[29] Katsuki T (2002) Adv Synth Catal 344:131–147

Activation of Cycloolefin Metathesis by Ultrasonic Irradiation

Ileana Dragutan,[1*] Valerian Dragutan,[1] Petru Filip,[1] Albert Demonceau[2]

[1]Institute of Organic Chemistry of the Romanian Academy 202B Spl. Independentei, 060023 Bucharest, Romania

[2]University of Liège, Sart Tilman, Liège, Belgium

[*]E-mail: idragutan@yahoo.com

Abstract The present research focuses on the impact of power ultrasound on the synthesis of the tungsten-based metathesis catalytic system WCl_6/Me_4Sn and its activity in ring-opening metathesis polymerization of cyclooctene and cyclododecene. As compared to corresponding silent ROMP reactions with this mild catalytic system, altered reaction kinetics and different product selectivity have been found. Rate acceleration and an enhancement of oligomer formation have been clearly evidenced. The demonstrated possibility of employing technical grade solvents in ROMP induced by WCl_6/Me_4Sn is a further gain of the ultrasound strategy. Under the right conditions, ultrasound may thus promote greener, more cost effective and sustainable metathetic procedures.

Keywords Olefin metathesis · Tungsten-based catalysts · Power ultrasound · Sonochemical activation · ROMP · Cycloolefins

1 Introduction

As an efficient and virtually innocuous means of chemical activation, generation of products under sonochemical conditions is attracting considerable interest in synthetic organic, organometallic and inorganic chemistry [1, 2]. Strategies involving sonication are rapidly expanding to chemical processes, materials (nanosynthesis, nanomaterials) and life sciences (medical scanning ultrasonic therapy, ultrasonic disinfection, diagnostic ultrasound [1–10 MHz] et al.) but have also raised to commercial exploitation in an important range of industrial and technical fields such as ultrasonic impact treatment of surface layers of alloys to improve mechanical properties, surface cleaning, mineral processing, ultrasonic plastic and metal welding, spraying, ultrasound-assisted polymer recycling, food and beverage technology etc. Besides, ultrasound is currently employed in environmental chemistry for non-destructive remediation of water, land and air (degradation of environmental

V. Dragutan et al. (eds.), *Green Metathesis Chemistry: Great Challenges in Synthesis, Catalysis and Nanotechnology*,

pollutants, sonolysis of organic pollutants in water, aerosol precipitation, destruction of micro organisms by power ultrasound) [3–5].

The striking effects of ultrasound are usually interpreted in terms of the characteristic physical phenomenon called acoustic cavitation (i.e. formation, growth and implosive collapse of gas and vapor filled, micrometer-sized bubbles or cavities). Two forms of cavitation are known: transient and stable. In transient cavitation, the bubbles grow over one, two or three acoustic cycles to double their initial size and finally collapse. The collapse of bubbles can be violent enough to generate transient hot spots with very high local temperatures and pressures, within nanoseconds and with extreme cooling rates. In stable cavitation, bubbles oscillate in a regular fashion for many acoustic cycles inducing micro streaming in the surrounding liquid. An important consequence of the fluid micro-convection induced by bubble collapse is a sharp increase in the mass transfer at liquid–solid interfaces.

Cavitation is accompanied by emission of light (sonoluminescence), mechanical effects and chemical effects [6]. Waves resulting from cavitation promote vibrational motion of molecules which alternately compress and stretch the molecular structure of the medium. As molecules oscillate, a point is reached at which intramolecular forces are not able to hold the molecular structure; the released kinetic energy suffices to break the chemical bonds. This is the basic explanation for the multitude of chemical effects which have been observed upon ultrasound treatment. Effects include the production of radicals and excited state species, enhancement of mass transfer and catalysis at solid surfaces and excellent mixing in multiphase systems. Depending on the sonication conditions, some chemical effects of power ultrasound, relevant for this study, may intervene: (i) ligand–metal bond cleavage in transition metal complexes to give coordinatively unsaturated species; (ii) acceleration of single electron transfer (SET) (if ionic and electron transfer pathways are possible the latter is preferred – "sonochemical switching"); (iii) improved mass transfer, emulsification, increase of the effect of phase transfer catalysts, in liquid–liquid systems; (iv) homolytic fragmentation to radicals, rupture of polymers, generation of excited states; (v) disrupture of the solvent structure, thus altering solvation of the reactants; (vi) modified properties of solid particles.

Consequently, ultrasound holds great promise for promoting and accelerating a whole range of chemical reactions. The increasing interest in specific uses of power ultrasound has resulted in the sustained research conducted in large academic and industrial groups. Many important applications [1, 2] pertain to organic chemistry (acceleration of solution or two-phase [solid–liquid] organic reactions [7–10], organometallic chemistry (preparation of organometallics of main group or transition metals) [11–16], catalysis (generation of activated metals by sonication, impregnation of metals or metal halides on supports), and inorganic chemistry (preparation of colloidal alkali metal and activated metal solutions, sonochemical reactions involving metals *in situ*, etc.). Ultrasonic action in specific heterogeneous systems has also been studied [17].

The present chapter is aimed at evaluating ultrasound-mediated ROMP of cycloolefins, in the context of the state-of-the-art metathesis reactions conducted in an ultrasonic field. While among diverse metathesis reactions sonochemical RCM and CM have been given some attention, to the best of our knowledge ultrasound-promoted ROMP has been first studied by our group [18]. In this research, we chose to explore the effects of power ultrasound in the preparation of tungsten-based metathesis catalysts and thereby catalyzed ring-opening metathesis polymerization of two common cycloolefins, cyclooctene and cyclododecene.

2 Ultrasound in Polymer Chemistry and Metathesis Reactions

Applying ultrasound in polymer chemistry is since long an established technique [19]. Initially, use of ultrasound in this realm was limited to directions such as preparation of anionic initiators and anionic polymerizations [20] or polymerizations initiated through radicals formed by decomposition of solvents and breakage of polymer chains. More recently, the power of ultrasound has been selectively focused to weak linkages within the polymer resulting in well-defined macroradicals [21] and a reduction of the molecular weight in a predictable manner. The chance of synthesizing polymers and co-polymers with controlled structure was thus offered [22]. In comparison to silent counterparts, sonochemically-assisted polymerizations proceed faster and with higher yields. Also, by variation of ultrasound properties, control of the molecular weight distribution can be achieved.

A distinct area in sonochemical polymer synthesis is the production of functionalized [23] and metal-containing polymers. Many other technically important utilizations actually include polymer recycling through ultrasound induced chain degradation [2, 19], preparation of molecularly imprinted polymers [24], metal–polymer composite materials or polymer–clay nanocomposites [25], ultrasound-aided extrusion processing, welding of thermoplastic materials, joining of plastics and plastic composites [26]. Producing such materials is of momentous consequence because of emerging applications founded thereupon [27].

Not surprisingly, in the context of contemporary trends for a "greener" chemistry, power ultrasound has also found valorization in metathesis reactions. The synergistic action of catalysts and sustainable physical techniques such as microwave or ultrasound irradiation is now considered as viable alternatives to conventional metathesis. However, in comparison to microwaves [28], the effect of ultrasound in metathesis reactions is much less studied. Pertinent examples are activation of ethenolysis of unsaturated esters of vegetable oils induced by supported rhenium oxide heterogeneous catalysts [29] and RCM of N,N-diallyltosylamide with homogeneous Ru-allenylidene complexes [30].

A remarkable recent study [31] focuses on olefin metathesis in water, as an ideal "green" replacement for conventional VOS because of its low cost and the absence of organic vapours. In this research, ultrasonication of water-insoluble reactants led to the formation of a stable emulsion, without the use of any surfactant (acoustic emulsification). After addition of commercially available catalysts, smooth metathesis (RCM, enyne or cross-metathesis) took place in the emulsion droplets where also the water insoluble catalyst could be solved. In this way aqueous metathesis of a variety of substrates proceeded with considerable efficiency, allegedly because the sensitive ruthenium intermediates are being "protected" inside the water-insoluble organic droplets thus allowing higher turnovers. Although environmentally significant aqueous olefin metathesis reactions have been successfully attempted before with water-soluble reagents and specially functionalized catalysts, it was now demonstrated that simple sonication enables high-yielding aqueous metathesis of even water insoluble substrates.

Very recently an innovative application of ultrasound in RCM has been proposed, i.e. the mechanochemical activation of latent ruthenium catalysts (A. Piermattei, S. Karthikeyan, R.P. Sijbesma, Nature Chem. 1, 133–137 [2009]). Specifically, sonication is efficiently used as a mechanical trigger to "switch on" catalytic activity of ruthenium bis(NHC) carbene complexes, appended by polymer chains, for the ring closing metathesis of diethyl diallyl malonate. It was proved that the anticipated dissociative metathesis mechanism comes into action due to transfer of mechanical forces from the polymeric substituent to the coordination bond, thus bringing about ligand dissociation. This strategy has bright perspectives for further practical or even industrial valorization considering the established Ru dormant precursors incorporating NHC, alcoxy, Schiff bases or S-containing ligands, so far rendered metathesis active only after thermal or chemical activation.

In the ultrasound-ROMP connection we would like to mention that an *in situ* pulse echo ultrasonic spectroscopy technique has been used to monitor the ring-opening metathesis polymerization of dicyclopentadiene induced by the Grubbs I catalyst, carried out in a reaction cell provided with a flexible poly(ethylene terephtalate) window [32]. However, with the exception of work in our group [18], no reports on ultrasound-assisted ring opening metathesis polymerization seem to have been published to date.

3 Results and Discussion

3.1 *Silent Versus Ultrasonic Preparation of the Binary Catalyst System, WCl_6/Me_4Sn*

In view of initiating ROMP of cycloolefins, the classical catalytic system WCl_6/ Me_4Sn, well-known for its applications in olefin metathesis under mild conditions [33], has been prepared either under or without ultrasonic treatment. The conventional synthetic protocol was followed throughout the study, i.e. addition of the organometallic compound (Me_4Sn) to a toluene solution of tungsten hexachloride, at 30°C, under inert atmosphere.

In the absence of ultrasound the binary catalytic initiator needs to be prepared in dry solvent. As evidenced in previous reports on the chemistry of the process [34], tungsten hexachloride is alkylated by tetramethyltin, initially to methyltungsten pentachloride and subsequently to the dimethyl derivative that supposedly generates the first very active methylene-tungsten tetrachloride species (I). Along parallel pathways, methyltungsten pentachloride generates (by reductive elimination followed by alkylation and a second elimination) other methylene-tungsten species, (II) and (III), in which the tungsten atom is in a lower oxidation state (Scheme 1). The overall process is accompanied by an obvious change in colour, from the initial brown to a bright reddish-brown characteristic for the active binary catalytic system. Generation of active species occurs typically in 15–30 min reaching a plateau within 1 h. Progress was monitored visually and by absorption spectroscopy methods (especially IR).

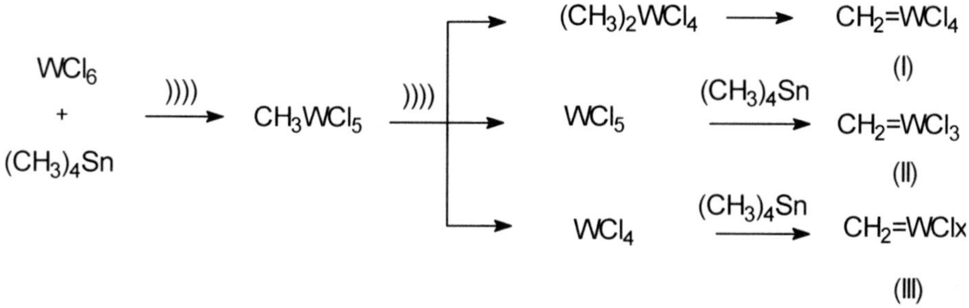

Scheme 1 Generation of the active species in $WCl_6/(CH_3)_4Sn$ system under ultrasound irradiation

In contrast to the silent experiments, a different behaviour is manifest under sonication. The active tungsten carbene species are detectable even after a few seconds after contacting the catalyst components. Such stimulation was to be expected considering the synergistic effects of ultrasound that come into force, namely rate enhancement through strip off of ligands to form highly reactive coordinatively unsaturated intermediates, rapid mass transfer due to cavitation, fast dispersion into the bulk solution of poisonous trace impurities and water. We have found indeed that ultrasound irradiaton brings about another striking benefit, i.e. an efficient preparation of the catalytic system in technical grade solvents (wet benzene or toluene).

3.2 *Polymerization of Cyclooctene and Cyclododecene*

ROMP of common cycloolefins was studied with the aim of testing a potential sonochemical switching. Two low ring-strain cycloolefins, cyclooctene and cyclododecene, reputedly slow in ROMP and greatly differing in reactivity from each other, have been selected as model substrates (Scheme 2).

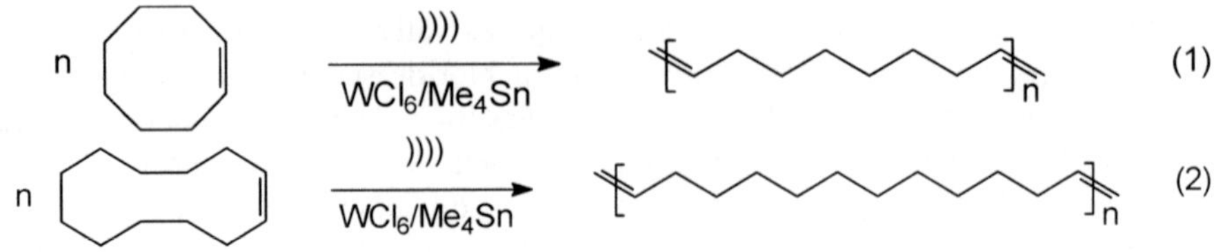

Scheme 2 ROMP of cyclooctene and cyclododecene under untrasound irradiation

Upon polymerizing cyclooctene with tungsten-based ROMP catalysts, quite varied conversions are attained depending mainly on the nature of the cocatalyst and the reaction parameters (Table 1). As could have been anticipated, results obtained under similar experimental conditions (Table 1) showcase the lowest cyclooctene conversion (Entry 6) for WCl_6/Me_4Sn, the least active catalyst in the studied range. For this reason and in view of a "greener" process, we decided to check if ultrasound could promote an enhancement of ROMP for this combination of a reluctant cycloolefin and a mild catalytic system. In spite of the low conversion recorded with the WCl_6/Me_4Sn initiating system, focusing on the Me_4Sn cocatalyst in tungsten-based systems is worthwhile because it affords a better functional group compatibility.

Table 1 ROMP of cyclooctene induced by tungsten-based catalytic systems[a]

Entry	Catalyst system	Temp. (°C)	Time (h)	Conversion (%)
1	WCl_6/i-Bu_3Al/chloranil	20	0.8	22
2	WCl_6/i-Bu_3Al/epichlorohydrin	20	0.8	22
3	WCl_6/ TIBAO[b,c]	25	0.8	48
4	WCl_6/ TIBAO[b,c]	12	0.8	78
5	$TPPWCl_4$/TIBAO[b,c]	20	1	21.5
6	WCl_6/Me_4Sn	25	1	7

[a] $[W] = 2.10^{-3}$ mol/l; [W]/ chloranil = 1:1; [W]/ epichlorohydrin = 1:1; [W]/ TIBAO = 1:2; [W]/ Sn = 1:2. Solvent: toluene.
[b] TIBAO = triisobutyl aluminoxane; TPP = tetraphenyl porphyrinate.
[c] This study and Ref. [35].

Ultrasound-assisted polymerization of cyclooctene triggered by the WCl_6/Me_4Sn catalyst system led to an interesting outcome (Table 2). A faster initiation of the reaction is evident. Monomer conversion, as inferred from GC analysis, was noticeable greater in the presence of ultrasound. Additionally, we found that the amount of oligomers was higher and the molecular weight distribution of the polymer broader for reactions conducted under ultrasound, especially for longer reaction times, pointing out an ultrasound-stimulated degradation of the polymer. It should be mentioned that under sonochemical conditions the temperature in the reaction vessel could be kept at 25°C with external ice-cooling but the experimental setup did not allow a further lowering of the reaction temperature which, very likely, would have led to increased conversions and a diminished polymer degradation (see Table 1, Entry 4 vs. Entry 3). ROMP of cyclododecene catalyzed by the system under scrutiny, WCl_6/Me_4Sn, was carried out either in the presence or in the absence of an ultrasonic field. The behaviour of cyclododecene parallels that of cyclooctene, with new results confirming earlier reports [18].

Table 2 Comparative ROMP of cyclooctene induced by the WCl_6/Me_4Sn catalytic system, with or without the action of an ultrasonic field (100 kHz)[a]

Time (h)	Conversion (%)	
	Silent	Ultrasound
1	7	10
2	9	12
3	11	13.5
4	13	15
10	26	33

[a] $[W] = 2.10^{-3}$ mol/l; W]/ Sn = 1:2. Temp. in the reaction flask: 25°C. Solvent: toluene.

Understanding the mechanism of ultrasound-assisted ROMP is difficult and not a clear-cut process. Although a series of general, empirical rules have been proposed to rationalize reactivity in an acoustic field, they should be used with caution [36]. However, data obtained in this study encourage us to consider that ultrasound facilitates a more efficient ROMP of cycloolefins. Improvement is due to ultrasound-associated effects intervening in the overall process. Superior homogeneization of the reaction mixture enables faster mass and charge transfer and finally translates into a more lively formation of the active sites which are also better protected from moisture and impurities traces. In addition, an easier access

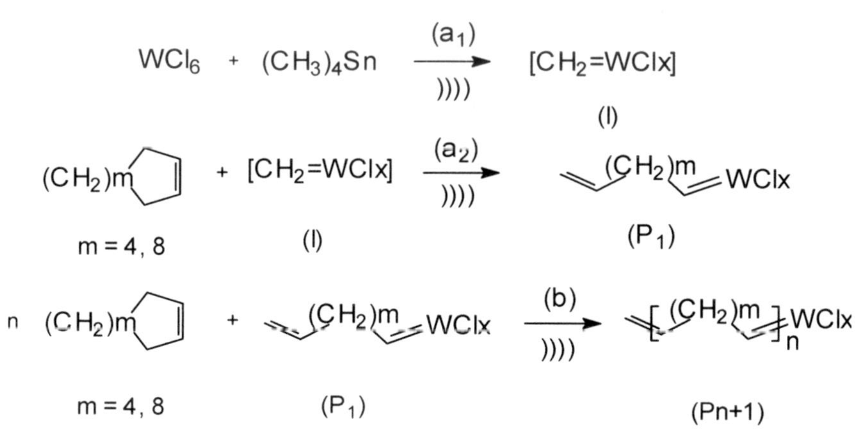

Scheme 3 Mechanism of ROMP of cycloolefins under ultrasound irradiation

of the monomer to the coordinatively unsaturated metal centre is ensured affording more rapid initiation and chain growth, in accordance to the general metallacarbene/metallacyclobutane metathesis mechanism [37–39] and reports on silent ROMP induced by the WCl_6/Me_4Sn system [34, 40 41]. Therefore we admit that ultrasound speeds up the generation (a_1/Step 1) of metallacarbene catalytic species (I), the formation of the first propagating species (P_1) (from (I) and the monomer; Step 2/a_2, initiation) and the propagation by successive monomer additions to the propagating species (Pn+1) (Step 3) (Scheme 3).

Gas-chromatography analyses of sonicated polymerizations suggest that power ultrasound has also promoted the back-biting metathesis route leading to formation of oligomers from the metallacarbene propagating species Pn + 1 (Scheme 4).

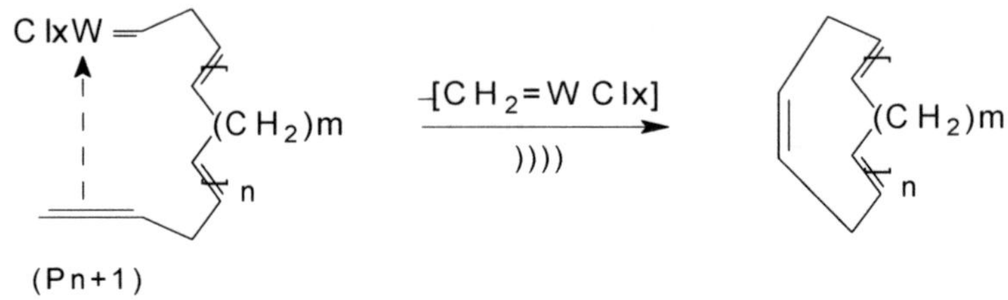

(3)

Scheme 4 Back-biting metathesis route to oligomers under ultrasound irradiation

4 Experimental

Starting materials. Tungsten(VI) chloride (Aldrich, 99.9%) was purified by repeated sublimation. Tetramethyltin (Schering) was used as such. Cyclooctene (Merck, 99.9%) and cyclododecene (Merck, 99.9%) have been employed as the monomers without further purification. Commercial grade toluene (Aldrich, 99.8%) was used in some of the sonicated reactions. For silent ROMP toluene was dried by refluxing on a Na-K alloy, under nitrogen, and distilled from the alloy just before use. Sensitive reagents (WCl_6, $SnMe_4$) were stored and handled under an extremely pure nitrogen atmosphere.

Catalyst preparation. Handling and dosage were performed under nitrogen of high purity. WCl_6 was stored in sealed vials. In some of the experiments the catalyst was preformed by complexation of WCl_6 with $SnMe_4$, in the presence or in the absence of ultrasound irradiation. Most frequently the catalytic system was prepared *in situ,* with addition of first the monomer and then the catalyst components, both under or without an ultrasound field (power ultrasound 50–100 kHz).

Polymerization process. The polymerization reactions were carried out in a standard installation provided with stirring, nitrogen inlet and sample collector, and immersed in an ultrasonic bath containing water [18]. Samples were collected at predetermined intervals. Finally, the catalyst was deactivated with a 2% NaOH methanol solution, water was added to remove the deactivation products, then phases separated. The organic phase, a wet toluene solution of oligomers and polymers, was concentrated in vacuum. The crude oily product was analyzed by gas-chromatography.

Gas-chromatography analyses. GC analyses were carried out with a Carlo Erba HR6C 5300 MEGA SERIES gas chromatograph using a fused-silica capillary column (L = 25 m; inner diameter = 0.35 mm) with a poly(methyl phenyl polysiloxane) stationary phase (SE 52, Carlo Erba; thickness of the deposited layer = 0.15 μm). The carrier gas was hydrogen (1 ml/min) and the splitting ratio, 1:50. The following temperature conditions afforded best results: FID detector, 250°C; vaporizer, 250°C; the column temperature was programmed from 50°C to 250°C with an increase of 10°/min.

IR and NMR analysis of polymers. Infrared spectra of the polymeric products were recorded on a Nicolet 10MX(FT) spectrophotometer. ^{1}H and ^{13}C NMR spectra of the polymers in $CDCl_3$ were collected using a Brucker (300 MHz) spectrophotometer with TMS as internal standard and confronted with previously reported data from the silent ROMP of the respective cycloolefins [42].

5 Conclusions

The present study shows, for the first time, that ultrasound irradiation in a ROMP reaction represents an interesting option for enhancing the rate of this kind of metathesis. Common chemical effects of ultrasound, such as reduction of the induction period and rate acceleration, have been identified in ROMP of the two investigated cycloolefins (cyclooctene and cyclododecene) induced by the mild catalytic system WCl_6/Me_4Sn. Comparison with silent reactions, carried out under the same conditions, clearly evidenced an enhancement of the catalytic activity in ultrasonic field, though rather modest under the experimental setting available. The superiority of the ultrasonically-assisted method is also demonstrated from an obvious additional benefit: the possibility of employing technical grade solvents in ROMP with WCl_6/Me_4Sn. Under the right conditions, ultrasound may promote greener, more cost effective and sustainable metathetic procedures.

References

[1] (a) Mason TJ, Larimer JP (1988) Sonochemistry: Theory, applications and uses of ultrasound in chemistry. Ellis Horwood, Chichester, England; (b) Mason TJ (ed.) (1990) Sonochemistry: The uses of ultrasound in chemistry. Royal Society of Chemistry, Cambridge, England; (c) Mason TJ, Peters D (2002) Power ultrasound. Uses and applications (2nd edn.). Horwood, Chichester; (d) Mason TJ, Lorimer JP (2002) Applied sonochemistry. Wiley-VCH, Weinheim

[2] (a) Suslick KS (ed.) (1987) High energy processes in organometallic chemistry. ACS Symposium Series 333, New York; (b) Suslick KS (ed.) (1988) Ultrasound, its chemical, physical and biological effects. VCH Publishers, New York; (c) Suslick KS (1994) The chemistry of ultrasound, in Encyclopedia Britannica Yearbook of science and the future

1994. Britannica, Chicago, IL, pp. 138–155; (d) Suslick KS (1988) Sonochemistry, in Kirk-Othmer Encyclopedia of Chemical Technology (4th edn.). Wiley, New York, vol. 26, pp. 517–541

[3] Price GJ, Matthias P, Lenz EJ (1994) Process Safety Environ Prot 72B1:27

[4] (a) Mason TJ (2007) Ultrasonics Sonochemistry 14:476–483; (b) Joyce EM, Mason TJ (2008) Chimica Oggi • Chem Today 26:22–26

[5] (a) Cravotto G, Cintas P (2006) Chem Soc Rev 35:180–196; (b) Cravotto G, Garella D, Gaudino CE, Leveque JM (2008) Chimica Oggi • Chem Today 26:44–47

[6] Price GJ, Ashokkumar M, Hodnett M, Zequiri B, Grieser F (2005) J Phys Chem B 109:17799–17801

[7] Caulier TP, Maeck M, Reisse J (1995) J Org Chem 60:272

[8] Peters D, Pautet F, Fakih H-El, Luche J-L, Fillion H (1995) J Prakt Chem 337:363

[9] Dauben WG, Bridon DP, Kowalczyk BA (1989) J Org Chem 54:6101

[10] Eshuis JJW (1994) Tetrahedron Lett 35:7833

[11] Luche J-L, Damiano J-C (1980) J Am Chem Soc 102:7926

[12] Hyeon T, Fang M, Cichowlas AA, Suslick KS (1995) Prepr (Am Chem Soc Div Fuel Chem) 40:365

[13] Ley SV, Low CMR (1989) Ultrasound in synthesis. Springer, Berlin

[14] Cheng J, Luo F (1989) Bull Inst Chem Acad Sin 36:9

[15] Billington DC, Helps IM, Paulson PL, Thompson, Willison D (1988) J Organomet Chem 354:233

[16] Harrity JPA, Kerr WJ, Middleniss D (1993) Tetrahedron 49:5565

[17] (a) Suslick KS, Goodale JW, Schubert PF, Wang HH (1983) J Am Chem Soc 105:5781; (b) Suslick KS, Skrabalak SE (2008) Sonocatalysis. In: Handbook of heterogeneous catalysis, vol. 4, Ertl G, Knözinger H, Schüth F, Weitkamp J (eds.) Wiley-VCH, Weinheim. Weinheim. pp. 2006–2017

[18] Dragutan I, Dragutan V, Petride A, Vanatoru M, Filip P (2002) Ultrasound assisted metathesis of monocyclic olefins with tungsten-based catalysts. In: Khosravi E, Szymanska-Buzar T (eds.) Ring-opening metathesis polymerization and related chemistry: State of the art and visions for the new century. NATO Science Series II. Mathematics, Physics and Chemistry, Kluwer, Dordrecht, The Netherlands, vol. 56, pp. 477–482

[19] (a) Paulusse JMJ, Sijbesma RP (2006) J Polym Sci Part A: Polym Chem 44:5445–5453; (b) Akyuz A, Catalgil-Giz H, Giz AT (2008) Macromol Chem Phys 209:801–809

[20] Schulz DN, Sissano JA, Costello CA (1994) Polym Prepr (Am Chem Soc Div Polym Chem) 35:514

[21] Price GJ, Garland L, Comina J, Davis M, Snell DJ, West PJ (2004) Res Chem Intermed 30:807–827

[22] (a) Price GJ (2003) Ultrasonics Sonochem 10:277–283; (b) Price GJ (1993) Chem Ind 3:75–78

[23] (a) Price GJ, Lenz EJ, Ansell CWG (2002) Eur Polym J 38:1753–1760; (b) Price GJ, Lenz EJ, Ansell CWG (2002) Eur Polym J 38:1531–1536

[24] Svenson J (2006) Anal Lett 39:2749–2760

[25] (a) Wang JZ, Hu Y, Chen ZY (2003) Rare Metal Mater Eng 32:585–590; (b) Sonawane SH, Chaudhari PL, Ghodke SA, Parande MG, Bhandari VM, Mishra S, Kulkarni RD (2009) Ultrasonics Sonochem 16:351–355

[26] Stokes VK (1989) Polym Eng Sci 29:1310–1324

[27] (a) Peters D (1996) J Mater Chem 6:1605–1618; (b) Suslick KS, Price GJ (1999) Ann Rev Mater Sci 29:295–326; (c) Dhas NA, Suslick KS (2005) J Am Chem Soc 127:2368–2369

[28] (a) Coquerel Y, Rodriguez J (2008) Eur J Org Chem 1125–1132 (b) Gebauer J, Arseniyadis S, Cossy J (2008) Eur J Org Chem 2701–2704

[29] Mandelli D, Jannini MJDM, Buffon R, Schuchardt U (1996) J Am Oil Chemists' Soc 73:229–232

[30] Furstner A, Dixneuf P, Bruneau C, Picquet M (2003) US Patent 6590048

[31] Gulajski L, Sledz P, Lupa A, Grela K (2008) Green Chem 10 :271–274

[32] Constable GS, Lesser AJ, Coughlin EB (2003) J Polym Sci Part B – Polym Phys 41:1323–1333
[33] Boelhower C, Mol JC (1985) Progr Lipid Res 24:243
[34] Thorn-Csanyi E, Kessler M (1991) J Mol Catal 65:253–260
[35] Dragutan V, Dragutan I, Dimonie M (2001) Polym Prepr (Am Chem Soc Div Polym Chem) 42:362–363
[36] (a) Luche J-L, Einhorn C, Einhorn J, Sinisterra-Gago JV (1990) Tetrahedron Lett 31:4125–4128; (b) Chanon M, Luche J-L (1998) In Luche J-L (ed.) Synthetic organic sonochemistry. Plenum, New York, pp. 377–392
[37] Dragutan V, Balaban AT, Dimonie M (1985) Olefin metathesis and ring-opening polymerization of cycloolefins. Wiley, New York
[38] Ivin KJ, Mol JC (1997) Olefin metathesis and metathesis polymerization. Academic Press, London
[39] Dragutan V, Streck R (2000) Catalytic polymerization of cycloolefins. Elsevier, Amsterdam
[40] Thorn-Csanyi E, Timm H (1985) J Mol Catal 28:37
[41] Thorn-Csanyi E, Kessler M (1986) J Mol Catal 36:31
[42] Dimonie M, Coca S, Teodorescu M, Popescu R, Chipara M, Dragutan V (1994) J Mol Catal 90:117–124

Microwave-Assisted Olefin Metathesis

François Nicks, Yannick Borguet, Xavier Sauvage, Dario Bicchielli, Sébastien Delfosse, Lionel Delaude, Albert Demonceau*

Laboratory of Macromolecular Chemistry and Organic Catalysis, University of Liège, Sart-Tilman (B.6a), B-4000 Liège, Belgium

*E-mail: A.Demonceau@ulg.ac.be

Abstract Since the first reports on the use of microwave irradiation to accelerate organic chemical transformations, a plethora of papers have been published in this field. In most examples, microwave heating has been shown to dramatically reduce reaction times, increase product yields, and enhance product purity by reducing unwanted side reactions compared to conventional heating methods. The present contribution aims at illustrating the advantages of this technology in olefin metathesis and, when data are available, at comparing microwave-heated and conventionally heated experiments.

Keywords Cross-metathesis (CM) · Homogeneous catalysis · Microwaves (MW) · Olefin metathesis · Ring-closing metathesis (RCM) · Ruthenium

1 Introduction

Microwave-assisted organic synthesis is a relatively new technique in the present-day synthetic world and has grown extensively in the last decade from a mere academic tool to a heartily used technique in the industrial research laboratories. It is apparent from the recent literature that, compared to conventional heating methods, microwave irradiation (MW) mostly results in a dramatic acceleration of reactions, most often resulting in cleaner outcomes and increased yields [1, 2].

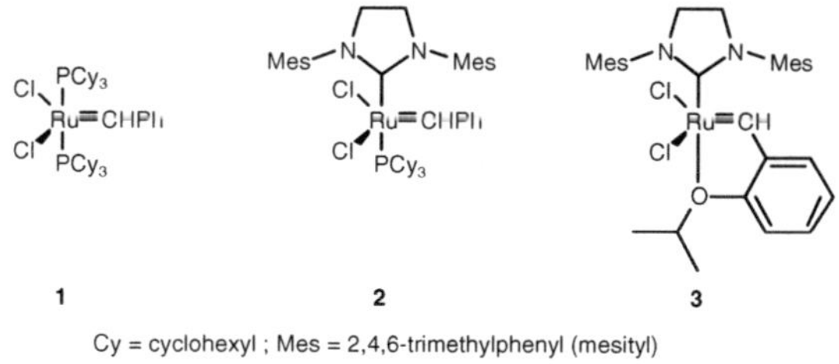

Scheme 1 The classical ruthenium catalyst precursors for olefin metathesis

V. Dragutan et al. (eds.), *Green Metathesis Chemistry: Great Challenges in Synthesis, Catalysis and Nanotechnology*,

In recent years, the olefin metathesis reaction has attracted widespread attention as a versatile carbon–carbon bond-forming method. Many new applications have become possible because of major advances in catalyst design. In particular, the advent of well-defined ruthenium-based metathesis pre-catalysts (Scheme 1) has triggered an explosive growth of interest in this reaction both from the organic and polymer chemist communities. Despite these efforts, some substrates are still reluctant to undergo efficient metathesis reactions under classical thermal conditions, which translates into prolonged reaction times, low yields, and/or large amounts of catalyst. Not surprisingly, microwave irradiation has also a pronounced beneficial impact on olefin metathesis reactions. The first report on microwave-assisted olefin metathesis appeared in 2000 [3]. Since then, this method has gained increasing popularity (Figure 1) and, in particular, has been very recently reviewed by Coquerel and Rodriguez [4]. This chapter aims at summarising the most recent contributions in microwave-assisted olefin metathesis. Accordingly, the manuscript will be divided into three sections dealing with ring-closing metathesis, cross-metathesis, and enyne metathesis. Recent progress in microwave-assisted alkyne metathesis will also be summarised in the last part of this chapter.

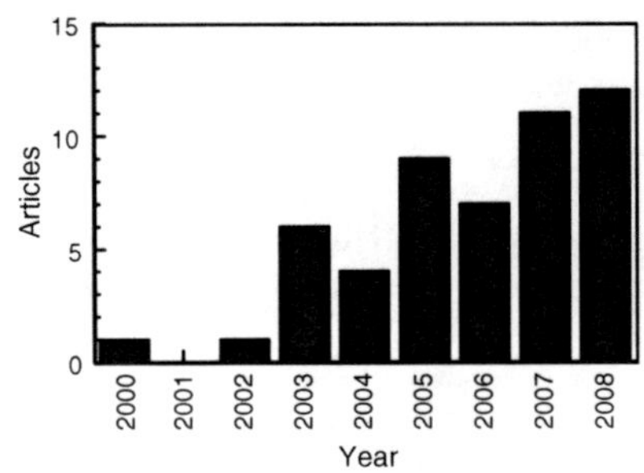

Figure 1 Number of articles on microwave-assisted olefin metathesis published every year from 2000 (November 2008)

2 Microwave-Assisted Ring-Closing Metathesis

The first report on microwave-assisted olefin metathesis appeared in 2000 and described the preparation of a poly(ethylene glycol)-supported cyclic amino acid derivative using the ring-closing metathesis (RCM) of diene **4** catalysed by the first-generation Grubbs' complex (**1**) (Scheme 2) [3] In this case, the microwave irradiation allowed the reaction time to be reduced from 8 h under thermal conditions to 10 min (Table 1). By contrast, with diene **4** bearing a methyl ester functionality, the RCM reaction took place equally well under both conventional and microwave conditions. In particular, under microwave irradiation in the absence of solvent, cyclisation yielded 91% of isolated **5**. Furthermore, it should be noted that, in this initial study, a domestic multimode microwave oven was used and the temperature of the irradiated reaction mixture was not measured [3].

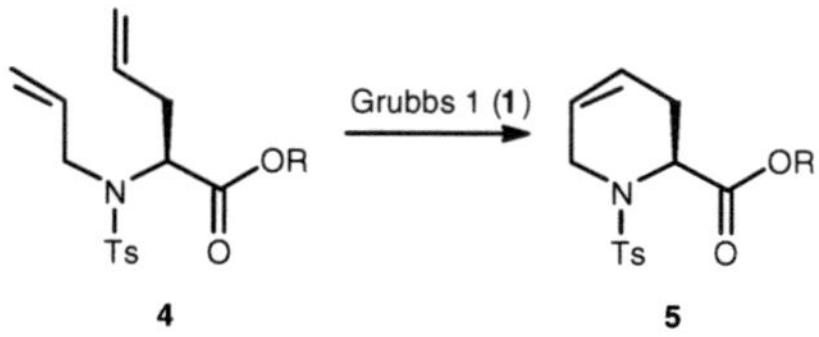

Ts = *p*-toluenesulfonyl (tosyl)

Scheme 2 The first microwave-assisted olefin metathesis reaction

Table 1 Thermal versus microwave-assisted ring-closing metathesis of diene **4**

R	1 (mol%)	Solvent	Reaction conditions	Conversion (%)
Me	10	CH_2Cl_2	Δ, 20°C, 10 min	100
	10	–	MW, 850 W,[b] 10 min	100
PEG[a]	10	CH_2Cl_2	Δ, 20°C, 2 h	38
	10	CH_2Cl_2	Δ, 20°C, > 24 h	50
	40	CH_2Cl_2	Δ, 20°C, 8 h	100
	50	–	MW, 850 W,[b] 10 min	100

[a]PEG = poly(ethylene glycol).
[b]The temperature was not measured.

With the development of safe and reliable mono- or multimodal microwave reactors specifically designed for chemical applications, microwave-assisted organic synthesis has made great strides, which were profitable to olefin metathesis [4]. For instance, microwave-assisted olefin metathesis was recently applied successfully as a key transformation in the total synthesis of aigialomycin D (**6**) (Scheme 3), a natural compound possessing potent anti-tumour and anti-malarial activity. Thus, the macrocyclisation of **7** by ring-closing metathesis following a reported protocol [5]

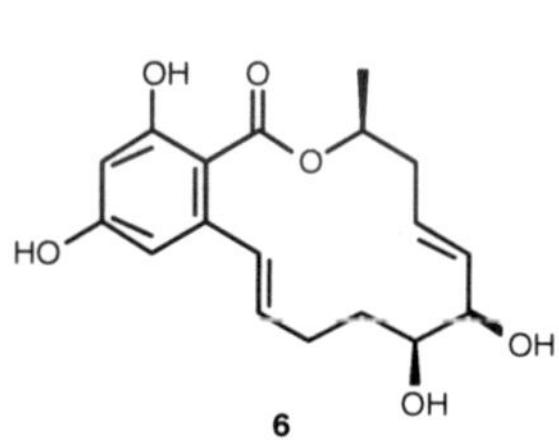

Scheme 3 Structure of aigialomycin D

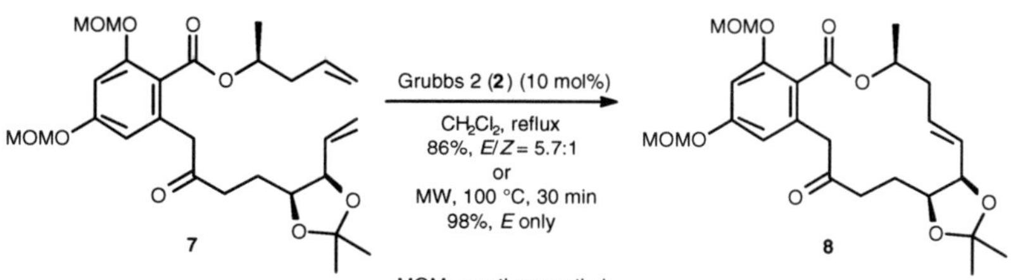

MOM = methoxymethyl

Scheme 4 The key RCM step in the synthesis of aigialomycin D afforded the cyclised product **8** in 86% yield and with an *E*/*Z* ratio of 5.7:1 (Scheme 4). When the same reaction was performed under microwave irradiation at 100°C for only 30 min, the cyclised product **8** was obtained in 98% yield with a complete selectivity in favour of the thermodynamically more stable *E*-isomer [6].

In the course of an investigation toward the total synthesis of guaianolides such as (+)-arglabin (**9**) and ixerin Y (**10**) (Scheme 5), Reiser and coworkers reported a systematic study of the RCM of diene **11** to give the tetrasubstituted double bond in **12** (Scheme 6) [7].

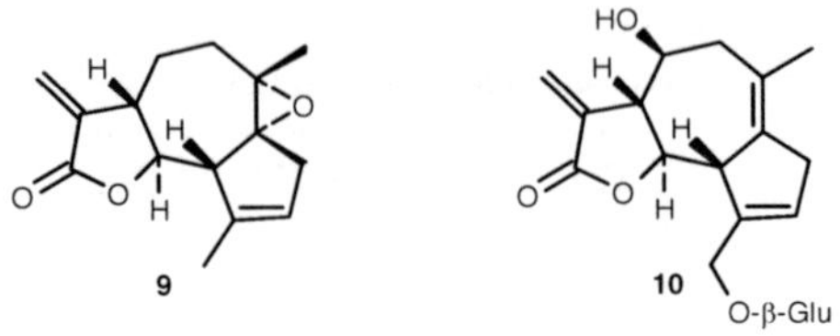

Scheme 5 Structure of (+)-arglabin and ixerin Y

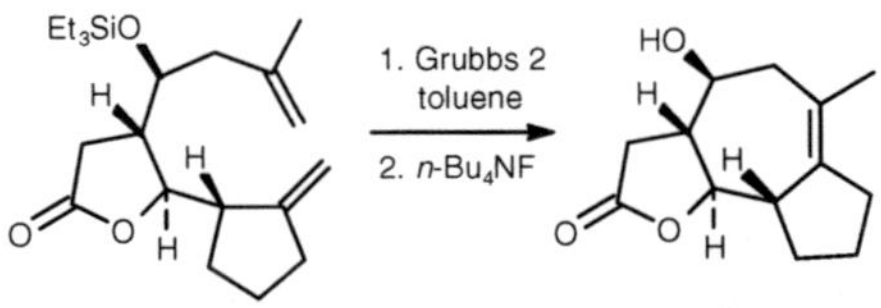

Scheme 6 The key RCM step in the synthesis of the 5.7.5-tricyclic ring systems **12**

By combining microwave heating and sparging an inert gas through the reaction mixture, a high-yielding RCM was achieved with this challenging substrate. Thus, by applying a 15 mol% loading of the Grubbs 2 catalyst, the RCM of **11** could be accomplished with an isolated yield of 98%, which, to the best of our knowledge, represents one of the most efficient examples for an RCM towards a highly substituted cycloalkene (Table 2). Conventional heating using a preheated oil bath set at 120°C gave much slower conversion (60% yield), and even at prolonged reaction times the yield did not reach the value obtained under microwave conditions. A key aspect seems to be that rapid microwave irradiation diminishes catalyst decay by allowing the required high reaction temperature to be reached quickly and homogeneously and thereby providing enough energy for a successful metathesis reaction. On the other hand, passing an inert gas through the reaction solution purges off evolving ethylene to shift the equilibrium to the desired product. Noteworthy, attempts to achieve RCM in a closed reaction vessel under microwave irradiation failed completely and the starting material could be recovered almost quantitatively, demonstrating thereby the crucial role played by the inert-gas sparging.

In a new approach for the synthesis of substituted pyrrolo-[3,2-*c*]quinoline derivatives, the use of a microwave-assisted ring-closing metathesis resulted again in high yielding and short reaction times (Scheme 7) [8]. Indeed, while cyclisation of the diene **13** was complete within 12 h at room temperature using 10 mol% Grubbs' catalyst **2** with 87% yield, it could be conveniently accelerated by microwaves using only 4 mol% of the catalyst. Under these conditions, completion of the cyclisation was reached within 2 h yielding 84% of pyrroline **14**. Interestingly, the cyclisation of **13** was slower than in the case of the related dienes **18** (Scheme 7) and **20** (Scheme 8) devoid of nitro group [9]. Cyclisation of **18**, indeed, was complete within 12 h at room temperature, whereas 5 min at 100°C under microwave activation were sufficient to drive the reaction to completion (Table 3). The lower reactivity of diene **13** might be due to the coordinating effect of the nitro substituent onto the ruthenium centre.

Table 2 Thermal versus microwave-assisted ring-closing metathesis of diene **11**

2 (mol%)	Reaction conditions	Yield (%)
10	Δ, 80°C, 24 h, N_2 atmosphere	35
5	Δ, 80°C, 24 h, N_2 sparging	49
10	Δ, 80°C, 24 h, N_2 sparging	66
15	Δ, 80°C, 24 h, N_2 sparging	80
15	Δ, 110°C, 2 h, Ar sparging	60
15	Δ, 110°C, 4.17 h, Ar sparging	82
15	MW, 300 W, 110°C, 1.5 h, Ar sparging	98

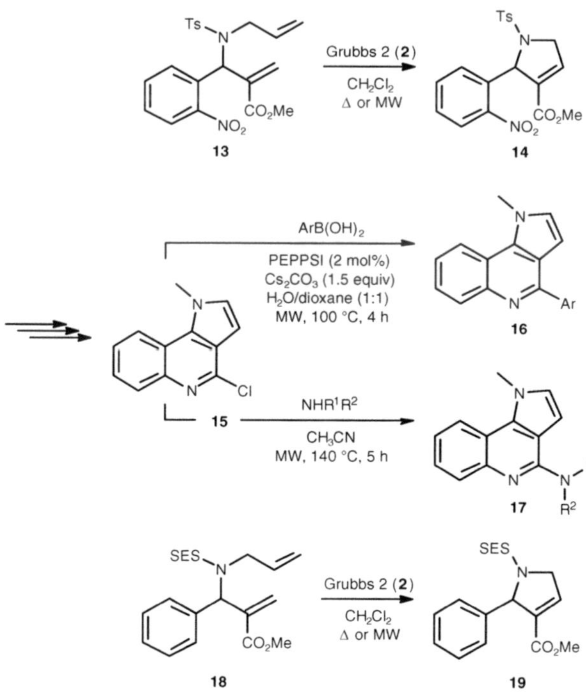

SES = 2-trimethylsilylethylsulfonyl [$(CH_3)_3SiCH_2CH_2SO_2$]

Scheme 7 Synthesis of pyrrolines **14** and **19** by ring-closing metathesis

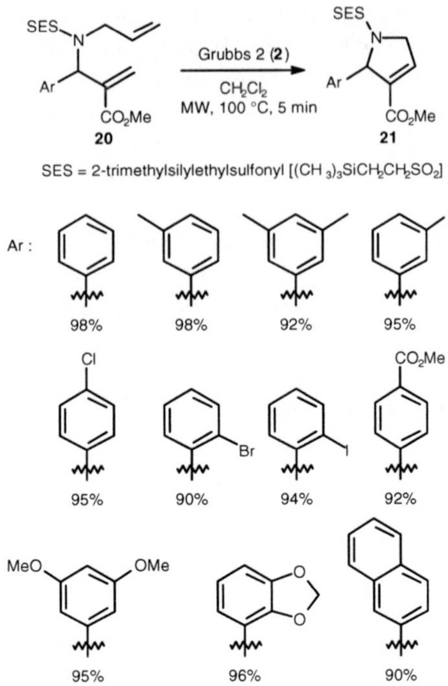

Scheme 8 Synthesis of pyrrolines **21** by ring-closing metathesis

Table 3 Thermal versus microwave-assisted ring-closing metathesis of dienes **13** and **18** catalysed by the Grubbs 2 complex (**2**)

Diene	2 (mol%)	Reaction conditions	Yield (%)
13	10	Δ, RT, 12 h	87
	4	MW, 100°C, 2 h	84
18	5	Δ, RT, 16 h	98
	5	MW, 100°C, 5 min	98

Five subsequent transformations, including two microwave-accelerated steps, led to pyrrolo-[3,2-*c*]quinoline **15**, which was then decorated under microwaves via either a Suzuki–Miyaura coupling reaction catalysed by the $Pd(OAc)_2/PPh_3$ system or, preferably, the PEPPSI (Pyridine-Enhanced Precatalyst Preparation, Stabilisation and Initiation) pre-catalyst (**22**) (Scheme 9) [10] or a nucleophilic substitution of the chlorine atom by amines (Scheme 7) [8].

Indolizidine derivatives could also be obtained via ring-closing metathesis, as illustrated in Scheme 10 [11]. Thus, when diene **23** was reacted with 10 mol% Grubbs 2 catalyst under microwave irradiation (Table 4), the expected indolizidine **24** bearing a tetrasubstituted double bond was obtained in excellent 97% isolated yield. Noteworthy, cyclisation of **23** was unsuccessful at room temperature.

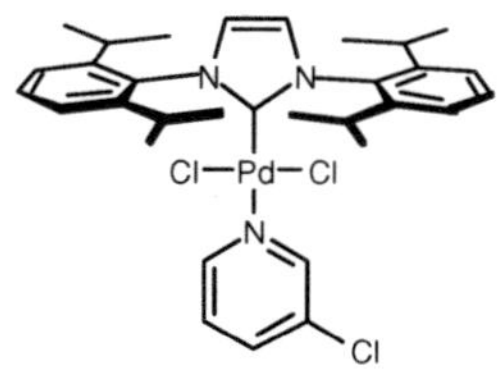

22

Scheme 9 Structure of the PEPPSI pre-catalyst

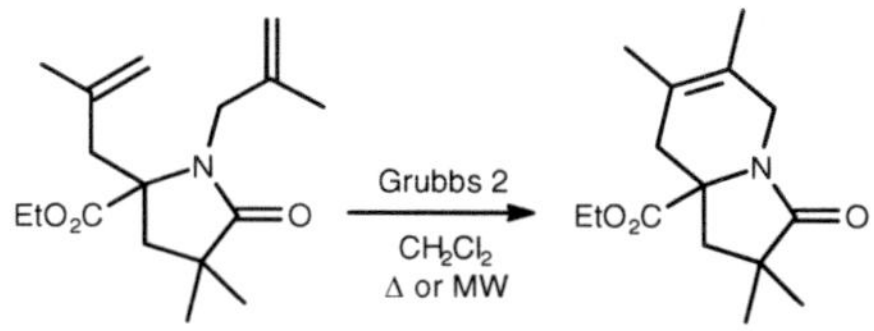

23 24

Scheme 10 Synthesis of indolizidine **24** via ring-closing metathesis

Table 4 Thermal versus microwave-assisted ring-closing metathesis of diene **23** catalysed by Grubbs 2 catalyst (**2**)

2 (mol%)	Reaction conditions	Yield (%)
20	Δ, RT, 4 d	0
10	MW, 100°C, 30 min	97

Even after 4 days and use of 20 mol% of catalyst, only starting material could be recovered.

Conformationally constrained spiro-amino acids **27** and spiro-2,5-diketo-piperazines **28** have also been prepared using a microwave-assisted RCM reaction as the key step (Scheme 11) [12]. In a preliminary investigation [13], the RCM reactions were run in benzene or toluene at slightly elevated temperatures (20–100°C) for 5–24 h, using the first-generation Grubbs' catalyst (**1**). This classical procedure furnished the desired spiro compounds **26** in 53–99% yields. Subsequently, the second-generation Grubbs' catalyst (**2**) was used in conjunction with microwave-assisted heating to improve the efficiency of the transformation. Gratifyingly, spiranes **26** were obtained in 63–99% isolated yields after 15 min at 140°C. In particular, the six- and seven-membered rings were obtained in higher yields (80–99%) than the five-membered counterpart (63%). Shorter reaction times or lower temperatures resulted in incomplete reactions, and longer reaction times, e.g., 30 min, gave no further increase in yields [12]. It should be noted, on the other hand, that all subsequent transformations to the spiro-2,5-diketopiperazines **28** were also run with microwave-assisted heating, resulting in high yields and short reaction times for all steps. Consequently, the total reaction times for the RCM reaction and the subsequent transformations were only 55 min using microwave heating!

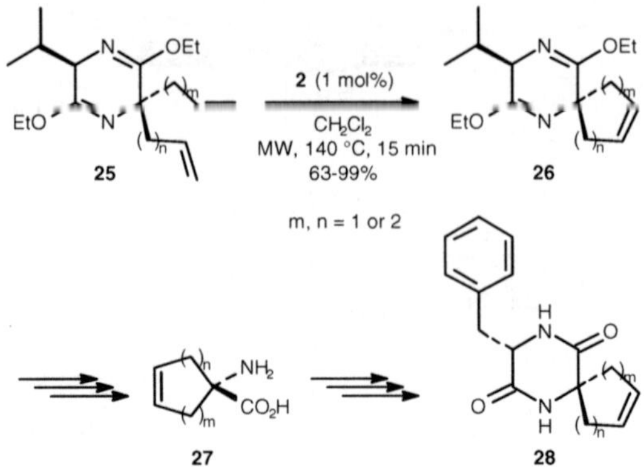

Scheme 11 The key RCM step in the synthesis of the conformationally constrained spiro-amino acids **27** and spiro-2,5-diketopiperazines **28**

On the basis of the reported data regarding the successful use of microwave in RCM, microwave irradiation instead of conventional heating was used for the synthesis of jaspamide analogues (Scheme 12) [14]. The cyclodepsipeptide jaspamide (**31**) (Scheme 13) is an interesting marine metabolite, possessing a potent inhibitory activity against breast and prostate cancer, as a consequence of its ability to disrupt actin cytoskeleton dynamics.

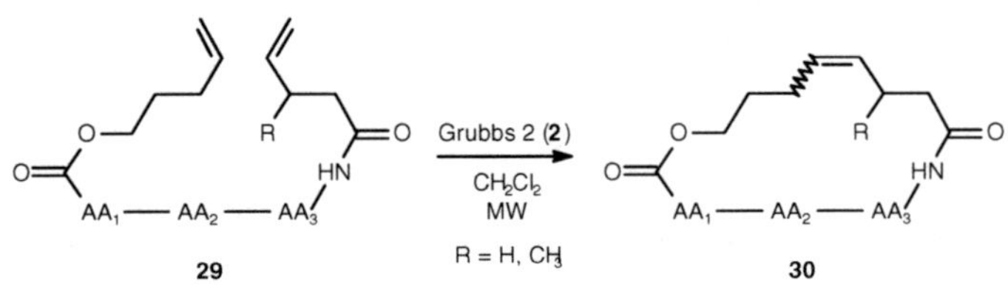

Scheme 12 The key RCM step in the synthesis of jaspamide analogues **30**

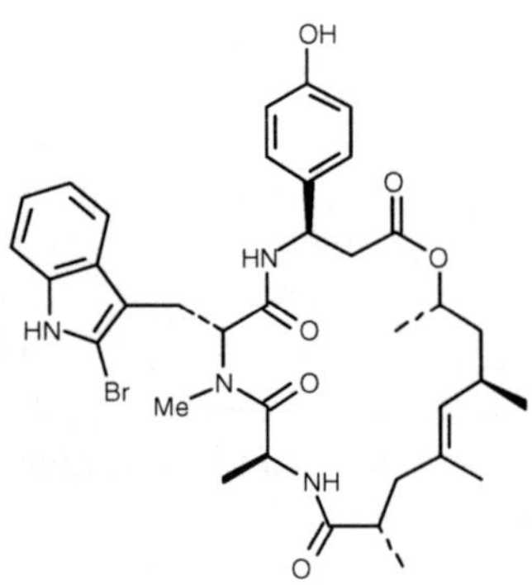

31

Scheme 13 Structure of jaspamide

The following optimised procedure was used for the synthesis of jaspamide analogues **30**: refluxing dichloromethane as solvent, 10 mol% of second-generation Grubbs' catalyst, microwave heating at 300 W for two 40 min periods (Scheme 12). A second addition of fresh 10 mol% catalyst was performed after the first 40 min in order to drive the RCM reaction to completion. In addition, a gentle stream of an inert gas was employed in order to remove the evolving ethylene during the RCM. After RCM, all cyclodepsipeptides were deprotected, affording the desired compounds in 20–33% yield after two HPLC purification steps.

A microwave-accelerated RCM reaction has also been used as the key step of the synthesis of resin-attached peptides and this methodology was illustrated by the highly selective synthesis of dicarba analogues of α-conotoxin IMI (**32**, Scheme 14), a disulphide-rich peptide isolated from the venom of the vermivorous conus species *Conus imperialis* [15].

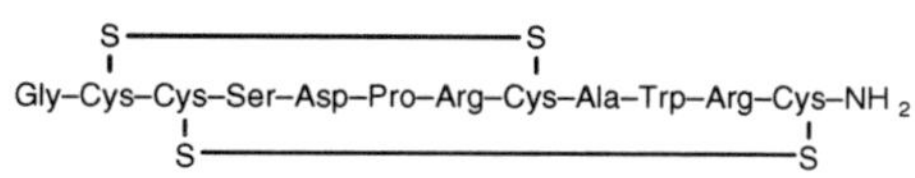

32

Scheme 14 Structure of α-conotoxin IMI

Microwave irradiation of a mixture of Rink amide bound-peptide **33** and second-generation Grubbs' catalyst (10 mol%) in dichloromethane containing 10% lithium chloride in dimethylformamide resulted in complete ring closure in only 1 h (Scheme 15). Complete conversion could also be achieved using a 5 mol% loading of catalyst and 2 h of microwave irradiation. Significantly, attempted RCM reactions without microwave irradiation were less successful. Indeed, exposure of peptide **33** to the first-generation Grubbs' catalyst (50 mol%) in CH_2Cl_2 at 50°C for 72 h gave only trace amounts (<10%) of cyclised product **34**. While RCM progressed further (~70%) using the more reactive Grubbs 2 catalyst, conditions could not be found to affect full cyclisation to **34** even under high catalyst loading (50 mol%). Changes in solvent, catalyst loading and reaction time had no positive effect on conversion. The addition of chaotropic salts to the reaction medium to disrupt aggregation also had no effect on RCM yield.

Hydrogenation of the carbocycle **34**, followed by Fmoc-deprotection, resin-cleavage, and aerial oxidation afforded the desired saturated peptide **35**. Analogous procedures led to the unsaturated counterpart **36**, as well as to the related **37**/**38** pair (Scheme 16) [16].

A similar methodology led to the resin-bound pentapeptide **39** (Scheme 17) bearing a pendant crotyl group. Subsequent microwave-assisted cross-metathesis between **39** and excess crotylglycine derivative **40** using 40 mol% Grubbs 2 catalyst in CH_2Cl_2 and 10% LiCl in DMF resulted in the formation of the unsaturated peptide **41** accompanied by dimer **43** (Scheme 18) [17]. Finally, a quantitative hydrogenation of **41** furnished the target peptide **42** containing two selectively constructed dicarba bridges [18].

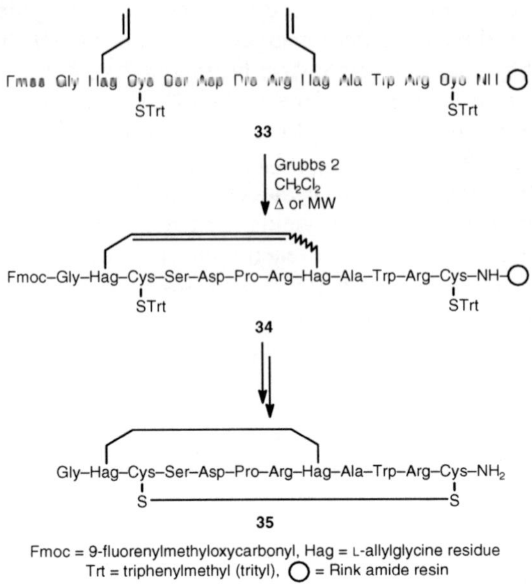

Scheme 15 Synthesis of the carbocyclic analogue **35** of α-conotoxin IMI

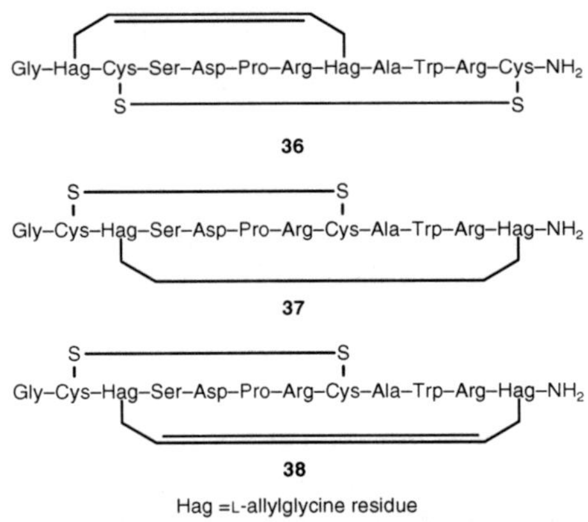

Scheme 16 Saturated and unsaturated analogues of α-conotoxin IMI

3 Cross-Metathesis

In connection with peptidomimetics and synthetic peptide conjugates, several α-substituted *N*-protected prolines containing various functionalities have been recently prepared by applying the cross-metathesis (CM) reaction between allylprolines **44** and **47** with terminal alkenes **45**, followed by catalytic hydrogenation of the intermediate compounds **46** and **48** (Scheme 19) [19].

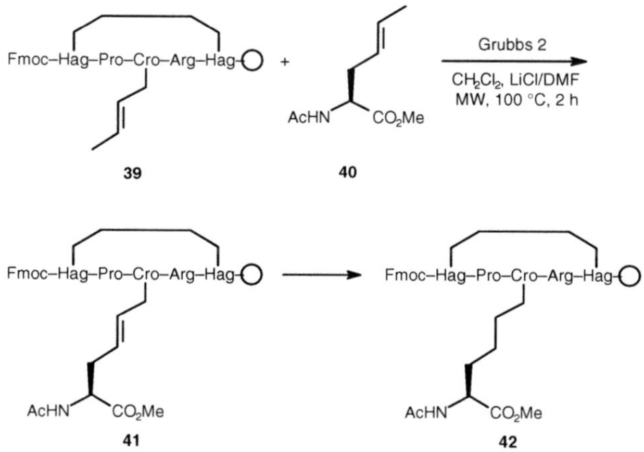

Fmoc = 9-fluorenylmethyloxycarbonyl
Hag = L-allylglycine residue, Cro = L-crotylglycine residue
○ = Wang resin

Scheme 17 Selective synthesis of peptide derivatives containing two dicarba bridges

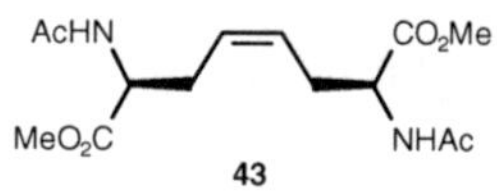

Scheme 18 Cross-metathesis product of crotylglycine derivative **40**

This versatile process allowed the achievement of various novel amino acids containing lipophilic and functionalised side chains from a common enantiomerically pure single precursor **44** or **47**. The cross-metathesis occurred with good selectivity and short reaction time under microwave heating conditions (150 W, 80°C with simultaneous cooling), affording yields in the range of 40–92%. Addition of $Ti(Oi\text{-}Pr)_4$ as a Lewis acid allowed a slight increase of the yield in the case of alkenes with Lewis basic substituents. In the test reaction between allylproline **44** and hept-1-ene, the desired heterocoupling product **46** was obtained in 60% isolated yield after 20 min of microwave irradiation at 80°C (Table 5). A slight increase of the yield (74%) was observed by doubling the amount of heptene with respect to **44**. By contrast, after 2 h in refluxing CH_2Cl_2, a 61% yield of **46** was reached when 2 eq. of heptene were used (Scheme 20).

Interestingly, this methodology led to the synthesis of the glucosyl-conjugated proline **49**, a new member of the class of glycosylated amino acids, which are nowadays the subject of intense synthetic efforts for their applications as biological tools in glycopeptidomimetics [19].

Another impressive application of microwave-assisted olefin metathesis was published by Barrett [20] with the synthesis of some viridiofungins **50** (viridiofungin A (Ar = 4-HOC_6H_4), B (Ar = Ph), and C (Ar = 3-indolyl)), members of

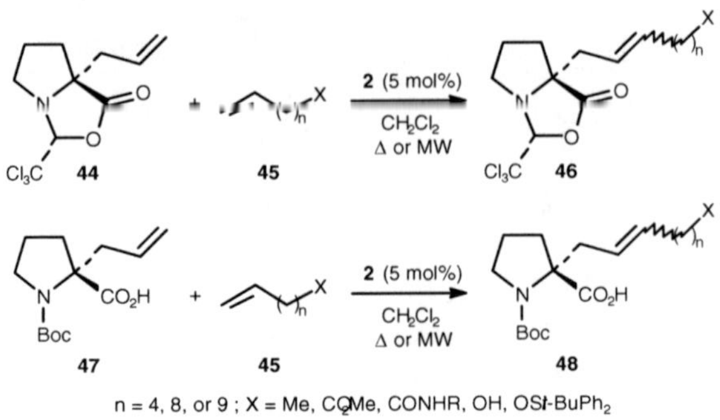

Scheme 19 The key CM step in the synthesis of α-substituted prolines

Table 5 Thermal versus microwave-assisted cross-metathesis between allylproline **44** and hept-1-ene (**45**, n = 4, X = Me) catalysed by Grubbs' complex **2**

Hept-1-ene	Reaction conditions	Yield (%)	*E*/*Z* ratio
1 eq.	Δ, 40°C, 2 h	53	4.8:1
	MW, 150 W, 80°C,[a] 20 min	60	5.0:1
2 eq.	Δ, 40°C, 2 h	61	6.3:1
	MW, 150 W, 80°C,[a] 20 min	74	4.4:1

[a] With simultaneous cooling.

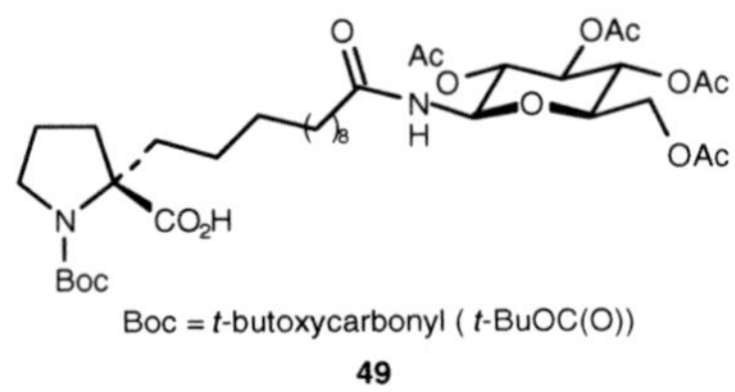

Scheme 20 Structure of a *N*-Boc-protected α-alkylated proline

a family of aminoacyl vinyl citrate antibiotics (Scheme 21). In a key C–C bond-forming step, cross-metathesis of alkene **51** and citrate **52** was examined using the second-generation Grubbs' catalyst (**2**), the second-generation Hoveyda–Grubbs' catalyst (**3**), the Blechert's catalyst (**56**), and the Grela's catalyst (**57**) (Scheme 22). Neither catalyst **2** nor **3** was especially effective with slow and incomplete conversions. Both catalysts **56** and **57** were superior, with the Grela's catalyst **57** the most effective under microwave irradiation at 55°C for 1 h. Alternatively, amides **55** could be converted in high yield (87–92%) into the viridiofungin derivatives **54** by cross-metathesis with alkene **51**. Noteworthingly, the vinyl citrates **55** were less reactive than the parent compound **52**, presumably because of their increased steric hindrance, which necessitated a prolonged reaction time (4 h) at 65°C.

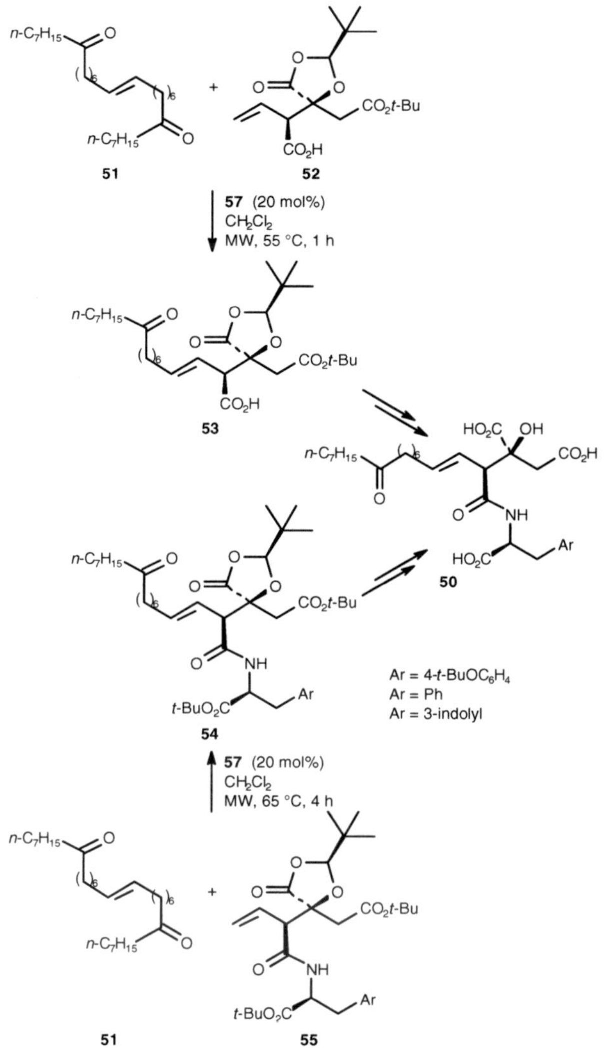

Scheme 21 The key CM steps in the synthesis of viridiofungin derivatives **50**

In a recent study of Lombardo et al., the importance of microwave irradiation in cross-metathesis has also been emphasised [21]. Thus, in their total synthesis of D-*ribo*-phytosphingosine **61**, a natural bioactive lipid ubiquitously distributed in many mammalian tissues, plants, fungi, as well as marine organisms, chain elongation of olefin **58** was based on a cross-metathesis with 1-tetradecene (**59**) (Scheme 23). Preliminary screening experiments in refluxing 1,2-dichloroethane with 2 eq. of **59** and first-generation Grubbs' catalyst (**1**) did not furnish any cross-coupled product, while second-generation Grubbs' catalyst (**2**) afforded the desired product **60** in 65% isolated yield after 24 h. Adopting the same reaction conditions for microwave heating resulted in 79% isolated yield after only 5 min of irradiation at 200 W (Table 6). Grubbs's catalyst **2** could be reduced from 10 mol% down to 3.5 mol% without appreciable reduction in overall isolated yields, while with lower amounts (e.g., 1 mol%) the yields dropped considerably (23%) [21].

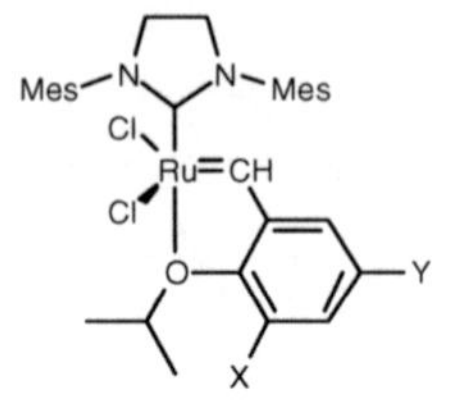

3 X = Y = H
56 X = Ph, Y = H
57 X = H, Y = NO_2

Scheme 22 The second-generation Hoveyda–Grubbs' catalyst (**3**) and derivatives

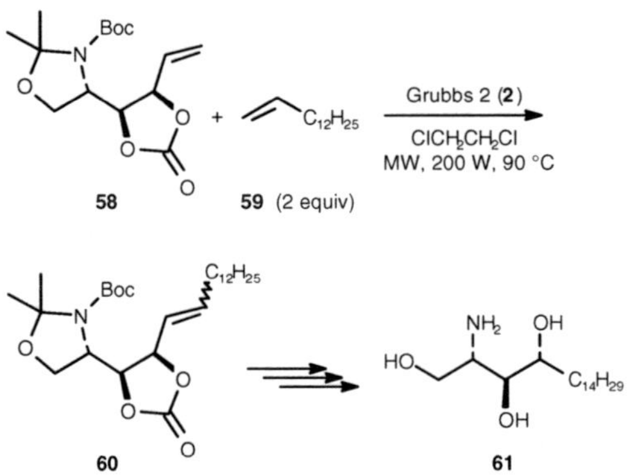

Boc = *t*-butoxycarbonyl (C(O)O *t*-Bu)

Scheme 23 The key CM step in the synthesis of D-*ribo*-phytosphingosine

Table 6 Microwave-assisted cross-metathesis between olefins **58** and **59** catalysed by Grubbs' complex **2**

2 (mol%)	Reaction time (min)	Yield (%)
10	5	79
5	5	76
3.5	5	77
3	5	70
1	15	23

Microwave-assisted cross-metathesis was also employed as the key step of the total synthesis of (+)-negamycin **66** (Scheme 24) [22], an unusual antibiotic containing a hydrazine peptide bond and exhibiting very low acute toxicity and strong inhibitory activity against multiple drug-resistant enteric Gram-negative bacteria. Negamycin has also considerable potential as a chemotherapeutic agent against genetic diseases. The cross-metathesis reaction between **62** and *t*-butyl acrylate (**63**) was investigated in the presence of the first- and second-generation Grubbs' catalysts (Scheme 24 and Table 7). It turned out that the conversion of olefin **62** and chemoselectivity (**64**/**65** ratio) enhancements were definitely more pronounced for catalyst **2** than **1**. With the latter catalyst, indeed, the conversions were modest and the unwanted dimer **65** prevailed over the target compound **64**. The opposite trend was found with catalyst **2**: the conversions were (almost) quantitative and the desired product **64** was formed selectively. Furthermore, microwave irradiation drastically shortened the cross-metathesis reaction time by 20-fold. For the conventional heating, indeed, 5 h were needed to bring the reaction to completion, whereas 15 min were sufficient when the cross-metathesis was conducted under microwaves. As a result, **64** was isolated with 83% yield [22].

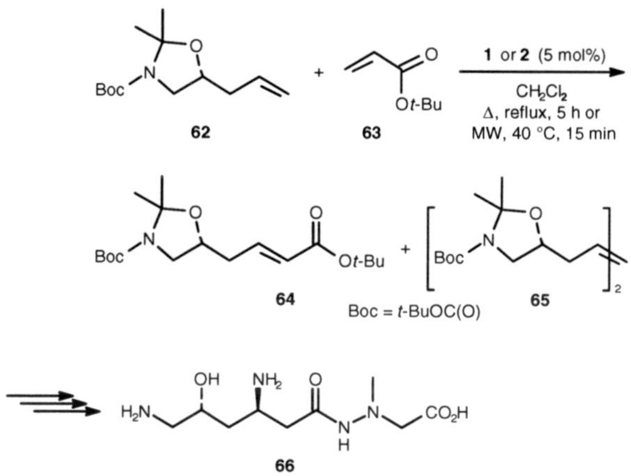

Scheme 24 The key CM step in the synthesis of (+)-negamycin

Table 7 Thermal versus microwave-assisted cross-metathesis between olefin **62** and *t*-butyl acrylate (**63**) catalysed by Grubbs' complexes **1** and **2**

Catalyst	Reaction conditions	62, conversion (%)	64/65 ratio
1	Δ, reflux, 5 h	36	41/59
	MW, 40°C, 15 min	47	24/76
2	Δ, reflux, 5 h	96	>99/traces
	MW, 40°C, 15 min	100	100/0

A recent, short, and convergent synthesis of cystothiazole A also relies on a microwave-assisted olefin cross-metathesis reaction [23]. Cystothiazole A (**69**, Scheme 25) is the major and most active member of a family of bithiazole metabolites. The cystothiazoles belong to the class of complex III inhibitors, which selectively bind to the cytochrome bc_1 complex of the mitochondrial respiratory chain. Cystothiazole A also displays significant activity against a broad range of fungi and in vitro cytotoxicity towards human colon carcinoma HCT-116 and leukemia K562 cells. When the cross-metathesis of **67** with the β-methoxyacrylate

68 by using 30 mol% of the second-generation Grubbs' catalyst (**2**) was performed under microwave irradiation at 100°C for 5.5 h with simultaneous cooling, cystothiazole A (**69**) was obtained in 25% isolated yield, together with an unidentified side-product and homodimers of both olefins **67** and **68**. By contrast, attempted cross-metatheses under various conventional heating conditions resulted in less than 5% conversion to the target compound **69**, and mainly homodimerisation and decomposition of the vinylbithiazole **67** were observed. In an alternative synthesis of **69** (Scheme 25), cross-metathesis of **68** with **71** under the optimised microwave conditions furnished the key intermediate **70** in an acceptable isolated yield of 55%. As expected, the final arylation of **70** using the Stille coupling protocol proceeded smoothly and afforded cystothiazole A in 83% yield [23].

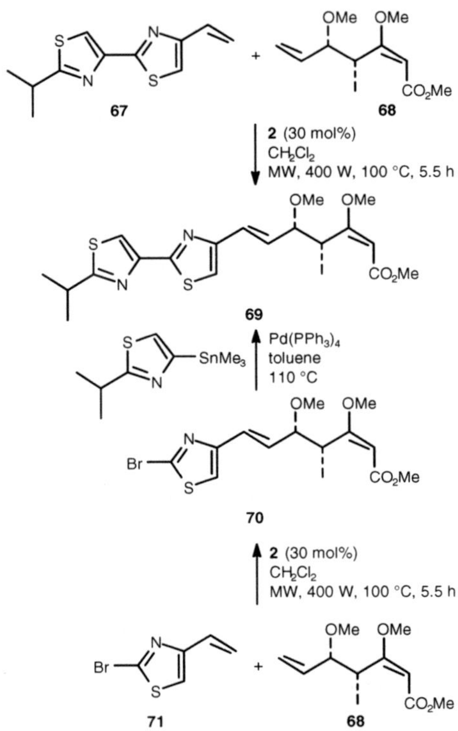

Scheme 25 The key CM step in the syntheses of cystothiazole A

Very recently, a general, two-step method has been described for the synthesis of azaspirocycles **75** (Scheme 26), involving the microwave-assisted cross-metathesis of methylenecycloalkanes **72** with protected allyl- and 3-butenylglycines **73** to afford the intermediates **74**, followed by treatment with triflic acid [24].

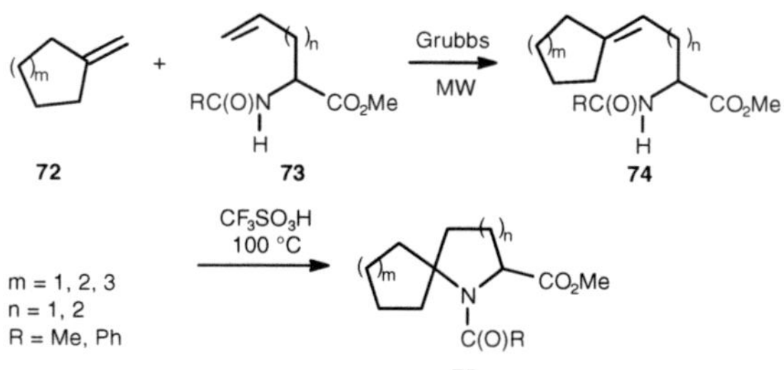

Scheme 26 Cross-metathesis route to azaspirocycles **75**

Furthermore, a recent, stereocontrolled total synthesis of the fungal metabolites stephacidin A (**76**), stephacidin B, and notoamide B (**77**) (Scheme 27) has also extensively taken advantage of microwave technology, which reduced reaction times from hours to minutes as well as increased the yields of several key transformations [25]. The stephacidins A and B are potent inhibitors of several human tumour cell lines with stephacidin B exhibiting a high cytotoxic potency against testosterone-dependent prostate LNCaP lymphoma.

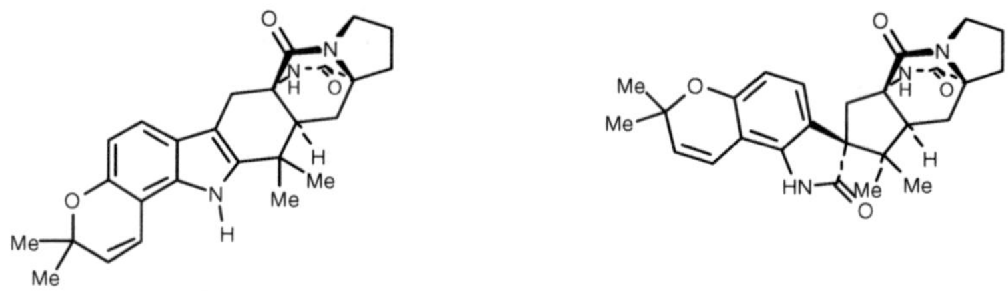

Scheme 27 Structure of stephacidin A (**76**) and notoamide B (**77**)

In a quite classical approach, cross-metathesis of the terminal olefin of **78** (Scheme 28) with methacrolein (**79**) using catalytic amounts of the Grubbs' second-generation catalyst in refluxing CH_2Cl_2 for 24 h readily afforded the aldehyde **80** in 65% yield with 15% recovered starting material. Unfortunately, additional quantities of the Grubbs 2 catalyst had to be added during the reaction raising the catalyst loading from the initial 5 to 20 mol%. Switching to the second-generation Hoveyda–Grubbs' catalyst **3** reduced the catalyst loading to 5 mol% and increased the yield for **80** to 71% with 10% recovered **78** after 48 h of reflux. Once again, recourse to microwave irradiation facilitated the heating of the metathesis reaction and reduced the reaction time. In the event, heating of the olefin **78** with methacrolein (**79**) in the presence of 5 mol% of Hoveyda–Grubbs 2 in CH_2Cl_2 at 100°C for 20 min generated the aldehyde **80** in 73% with 10% recovered starting material [25].

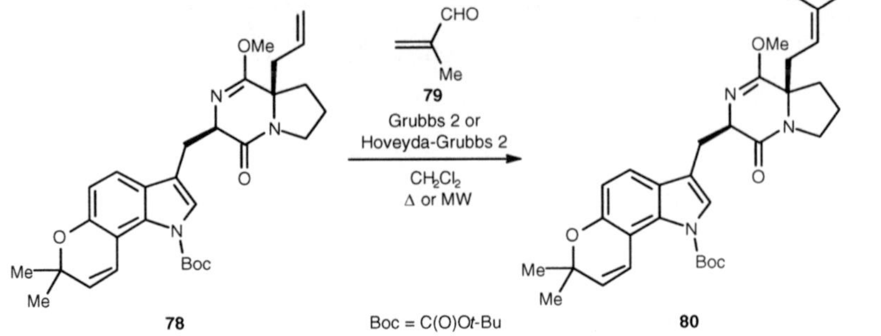

Scheme 28 Cross-metathesis route to stephacidin A (**76**) and notoamide B (**77**)

4 Enyne Metathesis

Recent efforts led to the synthesis of cyclobutene derivatives through the enyne metathesis of 1,5-enynes using the Hoveyda–Grubbs' catalyst **3** (Scheme 29) [26]. Initially, the reactivity of enyne **81** was explored in refluxing dichloromethane. After 36 h of reaction, the cyclobutene **82** was obtained in 35% yield. As previously observed in other investigations [4], the use of microwave irradiation proved beneficial, and the desired cyclobutene **82** was formed in 58% yield after only 35 min at 70°C (Table 8). The reaction was rather clean and, along with the expected cyclobutene **82**, only a small amount of the cross-metathesis product (5%) and the starting 1,5-enyne **81** (10–15%) were isolated. The beneficial effect of microwaves, however, was not confirmed in reactions carried out under an ethylene atmosphere. Under both classical and microwave heating conditions, cyclobutene **82** culminated at 20–22% yield, and extensive by-product formation was observed.

Table 8 Thermal versus microwave-assisted metathesis of enyne **81** catalysed by Hoveyda–Grubbs' complex **3**

Reaction conditions			Yield (%)
Δ, CH_2Cl_2, reflux		36 h	35
MW, CH_2Cl_2, 70°C		35 min	58
Δ, CH_2Cl_2, reflux	Ethylene (1 atm)	36 h	22
MW, CH_2Cl_2, 70°C	Ethylene (1 atm)	35 min	20

This methodology was successfully applied to a variety of 1,5-enynes **83** (Scheme 29). Likewise, the double cyclisation of the *para*-disubstituted phenyl derivative **85** led to the bis-cyclobutene **86**, albeit in a modest 19% yield (Scheme 29).

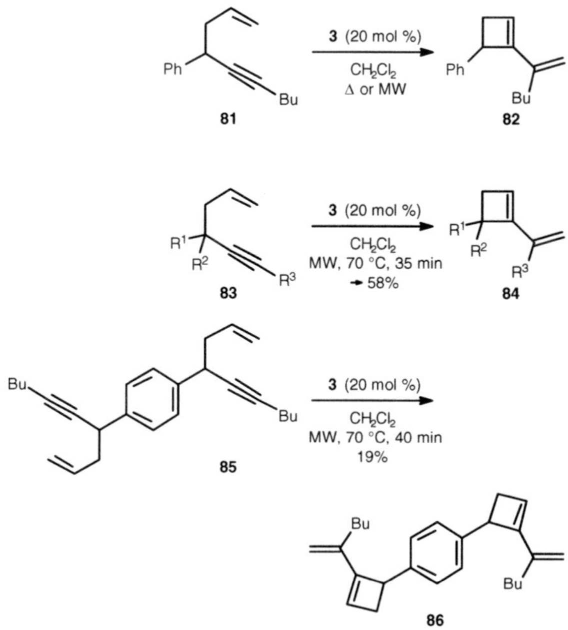

Scheme 29 Synthesis of cyclobutenes via enyne metathesis

Microwave-assisted olefin metathesis has also been employed for the projected total synthesis of the phenomenally potent anti-tumour macrolide amphidinolide N (**87**) (Scheme 30).

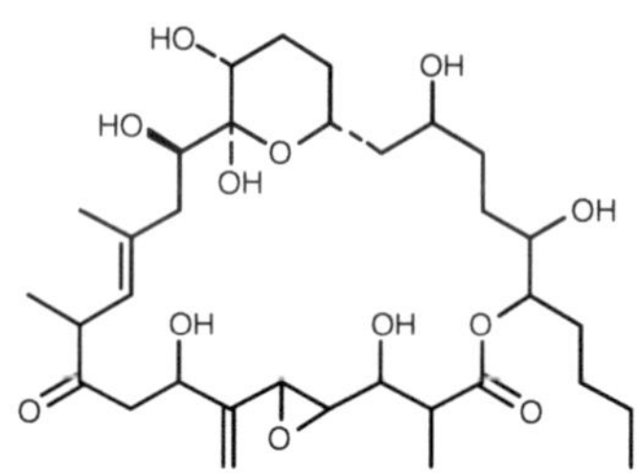

Scheme 30 Structure of amphidinolide N

Despite numerous attempts and extensive variation of the reaction parameters, enyne **88** could not be cyclised directly to generate macrocyclic compound **90**, but was cleanly converted into diene **89** in 60% yield upon microwave irradiation in the presence of the second-generation Grubbs' ruthenium carbene complex (**2**) under an ethylene atmosphere (Scheme 31). Diene **89** was apparently all but inert to further productive metathesis events, failing either to cyclise to the corresponding macrocycle (**90**) or to undergo a significant degree of oligomerisation, despite

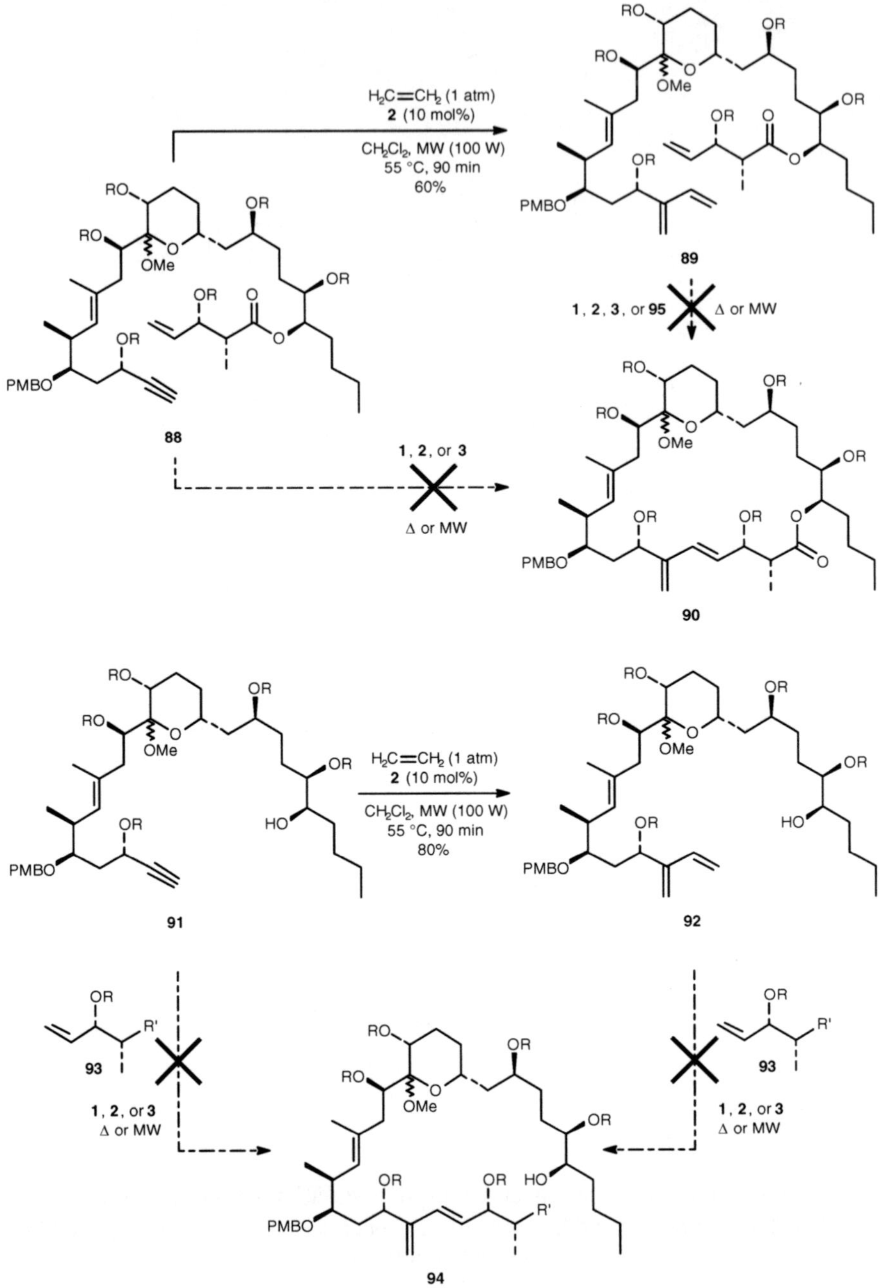

PMB = $CH_2C_6H_4OCH_3$; R = Si*t*-BuMe$_2$; R' = CO_2H, CO_2Me, CH_2OH or $CH_2OC(O)CMe_3$

Scheme 31 The key enyne and cross-metathesis steps in the synthesis of amphidinolide N

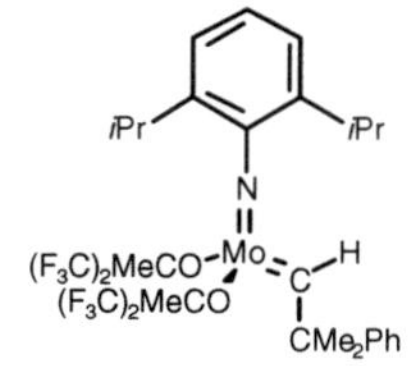
95

Scheme 32 Structure of the Schrock catalyst **95**

prolonged exposure (under purely thermal conditions or microwave irradiation) to the ruthenium-based catalysts **1**, **2**, and **3** or the highly active Schrock molybdenum-based catalyst **95** (Scheme 32) [27]. In an alternative strategy to amphidinolide N, it was found that, when alkyne **91** was exposed to catalyst **2** in CH_2Cl_2 saturated with ethylene under microwave irradiation, enyne cross-metathesis did occur cleanly, to give diene **92** in 80% yield. However, and for reasons that are presently unclear, alkyne **91** and diene **92** proved to be resistant to cross-metathesis with any of the terminal alkene coupling partners **93** (Scheme 31).

Microwave-assisted alkene–alkyne cross-metathesis (Scheme 33) and alkyne–diene methylene-free tandem-metathesis reactions (Scheme 34) also represented the key steps of the synthesis of benzhydrylamine derivatives [28]. Taking advantage of microwave irradiation, both reactions were performed in a few

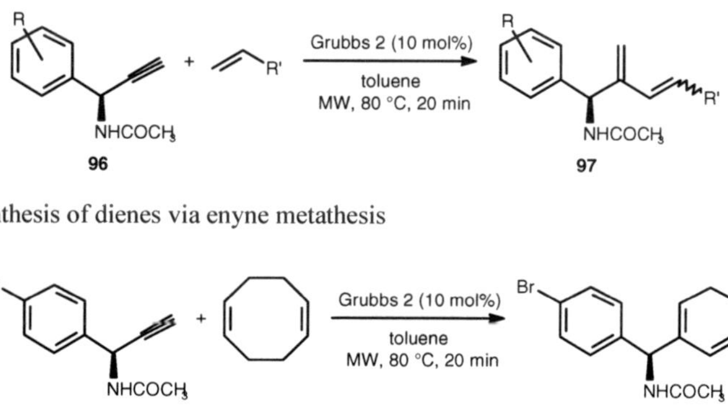

96 97

Scheme 33 Synthesis of dienes via enyne metathesis

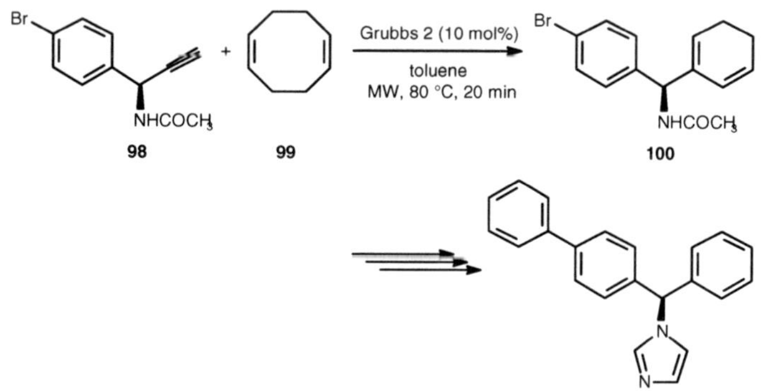

98 99 100 101

Scheme 34 Synthesis of (R)-bifonazole (**101**) via tandem-metathesis minutes and high yields. Thus, treatment of alkyne **96** (Scheme 33) with different alkenes in the presence of the Grubbs 2 catalyst in toluene at 80°C afforded dienes **97** in moderate to good yields (45–88%) after only 20 min of microwave irradiation. On the other hand, en route to bifonazole (**101**) (Scheme 34), an anti-fungal agent, the preparation of the 1,3-cyclohexadiene motif in compound **100** took place with 60% yield via a tandem-metathesis between alkyne **98** and *cis,cis*-1,5-cyclooctadiene (**99**), following the same microwave-assisted method [28].

Indolizidine derivatives could also be obtained via ring-closing enyne metathesis, as illustrated in Scheme 35 [11]. Diene **103**, for instance, was formed from enyne **102** in good 72% isolated yield, using either conventional heating or microwave irradiation. Noteworthy, the reaction time was strongly reduced from 2 days at room temperature to 30 min at 100°C under microwave irradiation.

102 103

Scheme 35 Synthesis of indolizidine **103** via enyne metathesis of **102**

5 Alkyne Metathesis

The metathesis of alkynes has received less attention than that of alkenes because alkynes and active alkyne metathesis catalysts are less common than their alkene counterparts. Metathesis of alkenes generally results in a mixture of *E*- and *Z*-isomers and the stereochemistry cannot at present be properly controlled. Alkynes, however, are readily available and their stereoselective semi-reduction into *E*- or *Z*-alkenes is routinely employed in organic synthesis.

The first homogeneous catalyst for alkyne metathesis was generated from $[Mo(CO)_6]$ and phenol as the co-catalyst [29]. Afterwards, phenol was advent-ageously replaced by (poly)halophenols and silanols [30]. In particular, with triphenylsilanol and diphenylsilanediol as co-catalysts for the cross-metathesis reaction of methoxytolan (**104**) (Scheme 36), the metathesis equilibrium was obtained after 16 h of reflux in octane under argon. In the case of the trisilanol **107** (Scheme 37), the equilibrium was reached only after 20 h of reflux. Not surprisingly, microwave irradiation significantly accelerated the reaction. Thus, with Ph_3SiOH and $Ph_2Si(OH)_2$, the equilibrium was observed after only 10 min of microwave irradiation in a resonance cavity, whereas with trisilanol **107**, the equilibrium was reached after 20 min of irradiation instead of 20 h under reflux. Noteworthy, this methodology was successfully extended to a range of functionalised alkynes and, using the catalyst system $[Mo(CO)_6]$–Ph_3SiOH, the equilibrium was attained in all cases after less than 15 min of irradiation. With 1-phenyl-2-trimethylsilylacetylene (**108**), the equilibrium was shifted to the desired 1,2-diphenylacetylene (**109**) (100% yield) due to the escape of the volatile 1,2-bis(trimethylsilyl)acetylene (**110**) (Scheme 38) [30].

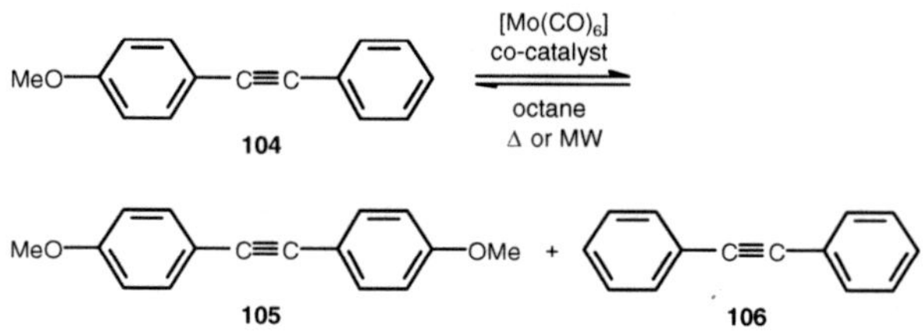

Scheme 36 Metathesis of 4-methoxytolan

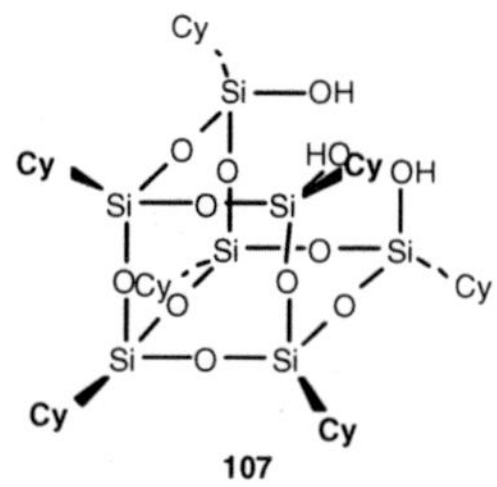

Scheme 37 The trisilanol used as co-catalyst with [Mo(CO)$_6$]

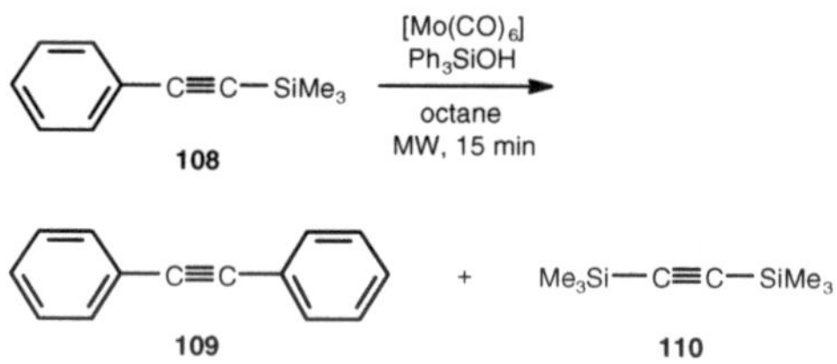

Scheme 38 Metathesis of 1-phenyl-2-trimethylsilylacetylene

Recourse to both ring-closing alkyne metathesis (RCAM) and microwave heating proved to be particularly powerful and flexible for the total synthesis of three members of the turriane family of natural products (**111**–**113**) (Scheme 39) isolated from the stem wood of an Australian tree [31].

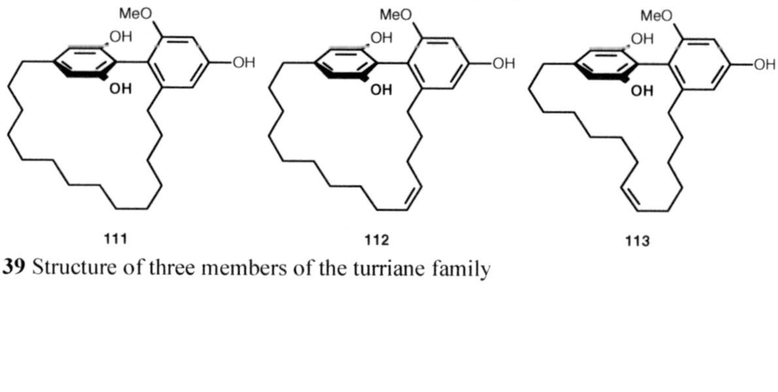

Scheme 39 Structure of three members of the turriane family

The total synthesis of turrianes **112** and **113** as reported by Fürstner et al. involves two key steps: the formation of (*i*) the sterically hindered biaryl entity and (*ii*) the 20-membered macrocycle containing a *Z*-configured double bond. For the latter reaction, the conventional RCM catalysed by various ruthenium–carbene complexes invariably led to the formation of mixtures of both stereoisomers with the undesirable *E*-cycloalkene prevailing. To circumvent this severe limitation, ring-closing alkyne metathesis (Scheme 40) followed by Lindlar hydrogenation of the resulting cycloalkynes was considered. It turned out that, under conventional heating, the well-defined tungsten–alkylidyne complex [(*t*-BuO)$_3$W≡C*t*-Bu] converted diynes **114** and **116** into the desired cycloalkynes **115** and **117** in reasonable yields (Table 9), whereas the use of [Mo(CO)$_6$] activated with 4-tri-fluoromethylphenol in chlorobenzene at 135°C provided even better results. Furthermore, application of microwave technology instead of conventional heating allowed to reduce the reaction time from 4–6 h to 5 min [31].

Table 9 Thermal versus microwave-assisted formation of the cyclophane core by RCAM

Substrate	Catalyst[a]	Reaction conditions	Yield (%)
114	[(*t*-BuO)$_3$W≡C*t*-Bu]	Δ, toluene, 80°C, 16 h	64
	[Mo(CO)$_6$], $F_3CC_6H_4OH$	Δ, chlorobenzene, 135°C, 4 h	83
	[Mo(CO)$_6$], $F_3CC_6H_4OH$	MW, chlorobenzene, 150°C, 5 min	69
116	[(*t*-BuO)$_3$W≡C*t*-Bu]	Δ, toluene, 80°C, 16 h	61
	[Mo(CO)$_6$], $F_3CC_6H_4OH$	Δ, chlorobenzene, 135°C, 6 h	76
	[Mo(CO)$_6$], $F_3CC_6H_4OH$	MW, chlorobenzene, 150°C, 5 min	71

[a] 10 mol%.

In sharp contrast with internal alkynes, terminal alkynes generally undergo a [2 + 2 + 2] cyclotrimerisation reaction, as illustrated in Scheme 41 by the cyclotrimerisation of triyne **118** in the presence of both the first-generation Grubbs' catalyst (**1**) and the second-generation catalyst system (**2**) [32].

Not surprisingly, when dienyne **120** was subjected to the Grubbs' complexes, a domino ring-closing metathesis reaction took place (Scheme 42). With the first-generation Grubbs' catalyst (**1**), optimum conditions resulted with reactions performed under microwave irradiation at 160°C in which case 76% conversion was observed after 45 min. For comparison, in the conventional preparative work, very little RCM product was obtained, even after prolonged reaction times (Table 10). On the other hand, in the conventional chemistry, the second-generation Grubbs' catalyst (**2**) dramatically improved the RCM reaction from an almost zero yielding reaction with catalyst **1** to a high yielding process of *ca.* 92% when the catalyst **2** was added at intervals. The change of reactivity was also seen during microwave heating. At *ca*. 5 mol% catalyst loading, full conversion was observed in toluene at 160°C after only 10 min [32].

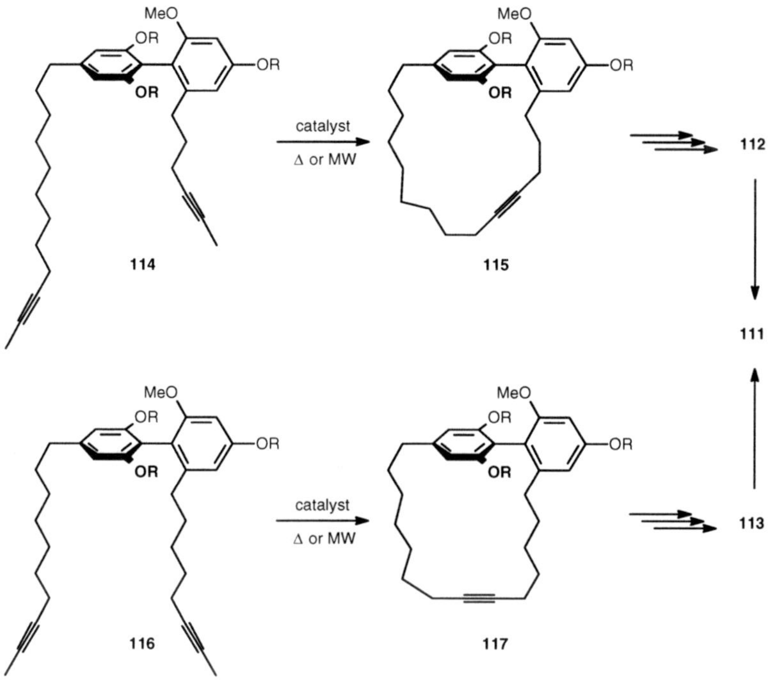

Scheme 40 Formation of the cyclophane core of the turrianes **111–113** by ring-closing alkyne metathesis

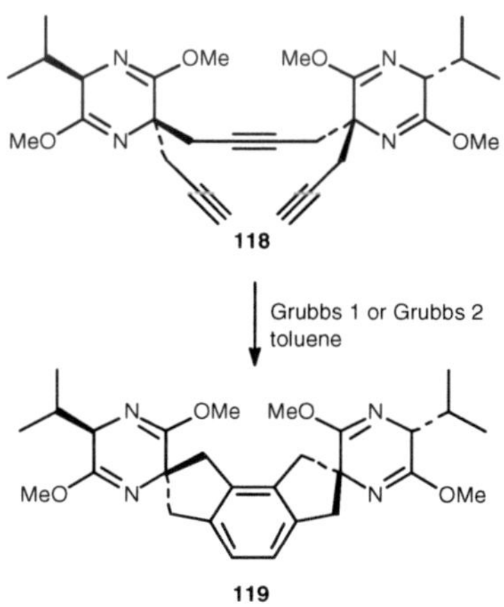

Scheme 41 [2 + 2 + 2] Cyclotrimerisation of triynes **118**

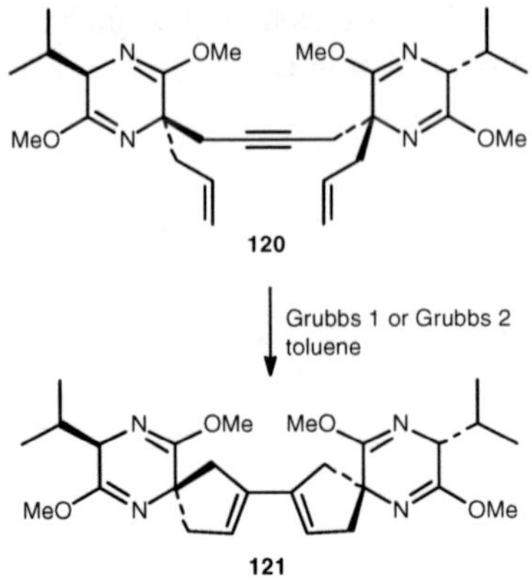

Scheme 42 RCM of dienyne **120**

Table 10 Thermal versus microwave-assisted cyclisation of dienyne **120** catalysed by Grubbs' complexes **1** and **2**

Catalyst	Reaction conditions	Yield (%)
1 (2 × 10 mol%)	Δ, 85°C, 2 × 5 h	Traces
1 (15 mol%)	MW, 160°C, 45 min	76
2 (3 × 10 mol%)	Δ, 85°C, 3 × 3 h	92
2 (5 mol%)	MW, 160°C, 10 min	100

Microwave-assisted alkyne metathesis has also been exploited in macromolecular chemistry. The first metathesis polymerisation of phenylacetylenes using $[W(CO)_6]$ and $[Mo(CO)_6]$ as catalyst precursors under microwave heating was reported by Dhanalakshmi and Sundararajan (Scheme 43) [33]. They utilised a kitchen microwave oven and specially designed high-pressure glassware, and conducted their reactions in a mixture of solvents for 1 h at high pressures and temperatures. However, they did not report the exact thermal conditions used. Similar polymerisations were performed under conventional heating at reflux, but 24 h were needed to complete the reaction.

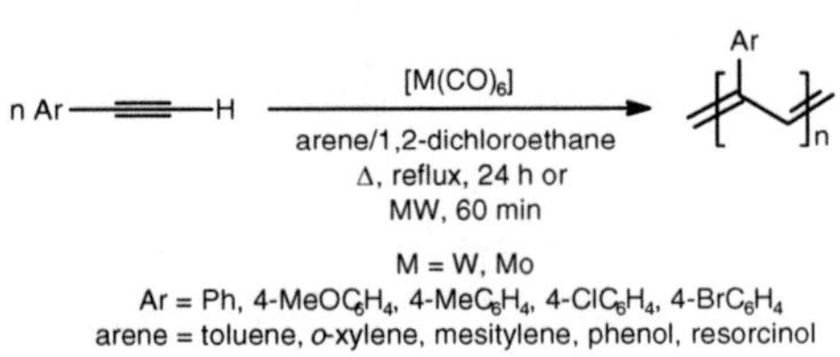

Scheme 43 Polymerisation of phenylacetylenes

Recently, disubstituted polyacetylene brushes were grown from modified silicon and quartz surfaces using the metathesis polymerisation technique employing tungsten hexachloride/tetraphenyl tin as the catalyst system (Scheme 44) [34]. The substrate surfaces were initially functionalised with terminal alkyne functional groups by using an alkyne-functionalised silane as a surface coupling agent. Surface polymerisation of 5-decyne under microwave irradiation at 150°C for 30 min was performed on the functional surfaces (**122**) to produce surfaces consisting of grafted poly(1,2-dibutylacetylene) brushes (**123**).

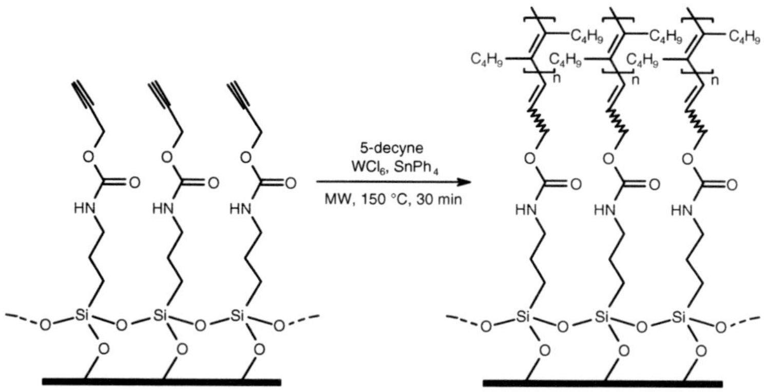

Scheme 44 Preparation of polymer brushes from alkyne-functionalised silicon substrate using tungsten-catalysed polymerisation of 5-decyne

The advantages of utilising microwave heating included rapid, highly efficient heating and the observation that it was not necessary to pre-age the catalyst solution or use longer time periods of heating (~24 h), which are conditions generally described for the polymerisation of alkynes during $WCl_6/SnPh_4$-catalysed polymerisations of disubstituted alkynes. Microwave-heated polymerisation commences at the onset of heating with no observed activation period.

6 Non-thermal Microwave Effects?

In the majority of the papers on microwave-assisted organic synthesis, microwave heating has been shown to dramatically reduce reaction times, increase product yields, and enhance product purity by reducing unwanted side reactions compared to conventional heating methods. The exact reasons why microwave irradiation is able to enhance chemical processes are still unknown. Since the early days of microwave synthesis, the observed rate accelerations and, sometimes, altered product distributions compared to conventionally heated experiments have led to speculation on the existence of so-called "specific" or "non-thermal" microwave effects. This concept has received considerable attention and is the subject of intense debate in the scientific community [35]. Non-thermal microwave effects have been postulated to result from a direct stabilising interaction of the electric field with specific molecules in the reaction medium that is not related to a macroscopic temperature effect.

The question of the involvement of an eventual non-thermal microwave effect in olefin metathesis has been raised by Lavastre, Kappe and coworkers, who investigated the RCM reaction of standard dienes **124** to produce five-, six-, or seven-membered carbo- or heterocycles **125** (Scheme 45) [36]. Careful comparison studies indicated that the rate enhancements observed under controlled microwave irradiation were merely due to thermal effects and not to the so-called "microwave effect".

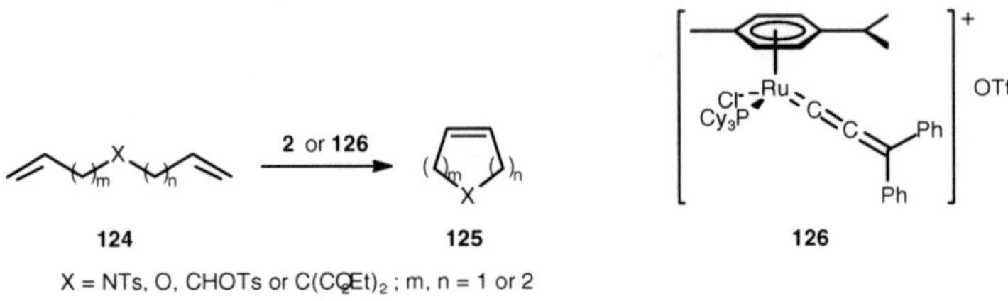

Scheme 45 RCM of standard dienes **124**

Furthermore, in order to probe the existence of non-thermal microwave effects, four synthetic transformations (Diels–Alder cycloaddition (Scheme 46a), alkylation of triphenylphosphine (Scheme 46b) and 1,2,4-triazole (Scheme 46c), and direct amidation (Scheme 47)) were re-evaluated under both microwave dielectric heating and conventional thermal heating [37]. Whereas previous studies have claimed the existence of non-thermal microwave effects in these reactions, a critical re-evaluation revealed that the observed effects were purely thermal and not related to the microwave field. Likewise, a detailed investigation of the kinetic profile of the Newman–Kwart rearrangement (Scheme 48) established that conventional and microwave heating were equivalent [38].

The enhancing effects of microwave irradiation, however, has been clearly demonstrated for the Ni-catalysed [2 + 2 + 2] cyclotrimerisation reaction.

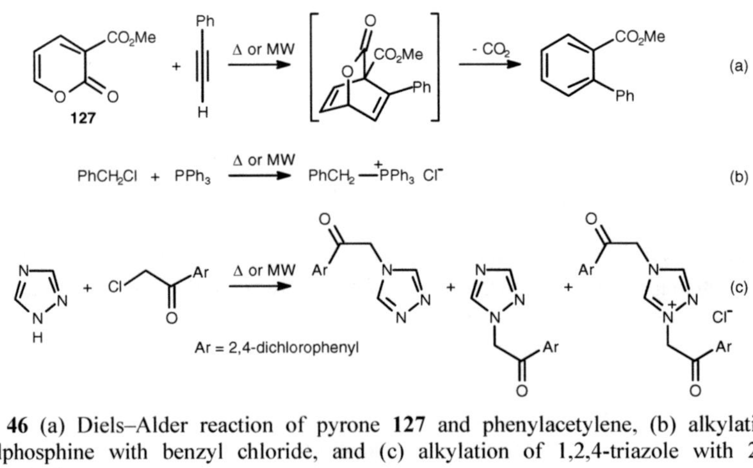

Scheme 46 (a) Diels–Alder reaction of pyrone **127** and phenylacetylene, (b) alkylation of triphenylphosphine with benzyl chloride, and (c) alkylation of 1,2,4-triazole with 2,2′,4′-trichloroacetophenone

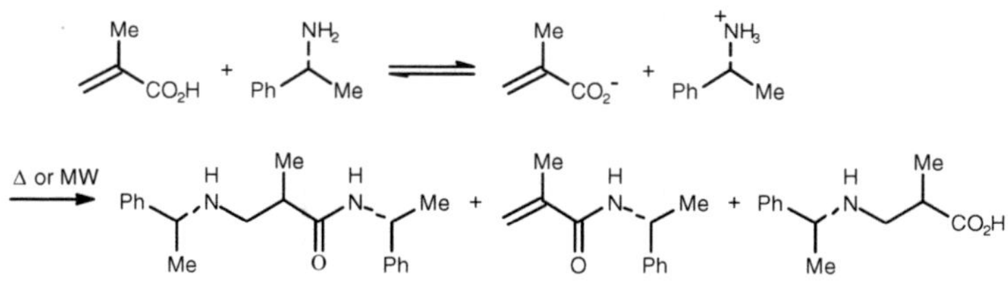

Scheme 47 Amidation of methacrylic acid with (R)-1-phenylethylamine

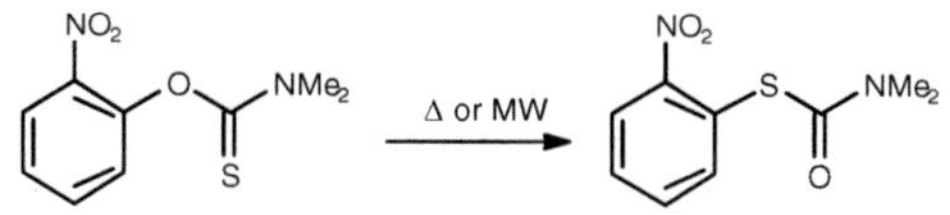

Scheme 48 The Newman–Kwart rearrangement

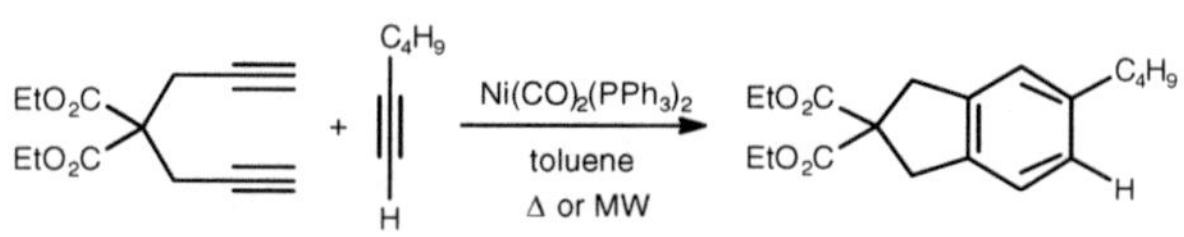

Scheme 49 Nickel-catalysed [2 + 2 + 2] cyclotrimerisation of diyne **128** and 1-hexyne

In particular, microwave irradiation of a mixture of diethyl dipropargyl malonate (**128**) and 1-hexyne **129** in toluene for only 2 min led to complete conversion of the diyne **128** and furnished the desired product **130** in 78% yield, with a final reaction temperature of 82°C (Scheme 49) [39]. In sharp contrast, conducting the same reaction without microwave irradiation in an oil bath heated at 92°C so as to reach a final temperature of 82°C inside the reaction vessel, did not yield any cyclotrimerisation product, holding thereby the controversy on the existence or non-existence of specific microwave effects in organic synthesis. In olefin metathesis, all the investigations –the one of Lavastre and Kappe excepted– [36] were performed under different reaction conditions rendering therefore impossible the comparison of results obtained under conventional and microwave heating.

7 Conclusions

Microwave-assisted organic synthesis has received increasing attention in recent years as a valuable technique for accelerating olefin metathesis reactions. Reductions in reaction time, increases in yield, and suppression of side-product formation are often observed when switching from conductive to microwave heating. In this regard, microwave-assisted olefin metathesis contributes to green chemistry, particularly towards developing environment-friendly synthetic procedures. Furthermore, by reducing reaction times from hours to minutes, microwave heating also allows significant savings in energy [40]. Although there is still considerable debate and speculation on the nature and/or the existence of so-called "non-thermal" microwave effects that could provide a rationalisation for the often observed significant rate and yield enhancements, there is little doubt that microwave heating will become a standard technique in most laboratories within a few years.

Acknowledgements The authors gratefully acknowledge the "Fonds national de la Recherche scientifique" (F.N.R.S.), Brussels, and the "Fonds spéciaux pour la Recherche" (University of Liège) for the purchase of a microwave reactor. The authors also thank the "Fonds pour la Formation à la Recherche dans l'Industrie et dans l'Agriculture" (F.R.I.A.) for a fellowship to F. Nicks and D. Bicchielli.

References

[1] For some recent monographs on microwave-assisted reactions, see: (a) Hayes BL (2002) Microwave synthesis: chemistry at the speed of light, 1st edn., CEM, Matthews, NC; (b) Microwaves in organic and medicinal chemistry. In Kappe CO, Stadler A (eds.) (2005) Methods and principles in medicinal chemistry, vol. 25, Wiley-VCH, Weinheim, Germany; (c) Loupy A (2006) Microwaves in organic synthesis, 2nd edn., Wiley-VCH, Weinheim, Germany

[2] For some recent reviews on microwave-assisted reactions, see: (a) Perreux L, Loupy A (2001) A tentative rationalization of microwave effects in organic synthesis according to the reaction medium, and mechanistic considerations. Tetrahedron 57: 9199–9223; (b) Lidström P, Tierney J, Wathey B, Westman J (2001) Microwave-assisted organic synthesis – a review. Tetrahedron 57: 9225–9283; (c) Larhed M, Moberg C, Hallberg A (2002) Microwave-accelerated homogeneous catalysis in organic chemistry. Acc Chem Res 35: 717–717; (d) Kappe CO (2004) Controlled microwave heating in modern organic synthesis. Angew Chem Int Ed 43: 6250–6284; (e) Hayes BL (2004) Recent advances in microwave-assisted synthesis. Aldrichim Acta 37: 66–76; (f) Nüchter M, Ondruschka B, Bonrath W, Gum A (2004) Microwave-assisted synthesis – a critical technology overview. Green Chem 6: 128–141; (g) Xu Y, Guo QX. (2004) Syntheses of heterocyclic compounds under microwave irradiation. Heterocycles 63: 903–974; (h) Desai B, Kappe CO (2004) Microwave-assisted synthesis involving immobilized catalysts. Top Curr Chem 242: 177–208; (i) de la Hoz A, Diaz-Ortiz A, Moreno A (2005) Activation of organic reactions by microwaves. Adv Org Synth 1: 119–171; (j) Dallinger D, Kappe CO (2007) Microwave-assisted synthesis in water as solvent. Chem Rev 107: 2563–2591; (k) Baxendale IR, Hayward JJ, Ley SV (2007) Microwave reactions under continuous flow conditions. Comb Chem High Throughput Screen 10: 802–836; (l) Kappe CO (2008) Microwave dielectric heating in synthetic organic chemistry. Chem Soc Rev 37: 1127–1139; (m) Polshettiwar V, Varma RS (2008) Aqueous microwave chemistry: a clean and green synthetic tool for rapid drug discovery. Chem Soc Rev 37: 1546–1557; (n) Appukkuttan P, Van der Eycken E (2008) Recent developments in microwave-assisted, transition-metal-catalysed C–C and C–N bond-forming reactions. Eur J Org Chem 1133–1155

[3] Varray S, Gauzy C, Lamaty F, Lazaro R, Martinez J (2000) Synthesis of cyclic amino acid derivatives via ring closing metathesis on a poly(ethylene glycol) supported substrate. J Org Chem 65: 6787–6790

[4] Coquerel Y, Rodriguez J (2008) Microwave-assisted olefin metathesis. Eur J Org Chem 1125–1132

[5] (a) Geng X, Danishefsky SJ (2004) Total synthesis of aigialomycin D. Org Lett 6: 413–416; (b) Yang ZQ, Geng X, Solit D, Pratilas CA, Rosen N, Danishefsky SJ (2004) New efficient synthesis of resorcinylic macrolides via ynolides: establishment of cycloproparadicicol as synthetically feasible preclinical anticancer agent based on Hsp90 as the target. J Am Chem Soc 126: 7881–7889

[6] Vu NQ, Chai CLL, Lim KP, Chia SC, Chen A (2007) An efficient and practical total synthesis of aigialomycin D. Tetrahedron 63: 7053–7058

[7] Nosse B, Schall A, Jeong WB, Reiser O (2005) Optimization of ring-closing metathesis: inert gas sparging and microwave irradiation. Adv Synth Catal 347: 1869–1874

[8] Benakki H, Colacino E, André C, Guenoun F, Martinez J, Lamaty F (2008) Microwave-assisted multi-step synthesis of novel pyrrolo-[3,2-*c*]quinoline derivatives. Tetrahedron 64: 5949–5955

[9] Declerck V, Ribière P, Martinez J, Lamaty F (2004) Sequential *aza*-Baylis–Hillman/ring closing metathesis/aromatization as a novel route for the synthesis of substituted pyrroles. J Org Chem 69: 8372–8381

[10] (a) O'Brien CJ, Kantchev EAB, Valente C, Hadei N, Chass GA, Lough A, Hopkinson AC, Organ MG (2006) Easily prepared air- and moisture-stable Pd–NHC (NHC = N-heterocyclic carbene) complexes: a reliable, user-friendly, highly active palladium precatalyst for the Suzuki–Miyaura reaction. Chem Eur J 12: 4743–4748; (b) Organ MG, Avola S, Dubovyk I, Hadei N, Kantchev EAB, O'Brien CJ, Valente C (2006) A user-friendly, all-purpose Pd–NHC (NHC = N-heterocyclic carbene) precatalyst for the Negishi reaction: a step towards a universal cross-coupling catalyst. Chem Eur J 12: 4749–4755

[11] Lamberto M, Kilburn JD (2008) Synthesis of indolizidines from dialkylated isocyanides: a novel radical cyclisation/N-alkylation/ring closing metathesis approach. Tetrahedron Lett 49: 6364–6367

[12] Jam F, Tullberg M, Luthman K, Grøtli M (2007) Microwave assisted synthesis of spiro-2,5-diketopiperazines. Tetrahedron 63: 9881–9889

[13] Hammer K, Undheim K (1997) Ruthenium(II) in ring closing metathesis for the stereoselective preparation of cyclic 1-amino-1-carboxylic acids. Tetrahedron 53: 2309–2322

[14] Terracciano S, Bruno I, D'Amico E, Bifulco G, Zampella A, Sepe V, Smith CD, Riccio R (2008) Synthetic and pharmacological studies on new simplified analogues of the potent actin-targeting jaspamide. Bioorg Med Chem 16: 6580–6588

[15] Robinson AJ, Elaridi J, van Lierop BJ, Mujcinovic S, Jackson WR (2007) Microwave-assisted RCM for the synthesis of carbocyclic peptides. J Pept Sci 13: 280–285

[16] See also: (a) Robinson AJ, Garland R, Illesinghe J, van Lierop B, Gooding S, Whelan A, Elaridi J, Teoh E, Jackson WR, Applying homogeneous catalysis to the synthesis of peptidomimetics, 16th International Symposium on Homogeneous Catalysis, Florence (Italy), July 6–11, 2008, abstract P 373; (b) Illesinghe J, Garland R, Xing Guo C, Ahmed A, van Lierop B, Jackson WR, Robinson AJ, Metathesis-assisted synthesis of cyclic peptides, 16th International Symposium on Homogeneous Catalysis, Florence (Italy), July 6–11, 2008, abstract P 374

[17] Elaridi J, Patel J, Jackson WR, Robinson AJ (2006) Controlled synthesis of (*S*,*S*)-2,7-diaminosuberic acid: a method for regioselective construction of dicarba analogues of multicystine-containing peptides. J Org Chem 71: 7538–7545

[18] See also: Teoh E, Garland R, Illesinghe J, Whelan A, Kozowski Z, Jackson WR, Robinson AJ, Using cross metathesis to generate peptide-polymer hybrids, 16th International Symposium on Homogeneous Catalysis, Florence (Italy), July 6–11, 2008, abstract P 375

[19] Lumini M, Cordero FM, Pisaneschi F, Brandi A (2008) Straightforward synthesis of α-substituted prolines by cross-metathesis. Eur J Org Chem 2817–2824

[20] Goldup SM, Pilkington CJ, White AJP, Burton A, Barrett AGM (2006) A simple, short, and flexible synthesis of viridiofungin derivatives. J Org Chem 71: 6185–6191

[21] Lombardo M, Capdevila MG, Pasi F, Trombini C (2006) An efficient high-yield synthesis of D-*ribo*-phytosphingosine. Org Lett 8: 3303–3305

[22] Hayashi Y, Regnier T, Nishiguchi S, Sydnes MO, Hashimoto D, Hasegawa J, Katoh T, Kajimoto T, Shiozuka M, Matsuda R, Node M, Kiso Y (2008) Efficient total synthesis of (+)-negamycin, a potential chemotherapeutic agent for genetic diseases. Chem Commun 2379–2381

[23] Gebauer J, Arseniyadis S, Cossy J (2008) Total synthesis of cystothiazole A by microwave-assisted olefin cross-metathesis. Eur J Org Chem 2701–2704

[24] Jackson R, Gartshore C, Illesinghe J, Robinson AJ, A metathesis route to spirocyclic alkaloid precursors, 16th International Symposium on Homogeneous Catalysis, Florence (Italy), July 6–11, 2008, abstract P 378

[25] Artman GD III, Grubbs AW, Williams RM (2007) Concise, asymmetric, stereocontrolled total synthesis of stephacidins A, B and notoamide B. J Am Chem Soc 129: 6336–6342

[26] Debleds O, Campagne JM (2008) 1,5-Enyne metathesis. J Am Chem Soc 130: 1562–1563

[27] Nicolaou KC, Brenzovich WE, Bulger PG, Francis TM (2006) Synthesis of *iso*-epoxy-amphidinolide N and *des*-epoxy-carabenolide I structures. Initial forays. Org Biomol Chem 4: 2119–2157

[28] (a) Castagnolo D, Renzulli ML, Galletti E, Corelli F, Botta M (2005) Microwave-assisted ethylene–alkyne cross-metathesis: synthesis of chiral 2-(*N*-1-acetyl-1-arylmethyl)-1,3-butadienes. Tetrahedron: Asymmetry 16: 2893–2896; (b) Castagnolo D, Giorgi G, Spinosa R, Corelli F, Botta M (2007) Practical syntheses of enantiomerically pure *N*-acetylbenzhydrylamines. Eur J Org Chem 3676–3686

[29] Mortreux A, Blanchard M (1974) Metathesis of alkynes by a molybdenum hexacarbonyl–resorcinol catalyst. J Chem Soc Chem Commun 786–787

[30] Villemin D, Héroux M, Blot V (2001) Silanol–molybdenum hexacarbonyl as a new efficient catalyst for metathesis of functionalised alkynes under microwave irradiation. Tetrahedron Lett 42: 3701–3703

[31] Fürstner A, Stelzer F, Rumbo A, Krause H (2002) Total synthesis of the turrianes and evaluation of their DNA-cleaving properties. Chem Eur J 8: 1856–1871

[32] Efskind J, Undheim K (2003) High temperature microwave-accelerated ruthenium-catalysed domino RCM reactions. Tetrahedron Lett 44: 2837–2839

[33] Dhanalakshmi K, Sundararajan G (1997) Microwave-assisted polymerisation of phenylacetylenes. Polym Bull 39: 333–337

[34] Jhaveri SB, Carter KR (2007) Disubstituted polyacetylene brushes grown via surface-directed tungsten-catalyzed polymerization. Langmuir 23: 8288–8290

[35] de la Hoz A, Diaz-Ortiz Á, Moreno A (2005) Microwaves in organic synthesis. Thermal and non-thermal microwave effects. Chem Soc Rev 34: 164–178

[36] Garbacia S, Desai B, Lavastre O, Kappe CO (2003) Microwave-assisted ring-closing metathesis revisited. On the question of the nonthermal microwave effect. J Org Chem 68: 9136–9139

[37] Herrero MA, Kremsner JM, Kappe CO (2008) Nonthermal microwave effects revisited: on the importance of internal temperature monitoring and agitation in microwave chemistry. J Org Chem 73: 36–47

[38] Gilday JP, Lenden P, Moseley JD, Cox BG (2008) The Newman–Kwart rearrangement: a microwave kinetic study. J Org Chem 73: 3130–3134

[39] Teske JA, Deiters A (2008) Microwave-mediated nickel-catalyzed cyclo-trimerization reactions: total synthesis of illudine. J Org Chem 73: 342–345

[40] Razzaq T, Kappe CO (2008) On the energy efficiency of microwave-assisted organic reactions. ChemSusChem 1: 123–132

PART III. NEW MATERIALS BY METATHESIS POLYMERIZATION AND RELATED CHEMISTRY

Acyclic Diene Metathesis (ADMET) Polymerization of Bis(4-pentenyl) dimethylstannane and Bis(4-pentenyl) diphenylstannane with an Electrochemically Activated Catalyst System

Solmaz Karabulut, Yavuz İmamoğlu

Department of Chemistry, Hacettepe University 06800 Beytepe, Ankara, Turkey

Tel: +903122977955, +903122976082; fax: +903122992163;
e-mails: imamoglu@hacettepe.edu.tr, solmazk@hacettepe.edu.tr

Abstract The microstructure and thermal analysis of two polycarbostannanes, obtained by acyclic diene metathesis (ADMET) polymerization in the presence of an electrochemically reduced tungsten-based catalyst system, were investigated in this study by NMR, DSC, and TGA techniques.

Keywords WCl_6 · Tin · Electrochemically reduced catalyst · Hybrid polymers · Polycarbostannanes · Acyclic diene metathesis (ADMET) polymerization

1 Introduction

Organic–inorganic hybrid polymers are macromolecules in which inorganic groups, such as transition or main group metal atoms, are linked to organic groups. Such polymers are desirable because they combine the properties of inorganic polymers, such as heat resistance and ionic conductivity, with the processability, flexibility and lower cost of organic polymers [1]. They are also useful as processible precursors for ceramic materials [2, 3]. A variety of organic–inorganic hybrid polymers containing silane, siloxane, tin and germanium components has been recently made available using ADMET [4–10]. The ADMET polymerization route possesses a significant advantage as compared to other systems where prior monomer functionalization can lead to catalyst deactivation or thermodynamically unfavorable polymerization conditions. Furthermore, this methodology is versatile and flexible, since one parent monomer can be substituted rather simply using various nucleophiles thus creating a large family of derived polymers with vastly different properties. Well-defined catalysts such as Schrock's [Mo] and [W] catalysts and Grubbs' first and second generation Ru catalysts, as well as classical catalyst systems based on phenoxide derivatives of tungsten halides were used as catalyst precursors for the ADMET polymerization [5–8].

This study describes polymerization of dienes containing tin via ADMET chemistry by an alternative catalyst system (WCl_6-e^--Al-CH_2Cl_2) that is prepared easier and with lower cost than other catalysts. The microstructure and thermal behaviour of the resulted polycarbostannanes were examined by ^{13}C NMR, ^1H NMR and DSC, TGA. Gilet and coworkers reported that the electrochemical reduction of WCl_6 and $MoCl_5$ produces metathetical active species [11, 12]. A WCl_6-e^--Al-CH_2Cl_2 system, which has been electrochemically reduced, catalyzes the metathesis of α- and β-olefins with good activity and selectivity. A facile route for the electrochemically generation of an alkene metathesis catalyst from a methylene chloride solution of WCl_6 was described by Düz et al. [13]. The WCl_6–e^-–Al–CH_2Cl_2 system has been previously proven to catalyze the acyclic diene metathesis (ADMET) polymerization with good activity and selectivity [14–17].

2 Experimental

The synthesis of monomers was reported in literature [16]. The catalyst was prepared according to previously described procedures [15]. All polymerization reactions were performed using standard Schlenk techniques, at room temperature and under nitrogen atmosphere. A typical polymerization reaction is as follows: Monomer (1.6 mmol) was charged to a round bottom flask equipped with a Teflon Roto-flow valve and a magnetic stirring bar in a nitrogen-purged dry box. Then 1 ml (0.02 M) of catalytic solution was added to the monomer from a Schlenk tube. The valve was sealed and the flask taken from the dry box to a high-vacuum Schlenk line and evacuated while stirring via magnetic agitation. A slow gelation was observed and stirring was continued for about 24–36 h until ethylene evolution stopped. Purification of the polymer from the dry chloroform solution was achieved by transfer to dry methanol using standard Schlenk techniques. The viscous liquid polymers were dried overnight in vacuum, at room temperature. Yields of the polymerizations were determined gravimetrically. The purified polymers were analyzed by ^1H and ^{13}C-NMR, GPC, DSC and TGA techniques. ^1H and ^{13}C NMR spectra were recorded with a Bruker GmbH 400 MHz high-performance digital FT-NMR spectrometer using $CDCl_3$ as the solvent and tetramethylsilane as the reference. Average molecular weight (M_w) was determined by gel permeation chromatography. GPC analyses were performed with a Shimadzu LC-10AD/VP liquid chromatograph equipped with a Shimadzu SPD-10A/VP UV detector, relative to polystyrene standards. Samples prepared in THF (1%) were passed through a μ-styragel column. A constant flow rate of 1 ml/min was maintained at 25°C. Glass transition temperatures were measured at a heating rate of 10°C/min in the range −100°C to +50°C by Shimadzu DSC-60. Thermogravimetric measurements were made on a SETARAM 92 apparatus under an argon with a flow rate of 40 ml/min and programmed heating from 25–800°C, at 10°C/min. Elemental analyses were performed by CHNS-932 (LECO) Elementel Analyzer.

3 Results and Discussion

In this work, ADMET polymerization of tin containing monomers (**1a** and **1b**), catalyzed by electrochemically produced W-based active species, were studied (Equation 1). All reactions were initiated in bulk, at room temperature and under nitrogen atmosphere. Characterization using NMR, GPC and elemental analysis showed products (**2a**, **2b**) to be consistent with the assigned structures, typical for ADMET polymers. Evolution of ethylene occurs slowly, upon exposure of the dienes (**1a**, **1b**) to the electrochemically reduced W-based catalyst. Viscosity changes are apparent over a 24 h period while magnetic agititation becomes difficult.

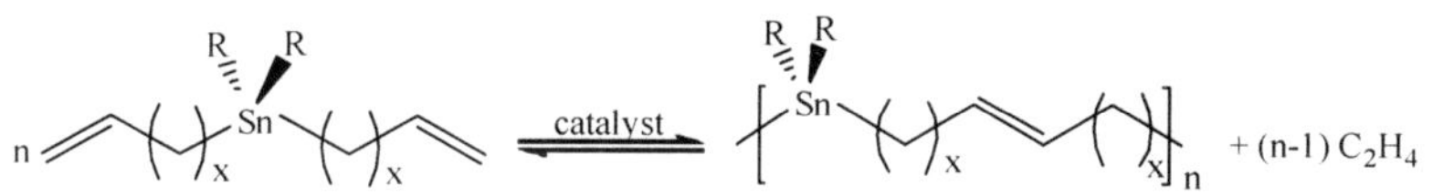

(1)

(**1a**) x= 3 R=CH_3 (**2a**) x= 3 R=CH_3

(**1b**) x= 3 R=C_6H_5 (**2b**) x= 3 R=C_6H_5

All visual characteristics of the reaction indicate that successful polymerization is taking place. Relative molecular weight values and polydispersities, determined by gel permation chromatography (GPC), approached values of 15,200–15,300 and of 2.10–2.80, respectively, consistent with the ADMET stepwise condensation mechanism (Table 1). The polymers (**2a**, **2b**) obtained are soluble in several common organic solvents and exist as highly viscous, transparent materials at room temperature.

The ^{1}H-NMR and ^{13}C-NMR spectra of the polymers (**2a**, **2b**) obtained with electrochemically reduced tungsten based active species are shown in Figures 1 and 2, respectively. In the ^{1}H NMR spectrum of the polymers (**2a** and **2b**), a minor terminal olefin signal can be observed at around 5.0 and 5.9 ppm. The ^{13}C-NMR spectrum of the product **2a** illustrates that metathesis has yielded both *trans* and *cis* internal olefins whose signals appear at 130.68 and 129.56 ppm, respectively (Figure 1). Based on the intensities of these peaks, the polymer was assigned to have a higher amount of *trans* configuration (Table 1). As evidenced in Figure 2, the terminal C=C double bond in each polymer unit arises as a consequence of metathesis condensation. The respective signals (at 129.96 and 130.11 ppm) for the polymer **2b** represent respectively the *cis* and *trans* geometric isomers of olefins located along the unsaturated polymer backbone.

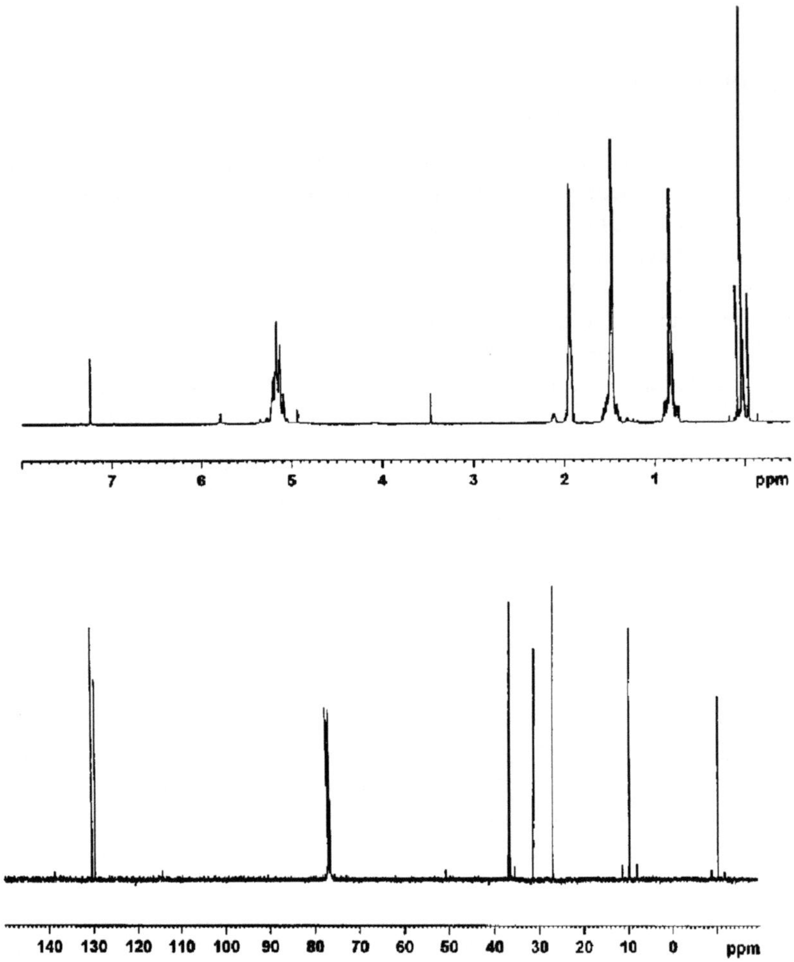

Figure 1 ^{1}H and ^{13}C NMR spectrum of the polymer of bis(4-pentenyl)dimethylstannane (**2a**)

Experiments to thoroughly examine thermal transitions for each polymer by means of differential scanning calorimetry (DSC) have been pursued. DSC curves for the polymers **2a** and **2b** show no observable melting transitions (for scans running from -100°C to +100°C, at a heating rate of 10°C/min). This would indicate that the polymers are completely amorphous at least within the detectable limits of the instrument. However, polymer **2a**, as well as polymer **2b**, do indeed exibit glass transition temperatures (T_g) at approximately -32°C and -11°C, respectively (Figure 3 and Table 1). The polymers (**2a**, **2b**) have also been investigated using thermal gravimetric analysis (TGA) to determine their thermal stability under high temperatures, in a argon atmosphere. Figure 4 displays the TGA scans obtained for the oligomers produced from ADMET polymerization of monomers **1a, 1b**. In the case of the polymer **2a**, the initial onset of decomposition does not ocur until 215°C, after which the polymer decomposes rapidly and loses approximately 66% of its weight between 240°C and 400°C.

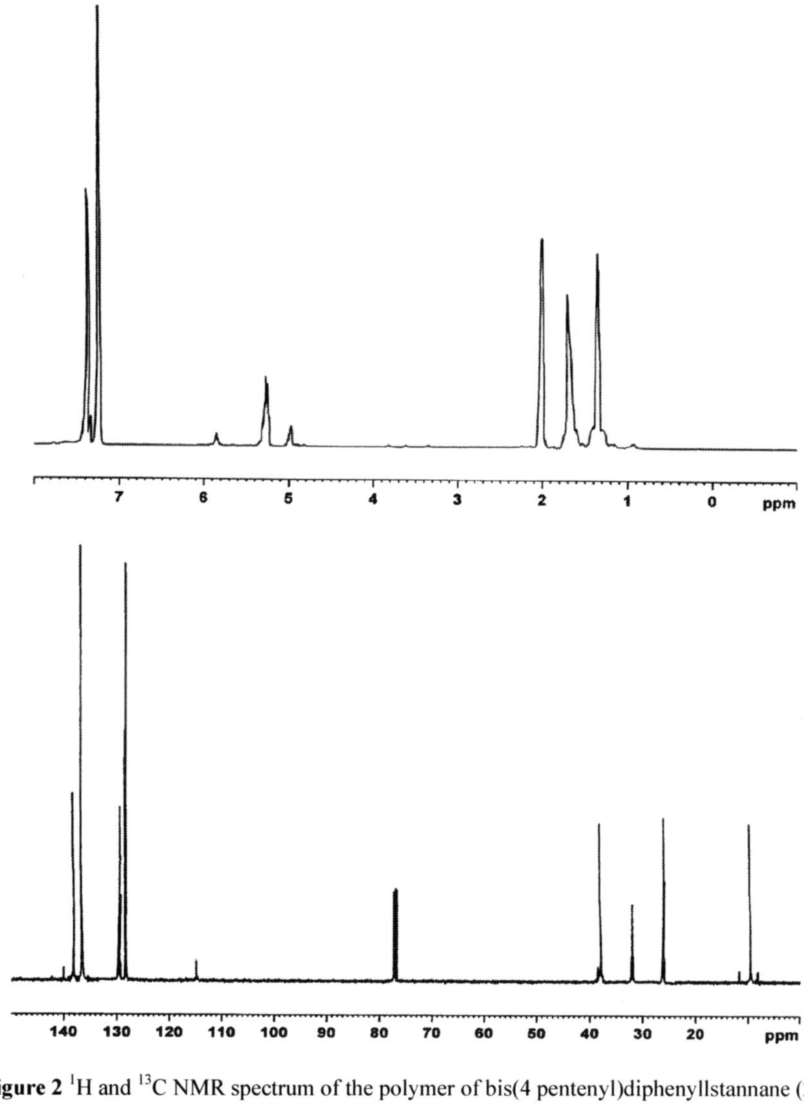

Figure 2 ^{1}H and ^{13}C NMR spectrum of the polymer of bis(4 pentenyl)diphenyllstannane (**2b**)

Table 1 Results for polycarbostannanes **2a** and **2b** characterization

				Elemental Analysis						
				Theor		Experim				
Diene	**Yield[a] (%)**	**M_n[b]**	**PDI[b]**	**%H**	**%C**	**%H**	**%C**	**Trans[c] (%)**	**T_g[d] (°C)**	**TDT[e] (°C)**
2a	92	15,300	2.14	8.10	48.84	8.11	49.34	56	−32	362
2b	76	15,200	2.80	6.37	62.74	6.12	62.86	67	−11	293

[a]Determined by gravimetrically.
[b]Determined by GPC, relative to polystyrene standard.
[c]Calculated from ^{13}C-NMR spectra.
[d]Determined by DSC in Ar atmosphere (heating rate: 10°C/min).
[e]Determined by TGA in Ar atmosphere (10% weight loss temperature).

The decomposition then ceases at approximately 420°C and the material remains stable up to 700°C with no further decomposition. An important feature of the curve is the constant weight of the sample, at approximately 34%, which is maintained after 420°C. By weight, the sample itself contains 38% tin. These results are supported by Tilley and coworkers [18] in whose research the tin content in the polystannanes synthesized *via* dehydropolymerization could be monitored using TGA thermal data [19]. The thermal stability of the polymer **2b** showing the same type of behavior is illustrated in Figure 4. Onset of decomposition at 250°C is observed, with a rapid decrease in weight of the polymer **2b**. The decomposition finally ceases around 420°C **(2b)** and then reaches a plateau. The decomposition stabilizes itself at around 30–33% weight, again reflecting the tin content of the polymer.

4 Conclusion

We conclude that in the production of polycarbostannanes the electrochemically reduced tungsten-based system leads to mainly *trans* polymer (**2a** or **2b**) (56–80% trans). Polymers (**2a** and **2b**) showed glass transition temperatures at approximately −32°C and −11°C, respectively. DSC thermograms for the polymer **2a** and **2b** displayed no observable transitions between −100°C to +100°C leading to the conclusion that the polymers are completely amorphous. Thermal analysis of the products (**2a**, **2b**) indicated that these polymers exhibit good thermal stability. Thermal decomposition temperatures (TDT) of polymers **2a** and **2b** were found to be 362°C and 293°C, respectively.

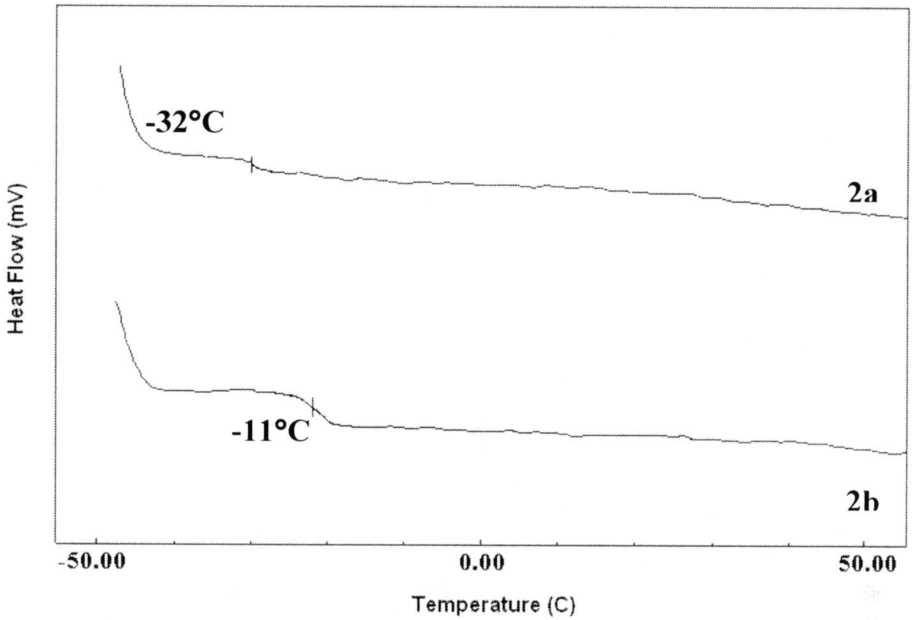

Figure 3 DSC curves for the polycarbostannanes **2a** and **2b**

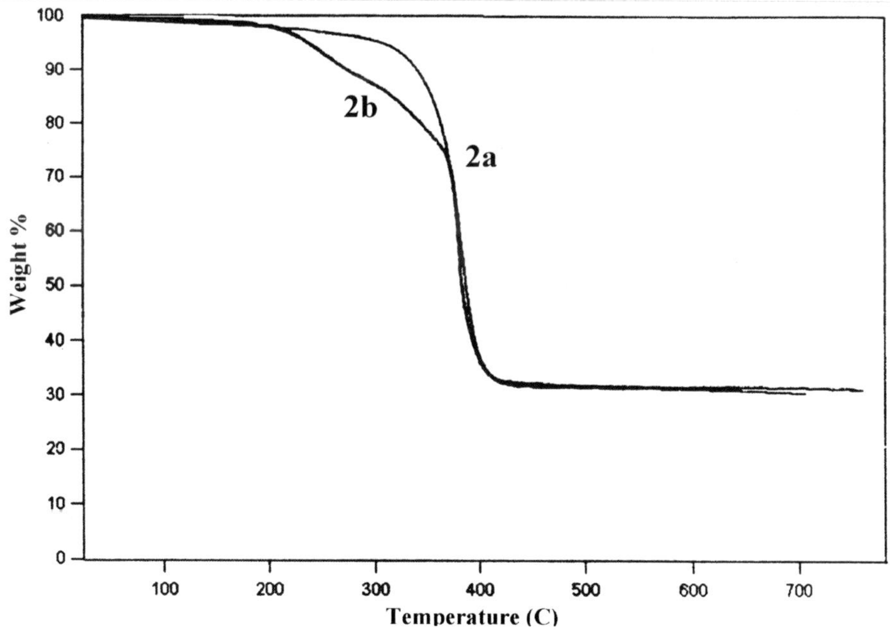

Figure 4 TGA thermograms of polycarbostannanes **2a** and **2b**

Acknowledgements We would like to thank the Hacettepe University Research Fund and Scientific and the Technological Research Council of Turkey (TUBITAK, 104T344) for their financial support of this work.

References

[1] Zeldin M, Wynne KJ, Allcock HR (eds.) (1998), *Inorganic and Organometallic Polymers: Macromolecules Containing Silicon, Phosphorous, and Other Inorganic Elements*, American Chemical Society, Washington, DC
[2] Zeigler JM, Fearon FWG (eds.) (1990) *Silicon-Based Polymer Science*, American Chemical Society, Washington, DC
[3] Brook MA (ed.) (2000) *Silicon in Organic, Organometallic, and Polymer Chemistry*, Wiley, New York
[4] Gomez FJ, Wagener KB (1999) J Organomet Chem 592: 271
[5] Smith DW, Wagener KB (1991) Macromolecules 24: 6073
[6] Smith DW, Wagener KB (1993) Macromolecules 26: 1633
[7] Gomez FJ, Wagener KB (1998) Macromol Chem Phys 199: 1581
[8] Wolfe PS, Gomez FJ, Wagener KB (1997) Abstracts of Papers of the American Chemical Society 213: 156-PMSE
[9] Wagener KB, Boncella JM, Nel JG (1991) Macromolecules 24: 2694
[10] Wagener KB, Boncella JM, Nel JG, Duttweiler RP, Hillmyer MA (1990) Makromol Chem 191: 365
[11] Gilet M, Mortreux A, Folest J C, Petit F (1983) J Am Chem Soc 105: 3876
[12] Gilet M, Mortreux A, Nicole J, Petit F (1979) J Chem Soc Chem Commun 521
[13] Düz B, Pekmez K, İmamoğlu Y, Süzer Ş, Yıldız A (2003) J Organomet Chem 77: 684
[14] Dereli O, Düz B, Zümreoğlu B K, İmamoğlu Y (2004) Appl Organomet Chem 18: 130
[15] Karabulut S, Aydogdu C, Düz B, İmamoğlu Y (2006) J Inorg Organomet Polym Mater 16: 115
[16] Karabulut S, Aydogdu C, Düz B, İmamoğlu Y (2006) J Mol Catal 254: 186
[17] Karabulut S, Aydogdu C, Düz B, İmamoğlu Y (2007) J Inorg Organomet Polym Mater 17: 517
[18] Tilley TD (1993) Acc Chem Res 26: 22
[19] Imori T, Lu V, Cai H, Tilley TD (1995) J Am Chem Soc 117: 9931

A Selective Route for Synthesis of Linear Polydicyclopentadiene

Valerian Dragutan,[1*] Ileana Dragutan,[1] Mihai Dimonie[2]

[1]Romanian Academy, Institute of Organic Chemistry, 202B Spl. Independentei, 060023 Bucharest, Romania

[2]Polytechnic University of Bucharest, Romania

[*]E-mail: vdragutan@yahoo.com

Abstract The present work looks at results recently obtained in polymerization of dicyclopentadiene to linear polydicyclopentadiene (LPDCPD) using two families of highly active and selective tungsten-based catalytic systems. In a first approach LPDCPD was obtained from *endo*-DCPD in excellent yields using catalytic systems consisting of WCl_6 or $WOCl_4$ and organosilicon compounds. IR and ^{13}C NMR microstructure investigations indicated a prevailingly *cis* double bond configuration. The linear polymer displayed quite low glass-transition temperature and good thermal and electrical properties, combined with appreciable solubility in various organic solvents. From solutions of the linear polymer, elastic and resistant films having superior adhesion on many solid surfaces (wood, metal, plastic materials) could be produced. Conversely, the catalytic system derived from tungsten tetraphenyl-porphyrinate and diisobutylaluminoxane led to linear polydicyclopentadiene with a predominantly *trans* configuration of the backbone double bonds. This latter system behaved in a "living" fashion opening access to polymers with monomodal and narrow molecular weight distribution. Applying the examined strategy in copolymerization of dicyclopentadiene with other cycloolefins (cyclopentene and cyclooctene) afforded new copolymers with incorporated LPDCPD blocks recommending them as potentially valuable materials.

Keywords Dicyclopentadiene · WCl_6 · $WOCl_4$ · Diisobutylaluminoxane · Triisobutyl-aluminoxane (TIBA) · Tungsten tetraphenylporphyrinate · Dimethylallylsilane · Tetraallylsilane · ROMP · Homopolymers · Copolymers

1 Introduction

Owing to accessibility of the hydrocarbon C_5 fraction as the main feedstock but also to the commendable properties of the final product, polymerization of dicyclopentadiene has become a technologically important process. Ring-opening

V. Dragutan et al. (eds.), *Green Metathesis Chemistry: Great Challenges in Synthesis, Catalysis and Nanotechnology*,

metathesis polymerization (ROMP) of dicyclopentadiene can proceed by two different pathways: in the first, the reaction occurs with opening of the norbornene unit to give linear polydicyclopentadiene while in the second both the norbornene and cyclopentene structural entities in the monomer are successively opened forming a cross-linked polymer [1–3]. Because of the substantial difference in reactivity of the double bond in the strained norbornene, relative to that in the cyclopentene ring, formation of the linear polyalkenamer by opening of the norbornene moiety is favored. However, depending on the catalyst activity and selectivity, the architecture of the main chain in the linear polymer may be more complex, namely when the norbornene and cyclopentene units become competitive giving either distinct polymer chains with the two possible linear structures (Scheme 1, Route a) or a product containing both these structural blocks in the same polymer chain [4] (Scheme 1, b).

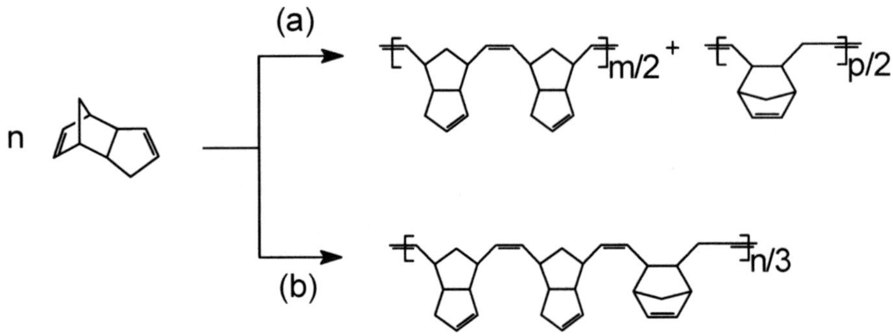

Scheme 1 Distinct routes in ROMP of dicyclopentadiene

These products also greatly differ in respect of their physical and chemical properties and accordingly can be used in various fields.

Ring-opening polymerization of dicyclopentadiene has been extensively studied as promoted by a broad range of catalysts mainly including systems based on $TiCl_4$ [5], titanacyclobutane complexes [6], WCl_6 [7–10], $WOCl_4$ [11], $MoCl_5$ [12], $ReCl_5$ [11, 13], $IrCl_3$, $OsCl_3$ and $RuCl_3$ [11], (mesitylene)$W(CO)_3$ [11], $Mo(CO)_5(Py)$ [14], various $WCl_n(OR)_{6-n}$ compounds [14–20], organoammonium molybdates and tungstates [21, 22], the molybdenum carbene precursor Mo(=NAr)(=NC*t*Bu)(CH_2*t*Bu)$_2$ in conjunction with a phenolic activator [23], tungsten tetraphenylporphyrinates associated with aluminoxane [24, 25], and various polymetallates [26]. Patent [27–29] and open literature [30–35] also describe application of binary catalysts that induce ROMP at both the norbornene and cyclopentene units of DCPD leading to cross-linked structures with outstanding physical and mechanical properties.

Important data have been reported on the polymerization of *endo*- and *exo*-dicyclopentadiene with tungsten alkoxide or phenoxide complexes, as such or associated with other organometallic compounds [16–20]. The latter systems constitute the most efficient catalysts for poly(dicyclopentadiene) manufacturing through RIM processes. Some of the $WOCl_{4-x}(OAr)_x$ complexes ($WOCl_3(OAr)$, $WOCl_2(OAr)_2$, and $WOCl(OAr)_3$), in combination with trialkyltin hydrides or triaryltin hydrides, showed to be quite stable, and useful to polymerize dicyclopentadiene in bulk, in high yield [19, 20].

It should be mentioned that ROMP of DCPD with selective ring-opening of the norbornene moiety to yield linear polymer has also been examined. Either well-defined one-component catalysts [36–38] or reactive binary catalysts [13] have been employed. In the latter case, the linear polydicyclopentadiene is accompanied by some cross-linked polymer, when working at high monomer concentration and with high catalyst loadings. In polymerization reactions of *endo*-dicyclopentadiene carried out by Pacreau and Fontanille [13] with the binary catalyst $ReCl_5/Me_4Sn$, a substantial amount of linear polymer with high molecular weight and having a large content of *cis* double bonds has been obtained. Interestingly, kinetic studies of early stages of the reaction showed that the polymerization proceeds *via* oligomerization, followed by the formation of polymer when an equilibrium between oligomers and polymer has been established. The influence of temperature, monomer concentration, and the catalyst/cocatalyst and catalyst/monomer ratios on the activity of the system and the polymer yield has been investigated. Furthermore, insoluble polymers were obtained with $ReCl_5$ and $EtAlCl_2$ or Et_2AlCl. Studies by DSC and TGA of linear polydicyclopentadiene synthesized with WCl_6/allylsilanes showed a thermal behaviour of the polymer characterized by an endothermic (glass transition) and two exothermic stages (the first being a thermal polymerization of double bonds of the polymer chains and the second a thermo-oxidation) [39].

As an area of great scientific endeavour ring-opening polymerization has lately been applied to dicyclopentadiene using newly disclosed catalytic systems based on ruthenium [40]. Although the *endo*-isomer of dicyclopentadiene (DCPD) is commercially available, the *exo*-stereoisomer has much faster reaction rates in olefin metathesis with first-generation Grubbs' catalyst.

In the light of the green chemistry concept, traditional systems continue today to prove their value because they are more efficient and stable, while reasonably selective, and better recommended for industrial applications.

ROMP of DCPD is amenable for autonomic healing applications of polymeric materials [41]. Using catalyst loading levels previously reported to be effective for *endo*-DCPD, *exo*-DCPD was found to heal approximately 20 times faster than the *endo*-isomer, but with a lower healing efficiency.

In continuation of previous research where we carried out polymerization of DCPD to linear polydicyclopentadiene using two types of very active and selective binary catalytic systems (tungsten tetraphenylporphyrinate and aluminoxane [25, 42]; WCl_6 or $WOCl_4$ and organosilicon compounds [43]), the present work highlights relevant aspects of the synthesis of linear poly-DCPD under the influence of these two catalytic systems providing useful additional data on the reaction products, mechanism and stereochemistry. Under controlled conditions, high molecular weight poly-DCPD has been prepared. In the case of the first catalytic system, a direct correlation between monomer conversion and molecular weight has been observed.

2 Experimental

Starting materials. Commercially available dicyclopentadiene, DCPD (Aldrich, 95%) was used as the starting raw material. Pure CPD was obtained by thermal cracking (170°C) of the commercial dimer, followed by distillation through a Vigreux column. It was then stored for several days at 30°C for the monomer to dimerize to DCPD. Finally, the resulted mixture of cyclopentadiene and dicyclopentadiene was distilled under vacuum (p = 15 mmHg, Vigreux column), immediately before carrying out the polymerization reaction. Toluene (Aldrich, 99.8%) was refluxed on a Na-K alloy, under nitrogen and distilled before being used. All other reagents were handled and stored under an extremely pure nitrogen atmosphere.

Catalyst preparation. Tungsten(VI) chloride (Aldrich, 99.9%) and tungsten(VI) oxychloride (Aldrich, 98%) were used without further purification. Handling and dosage were performed under nitrogen of high purity. These reagents were stored in sealed vials. Catalysts were prepared on adding the components either by pre-complexation or "*in situ*". Dimethylallylsilane (Aldrich) and tetraallylsilane (Aldrich) were used without further purification. Handling, dosage and storage were performed under ultra-pure nitrogen atmosphere. Tungsten oxychloride was also prepared "*in situ*" starting from WCl_6 and traces of water. In this case, the content of water in the reaction mixture was carefully controlled by using a toluene solution (wet toluene). Tungsten tetraphenylporphyrinate chloride ($TPPWCl_4$) was synthesized as previously described [25] by treating equimolar amounts of WCl_6 with tetraphenylporphyrin ($TPPH_2$) in carbon tetrachloride, under inert atmosphere, and further refluxing for 24 h. After separation and drying under vacuum, the product was characterized by absorption spectroscopy in the UV–VIS and IR regions. Triisobutylaluminoxane (TIBA) of commercial grade was distilled and sealed under high vacuum. Diisobutylaluminoxane was prepared from TIBA by a rigorously controlled reaction with water using the special technique described elsewhere [25].

Polymerization process. The polymerization reactions were carried out in a 100 ml, one-necked round-bottomed flask equipped with a magnetic stirrer, a N_2 purge device and a sample collector. After the polymerization reaction had ended the catalyst was deactivated using a 2% sodium hydroxide solution in methanol. The reaction mixture was washed with water to remove the deactivation products and then the organic phase was separated from the aqueous phase. This special procedure was necessary to totally eliminate traces of the deactivated catalyst from the polymer. (It was observed that traces of catalyst in the product promote an advanced cross-linking of the polymer within several days). After work-up a wet toluene solution of polymer resulted from which traces of water were carefully removed by vacuum distillation. The polymer was fully structurally characterized by spectroscopic techniques supplemented by DSC and TGA measurements.

IR and NMR analyses of polymers. Infrared spectra of polydicyclopentadiene were recorded on a Nicolet 10MX(FT) spectrophotometer. Polymer samples for this analysis were films prepared by evaporating a thin layer of polymer solution. ^{13}C NMR spectra of linear PDCPD were obtained in $CDCl_3$ using a Brucker (300 MHz) spectrophotometer with TMS as internal standard.

3 Results and Discussion

Two types of tungsten-based catalytic systems (WCl_6 or $WOCl_4$ in conjunction with organosilicon, and tungsten tetraphenylporphyrinate associated with diisobutylaluminoxane) have been used in this work for polymerization of DCPD. Both catalyst types allowed synthesis of linear polydicyclopentadiene in high yields but having different steric configurations at the carbon–carbon double bonds, depending essentially on the nature of the catalyst. The polymerization behaved in a "living" manner, allowing block copolymers with cyclopentene and cyclooctene to be prepared. Monomodal molecular weight distributions of low to high molecular weight polymers have been obtained using these catalysts. The linear polydicyclopentadiene displayed a wide range of physicochemical properties. The polymer was soluble in common organic solvents and had no tendency to cross-link.

The catalytic system WCl_6/All_4Si (All = allyl group), in a narrow range of WCl_6 concentrations ($[WCl_6] = 0.5 \times 10^{-3}$–$1.5 \times 10^{-3}$ mol/l), gave high yields (98–100%) of linear poly-DCPD. The reactions were carried out in toluene, mostly at a monomer concentration between 0.68 and 1.67 mol/l (8–22 wt%); best results were obtained with a monomer concentration of 1.6 mol/l (21.8 wt%). At higher catalyst concentration ($[WCl_6] > 1.5 \times 10^{-3}$ mol/l) the reaction led to cross-linked polymer. At lower catalyst concentration non-reproducible results were registered. For practical reasons, still higher monomer concentrations have not been explored.

Polymerization gave also good yields in linear polymer when WCl_6 was replaced with $WOCl_4$. Furthermore, $WOCl_4/Me_2All_2Si$, $WOCl_4/All_4Si$, $WOCl_4/H_2O/Me_2All_2Si$ and $WCl_6/H_2O/All_4Si$ (Me = methyl group) acted as suitable catalytic systems in polymerization to linear polymer but, at variance with the first group of catalysts, they showed quite high activity and selectivity within much larger limits of catalyst concentrations ($[WOCl_4] = 0.5 \times 10^{-3}$–$6 \times 10^{-3}$ mol/l). This wider concentration range ensures a better control of the polymerization, as compared with reactions employing WCl_6. Data obtained are summarized in Table 1.

As can be seen from Table 1, at high monomer conversions, the polymer molecular weight and molecular weight distribution varied as a function of the catalyst and monomer concentrations, the ratio between the catalyst components and the time elapsed between the reaction completion (100% monomer conversion) and the deactivation of the catalyst.

We assume that the different behavior of the two catalytic systems based on WCl_6 or $WOCl_4$ is to be assigned to the higher electrophilicity of WCl_6 as compared to $WOCl_4$. Our results seem to be in agreement with other work [44] where cross-linked polydicyclopentadiene apparently forms from the initially linear polydicyclopentadiene, by a cationic reaction implying the double bonds of the polymer chain. Further support for this assumption comes from data concerning the unexpected molecular weight variation of the linear polydicyclopentadiene observed when the catalyst ($WCl_6/H_2O/Me_2All_2Si$) was deactivated 20 h after the reaction completion (Figure 1).

Table 1 Polymerization of dicyclopentadiene to linear polymer in the presence of WCl_6 and $WOCl_4$ associated with organosilicon compounds[a,b]

Catalytic system	[W] (mol/l × 10^3)	Si/W	Molecular weight	M_w/M_n
$WC_6/SiAllyl_4$	0.5	1	M_n = 22.000 M_w = 39.000	1.77
$WC_6/SiAllyl_4$	1.12	1	M_n = 17.000 M_w = 35.500	2.1
$WC_6/SiAllyl_4$	1.7	1	–	–
$WCl_6/SiMe_2Allyl_2$	1.12	1	M_n = 19.000 M_w = 39.000	2.05
$WOCl_4/SiMe_2Allyl_2$	3.4	2	M_n = 30.000 M_w = 57.000	1.9
$WOCl_4/SiAllyl_4$	3.4	1	M_n = 22.000 M_w = 46.000	2.1
$WOCl_4/SiAllyl_4$	5	2	M_n = 18.000 M_w = 36.000	2
$WCl_6/H_2O/SiAllyl_4$	3.0	2	M_n = 28.500 M_w = 83.000	2.9
$WCl_6/H_2O/SiMe_2Allyl_2$ H_2O / WCl_6 = 0.7[d]	3.4	2	M_n = 26.000 M_w = 50.100	1.92
$WCl_6/H_2O/SiMe_2Allyl_2$ H_2O / WCl_6 = 0.7[e]	3.4	2	M_n = 41.400 M_w = 234.000	5.65
$WCl_6/H_2O/$ $(iBu)_2Al$-O-$Al(iBu)_2$	4	1	–	–
$WCl6/H2O/$ $(iBu)_2Al$-O-$Al(iBu)_2$	4	0.1	$M_n = 2.9 \times 10^6$ $M_w = 1.2 \times 10^7$	4.17
WCl_6/ECH/$SiAllyl_4$	3.5	1	M_n = 52.600 M_w = 94.300	1.79
WCl_6/ECH/ $SiMe_2Allyl_2$	3.5	1	M_n = 76.500 M_w = 248.200	3.24
WCl_6/ECH/ $(iBu)_2Al$-O-$Al(iBu)_2$	3.5	0,1	$M_n = 1.7 \times 10^6$ $M_w = 1.1 \times 10^7$	6.08

[a]Monomer concentration 1.67 mol/l.
[b]Solvent toluene.
[c]Polymerization temperature = 25°C.
[d]Deactivated immediately after the end of the polymerization reaction (H_2O/WCl_6 = 0.7).
[e]Deactivated after 20 h after the end of polymerization reaction ($H_2O:WCl_6$ = 0.7).

The substantial increase of the polymerization degree and the significant widening of the molecular weight distribution when the catalyst deactivation was performed after 20 h, illustrated in Figure 1, are ascribed to the cross-linking reaction through cationic active centers. The polymer obtained was characterized by IR (Figure 2), ^{13}C NMR (Figure 3), DSC and TGA analyses. The assignment of these absorption bands is presented in Table 2.

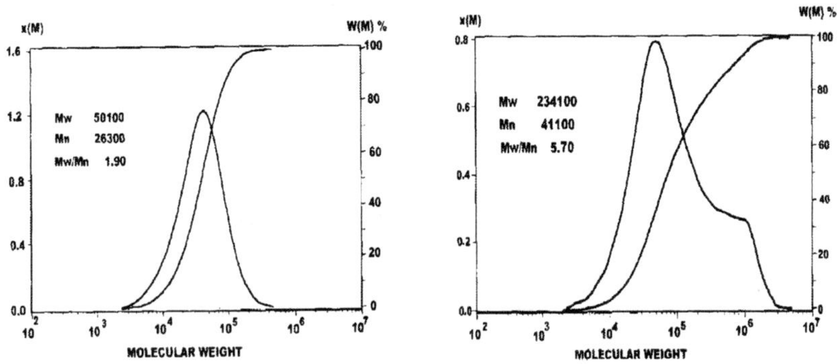

(A) (B)

Figure 1 Differential and cumulative log molecular weight distribution of linear poly-dicyclopentadiene prepared with the $WCl_6/H_2O/Me_2All_2Si$ system (**a**) catalyst deactivation immediately after the end of polymerization; (**b**) catalyst deactivation 20 h after the end of polymerization)

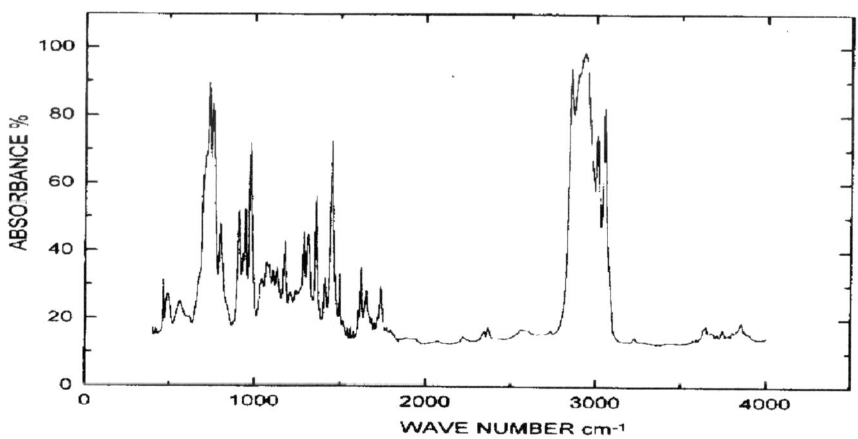

Figure 2 Infrared spectrum of linear polydicyclopentadiene

Table 2 Assignment of the main absorption bands in the IR spectrum of the linear poly-dicyclopentadiene

Wave number (cm^{-1})	Assignment
3,050	ν (= C–H), cyclic double bond
3,010	ν (= C–H), acyclic double bond
2,929	ν ($-CH_2-$), asym.
2,851	ν ($-CH_2-$), sym.
1,655	ν (C=C), acyclic *cis*
1,620	ν (C=C), cyclic *cis*
1,451	δ ($-CH_2-$)
975	γ (=C–H), acyclic *trans*
695	γ (=C–H), cyclic *cis*
686	γ (=C–H), acyclic *cis*

The infrared data of LPDCPD indicate the presence of both acyclic >C=C< (*cis* and *trans*) and cyclic >C=C< moieties in the cyclopentene rings. The relative intensity of the IR absorption bands suggests that the polymer stereoconfiguration of the carbon-carbon double bonds is prevailingly *cis*. These results are also confirmed by the ^{13}C NMR spectrum. Assessments were made from off-resonance spectrum and signal intensity measurements (Figure 3).

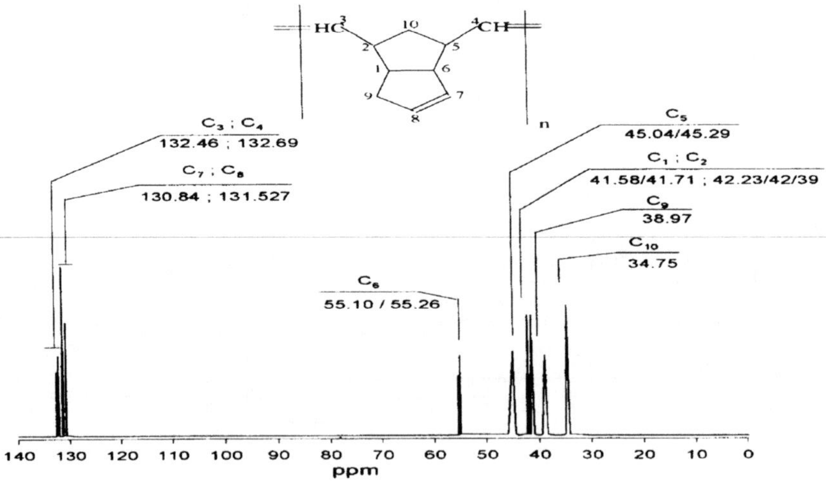

Figure 3 ^{13}C NMR Spectrum of linear polydicyclopentadiene

Figure 3 shows that the methylene carbons appear as singlets C_9 (38.97), C_{10} (34.75) while the methine carbons as doublets C_1(41.58/41.71), C_2(42.23/42.39), C_5(45.04/45.29), C_6(55.10/55.26). The following signals were assigned to the olefinic carbons: C_7 (130.84) >C=C< , C_8 (131.52), C_3 (132.46), C_4 (132.69). Noteworthy, the off-resonance ^{13}C NMR spectrum illustrated in Figure 3 is similar to that published by Ivin et al. [19] for *cis*-polydicyclopentadiene obtained by the polymerization of *endo*-DCPD with a $ReCl_5$ catalyst resulting in highly *cis*-LPDCPD.

Measurements by differential scanning calorimetry (DSC) analysis of LPDCPD indicated the existence of a T_g at 53°C. This value of T_g is reasonable for linear polydicyclopentadiene taking into account the presence of the double bonds and, at the same time, of bulky repeat units in the main chain. The thermostability of the linear PDCPD in nitrogen and air has been further followed by thermogravimetric analysis (TGA). It is important to note that under nitrogen atmosphere the weight loss is practically zero up to ca. 450°C whereas between 450–520°C the polymer rapidly loses about 30% of its weight. From 500°C to temperatures close to 1,000°C, the weight of the polymer remains invariant. This particular thermostability suggests the formation of graphitic structures between 500°C and 1,000°C. On the other hand, in the presence of air, the behavior of the polymer is totally different. In this case, no weight loss is observed up to 380°C. However, over 400°C a rapid degradation of the polymer is evidenced while at 600°C the residual product is only 10%. This totally unexpected thermal stability of a polymer with such elevated unsaturation like linear PDCPD is probably due to thermal polymerization of the carbon–carbon double bonds during heating at high temperatures, with formation of a highly cross-linked polymer. Relevant electrical properties of the linear polydicyclopentadiene prepared with the above catalytic systems are summarized in Table 3.

Table 3 Main electrical parameters of linear polydicyclopentadiene prepared with tungsten-based catalytic systems

Properties	Values			
Volume resistivity (Ω.cm)	$6.6 \cdot 10^{14}$			
Frequency (kHz)	1	5	10	50
Dielectric constant (t = 20°C)	2.98	2.69	2.63	2.57
Tangent of dielectric loss	$8.01 \cdot 10^{-4}$	$3.05 \cdot 10^{-4}$	$2.03 \cdot 10^{-4}$	$2.0 \cdot 10^{-4}$

These data suggest a good application profile of this polymer for various electrical domains.

Further studies on dicyclopentadiene polymerizations were conducted in the presence of the catalytic system consisting of tungsten tetraphenylporphyrinate and diisobutylaluminoxane. Under these conditions, high monomer conversions and high polymer yields were attained, when working in toluene solutions at room temperature. The resulted polydicyclopentadiene displayed a monomodal and narrow molecular weight distribution (Figure 4).

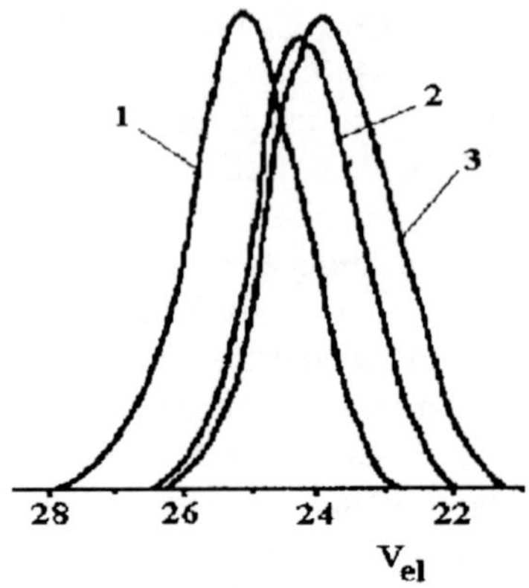

Figure 4 Catalytic precursor (A) of the tungsten tetraphenylporphyrinate catalyst and molecular weight distribution (B) of homopolymers (1) and copolymers prepared from dicyclopentadiene with cyclopentene (2) and cyclooctene (3)

It should be noted that the catalytic system derived from tungsten tetraphenylporphyrinate and diisobutylaluminoxane behaved in a “living” fashion showing a linear relationship between the molecular weight and the monomer conversion. In other experiments block copolymers of dicyclopentadiene with cyclopentene and cyclooctene have also been prepared. Investigation of polymer microstructure by both IR and ^{13}C NMR spectroscopy indicated a prevailingly *trans* stereoconfiguration of the carbon–carbon double bonds in the polymer chain (Figure 5).

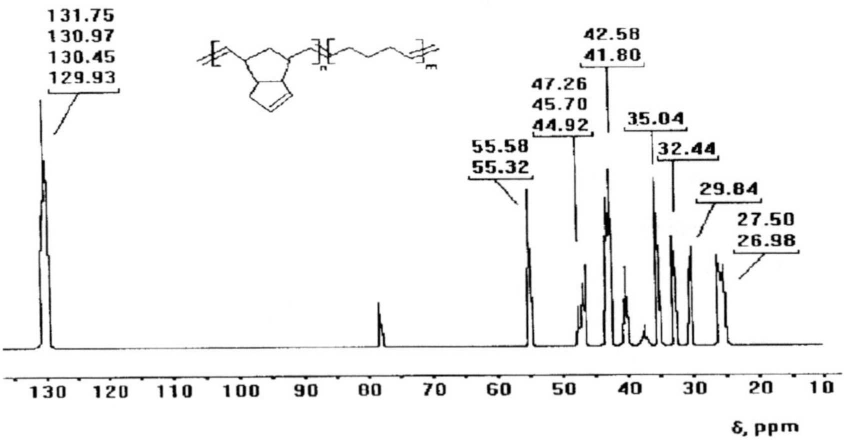

Figure 5 ^{13}C NMR Spectrum of copolymer prepared from dicyclopentadiene and cyclopentene with tungsten tetraphenylporphyrinate/diisobutylaluminoxane catalyst

The different stereoselectivity of the two types of catalysts taken into consideration in this work for the polymerization of dicyclopentadiene is striking. We assume that two distinct kinetic species arise at the catalyst active site leading to two sterically different, metallacyclic intermediates: a *cis* metallacyclobutane in the case of the WCl_6 or $WOCl_4$ systems and a *trans* metallacyclobutane from the system consisting of tungsten tetraphenylporphyrinate and diisobutylaluminoxane. The favoured pathway is governed by the configuration of these initiating and propagating tungsten-carbene species; consequently, these metallacyclobutane intermediates will give preferentially the *cis* and *trans* polymer, respectively.

4 Conclusions

The ring-opening polymerization of *endo*-dicyclopentadiene (*endo*-DCPD) yielding linear polydicyclopentadiene (LPDCPD) was performed effectively using two types of tungsten-based catalytic systems. With the catalysts derived from WCl_6 or $WOCl_4$ in conjunction with organosilicon compounds, the best results have been attained working in solution and under severely controlled monomer and catalyst concentrations ([DCPD] = 1.6 mol/l (21.8 wt%); [W] = 0.5×10^{-3}–1.5×10^{-3} mol/l). The polymer had a prevailingly *cis* stereochemistry at the carbon–carbon double bonds. The linear polydicyclopentadiene obtained under optimum conditions exhibited a T_g at 53°C. The polymer showed a good thermostability up to 450°C but above this temperature the thermal behavior strongly depends on whether heating is run in an inert atmosphere or in air. Systems derived from tungsten tetraphenylporphyrinate and diisobutylaluminoxane led, in a "living" fashion, to linear polydicyclopentadiene with high molecular mass. The polymer displayed a predominantly *trans* configuration of the carbon–carbon double bonds. Using this catalytic system, copolymers with cyclopentene and cyclooctene have also been prepared.

References

[1] Dragutan V, Streck R (2000) Catalytic polymerization of cycloolefins. Elsevier, Amsterdam

[2] Ivin KJ, Mol JC (1997) Olefin metathesis and metathesis polymerization. Academic Press, London

[3] Dimonie M, Coca S, Teodorescu M, Popescu L, Chipara M, Dragutan V (1994) J Mol Catal 90:117–124; Dimonie M, Coca S, Teodorescu M, Popescu L, Chipara M, Dragutan V (1994) J Mol Catal 90:117–124; Dimonie M, Coca S, Dragutan V (1992) J Mol Catal 76:79–91; Breslow DS (1990) Polymer Preprints (Am Chem Soc Div Polymer Chem) 31:410

[4] Johnston JA, Farona MF (1991) Polymer Bull 25:625

[5] Winstein CZ (1977) Chem Abstr 86:122050

[6] (a) Cannizzo LF, Grubbs RH (1988) Macromolecules 21:1961; (b) Risse W, Grubbs RH (1989) Macromolecules 22:1558; (c) Fischer RA, Grubbs RH (1992) Makromol Chem Macromol Symp 63:271

[7] (a) Klosiewitz DW (1983) US Patent 4,400,340; (b) Klosiewitz DW (1984) US Patent 4,469,809

[8] (a) Sjardjin W, Kramer AH (1986) US Patent 4,729,976; (b) Sjardjin W, Kramer AH (1987) US Patent 4,810,762

[9] Nelson LL (1989) US Patent 4,826,942

[10] Hamilton JG, Ivin KJ, Rooney JJ (1986) J Mol Catal 36:115; Hamilton JG (1998) Polymer 39:1669–1689

[11] Martin EA (1990) Eur Patent 360,262

[12] Goodrich BF (1988) US Patent 4,923,734

[13] Pacreau A, Fontanille M (1987) Makromol Chem 188:2585

[14] (a) Basset JM, Leconte M, Ollivier J, Quignard F (1985) US Patent 4,550,216; (b) Basset JM, Leconte M, Ollivier J, Quignard F (1990) Eur Patent 0,259,215

[15] (a) Quignard F, Leconte M, Basset JM (1985) J Chem Soc Chem Commun: 13; (b) Quignard F, Leconte M, Basset JM (1986) J Mol Catal 36:13

[16] Boutarfa D, Paillet C, Leconte M, Basset JM (1991) J Mol Catal 69:157

[17] Heroguez V, Soum A, Fontanille M (1992) Polymer 33:3302

[18] Balcar H, Dosedlova A, Petrusova L (1992) J Mol Catal 77:289

[19] Bell A (1991) Polym Prepr (Am Chem Soc Div Polym Mater Sci Eng) 64:102

[20] Bell A (1992) Polym Prepr (Am Chem Soc Div Polym Mater Sci Eng) 67:39

[21] Minchak RJ (1983) US Patent 4,380,617

[22] Minchak RJ (1984) US Patent 4,426,502

[23] (a) Bell A, Clegg W, Dyer PW, Elsegood MRJ, Gibson VC, Marshall EL (1994) J Chem Soc Chem Commun: 2247; (b) Bell A, Clegg W, Dyer PW, Elsegood MRJ, Gibson VC, Marshall EL (1994) J Chem Soc Chem Commun: 2547

[24] Dragutan V, Dimonie M, Coca S (1994) Polym Prepr (Am Chem Soc Div Polym Chem) 35:698

[25] Coca S, Dimonie M, Dragutan V, Ion R, Popescu L, Teodorescu M, Moise F, Vasilescu A (1994) J Mol Catal 90:101

[26] (a) Goodall BL, Kroenke WJ, Minchak RJ, Rhodes LF (1993) J Appl Polym Sci 47:607; (b) Goodall BL, McIntosch LH, Rhodes LF (1995) Makromol Chem Macromol Symp 89:421; (c) Goodrich BF (1989) US Patent 4,923,936

[27] (a) Tom GM (1985) US Patent 4,507,453; (b) Tom GM (1987) US Patent 4,661,575

[28] Klosiewicz DW (1987) US Patent 4,657,981

[29] Martin AE (1987) U.S. Patent 4,696,985

[30] Mateika L, Houtman C, Makosko W (1985) J Appl Polym Sci 30:2787

[31] Breslow DS (1998) Polym Mater Sci Eng (Am Chem Soc Div Polym Chem) 58:223

[32] Breslow DS (1980) Chemtech 10:540

[33] Bell A (1992) J Mol Catal 76:165

[34] (a) Dragutan I, Dragutan V, Drozdzak R, Verpoort F (2007) Ruthenium vinylidene complexes – An efficient class of homogeneous metathesis catalysts. In: Imamoglu Y, Dragutan V (eds.) Metathesis chemistry: From nanostructure design to synthesis of advanced materials. NATO Science Series II. Mathematics, Physics and Chemistry, vol. 243, pp. 137–150. Springer, Dordrecht; (b) Dragutan V, Balaban AT, Dimonie M (1985) Olefin metathesis and ring-opening polymerization of cycloolefins. Wiley, New York

[35] Ivin KJ (1983) Olefin metathesis. Academic Press, London

[36] Fisher RA, Grubbs RH (1992) Makromol Chem Macromol Symp 63:271

[37] Dall'Asta G, Motroni G, Manetti G, Tossi R (1969) Makromol Chem 130:153

[38] Marshall PR, Ridgewell BJ (1969) Eur Polym J 5:29

[39] Dimonie D, Dimonie M, Munteanu V, Iovu H, Couve J, Abadie MJ (2000) Polym Degrad Stab 70:319–324

[40] (a) Hejl A, Day MW, Grubbs RH (2006) Organometallics 25:6149–6154; (b) Allaert B, Dieltiens N, Ledoux N, Vercaemst C, Van der Voort P, Stevens CV, Linden A, Verpoort F (2006) J Mol Catal A-Chemical 260(1–2):221–226; (c) Dragutan I Dragutan V, Drozdzak R, Verpoort F (2007). In: Imamoglu Y; Dragutan V (eds.) Metathesis chemistry: From nanostructure design to synthesis of advanced materials Book Series: NATO Science Series, Series II: Mathematics, Physics and Chemistry, vol. 243, pp. 137–150. Springer, Dordrecht, The Netherlands

[41] (a) Mauldin TC, Rule JD, Sottos NR, White SR, Moore JS (2007) J Royal Soc Interface 4(13):389–393; (b) Wilson GO, Moore JS, White SR, Sottos NR, Andersson HM (2008) Adv Funct Mater 18(1):44–52; (c) Sheng X, Lee JK, Kessler MR (2009) Polymer 50:1264–1269

[42] Dragutan V, Popescu L, Coca S, Dimonie M (1998) Ring-opening metathesis polymerization of cycloolefins using tungsten-tetraphenylporphyrinate catalysys. In: Y Imamoglu (ed.) Metathesis polymerization of olefins and polymerization of alkynes. NATO Science Series II. Mathematics, Physics and Chemistry, vol. 506, pp. 103–115. Kluwer, Dordrecht, The Netherlands

[43] (a) Dragutan V, Dragutan I, Dimonie M, Couve C, Abadie MJ (2002) Catalyst activity and selectivity in ROMP of dicyclopentadiene induced by some tungsten systems. In: Khosravi E, Szymanska-Buzar T (eds.) Ring-opening metathesis polymerization and related chemistry: State of the art and visions for the new century. NATO Science Series II. Mathematics, Physics and Chemistry, vol. 56, pp. 465–476. Kluwer, Dordrecht/Boston/London; (b) Abadie MJ, Dimonie M, Couve C, Dragutan V (2000) Eur Polym J 36:1213–1219

[44] Davidson TA, Wagener KB, Priddy DB (1996) Macromolecules 29:786

Tuning Product Selectivity in ROMP of Cycloolefins with W-Based Catalytic Systems

Valerian Dragutan,[1*] Ileana Dragutan,[1] Mihai Dimonie[2]

[1]Institute of Organic Chemistry, 202B Spl. Independentei P.O. Box 35-108, 060023 Bucharest, Romania

[2]Polytechnic University of Bucharest, Bucharest, Romania

[*]E-mail: vdragutan@yahoo.com

Abstract Tuning product stereoselectivity in ROMP of a range of cycloolefins (e.g. cyclopentene, cyclooctadiene, cyclododecene, dicyclopentadiene) using different W-based catalytic systems (WCl_6/organoaluminium compounds, WCl_6/ organotin compounds, WCl_6 or $WOCl_4$/organosilicon compounds, WCl_4 (tetraphenyl-porphyrin)/isobutylaluminoxane) is presented and the role of the organometallic cocatalysts, electron donor and acceptor compounds and ligand associated with the W atom is particularly pointed out. The effect of these catalyst components on the *anti-syn* metallacarbene interconversion and the preferential formation of *cis* or *trans* metallacyclobutane intermediates are discussed in connection with the reaction stereoselectivity. In addition, the influence of relevant reaction parameters such as reaction temperature, reaction time, nature of solvent and nature of monomer on the product stereoselectivity is rationalized in terms of metallacarbene/ metallacyclobutane mechanism.

Keywords Cycloolefins · W-based catalysts · Tungsten tetraphenylporphyrinate · Organosilicon compounds · Diisobutylaluminoxane · Triisobutylaluminoxane (TIBA) · Dimethylallylsilane · Tetraallylsilane · ROMP · Metallacarbene · Metallacyclobutane

1 Introduction

Abundant data evidence that ring-opening metathesis polymerization (ROMP) of cycloolefins is a highly stereospecific reaction, when appropriately conducted [1]. This feature is essential in industrial ROMP applications having in mind the damaging effect of waste materials deposited in the environment. Depending mainly on the reaction conditions, polyalkenamers with varying stereoconfigurations at the carbon–carbon double bonds from all-*cis* to all-*trans* structures can be obtained [2]. Moreover, products differing in tacticities result as a function of the

V. Dragutan et al. (eds.), *Green Metathesis Chemistry: Great Challenges in Synthesis, Catalysis and Nanotechnology*,

nature of the catalytic system, monomer and reaction parameters such as monomer concentration, temperature, solvent and reaction time [3]. Both the steric configuration of the polyalkenamer and its tacticity are of major significance for practical applications because they impart special properties to the final product [4]. The present work deals with the influence that most relevant reaction parameters exert on the stereoselectivity in the ring-opening metathesis polymerization of monocyclic olefins catalyzed by an array of tungsten-based systems. Partial results on the stereoselectivity of ROMP of this class of cycloolefins have been reported previously [5].

2 Experimental

Starting materials. Cyclopentene was synthesized by dehydration of cyclopentanol with H_3PO_4, followed by distillation on a highly efficient column and storage over K-Na alloy. Cyclooctene (Merck), *cis,cis*-cyclooctadiene (Merck) and cyclododecene (Fluka) of commercial grade were first stored over K-Na alloy, then distilled under high vacuum. Toluene (Aldrich, 99.8%) was refluxed on Na-K alloy under nitrogen and distilled before use. WCl_6 (Merck, 99%), purified by carefully removing volatile impurities through sublimation, was sealed under argon or high vacuum. Et_3Al, Et_2AlCl and iBu_3Al (Merck) were distilled and sealed under vacuum. Commercial grade tetraethyltin, tetrabutyltin and tetraphenyltin (Merck) were used without further purification. Diisobutylaluminoxane was prepared from triisobutyl aluminium by reaction with a controlled amount of water (wet toluene). Epichlorohydrin (commercial grade) was distilled on a highly effective column, stored over CaH_2 and sealed under vacuum. Chloranil (Merck) was used as such. Dibenzoquinone was purified by recrystallization from benzene. Salicyl aldehyde was distilled under high vacuum. Maleic anhydride, cyanuric acid and cyanuric chloride of commercial grade were purified by recrystallization from benzene.

Analysis procedure. Infrared spectra of the polymer were recorded on a Nicolet 10MX(FT) spectrophotometer. ^{13}C NMR spectra were obtained on a Bruker (300 MHz) spectrometer using TMS as internal reference. Mass spectroscopic measurements were performed on a DP-102 type spectrometer. ESR spectra were recorded using a JES-ME-3X instrument, operating in the X-band frequency (9 GHz), with DPPH as the internal standard. Gel permeation chromatography (GPC) was used to determine molecular weights and molecular weight distributions, M_w/M_n, of polymer samples, with respect to polystyrene standards.

Polymerization method. The polymerization reactions were carried out in glass installations either using the high vacuum technique (10^{-3} torr) for handling and contacting the reactants or working under highly purified argon atmosphere as previously described [6]. Toluene was used as the solvent. Reactions were conducted at temperatures between −20°C and +20°C for times varying from 1 to 480 min. Reaction quenching was effected with methanol solutions. Yields of polymer from 10% to 98% were recorded.

3 Results and Discussion

Two-component catalytic systems. Ring-opening metathesis polymerization of cyclopentene using the binary catalysts WCl_6/Et_3Al, WCl_6/Et_2AlCl and WCl_6/iBu_3Al gives polypentenamers with prevailingly *trans*-configuration (60–80%), whereas on using binary systems containing organotin compounds associated with WCl_6, polypentenamers with mainly *cis* stereoconfiguration (70–80%) have been produced [5]. The *trans/cis* content of the obtained polypentenamers could be gradually tuned by changing the organoaluminium compound; the reducing ability of the latter was evidenced by the ESR spectra of the paramagnetic species (W(V) and W(III)) produced in the system. The two families of binary catalysts, containing either organoaluminiums or organotins, lead also to totally different polymer microstructures, the first to a random distribution and the latter to a blocky distribution of the carbon–carbon double bonds in the polymer chain. Consequently, on using these catalysts, both the steric configuration (*cis* vs. *trans*) and the polymer microstructure (random vs. blocky) could be rigorously controlled.

When organoaluminium or organotin compounds in the above catalytic systems were replaced by diisobutylaluminoxane, the obtained polymers had a high *cis* stereoconfiguration (Table 1). As can be readily observed, the polymer stereoconfiguration depends substantially on the molar ratio catalyst/monomer and monomer conversion.

Table 1 Stereoconfiguration of polypentenamer obtained with the binary catalytic system $WCl_6/iBu_2AlOAliBu_2$

Molar ratio WCl_6/CP (mmol/mol)	Reaction Temp (°C)	Conv. (%)	*cis*-Polypentenamer (%)	*trans*-Polypentenamer (%)
0.18	−10	2.0	56.0	44.0
0.2	−10	1.0	58.0	42.0
0.3	−10	37.5	90.9	9.1
0.3	0	48.0	88.8	11.2
0.5	−10	35.0	85.4	14.6

Noteworthy, the polymerization of cyclopentene in the presence of the binary catalyst tetraphenylporphyrinate tungsten/diisobutylaluminoxane gave polypentenamers with a high *trans* stereoconfiguration, in spite of the bulky ligand at the tungsten atom. Also with this system, the polymer configuration and microstructure were dependent on the monomer conversion (Table 2).

Table 2 Stereoconfiguration and microstructure of polypentenamer obtained with tetraphenylporphyrinate tungsten/diisobutylaluminoxane

Monomerconversion (%)	Polymermicrostructure[a] R	*cis*-Polypentenamer (%)	*trans*-Polypentenamer (%)
10.0	0.68	23.07	76.93
18.0	0.73	19.88	80.12
23.0	0.73	19.21	80.79
45.0	0.74	18.68	81.32

[a] $R = (P_{cc} + P_{tt})(P_{tc} + P_{ct})$ where P_{cc}, P_{tt}, P_{ct} and P_{tc} are the population of *cis–cis*, *trans–trans*, *cis–trans* and *trans–cis* dyads from the CH_2 region of the ^{13}C NMR spectrum.

By contrast, polymerization of cyclododecene in the presence of the $WCl_6/iBu_2AlOAliBu_2$ catalyst led mainly to *trans*-polydodecenamer, under most circumstances (Table 3).

Table 3 Stereoconfiguration of polydodecenamer prepared with the catalytic system $WCl_6/iBu_2AlOAliBu_2$

Reaction temperature (°C)	Monomer conversion (%)	*cis*-Polydodecenamer (%)	*trans*-Polydodecenamer (%)
+20	24.0	42.5	57.5
+20	42.0	38.5	61.5
+20	68.0	33.3	66.7
+20	78.0	31.2	68.8
−20	8.0	50.0	49.8
−20	10.0	45.4	54.6
−20	12.0	44.4	55.6

It can be easily observed that in the above conditions the content of *trans* stereoconfiguration did not vary drastically with reaction temperature or monomer conversion. Higher *cis*-polydodecenamer has been produced only at low temperatures and short reaction times but it readily turned into the *trans* stereoisomer at longer reaction times.

Three-component catalytic systems. On additionally using an electron donor/acceptor component or a bulky ligand, in ternary catalytic systems, the steric configuration and microstructure of polyalkenamers could be widely varied, besides the catalyst activity and stability. Some relevant examples for cyclopentene polymerization with ternary systems derived from WCl_6/organoaluminium compounds are given in Table 4. Obviously, a significant variation in the *trans* content of the polypentenamer is recorded for every ternary component: the *trans* stereoconfiguration increased from 72.0% to 85.5% when replacing chloranil with epichlorohydrin whereas with dibenzoquinone, instead of cyanuric chloride, the steric configuration changed from 73.0% to 83.0% *trans*.

Table 4 Steric configuration of polypentenamer prepared with ternary catalysts derived from WCl_6 and organoaluminium compounds[a]

Catalytic system	Reaction temp. (°C)	*cis*-Polypentenamer (%)	*trans*-Polypentenamer (%)
$WCl_6/iBu_3Al/EP$	0	14.5	85.5
$WCl_6/Et_3Al/EP$	0	18.4	81.6
$WCl_6/Et_3Al_2Cl_3/EP$	0	18.1	81.9
$WCl_6/Et_2AlCl/EP$	0	21.6	78.4
$WCl_6/iBu_3Al/CA$	0	28.0	72.0
$WCl_6/iBu_3Al/DBQ$	−15	17.0	83.0
$WCl_6/iBu_3Al/MalAnh$	0	28.0	72.0
$WCl_6/iBu_3Al/SalAld$	0	19.6	80.4
$WCl_6/iBu_3Al/CyanAc$	0	25.8	74.2
$WCl_6/iBu_3Al/CyanCl$	0	27.0	73.0

[a] EP = Epichlorohydrin, CA = chloranil, DBQ = dibenzoquinone, MalAnh = maleic anhydride, SalAld = salicyl aldehyde, CyanAc = cyanuric acid, CyanCl = cyanuric chloride.

Table 5 Stereoconfiguration of polyoctenamer obtained with WCl_6/iBu_3Al (I) and epichlorohydrin (EP) or chloranil (CA)

Reaction temperature (°C)	Monomer conversion (%)	*trans*-Polymer (%)	*cis*-Polymer (%)
Catalyst I+EP			
+20	8.0	37.9	62.1
+20	15.0	41.5	58.5
+20	19.0	46.8	53.2
+20	26.5	48.7	54.3
+20	27.0	49.9	50.1
+20	30.0	58.0	42.0
−20	7.5	34.9	65.1
−20	8.0	39.9	60.1
−20	9.0	44.7	55.3
−20	10.0	54.5	45.5
Catalyst I+CA			
+20	25.0	42.0	58.0
+20	55.0	53.0	47.0
+20	80.0	54.5	45.5

Differing from cyclopentene, the polymerization of cyclooctene induced by WCl_6/iBu_3Al in association with epichlorohydrin or chloranil yields mainly *cis*-polyoctenamer at both high and low reaction temperatures (Table 5).

These data for cyclooctene are quite unexpected considering the behavior of cyclopentene (mainly *trans*- and *cis*-polymer, respectively, with epichlorohydrin vs. chloranil; Table 4). Furthermore, as illustrated in Table 5, the *cis* content of polyoctenamer varied gradually with conversion. This result is very important for industrial applications where the stereoconfiguration of polyalkenamer has to be rigorously controlled. Similar data have also been recorded for cyclooctene polymerization with the WCl_6/diisobutylaluminoxane catalyst, a well-known high *cis*-directing system for cyclopentene polymerization [6]. Additionally, highly *cis*-polyoctenamers (*cis* content 85.8–91.5%) have been produced using WCl_6 in association with organosilicon compounds in binary and ternary catalytic systems, e.g. WCl_6/Me_2All_2Si and $WCl_6/Me_2All_2Si/H_2O$.

Of a special interest is the polymerization of 1,5-cyclooctadiene in the presence of the ternary catalytic system $WCl_6/Me_2All_2Si/H_2O$, leading to polybutenamer (1,4-polybutadiene); at various temperatures, polymers with 81–84% *cis* stereo-configuration have been obtained.

Four-component catalytic systems. Most efficiently, four-component catalysts allowed a rigorous control of polyalkenamer stereoconfiguration. With such a system, e.g. WCl_6/Bu_4Sn/chloranil/piperylene, 97.5% *cis*-polypentenamer could be conveniently produced.

Results have been rationalized in terms of the well-accepted metallacarbene–metallacyclobutane mechanism. We assume that the ligands associated with the transition metal affect the equilibrium between the *syn* and *anti* rotamers (A and B) of the *in-situ* generated metallacarbene, by an ancilliary effect (Scheme 1; Mt = metal + ligands).

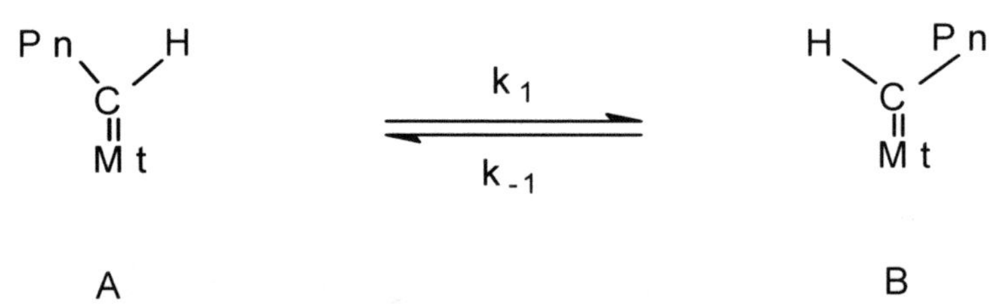

Scheme 1 *Syn* and *anti* rotamers in metallacarbene

A higher population of the *syn* rotamer, favored by electron-accepting ligands, will lead *via* a *cis*-metallacyclobutane to *cis*-polymer whereas a higher population of *anti* rotamer, promoted by electron-donating ligands, will form *trans*-polymer, *via* a *trans*-metallacyclobutane (Scheme 2).

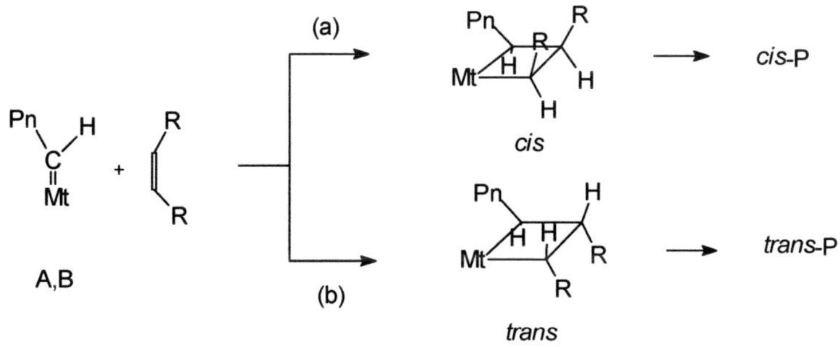

Scheme 2 *Cis* and *trans* metallacyclobutane routes to *cis* and *trans* polymers

4 Conclusions

Stereoselectivity in cycloolefin polymerization with tungsten-based ROMP catalysts is strongly influenced by the organometallic cocatalyst, the nature of the donor–acceptor ligands associated with the transition metal, the structure of the monomer, as well as by reaction temperature, conversion and molar ratios. In light of green chemistry, this feature is essential in industrial ROMP applications taking into account the damaging effect of waste materials on the environment. By monitoring these factors, the polyalkenamer stereoconfiguration could be easily tuned altering the physico-chemical properties of the products. The reaction pathway was rationalized in terms of the metallacarbene–metallacyclobutane mechanism evidencing the role played by the cocatalyst and the third component of the catalytic system.

Acknowledgements Support for this work from the Polytechnic University of Bucharest and the Romanian Academy is gratefully acknowledged.

References

[1] (a) Grubbs RH (ed.) (2003) Handbook of metathesis, vol. 3. Wiley-VCH, Weinheim; (b) Dimonie M, Coca S, Teodorescu M, Popescu L, Chipara M, Dragutan V (1994) J Mol Catal 90:117–124; (c) Amir-Ibrahimi V, Rooney JJ (2004) J Mol Catal A: Chem 208:103–108

[2] (a) Ivin KJ, Mol JC (1997) Olefin metathesis and metathesis polymerization, Academic Press, London; (b) Delaude L, Demonceau A, Noels AF (1999) Macromolecules 32:2091 2103; (c) Bokaris EP, Kosmas MM (2003) J Mol Catal A: Chem 192:263–273; (d) Fogg DE, Foucault H (2007) Ring-opening metathesis polymerization. In Crabtree RH, Mingos DMP (eds.) Comprehensive organometallic chemistry III, Elsevier, Oxford

[3] (a) Hamilton JG (1998) Polymer 39:1669–1689; (b) Mashima K, Kaidzu M, Tanaka Y, Nakayama Y, Nakamura A, Hamilton JG, Rooney JJ (1998) Organometallics 17:4183–4195; (c) Dimonie M, Coca S, Teodorescu M, Popescu L, Chipara M, Dragutan V (1994) J Mol Catal 90:117–124

[4] (a) Dragutan V, Streck R (2000) Catalytic polymerization of cycloolefins, Elsevier, Amsterdam; (b) Dragutan V, Balaban AT, Dimonie M (1985) Olefin metathesis and ring-opening polymerization of cycloolefins, Wiley, New York; (c) Dragutan V, Dragutan I, Balaban AT (2000) Platinum Metals Rev 44:168–172; (d) Vygodskii YS, Shaplov AS, Lozinskaya EI, Filippov OA, Shubina ES, Bandari R, Buchmeiser MR (2006) Macromolecules 39:7821–7830; (e) Dragutan I, Dragutan V, Delaude L, Demonceau A, Noels AF (2007) Rev Roumanie Chim 52:1013–1025; (f) Dragutan V, Verpoort F (2007) Rev Roumanie Chim 52:905–915; (g) Dragutan I, Dragutan V, Fischer H (2008) J Inorg Organomet Polym Mater 18:311–324; (h) Dragutan V, Dragutan I, Fischer H (2008) J Inorg Organomet Polym Mater 18:18–31; (i) Monsaert S, Drozdzak R, Dragutan V, Dragutan I, Verpoort F (2008) Eur J Inorg Chem 432–440

[5] Dragutan V, Coca S, Dimonie M (1998) Correlation between catalyst nature and polymer selectivity in ROMP of cycloolefins with WCl_6-based catalytic systems. In: Y Imamoglu (ed.) Metathesis polymerization of olefins and polymerization of alkynes. NATO Science Series II. Mathematics, Physics and Chemistry, vol. 506, pp. 89–102. Kluwer, Dordrecht, The Netherlands

[6] Dimonie M, Coca S, Dragutan V (1992) J Mol Catal 76:79–91

$[RuCl_2(p\text{-}Cymene)]_2$ Immobilized on Mesoporous Molecular Sieves SBA-15 as Catalyst for ROMP of Norbornene

David Bek,[1] Hynek Balcar,[1*] Jan Sedláček[2]

[1]J. Heyrovský Institute of Physical Chemistry AS CR, v.v.i, Dolejškova 3, 182 23 Prague 8, Czech Republic

[2]Department of Physical and Macromolecular Chemistry, Faculty of Science, Charles University, Albertov 2030, CZ-128 40, Prague 2, Czech Republic

*E-mail: balcar@jh-inst.cas.cz

Abstract A new heterogeneous catalyst for ring opening metathesis polymerization (ROMP) has been prepared by immobilization of $[RuCl_2(p\text{-cymene})]_2$ on siliceous mesoporous molecular sieves SBA-15. Activity of the catalyst was tested in ROMP of norbornene. Filtration test proved that the catalytic activity is bound to the solid phase. Catalyst could be easily separated from reaction mixture in contrast to the corresponding homogeneous system and therefore polymer with reduced amounts of catalyst residues was obtained.

Keywords Metathesis · ROMP · Norbornene · Ru dimer · Ru heterogeneous catalyst · Mesoporous molecular sieves

1 Introduction

After activation by trimethylsilyldiazomethane (TMSD), ruthenium(II) dimer $[RuCl_2(p\text{-cymene})]_2$ is active as catalyst in ROMP of norbornene [1]. This complex can be used as a starting compound for the synthesis of different catalyst precursors [2, 3] such as $RuCl_2(p\text{-cymene})(PCy_3)$, Cy = cyclohexane, which are used for ROMP of cycloolefins after reaction with a diazo compound forming ruthenium-carbene species. The advantage of these precursors is that they are relatively cheap compounds in contrast to the Grubbs catalysts.

Mesoporous molecular sieves are inorganic (mainly siliceous and aluminous) materials with well-defined regular architecture, described for the first time by Mobil Oil researchers [4] in 1992. Due to their properties such as large surface area, large void volume and narrow pore size distribution of mesopores, mesoporous molecular sieves represent advantageous support for heterogeneous catalysts [5]

V. Dragutan et al. (eds.), *Green Metathesis Chemistry: Great Challenges in Synthesis, Catalysis and Nanotechnology*,

including metathesis ones [6, 7]. They can be also used as a support for catalysts used in alkene and alkyne polymerizations [8, 9].

This contribution deals with the preparation of a new catalyst for ROMP by immobilization of $[RuCl_2(p\text{-cymene})]_2$ on siliceous mesoporous molecular sieves SBA-15 (one-dimensional mesoporous channels with hexagonal array) and testing its activity in ROMP of norbornene (Scheme 1).

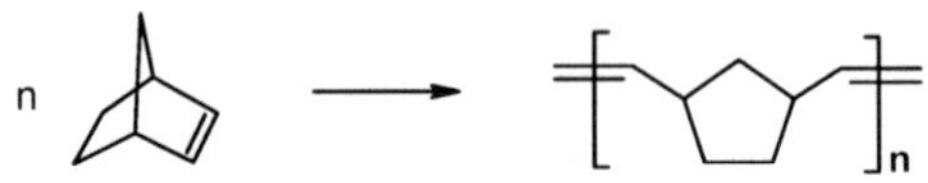

Scheme 1 ROMP of norbornene

2 Experimental

2.1 Materials

Toluene (Lach-Ner) was dried overnight by anhydrous Na_2SO_4, then distilled with Na and stored over drying molecular sieves of the type 4A. Dichloromethane (Lach-Ner) was dried overnight by anhydrous $CaCl_2$, then distilled with P_2O_5 and stored over drying molecular sieves type 4A. Other compounds were used as received, norbornene, NBE (Aldrich, purity 99%), $[RuCl_2(p\text{-cymene})]_2$ (Sigma-Aldrich), TMSD (Sigma-Aldrich, 2.0 M solution in hexanes).

2.2 Catalyst Preparation

Synthesis of siliceous SBA-15 was performed according to the procedure described in details in [10] using Pluronic PE 9400 (BASF) as a structure-directing agent and tetraethyl orthosilicate (Aldrich) as a silicon source. The structure-directing agent was removed by calcination in air carried out at 500°C for 6 h with a temperature ramp of 1°C/min. Sorption characteristics of the SBA-15 were determined from nitrogen adsorption isotherms: surface area S_{BET} = 915.1 m^2/g, average pore diameter d = 6.3 nm, volume of pores V = 1.101 cm^3/g. Silica gel 40, Merck, was taken as a conventional silica support with broad pore distribution: surface area S_{BET} = 559 m^2/g, average pore diameter d = 4.5 nm, volume of pores V = 0.473 cm^3/g.

Catalyst was prepared in Schlenk tube under Ar atmosphere. 3 g of dried SBA-15 was stirred with CH_2Cl_2 (60 ml), then 91 mg of $[RuCl_2(p\text{-cymene})]_2$ was added into this mixture. After 5 h of stirring at room temperature, the catalyst was settled down and the solvent above the yellow catalyst was colourless (colour of the parent $[RuCl_2(p\text{-cymene})]_2$ solution is orange). The prepared catalyst was washed three times by CH_2Cl_2 and then dried in vacuum at room temperature. Dried catalyst (denoted as Ru/SBA-15) was stored under Ar atmosphere.

2.3 *ROMP of Norbornene*

ROMP of NBE was performed under Ar atmosphere in a Schlenk tube equipped with a magnetic stirrer. In a typical experiment, 66.2 mg of Ru/SBA-15 catalyst was placed into the reactor, 12 ml of toluene was added and after several minutes of stirring 30 μl of TMSD (c = 2 mol/l in hexane) was added. After 10 min 139.5 mg of NBE was added (mole ratio NBE:Ru = 228) and the solution was warmed to 60°C in oil bath. After 3 h of reaction 2 ml of ethylvinyl ether (terminating agent) was added and after 5–10 min the reaction mixture was cooled, catalyst separated by centrifugation and polynorbornene (PNBE) precipitated by pouring the liquid part into 50 ml of methanol containing small amount of antioxidant 2,6-di-tert-butyl-*p*-cresol. Polymer was dried in vacuum oven at 60°C to the constant weight.

ROMP of NBE in homogeneous system was performed in the same way as heterogeneous polymerization, $[RuCl_2(p\text{-cymene})]_2$ was used as a catalyst.

2.4 *Methods*

Loading of Ru was determined by chemical analysis using an ICP-MS method (by ALS Czech Republic, s.r.o.). Size-exclusion chromatography (SEC) measurements were carried out on a Watrex Chromatograph fitted with a differential refractometer Shodex RI 101. A series of two PL-gel columns (mixed-B and mixed-C, Polymer Laboratories Bristol, UK) and THF (flow rate 0.7 ml/min) were used. Weight average molecular weight, M_w, and number average molecular weight, M_n, relative to polystyrene standards are reported. FTIR spectra of PNBE were recorded using KBr pellets on a FTIR spectrometer Nicolet Avatar 320 with DTGS-KBr detector. ^1H NMR spectra of PNBE in $CDCl_3$ were recorded on a Varian INOVA 400 instrument (referenced to the solvent line). Content of Ru in PNBE was determined by ICP-OES (by Research Institute of Inorganic Chemistry, a. s., Czech Republic).

3 Results and Discussion

Catalyst, Ru/SBA-15, containing 1 wt% of Ru was prepared. $[RuCl_2(p\text{-cymene})]_2$ was simply immobilized by stirring dried SBA-15 with the complex in CH_2Cl_2 for 5 h at room temperature. After the catalyst was settled down, the solvent above it was colourless (colour of original $[RuCl_2(p\text{-cymene})]_2$ solution is orange), which suggests that the immobilization proceeds quantitatively. The quantitative immobilization was also confirmed by elemental analysis. The mode of Ru complex immobilization on SBA-15 is not completely clear. The IR spectroscopy of Ru/SBA-15 confirmed presence of *p*-cymene moiety in the catalyst. We suppose that the immobilization proceeds via reaction of $[RuCl_2(p\text{-cymene})]_2$ with the surface OH groups of SBA-15 under formation of covalent bond Si-O-Ru (Scheme 2).

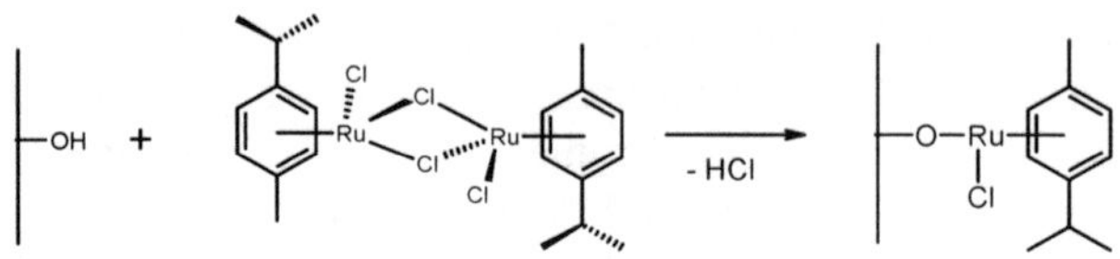

Scheme 2 Immobilization of $[RuCl_2(p\text{-cymene})]_2$ on SBA-15

According to Scheme 2, HCl should be formed as side product of this reaction and mole ratio Ru:Cl in Ru/SBA-15 should be 1:1. However, the elemental analysis of Ru/SBA-15 revealed the ratio Ru:Cl = 1:2, i.e. the same as in parent complex $[RuCl_2(p\text{-cymene})]_2$. Nevertheless, since HCl formed can be adsorbed on the surface of SBA-15, this result cannot be decisive.

Activity of Ru/SBA-15 was tested in ROMP of NBE. Ru/SBA-15 alone (i.e. without a cocatalyst) was inactive in this reaction, however, upon activation with TMSD (see Experimental) it provided high-molecular-weight PNBE, the yield of which was increasing with increasing TMSD/Ru mole ratio up to a limiting value of about 80% achieved at TMSD/Ru = 10 (Table 1). In the liquid part of the reaction system Ru/SBA-15/TMSD, free p-cymene was unambiguously detected by GC-MS method. The activation of Ru/SBA-15 with TMSD may thus proceed according to Scheme 3 under formation of metalocarbene species.

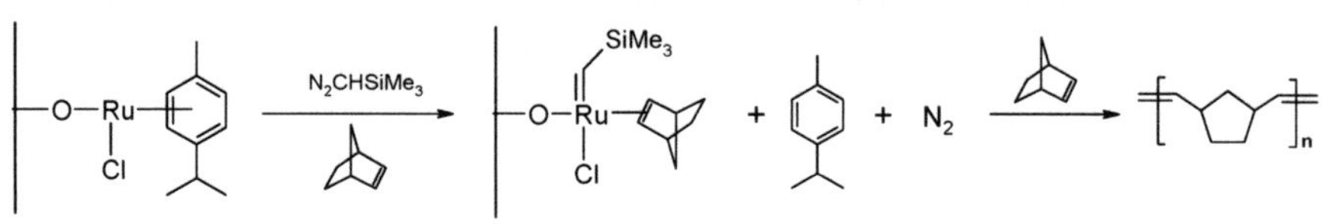

Scheme 3 ROMP of norbornene on Ru/SBA-15

Finding that TMSD had to be added in excess to Ru/SBA-15 for obtaining a good PNBE yield may reflect the preferential reaction of TMSD with surface OH groups of the support.

Table 1 Effect of mole ratio TMSD/Ru on PNBE yield. Initial concentrations [Ru] = 0.54 mmol/l, [NBE] = 123.5 mmol/l, toluene, reaction time 3 h

Mole ratio TMSD/Ru	PNBE yield (%)
0	0
5.2	12
10	78

The PNBE yield versus reaction time for NBE polymerization with Ru/SBA-15/TMSD is shown in Figure 1 together with the values of M_w of polymer isolated at various stages of the reaction.

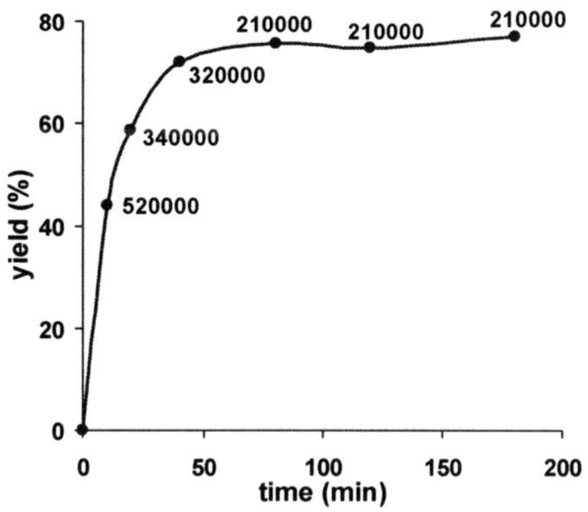

Figure 1 Dependence of yield and M_w of PNBE on reaction time. Initial concentrations: [Ru] = 0.54 mmol/l, [TMSD] = 5 mmol/l, [NBE] = 123.5 mmol/l. Toluene, reaction time 3 h, t = 60°C. Number at experimental points show corresponding values of M_w

Reaction was found to be rapid in the initial stage: within 10 min the yield of 44% was obtained and the reaction was completed approximately within 80 min (final yield of about 76%). Values of M_w were decreasing with increasing polymer yield. We suppose that this decrease can reflect (i) the formation of shorter polymer chains in the later stages of reaction due to the low monomer concentration, and/or (ii) the increasing extent of chain transfer reactions.

The effect of reaction temperature on PNBE yield is shown in Figure 2. It is evident that the optimum reaction temperature is 60°C for which the yield of (74 ± 4)% was obtained.

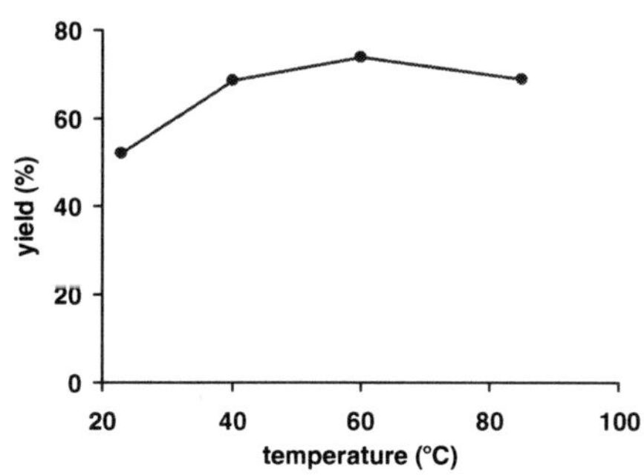

Figure 2 The effect of reaction temperature on yield of PNBE. Initial concentrations [Ru] = 0.54 mmol/l, [TMSD] = 5 mmol/l, [NBE] = 123.5 mmol/l, toluene, reaction time 3 h

For comparison, conventional silica with 1 wt% of Ru (prepared by immobilization of $[RuCl_2(p\text{-cymene})]_2$ on silica) was used in ROMP of NBE under the same conditions. Only 48% yield of PNBE was obtained in comparison with yield of 74% achieved on Ru/SBA-15 catalysts, which shows the advantage of the regular mesoporous support.

The effect of reaction temperature on weight average molecular weight, M_w, and number average molecular weight, M_n, of PNBE is shown in Figure 3. It was found that molecular weight of PNBE decreases with increasing reaction temperature, the polydispersity index, M_w/M_n, does not show any significant temperature dependence. The observed decrease in molecular weight of PNBE may reflect the increasing extent of transfer reactions with increasing temperature.

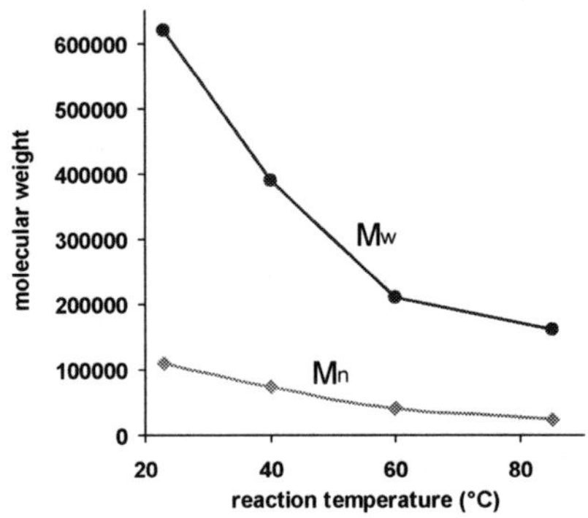

Figure 3 Effect of reaction temperature on M_w and M_n of PNBE. Initial concentrations: [Ru] = 0.54 mmol/l, [TMSD] = 5 mmol/l, [NBE] = 123.5 mmol/l. Toluene, reaction time 3 h

Three filtration tests were performed to verify whether metathesis activity is bound to the solid phase of the catalyst. In the first filtration test (Scheme 4), Ru/SBA-15 was stirred in toluene for 3 h at 60°C. Then the solid phase was filtered off and the colourless filtrate was obtained. TMSD and NBE were added into this filtrate and the resulting mixture was checked for the presence of PNBE after 3 h. Since no PNBE was detected it can be concluded that no releasing of active Ru species from Ru/SBA-15 due to its ageing in toluene at elevated temperature occurred within 3 h.

In the second filtration test (Scheme 5), NBE polymerization with Ru/SBA-15/TMSD (40°C) was started, then, after 5 min the reaction suspension was divided in two volume-equal portions by filtration. The first one, which contained solid catalyst and the second one, free from the solid catalyst. The both parts were allowed to continue to react under unchanged conditions for another 3 h. In the portion containing solid catalyst, the PNBE yield equal to 67% was determined after the termination (see Experimental). In the second portion, only 38% PNBE yield was determined.

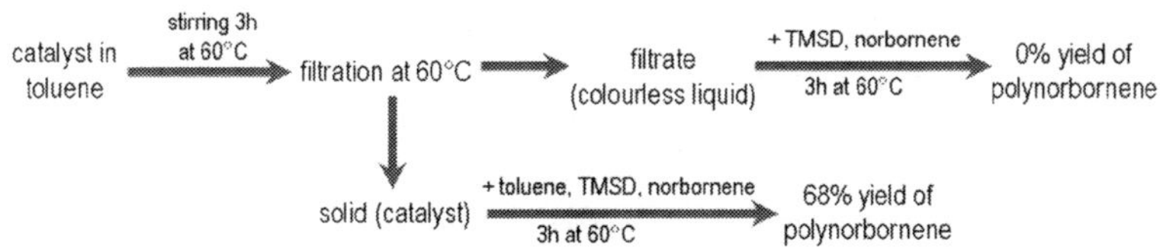

Scheme 4 First filtration test in NBE polymerization with Ru/SBA-15/TMSD

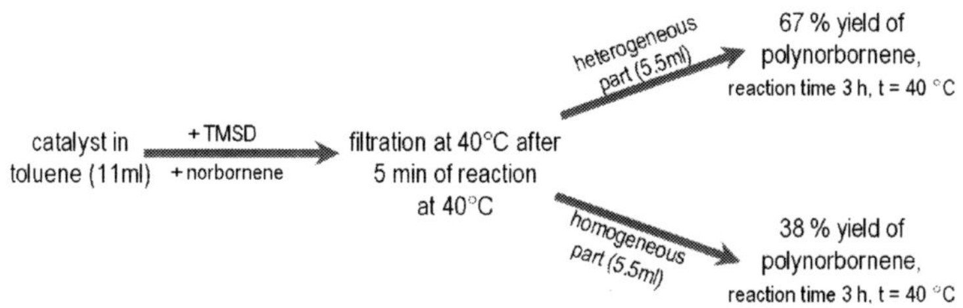

Scheme 5 Second filtration test in NBE polymerization with Ru/SBA-15/TMSD

Third filtration test (reaction temperature 40°C) was performed in the same way as the second one and it was aimed at determination of a difference in the yields of PNBE in homogeneous portion achieved (i) just after filtration, and (ii) after subsequent 3 h. We found this difference to be small, nevertheless it represented about 10–15% in PNBE yield. However, since there were significant difficulties in filtration of non-diluted viscous reaction mixture we cannot rule out a certain contamination of the filtrate with solid catalyst that might be a reason of the observed increase of the polymer yield in the separated liquid phase beside eventual small Ru leaching into the liquid phase. Thus on the base of the second and third filtration tests we can only conclude that as the yield of polymer in heterogeneous portion was always markedly higher than that in homogeneous one, the polymerization activity remains predominantly bound to the solid phase.

The low amount of catalyst residues in the isolated polymer is very important general feature, which is of the main advantages of heterogeneous catalysis in contrast to homogeneous one. Using the ICP-OES method we determined the amount of Ru residues in the PNBE prepared with Ru/SBA-15 (standard separation from the catalyst after the proper dilution of the reaction mixture, see Experimental) and in PNBE prepared and isolated under the same NBE/Ru initial mole ratio with homogeneous [RuCl$_2$(*p*-cymene)]$_2$ catalyst. In heterogeneously prepared PNBE sample (the polymer yield = 71%) the Ru content of 95 ppm was found while in a homogeneously prepared polymer (polymer yield = 85%) this value was 425 ppm. Significantly lower contamination of heterogeneously prepared PNBE with Ru residues is an unambiguous advantage of the polymerization performed with Ru/SBA-15. Nevertheless the Ru content in polymer (95 ppm) seems to be still relatively high, which indicates a certain extent of leaching of Ru compounds from Ru/SBA-15 into the liquid phase and/or certain contamination of polymer with microparticles of Ru/SBA-15.

PNBE prepared with Ru/SBA-15 was white solid well soluble in toluene, THF and $CHCl_3$. Samples prepared under various conditions were characterized by the IR and ^1H NMR spectroscopy. For all analyzed samples the polymer structure corresponding to the ROMP polymerization mode of NBE was confirmed. In Figure 4 the IR spectrum of PNBE prepared with Ru/SBA-15 (reaction temperature 60°C) is shown. In this spectrum the absorption bands of 741 and 966 cm^{-1}, correspond to the vibration of CH groups in the cis- and trans-configuration of the main chain double bonds, respectively [11].

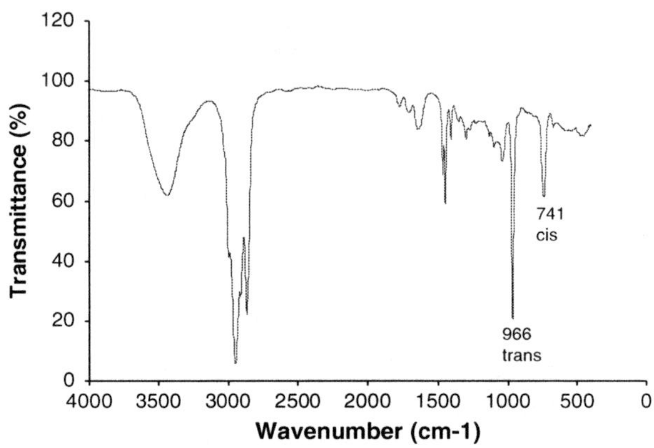

Figure 4 IR spectrum of PNBE prepared with Ru/ SBA-15

The quantification of the main chain double bonds *trans/cis* ratio was done on the base of ^1H NMR spectra from the signals at $\delta = 5.22$ ppm (*cis*-configuration) and $\delta = 5.36$ ppm (*trans*-configuration) [11]. In Table 2 the *trans/cis* ratio determined for PNBE samples prepared at various temperatures is given. The decrease in *trans/cis* ratio with increasing polymerization temperature is evident from Table 2.

Table 2 The effect of reaction temperature on the *trans/cis* ratio in PNBE

t (°C)	*Trans/cis*
23	1.13
40	1.05
60	0.89
85	0.68

4 Conclusions

We found that [$RuCl_2$(*p*-cymene)]$_2$ complex can be immobilized directly on SBA-15 without necessity of the support modification with a linker. Heterogeneous catalyst prepared exhibited, after its activation with TMSD, high activity in ROMP of NBE providing high molecular weight PNBE ($M_w = 2 \cdot 10^5 – 7 \cdot 10^5$). In the course of reaction, the PNBE formed was continuously released into the liquid phase of the reaction system probably by chain transfer. Filtration tests suggested that the catalytic activity was bound to the solid catalyst (at least predominantly). The PNBE yield achieved was slightly increasing with the increase in reaction temperature up to the optimum value of 60°C for which the PNBE yield of 74% was attained. On the other hand, the rise in reaction temperature caused the decrease in PNBE molecular weight. Polymer formed was easily separated from the catalyst and its contamination with Ru residues was significantly reduced in comparison to the contamination of PNBE prepared under the same conditions with [$RuCl_2$(*p*-cymene)]$_2$ homogeneous catalyst.

Acknowledgement We thank A. Zukal and M. Horáček (both J. Heyrovský Institute, Prague) for the preparation and characterization of SBA-15 and for GC-MS, respectively. The financial support from Grant Agency of the Academy of Science of the Czech Republic (project AA400400805), from the Czech Science Foundation (project no 203/08/H032) (D. Bek) and from the long-term research plan of the Ministry of Education of the Czech Republic (no. MSM0021620857) (J. Sedláček) is gratefully acknowledged.

References

[1] Demonceau A, Stumpf AW, Saive E, Noels AF (1997) Novel ruthenium-based catalyst systems for the ring-opening metathesis polymerization of low-strain cyclic olefins. Macromolecules 30:3127–3136

[2] Jan D, Delaude L, Simal F, Demonceau A, Noels AF (2000) Synthesis and evaluation of new $RuCl_2$(p-cymene)(ER_2R') and (η^1:η^6-phosphinoarene)$RuCl_2$ complexes as ring-opening metathesis polymerization catalysts. J Organomet Chem 606:55–64

[3] Sémeril D, Bruneau C, Dixneuf PH (2001) Ruthenium catalyst dichotomy: selective catalytic diene cycloisomerization or metathesis. Helv Chim Acta 84(11):3335–3341

[4] Beck JS, Vartuli JC, Roth WJ, Leonowicz ME, Kresge CTK, Schmitt D, Chu CTW, Olson DH, Sheppard EW (1992) A new family of mesoporous molecular sieves prepared with liquid crystal templates. J Am Chem Soc 114(27):10834–10843

[5] Schüth F (2001) Non-siliceous mesostructured and mesoporous materials. Chem Mater 13(10):3184–3195

[6] Balcar H, Čejka J (2007) Mesoporous molecular sieves as supports for metathesis catalysts. In: Imamoglu Y, Dragutan V (eds.) Metathesis chemistry: from nanostructure design to synthesis of advanced materials, NATO Science Ser II.243:151–166. Springer, Dordrecht

[7] Balcar H, Žilková N, Sedláček J, Zedník J (2005) MCM-41 anchored Schrock catalyst Mo(=$CHCMe_2Ph$)(=N-2,6-*i*-$Pr_2C_6H_3$)[$OCMe(CF_3)_2$]$_2$-activity in 1-heptene metathesis and cross-metathesis reactions. J Mol Catal A: Chem 232(1–2):53–58

[8] Balcar H, Čejka J, Sedláček J, Svoboda J, Zedník J, Bastl Z, Bosáček V, Vohlídal J (2003) $[Rh(cod)Cl]_2$ complex immobilized on mesoporous molecular sieves MCM-41- a new hybrid catalyst for polymerization of phenylacetylene. J Mol Catal A: Chem 203(1–2):287–298

[9] Abbenhuis HCL (1999) Heterogenization of metallocene catalysts for alkene polymerization. Angew Chem Int Ed 37:1058–1060

[10] Topka P, Balcar H, Rathouský J, Žilková N, Verpoort F, Čejka J (2006) Micropor Mesopor Mater 96:44

[11] Sakurai K, Kashiwagi T, Takahashi T (1993) Crystal structure of polynorbornene. J Appl Polym Sci 47(5):937–940

Behavior of Silyl-Containing Norbornenes in the Conditions of Addition Polymerization

Maria L. Gringolts,[1] Yulia V. Rogan,[1] Maxim V. Bermeshev,[1] Valentin G. Lakhtin,[2] Eugene Sh. Finkelshtein[1*]

[1]A.V. Topchiev Institute of Petrochemical Synthesis RAS, 119991 Moscow, Leninskii prospect 29, Russia

[2]M.V. Lomonosov Moscow State Academy of Fine Chemical Technology, 117571 Moscow, prospect Vernadskogo, 86, Russia

[*]Tel: +7(495)9554379; fax: +7(495)2302224; e-mail: fin@ips.ac.ru

Abstract Addition polymerization of norbornenes bearing Me_3Si-substituents was studied in the presence of Ni and Pd-containing catalyst systems. The main attention was paid to synthesis of bis(trimethylsilyl)norbornenes and their behaviour in the addition polymerization conditions. Unlike mono(trimethylsilyl)norbornene they were inactive in addition homopolymerization. However, their copolymerization with norbornene and its alkyl derivatives was successfully realized. Poly(5-(trimethylsilyl)norbornene) demonstrated high gas transport parameters in respect to hydrocarbon gases.

Keywords Norbornene · Me_3Si-norbornene · Betaines · Cyclooctene · Addition polymerization · Ru-based catalyst · Ni-containing catalyst · Pd-containing catalyst

1 Introduction

Earlier we have shown that norbornenes bearing Me_3Si-groups can be easily polymerized via metathesis route (ROMP) in the presence of different catalytic systems on the basis of Ru, W and Re compounds (Scheme 1) [1, 2].

Polycarbosilanes obtained had good film-forming and gas transport properties [2–5]. Systematic physico-chemical investigations of a series of metathesis polynorbornenes bearing different side substituents evidenced that exactly Me_3Si groups linked up directly with the main chain were responsible for high membrane parameters [2]. However, the presence of a double bond in each monomer unit imparts to these polymers some disadvantages in particular a rather high chemical activity and poor thermooxidative stability.

The goal of this work was a synthesis of completely saturated polynorbornenes having Me_3Si-substituents as side groups. For this aim we have used the addition polymerization of corresponding Me_3Si-containing norbornenes. In general addition

V. Dragutan et al. (eds.), *Green Metathesis Chemistry: Great Challenges in Synthesis, Catalysis and Nanotechnology*,

polynorbornenes are of interest as highly transparent materials suitable for various optical applications [6, 7]. Their gas transport parameters are also actively investigated [8].

2 Results and Discussion

We studied a behavior of silicon-containing norbornenes, such as mono Me_3Si-derivative (5-trimethylsilyl-norbornene) and bis Me_3Si-derivatives (*endo-,exo*-5,6-bis(trimethylsilyl)norbornene, 5,5-bis(trimethylsilyl)norbornene) as well as 2,3-bis(trimethylsilyl)norbornadiene under the conditions of addition polymerization initiated by some Ni- and Pd-containing catalytic systems. The catalysts of this type have already demonstrated high activity in addition polymerization of unsubstituted norbornene and its alkyl derivatives [6, 7].

The silicon-containing norbornenes mentioned above were prepared by using Diels–Alder reaction of cyclopentadiene with the corresponding silyl-ethylenes and acetylene. If necessary, diene condensation was followed by methylation of the corresponding chloro-silyl-norbornenes.

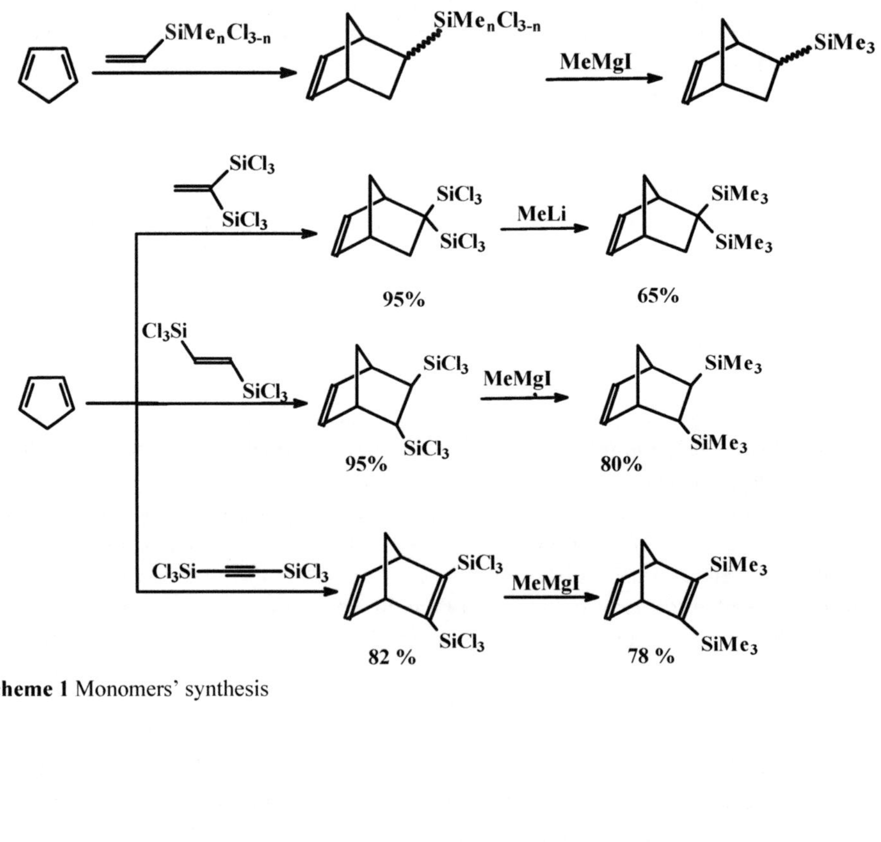

Scheme 1 Monomers' synthesis

Table 1 Activity of silyl-ethylenes in diene condensation

Olefin	I	II	III	IV	V	VI
Reaction T (°C)	50	50–60	100–120	210	25-65	50-60
Endo/exo in norbornene	76/24	73/27	60/40	52/48	–	–
Yield (%)	85–95	65–70	70–80	75–80	95	95

where I = $Cl_3SiCH{=}CH_2$; II = $Cl_2MeSiCH{=}CH_2$; III = $ClMe_2SiCH{=}CH_2$; IV = $(Cl_3Si)_2C{=}CH_2$; $Cl_3SiCH{=}CHSiCl_3$.

Table 1 demonstrates that activity of silyl-olefins in diene condensation increases with increase of number of chloro atoms in the initial silyl olefin: $Cl_3SiCH{=}CHSiCl_3 > (Cl_3Si)_2C{=}CH_2 > Cl_3SiCH{=}CH_2 > Cl_2MeSiCH{=}CH_2 > ClMe_2SiCH{=}CH_2 > Me_3SiCH{=}CH_2$. GLC analysis indicated that 5-trimethylsilyl-2-norbornene had 1:1 ratio of *endo-* and *exo-* conformers, whereas 5,6-bis(trimethylsilyl)-2-norbornene was defined as pure *endo-* , *exo-*isomer.

The known Pd-containing catalytic systems: $\{(\eta^3\text{-allyl})Pd(SbF_6)\}$ [9] and σ, π-bicyclic complex $[NB(OMe)PdCl]_2$ [10] turned out to be practically inactive in polymerization of 5-trimethylsilyl-2-norbornene. On the contrary, Ni-based complexes displayed a real activity in respect to this monomer. As a result saturated cyclolinear polymers were formed according to the Scheme 2 of addition polymerization. The absence of any unsaturation in these polymers was confirmed by both IR (no bands in 1,620–1,680 cm^{-1} region) and ^1H NMR spectroscopy (no signals at 5–6 ppm). Assignment of ^1H and ^{13}C NMR resonances of synthesized polymers was made with the help of assignments reported in [11] and model spectra.

GLC analysis of the final polymerization mixtures indicated that in the course of the reaction *exo*-conformer was consumed much faster than *endo*-form independently of the type of catalytic system employed.

Addition poly(5-trimethylsilyl-2-norbornenes) were obtained with the yields up to 80%. All of them were completely soluble in aromatic solvents. Among the Ni-based catalytic systems Ni(II) naphtenate – MAO and $(\pi\text{-}C_5H_9NiCl)_2$ – MAO are more active. Polymers prepared in the presence of Ni(II)naphtenate – MAO catalyst had the highest molecular weights and demonstrated good film-forming properties. They didn't show any glass transition up to 300°C (DSC).

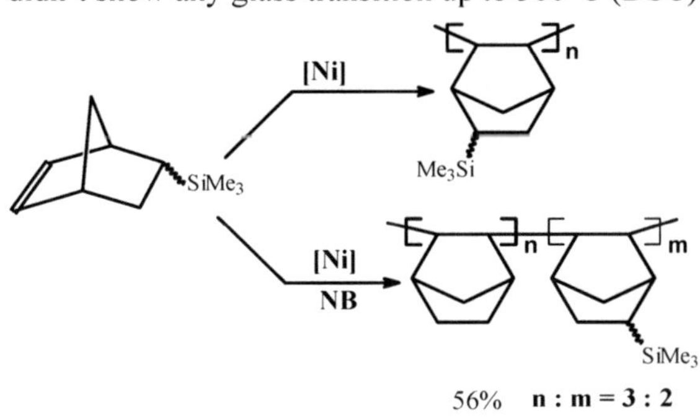

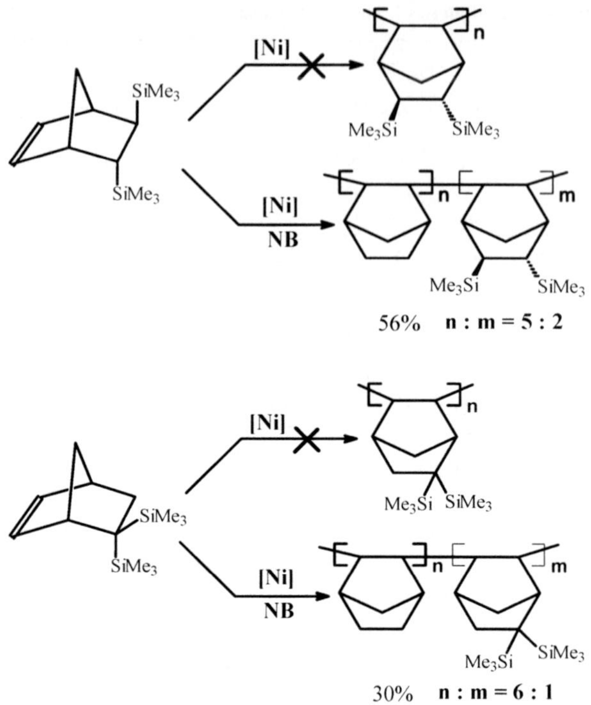

Scheme 2 Behavior of silyl-norbornenes under conditions of addition polymerization

It should be noted that polymerization of 5-trimethylsilyl-2-norbornene (5-NBSi) proceeded substantially slower than that of unsubstituted norbornene. At the same time *endo-,exo*-5,6-bis(trimethylsilyl)-2-norbornene (5,6-NBSi2), 5,5-bis(trimethylsilyl)-2-norbornene (5,5-NBSi2) and 2,3-bis(trimethylsilyl)norbornadiene didn't take part in addition homopolymerization (Table 2).

Table 2 Addition copolymerization of silyl-containing monomers (NBSi) with norbornene (NB)

Comonomers (NBSi)	[NBSi]/[NB]/[Cat[a]] (m/m/m)	Time(h)	Yield of copolymer (wt%)	Composition[b] (mol%)		Mw[c]	Mw/Mn
				NBSi	NB		
5-NBSi[d]	200/200/1	96	51	42	58	28,500	2.0
5,6-NBSi2	200/200/1	144	56	32	68	64,800	1.9
	200/200/1	166	43	25	75[e]	51,200	2.5
5,5-NBSi2	200/200/1	168	30	18	82	65,000	1.9

[a]$(Nph)_2Ni$:MAO = 1:100 m/m, RT, toluene.

[b]NMR data.

[c]Determined by GPC relative to polystyrene standards.

[d]$(\pi\text{-}C_5H_9NiCl)_2\text{-}Et_3Al_2Cl_3$, Al/Ni = 3.

[e]Copolymerization was performed with 5-n-hexyl-2-norbornene instead of norbornene.

Nevertheless, all the mentioned monomers turned out to be capable to take part in copolymerization with norbornene and 5-n-hexyl-2-norbornene. According to its NMR ^{1}H spectrum, the copolymer prepared from equimolar mixture of 5-trimethylsilyl-2-norbornene and norbornene contained 42 mol% units of silyl-derivative. In the case of bis(trimethylsilyl)-substituted-norbornenes (5,6-NBSi2 and 5,5-NBSi2) the content of silyl-containing units in copolymers was substantially lower (32 and 18 mol% respectively, Table 2).

GPC analysis confirmed the formation of copolymers by demonstrating unimodal and rather narrow molecular weight distribution for all the polymers obtained in this study. The data of T_g for the majority of synthesized polymers are not informative ($T_g \geq T_d$). However *endo-,exo*-5,6-bis(trimethylsilyl)-2-norbornene, incapable to give addition homopolymers, could copolymerize with 5-n-hexyl-2-norbornene giving polymer product with $T_g = 212°C$.

The structures of copolymers with hexylnorbornene can be depicted as follows:

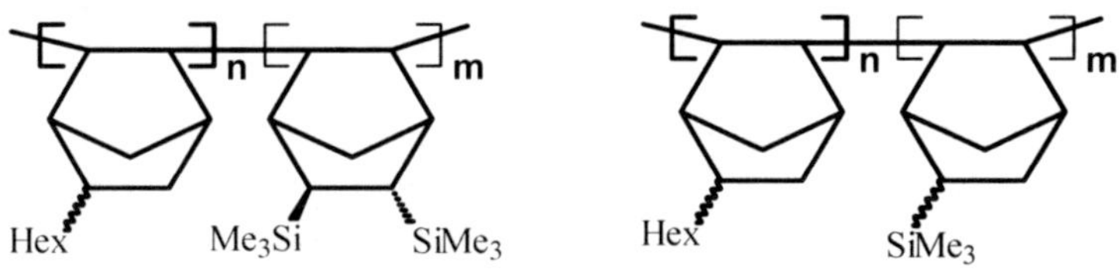

Lower activity of 5-trimethylsilyl-2-norbornene as compare with that of unsubstituted norbornene and a total inactivity of bis(trimethylsilyl)-2-norbornenes in addition polymerization can be explained by steric hindrances induced by Me_3Si-groups in *endo*-conformation. It is known that norbornene derivatives produced by Diels–Alder reaction are enriched, as a rule, by *endo*-forms. Usually the conformation of substituted norbornene does not have any considerable significance for ROMP process. However, for addition type process, the possibility of polymerization as well as its rate to a great degree depends on the presence of monomer in *exo*-conformation. The mechanistic reasons of this dependence were discussed by Sen et al. [12] on the example of simple 5-alkyl-2-norbornenes polymerization. It is possible that substituted norbornenes with 100% of endo-form are incapable in polymerizing via addition scheme at all. But when the mixture of different conformers is present, a part of *endo*-form could participate in copolymerization with *exo*-form. That's why we could obtain sufficiently high yields of poly(5-trimethylsilyl-2-norbornene) from a monomer with nearly equal amounts of *exo*- and *endo*-form. In the case of both bis (trimethylsilyl)-2-norbornenes one of substituents is always in the disadvantageous *endo*-conformation. This fact excludes the possibility of homopolymerization for this monomer but does not exclude the proceeding of its copolymerization with norbornene itself or with its substituted *exo*-derivative.

We have shown that ROMP Me_3Si-polynorbornenes can attract interest as potential materials for gas and vapor separation membranes [3]. However, ROMP polymers include double bonds, and this makes them relatively unstable materials: thus, unsubstituted polynorbornene shows obvious signs of deterioration (appearance of color and brittleness) after several weeks of storage in the ambient atmosphere. On the contrary, addition polynorbornenes are saturated polymers, so their chemical stability must be much better. According to contemporary wisdom on highly permeable polymers, among the prerequisites for high gas permeability and large free volume materials there are: presence of bulky nonpolar substituents, rigid main chains (great rotation barriers in them), and appearance of periodic kinks (frequent chain disruption) [13–15]. All of these features can be anticipated in addition polynorbornenes [16], so an appropriately selected catalytic system for the synthesis, chemical structure of a monomer, and molecular mass that would provide sufficiently good film forming properties can lead to addition polynorbornenes with attractive gas permeation properties.

Studies of the transport properties of unsubstituted addition type polynorbornene and polyalkylnorbornenes have been just started [8, 17]. Some results have been reported for norbornene–ethylene copolymers of different composition [18]. In the present work, we report membrane properties of the addition polynorbornene, bearing side $Si(CH_3)_3$ groups (Scheme 2). Appearance of this substituent, by analogy with other polymer classes, [13] would strongly enhance gas permeability and free volume. With this in mind, we tested gas permeation parameters of addition poly(trimethylsilyl)norbornene films for hydrocarbon gases. For comparison, we also prepared two other addition polynorbornenes: nonsubstituted polynorbornene and random copolymer of 5-trimethylsilyl-2-norbornene and 5-*n*-hexyl-2-norbornene.

Table 3 The most permeable polymers in respect to hydrocarbon gases

Polymer	Permeability, barrer				$\alpha(C_4H_{10}/CH_4)$
	CH_4	C_2H_6	C_3H_8	C_4H_{10}	
	790	1,430	1,740	17,500	22
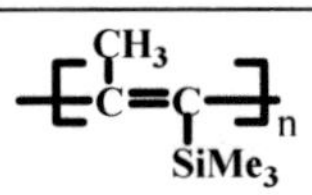	15,000	22,000	25,000	78,000	5,2
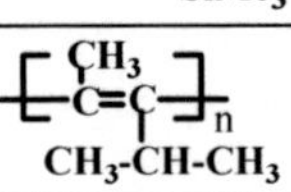	2,900	3,700	7,300	26,000	9
AF2400*	435	252	97	-	0.32

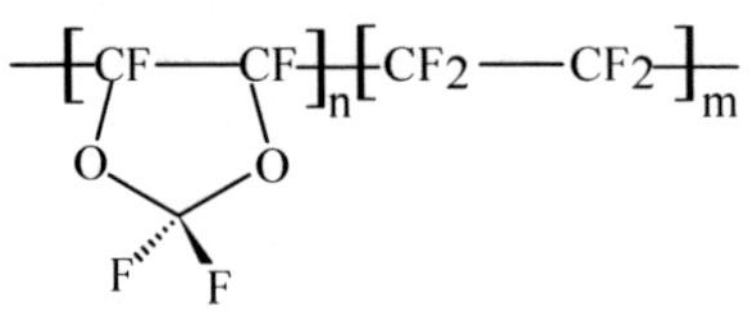

*

It turned out to be that the latters did not have high permeability in respect to methane (2.6 and 48). On the other hand the Me_3Si-derivative demonstrated very high permeability to all hydrocarbon gases of C1–C4 content (Table 3), especially n-butane. It belongs to a group of known the most permeable polymers.

3 Conclusions

First behavior of bis(trimethylsilyl)norbornenes under conditions of addition polymerization were studied in the presence of Ni-based catalytic systems: $(\pi\text{-}C_5H_9NiCl)_2 - Et_3Al_2Cl_3$, Ni(II) naphtenate – (MAO) and $(\pi\text{-}C_5H_9NiCl)_2$ – MAO. It was shown that unlike mono-5-(trimethylsilyl)derivative bis(trimethylsilyl)norbornenes was inactive in homopolymerization. On the other hand their copolymerization with norbornene and 5-hexylnorbornene was successfully realized.

Addition poly((monotrimethylsilyl)norbornene) demonstrated very high permeability and selectivity to hydrocarbon gases. It belongs to the family of known most permeable polymers.

Acknowledgements The authors would like to thank Prof. Yu. Yampolskii for the measurements of gas-transport parameters. This work was supported in part through the Russian Science Support Foundation (to M.B.).

References

[1] Finkelshtein ESh (1998). In: Imamoglu Y (ed.) Metathesis polymerization of olefins and polymerization of alkynes, pp. 201–224. Kluwer, Dordrecht

[2] Finkelshtein ESh, Gringolts ML, Ushakov NV, Lakhtin VG, Soloviev SA, Yampolskii YP (2003) Polymer 44:2843–2851

[3] Finkelshtein ESh, Makovetskii KL, Yampolskii YP, Portnykh EB, Ostrovskaya IY, Kaliuzhnyi NE, Pritula NA, Golberg AI, Yatsenko MS, Plate NA (1991) Makromol Chem 192:1

[4] Bondar VI, Kukharskii YM, Yampolskii, YP, Finkelshtein ESh, Makovetskii KL (1993) J Polym Sci Part B Polym Phys 31:1273

[5] Finkelshtein ESh (1998). In: Imamoglu Y (ed.) Metathesis polymerization of olefins and polymerization of alkynes, pp. 189–199. Kluwer, Dordrecht

[6] Janiak C, Lassahn PG (2001) J Mol Cat A 166:193–209

[7] Makovetsky KL (1999) J Polymer Sci Ser B 41:269

[8] (a) Zhao Ch, Ribeiro MR, de Pinho MN, Subrahmanyam VS, Gil CL, de Lima AP (2001) Polymer 42:2455–2462. (b) Wilks BR, Chung WJ, Ludovice PJ, Rezac MR, Meakin P, Hill AJ (2003) J Polym Sci Part B Polym Phys 41:2185–2199

[9] Mathew JP, Reinmuth A, Melia J, Swords N, Risse W (1996) Macromolecules 29:2755–2763

[10] Green M, Hancock R (1967) J Chem Soc A: 2054

[11] (a) Barnes DA, Benedikt GM, Goodall BL, Huang ShS, Kalamarides HA, Lenhard S, McIntosh LH, Selvy KT, Shick RA, Rhodes LF (2003) Macromolecules 36:2623–2632; (b) Myagmarsuren G, Lee KS, Jeong O-Y, Ihm S-K (2005) Polymer 46:3685–3692

[12] (a) Hennis AD, Polley JD, Long GS, Sen A, Yandulov D, Lipian J, Benedikt GM, Rhodes LF (2001) Organometallics 20:2802–2812; (b) Funk JK, Andes CE, Sen A (2004) Organometallics 23:1680–1683
[13] Yampolskii YP, Plate NA (1994). In: Paul DR, Yampolskii, YP (eds.) Polymeric gas separation membranes. CRC Press, Boca Raton, FL
[14] Budd PM, Msayib KJ, Tattershall CE, Ghanema BS, Kevin J, Reynolds KJ, McKeown NB, Fritsch DJ (2005) Membr Sci 251:263
[15] Yampolskii YP, Pinnau I, Freeman B (eds.) (2006) Materials science of membranes for gas and vapor separation. Wiley, Chichester
[16] Chung WJ (2003) Ph.D. dissertation, Georgia Technological Institute, Atlanta, GA
[17] Dorkenoo KD, Pfromm PH, Rezac ME (1998) J Polym Sci Part B Polym Phys 36:797
[18] Poulsen L, Zebger I, Klinger M, Eldrup M, Sommer-Larsen P, Ogilby PR (2003) Macromolecules 36:7189

New Applications of Ring-Opening Metathesis Polymerization for Grafting Alkylene Oxide-Based Copolymers

Bogdan Spurcaciu,[1] Emil Buzdugan,[1] Cristian Nicolae,[1] Paul Ghioca,[1] Lorena Iancu,[1] Valerian Dragutan,[2] Ileana Dragutan[2*]

[1]National Institute of Research & Development for Chemistry and Petrochemistry (ICECHIM), 202 Spl. Independentei, Bucharest, Romania; E-mail: bogdansssss@hotmail.com

[2]Institute of Organic Chemistry "C.D. Nenitescu" of the Romanian Academy, 202B Spl. Independentei, Bucharest, Romania

[*]E-mail: idragutan@yahoo.com

This research tackles the challenges of innovative modification of poly(allyl alkylene oxides) by ROMP to produce new materials. Firstly, binary and ternary copolymers, poly(epichlorohydrin-allyl glycidyl ether) (ECH-AGE) and poly (epichlorohydrin-propylene oxide-allyl glycidyl ether) (ECH-PO-AGE), have been prepared using as initiator a catalytic system consisting of an alkyl aluminium, controlled amounts of water and different compounds (ethers, diols, phosphines, salicylic acid derivatives, organozincs) acting as cocatalysts. Among catalysts explored in these copolymerizations most productive showed to be the systems triisobutylaluminium (TIBA), water and $Zn(DIPS)_2$ or $Zn(acac)_2$. Copolymers which have become thus available were subsequently grafted onto the pendent allylic groups by ROMP with cycloolefins (cyclooctene, norbornene, cyclododecene) involving ruthenium based catalysts.

Keywords Alkylene oxide · Allyl glycidyl ether · Binary and ternary copolymers · Cycloolefins · Epichlorihydrin · Graft copolymers · ROMP · Ruthenium alkylidene

1 Introduction

In macromolecular engineering, copolymerization is a technically important process providing access to new materials with properties tuned through adjustments to ratios and/or nature of individual monomer units of the copolymer [1]. Best approach to copolymers of choice is an inventive combination of two or more mechanistically distinct polymerization processes to obtain at reasonable cost products with optimized characteristics [2]. Polymerization reactions mostly applied in copolymer synthesis

V. Dragutan et al. (eds.), *Green Metathesis Chemistry: Great Challenges in Synthesis, Catalysis and Nanotechnology*,

involve anionic [3], cationic [4], coordination (Ziegler-Natta) [5], metathesis polymerization (e.g. ROMP, ADMET) [6], group transfer [7], and radical mechanisms (e.g. NMRP, ATRP) [8]. Modeling these fundamental processes in precisely controlled ways yields either advanced polymeric materials or well-defined macromonomers with targeted structures and function. When macromonomers arising from hydrophilic monomers are employed in a further polymerization step involving hydrocarbon monomers, amphiphilic copolymers are being created; the latter, or copolymers incorporating units differing in hydrophobicity, may organize by self-assembling [9] and thereby find particular practical applications.

Due to spectacular advances in well-defined metathesis initiating systems [10], ROMP has opened up enormous possibilities for commodity polymer synthesis. As a living process [11] it enables highly selective synthesis of well-characterized polymers with controlled microstructure and functionality, and narrow molecular weight distribution. By applying ROMP in tandem with one of the above polymerization techniques, copolymers with special architectures and displaying a range of nanoscale morphologies can be accessed [12–16].

In a previous study in our group [17] we have reported on an efficient synthetic route to high molecular weight copolymers using a sequence of anionic-coordinative ring-opening polymerization (ROP) of functionalized alkylene oxides, followed by ruthenium-catalyzed ring-opening metathesis polymerization (ROMP) of cycloolefins onto copolymers obtained in the first stage.

Materials based on copolymers of alkylene oxides, modified by ROMP, possess valuable properties recommending them for special applications in the automotive industry (hoses, fittings and complex assemblies), in the oil industry (drilling and oil processing), machine building, defence, aviation and aeronautics. Furthermore, a new paradigm in the design of polymer based composites is presently represented by the synthetic self-healing materials, i.e. materials that, when damaged, sense the failure and respond in an autonomous mode to restore the structural function [18]. Dicyclopentadiene and 5-ethylidene-2-norbornene have been considered as potential healing agents for self-healing composite materials, after ROMP triggered by first or second generation Grubbs catalysts [19]. Some systems for ROMP-based self-healing of epoxy resins also use Grubbs catalysts and exo-DCPD, embedded together in the resin matrix, to cure cracks in the resin by ROMP at room temperature, thus restoring the initial mechanical properties of the epoxy polymer [20].

However, as far as we know, with the exception of our earlier report [17], cycloctene, cyclododecene and norbornene have not been applied as ROMP-modifiers of allyl-substituted alkylene oxide copolymers. In the present work we extend our synthetic protocol to binary copolymers (from epichlorohydrin and allyl glycidyl ether) and ternary copolymers (from epichlorohydrin, propylene oxide and allyl glycidyl ether) and to their grafting via ROMP of cycloolefins occurring at the allyl units pendent from the main backbone to yield new comb-like polymers.

2 Results and Discussion

To open access to starting materials for ROMP, a first focus of this study was the synthesis of binary and ternary alkylene oxide copolymers. Poly(epichlorohydrin-allyl glycidyl ether)[poly(ECH-AGE)] and poly(epichlorohydrin – propylene oxide – allyl glycidyl ether)[poly(ECH-PO-AGE)], containing variable numbers of co-monomer units in the main chain, have been obtained by ROP of the respective comonomer mixtures, in the presence of an array of catalytic systems basically consisting of an alkyl aluminium (e.g. triisobutylaluminium (TIBA)), controlled amounts of water and different organic and organometallic compounds that act as cocatalysts, e.g. ethers, diols, phosphines, salicylic acid derivatives or organozincs (Table 1).

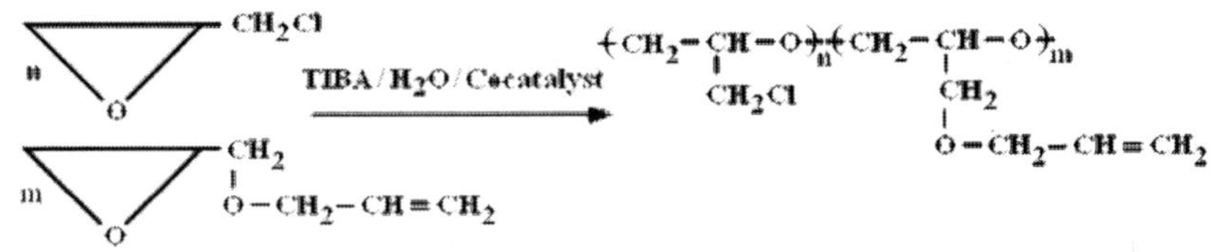

(1)

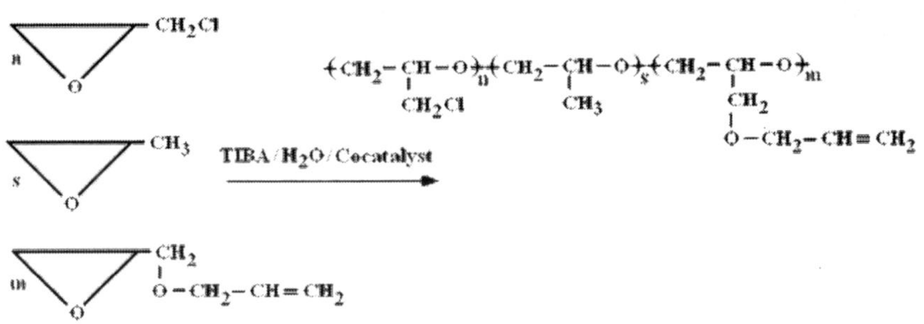

(2)

All binary (Equation 1) and ternary (Equation 2) copolymerizations were conducted in toluene, at 50°C and an overall monomer concentration of 100 g/800 ml solvent (Tables 2 and 3). Comonomer ratios were PO/AGE = 9:1 (wt/wt), ECH/AGE = 9:1, ECH/PO/AGE = 6/3/1 (wt/wt), with 30 mmol TIBA/100 g comonomers. The addition order of the catalytic components proved to be of essential importance for attaining high yields and reproducibility. For catalytic systems prepared in situ the optimal addition order was solvent, monomer, cocatyst, TIBA and water, whereas in the case of preformed catalysts best results were obtained by first diluting TIBA with the solvent up to a concentration of 20 wt% and then adding the cocatalyst and water. Reaction times were 6 and 8 h for binary and ternary copolymerizations, respectively.

Table 1 Catalytic systems used for synthesis of aplkylene oxide copolymers[a,b]

Catalyst	Catalyst components[c]
Cat 1	TIBA:H_2O:Et_2O = 1:0.4:0.25
Cat 2	TIBA:H_2O:MTBE = 1:0.4:0.25
Cat 3	TIBA:H_2O:DIPS Acid = 1:0.4:0.25
Cat 4	TIBA:H_2O:Zn(DIPS)$_2$ = 1:0.4:0.25
Cat 5	TIBA:H_2O:tris(di-n-butylamino)phosphine = 1:0.4:0.25
Cat 6	TIBA:H_2O:trimorpholide of phosphorous acid = 1:0.4:0.25
Cat 7	TIBA:H_2O:Zn(acac)$_2$ = 1:0.4:0.25

[a]This study and [17].

[b]TIBA = triisobutylaluminium, MTBE = methyl *tert*-butyl ether, DIPS = diisopropyl salicylic acid, acac = acetylacetonate.

[c]Molar ratios.

2.1 Binary Copolymerization of Epichlorohydrin and Allyl Glycidyl Ether

Binary copolymers poly(ECH-AGE) have been prepared with the catalytic systems listed in Table 1 and compared to the previously synthesized PO-AGE copolymers [17]. Considering both the copolymer yield and the molecular mass, among the catalysts tested in epichlorohydrin -allyl glycidyl ether copolymerizations the most efficient proved to be the ternary catalytic system TIBA:H_2O:Zn(DIPS)$_2$ (Cat. 4) (Table 2). This result parallels that observed for poly(PO-AGE). However, for most catalytic systems, binary copolymerizations of the new set of monomers (ECH-AGE) lead to higher molecular masses and conversions than in the case of PO-AGE.

Table 2 Synthesis of binary copolymers poly(ECH-AGE)[a]

Catalyst	Copolymer	Inh. visc. (dl/g)	Conv (%)
Cat 1 (TIBA:H_2O:Et_2O)	Poly(ECH-AGE) -1	1.17	69.6
	Poly(PO-AGE) – 1[a]	0.86	53.0
Cat 2 (TIBA:H_2O:MTBE)	Poly(ECHAGE) – 2	0.96	51.4
	Poly(PO-AGE) – 2[a]	0.79	27.2
Cat 3 (TIBA:H_2O:DIPS Acid)	Poly(ECHAGE) - 3	1.08	63.8
	Poly(PO-AGE) – 3[a]	–	11.7
Cat 4 (TIBA:H_2O:Zn(DIPS)$_2$)	Poly(ECHAGE) – 4	1.34	93.7
	Poly(PO-AGE) - 4[a]	1.03	83.7
Cat 5 (TIBA:H_2O:tris(di-n-butylamino)phosphine)	Poly(ECHAGE) – 5	1.22	84.2
	Poly(PO-AGE) – 5[a]	1.17	62.4
Cat 6 (TIBA:H_2O:trimorpholide of phosphorous acid)	Poly(ECHAGE) – 6	0.83	42.4
	Poly(PO-AGE) – 6[a]	0.93	37.1
Cat 7 (TIBA:H_2O:Zn(acac)$_2$)	Poly(ECHAGE) – 7	1.29	89.5
	Poly(PO-AGE) – 7[a]	1.21	94.5

[a]Data from [17].

Similarly to the poly(PO-AGE) spectra [17], ^{1}H-NMR spectra recorded for poly(ECH-AGE) indicate random copolymers and an AGE content of ca.10 mol%.

2.2 Ternary Copolymerization of Epichlorohydrin , Propylene Oxide and Allyl Glycidyl Ether

Conclusive results obtained in binary monomer systems encouraged us to also explore ternary copolymerizations of this type, e.g. that of epichlorohydrin – propylene oxide – allyl glycidyl ether (Table 3) using the same panel of multi-component catalysts given in Table 1. Again, the most efficient ROP promoter was TIBA:H_2O:Zn(DIPS)$_2$. It is apparent, nevertheless, that while conversions in ternary ROP are inferior to those obtained in binary copolymerizations induced by the same catalyst, the inherent viscosities tend to overpass values attained in reactions involving only two comonomers.

Table 3 Synthesis of ternary copolymers poly(ECH-PO-AGE)

Copolymer	Catalyst	Inherent viscosity (dl/g)	Conversion (%)
Poly(ECHPOAGE) – 1	Cat – 1	1.13	37.6
Poly(ECHPOAGE) – 2	Cat – 2	0.87	12.3
Poly(ECHPOAGE) – 3	Cat – 3	0.91	17.5
Poly(ECHPOAGE) – 4	Cat – 4	1.63	71.0
Poly(ECHPOAGE) – 5	Cat – 5	1.36	62.4
Poly(ECHPOAGE) – 6	Cat – 6	0.95	15.7
Poly(ECHPOAGE) – 7	Cat – 7	1.47	65.8

According to their ^{1}H-NMR spectra, ternary epichlorohydrin – propylene oxide – allyl glycidyl ether copolymers are also random and contain about 9 mol% AGE.

2.3 Subsequent Modification of Binary and Ternary Copolymers by Grafting via Ring-Opening Metathesis Polymerization of Cycloolefins

Modification of ECH–AGE binary copolymers and ECH–PO–AGE ternary copolymers by ring-opening metathesis polymerization is at the core of this study. Grafting was performed taking advantage of the pendent allyl groups of AGE units in the copolymers and using cyclooctene, norbornene or cyclododecene as reaction partners able to ROMP. Metathesis reactions were carried out in a toluene – THF mixture (5/1 vol/vol), in the presence of the first generation Grubbs catalyst and employing our earlier synthesis methodology [17]. The obtained new graft polymers are illustrated below and in Table 4.

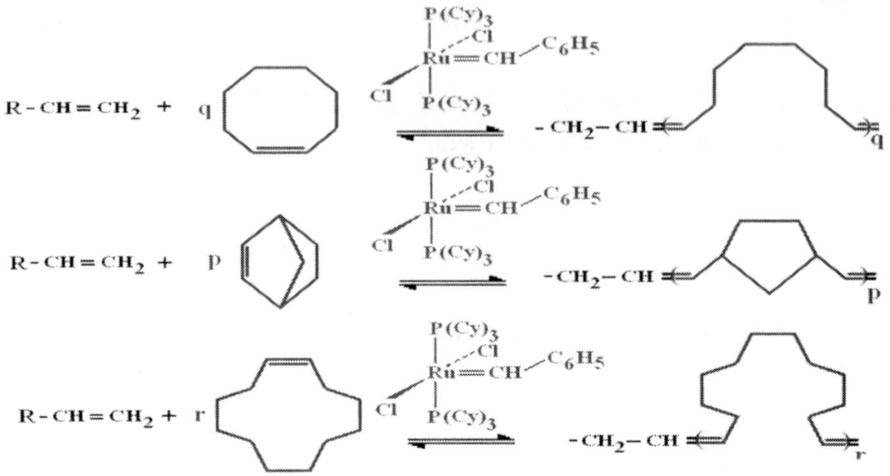

^1H - NMR spectra of the graft polymers enabled estimation of the extent to which modification has occurred. Based on the NMR data, grafting succeeded best when using norbornene as the cycloolefin subjected to ROMP, be it the case of binary or ternary alkylene oxide copolymers (Table 4). Cyclooctene and cyclododecene displayed similar, yet lower, performance as modification intercessors. Noteworthy, under similar conditions binary copolymers seem to be more modification-prone than ternary counterparts.

Table 4 Synthesis of graft copolymers alkylene oxide-cycloolefin by ROMP

Entry	Copolymer of	Cycloolefin	Percentage of modified allylic groups
1	ECH-AGE	Cyclooctene	52
2	ECH-AGE	Norbornene	65
3	ECH-AGE	Cyclododecene	51
4	ECH-PO-AGE	Cyclooctene	47
5	ECH-PO-AGE	Norbornene	61
6	ECH-PO-AGE	Cyclododecene	45

3 Conclusions

In the examined catalyst series, the system TIBA:H2O:Zn(DIPS)2 = 1:0.4:0.25 performed best in both binary ECH-GE and ternary ECH-PO-AGE copolymerizations.

Graft copolymers were successfully obtained by grafting cyclooctene, norbornene and cyclododecene via ring-opening metathesis polymerization onto the binary poly(ECH-AGE) and ternary poly(ECH-PO-AGE) copolymers, using the first generation Grubbs catalyst.

All newly synthesized copolymers are random. Modification of pendent allylic groups by metathesis took place in excess of 60 mol% with norbornene and only about 50 mol% with cyclooctene and cyclododecene.

References

[1] (a) Grubbs RH (ed.) (2003) Handbook of metathesis, vol. 3. Wiley-VCH, Weinheim; (b) Dragutan V, Streck R (2000) Catalytic polymerization of cycloolefins. Elsevier, Amsterdam

[2] (a) Khosravi E (2003), Ch 3.3 In: Grubbs RH (ed.) Handbook of metathesis, vol. 3. Wiley-VCH, Weinheim; (b) Enholm E, Joshi A, Wright DL (2005) Bioorg Med Chem Lett 15:5262–5265; (c) Matyjaszewski K (2005) Prog Polym Sci 30:858–875

[3] Szwarc M, Levy M, Milkovich R (1956) J Am Chem Soc 78:2656

[4] Matyjaszewski K, Gaynor S, Wang J-W (1995) Macromolecules 28:2093

[5] (a) Boor J Jr (1979) Ziegler-Natta catalysts and polymerizations. Academic Press, New York; (b) Kaminsky W, Bark A, Dake I (1990) In: Keii T, Soga K (eds.) Catalytic olefin polymerization, pp. 426–438. Elsevier, Amsterdam

[6] (a) Ivin KJ, Mol JC (1987) Olefin metathesis and metathesis polymerization. Academic Press, London; (b) Dragutan V, Balaban AT, Dimonie M (1985) Olefin metathesis and ring-opening polymerization of cycloolefins. Wiley, New York; (c) Slugovc C (2004) Macromol Rapid Commun 25:1283–1297; (d) Hilf S, Kilbinger AFM (2007) Macromol Rapid Commun 28:1225–1230; (e) Abd-El-Aziz AS, Manners I (2005) J Inorg Organomet Polym Mater 15:157–195; (f) Dragutan V, Dragutan I, Fischer H (2008) J Inorg Organomet Polym Mater 18:18–31; (g) Dragutan V, Dragutan I, Fischer H (2008) J Inorg Organomet Polym Mater 18:311–324; (h) Lee JC, Parker KA, Sampson NS (2006) J Am Chem Soc 128:4578–4579; (i) Song A, Parker KA, Sampson NS (2009) J Am Chem Soc 131:3444–3445; (j) Rojas G, Wagener KB (2007) Precision polyolefin structure: modelling polyethylene containing methyl and ethyl branches. In Imamoglu Y, Dragutan V (eds.) Metathesis chemistry: from nanostructure design to synthesis of advanced materials, NATO Science Series II. Mathematics, Physics and Chemistry, vol. 243, pp. 305–324. Springer, Dordrecht, The Netherlands

[7] (a) Webster OW, Sogah DY (1988) Group transfer and Aldol group transfer polymerization. In Comprehensive Polymer Science, vol. 4. Pergamon, New York; (b) Webster OW, Sogah DY (1987) Recent advances in the controlled synthesis of acrylic polymers by group transfer polymerization, pp. 3–21. In Fontanille M, Guyot A (eds.) Recent advances in mechanistic and synthetic aspects of polymerization. Reidel, Dordrecht, The Netherlands

[8] (a) (a) Matyjaszewski K, Davis TP (eds.) (2002) Handbook of radical polymerization. Wiley-Interscience, Hoboken, NJ; (b) Matyjaszewski K, Xia J (2001) Chem Rev 101:2921; (c) Patten TE, Matyjaszewski K (1999) Acc Chem Res 32:895; (d) Coca S, Paik H, Matyjaszewski K (1997) Macromolecules 30:6513; (e) Miura Y, Sakai Y, Taniguchi I (2003) Polymer 44:603

[9] (a) Miller AF, Richards RW (2000) Macromolecules 33:7618; (b) Ogata Y, Makita Y, Okaniwa M (2008) Polymer 49:4819–4825; (c) Liaw D-J, Wang K-L, Chen T-P, Lee K-R, Lai J-Y (2007) Polymer 48:3694–3702; (d) Akcora P, Liu H, Kumar SK, Moll J, Li Y, Benicewicz BC, Schadler LS, Acehan D, Panagiotopoulos AZ, Pryamitsyn V, Ganesan V, Ilavsky J, Thiyagarajan P, Colby RH, Douglas JF (2009) Nature Mater 8:354–359

[10] (a) Grubbs RH (ed.) (2003) Handbook of metathesis, vol. 1. Wiley-VCH, Weinheim; (b) Schrock RR, Hoveyda AH (2003) Angew Chem Int Ed 42:4592–4633; (c) Grela K, Harutyunyan S, Michrowska, A (2002) Angew Chem Int Ed 41:4038–4040; (d) Dragutan V, Dragutan I, Delaude L, Demonceau A (2007) Coord Chem Rev 251:765–794; (e) Ding F, Sun Y, Monsaert S, Drozdzak R, Dragutan I, Dragutan V, Verpoort F (2008) Curr Org Synth 5:291–304; (f) Monsaert S, Drozdzak R, Dragutan V, Dragutan I, Verpoort F (2008) Eur J Inorg Chem 3:432–440; (g) Dragutan I, Dragutan V, Delaude L, Demonceau A, Noels AF (2007) Rev Roumaine Chim 52:1013–1025; (h) Dragutan I, Dragutan V, Filip P (2005) ARKIVOC **2005** x:105–129; (i) Drozdzak R, Ledoux N, Allaert B, Dragutan I, Dragutan V, Verpoort F (2005) Central Eur J Chem 3:404–416; (j) Buchmeiser MR (2009) Chem Rev 109:303–321

[11] Bielawski CW, Grubbs RH (2007) Prog Polym Sci 32:1–29

[12] (a) Feast WJ, Gibson VC, Johnson AF, Khosravi E, Moshin MA (1994), Polymer 35:3542; (b) Feast WJ, Gibson VC, Johnson AF, Khosravi E, Moshin MA (1997) J Mol Catal A Chem 115:37

[13] Miller AF, Richards RW, Webster JRP (1997) Macromolecules 33:7618

[14] (a) Herouguez V, Gnanou Y, Fontanille M (1996) Macromolecules 29:4459; (b) Herouguez V, Gnanou Y, Fontanille M (1997) Macromolecules 30:4791

[15] Radano CP, Scherman OA, Stingelin-Stutzmann N, Müller C, Breiby DW, Smith P, Janssen RAJ, Meijer EW (2005) J Am Chem Soc 127:12502

[16] Allen MJ, Wangkanont K, Raines RT, Kiessling LL (2009) Macromolecules 42:4023–4027 DOI: 10.1021/ma900056b

[17] Spurcaciu B, Buzdugan E, Nicolae C, Dragutan V, Dragutan I (2007) Synthesis of new elastomers with controlled structures based on alkylene oxides, grafted by ROMP. In Imamoglu Y, Dragutan V (eds.) Metathesis chemistry: from nanostructure design to synthesis of advanced materials, NATO Science Series II. Mathematics, Physics and Chemistry, vol. 243, pp. 347–354. Springer, Dordrecht, The Netherlands

[18] Kessler MR (2007) Proc Inst Mech Eng Part G- J Aerosp Eng 221 (G4):479–495

[19] (a) Lee JK, Liu X, Yoon SH, Kessler MR (2007) J Polym Sci Part B Polym Phys 45:1771–1780; (b) Liu X, Sheng X, Lee JK, Kessler MR (2007) Intl Conf Smart Mater Nanotech Eng 1-3 Book Series: Proc Soc Photo Opt Instr Eng (SPIE) Du S, Leng J, Asundi AK (eds.) 64 (23) Part 1–3:42348–42348

[20] (a) Wilson GO, Moore JS, White SR, Sottos NR, Andersson HM (2008) Adv Funct Mater 18:44–52; (b) Mauldin TC, Rule JD, Sottos NR, White SR, Moore JS (2007) J Royal Soc Interf 4:389–393

Subject Index

Author Index

Lightning Source UK Ltd.
Milton Keynes UK
14 March 2010

151358UK00006BA/82/P

9 789048 134311